A Text Book on
NUCLEAR PHYSICS

For Graduate Students

Prof. S. Devanarayanan, Ph.D.; D.Sc.,
Dip(Uppsala)

(Revised Edition)

ISBN 9781540538451
CreateSpace Ind. Publ., USA.
(First Edition   January 2016)

Revised Edition
November 2016

DEDICATION

This work is dedicated to my teachers.

|| Gurur Brahma – Gurur Vishnu – Guru : devo Maheswara : |
Guru saakshaat – Param Brahma – Tasmai – Sree Guruve – Nama: ||

|| Yada tu sarva-bhuteshu Darushvagnimiva stitam
Pratichaksheeta Maam Loko jaythrhyeva kasmalam ||
(Srimad Bhagavatam: Sk III. Ch9. Sl 32)

	OM Namo Bhagavate Vaasudevaya	
	Ahamevaasamevaagre naanyadyat sadasatparam
Paschaadaham yadetaccha yo f vasishyeta so f smyaham	
	Riter f tham yatpratiyeta na pratiyeta	chaatmani
Tatvidyaatmano maayam yatha ff bhaaso yatha tama:	
	Yatha mahaanti bhutaani bhuteshoocchaavacheshvanu
Pravishtaanya-pravishtaani tatha teshu na teshavaham	
	Etaavadeva jigjnaasyam Tatva jigjnaasunaa ff tmana:
Anvaya vyatirekaabhayaam yatsyaat sarvatra sarvadaa	
*{|| Srimad Bhaagavatam (Sk II; Chapter 9: Verses 32-35)
Satya – Sanaatana – Dharmam vijayetetaram || }*
- Bhagavan the Supreme God Head, spoke to Brahma the Creator, *the* Bahaagavatam in a Nutshell.

PREFACE

Nuclear Physics has been occupying continuously an important place in any University course in physics at the graduate and PG levels in India. The main purpose of *A Text Book on Nuclear Physics* is to give a concise account of the fundamentals of the physics of the nuclei and particles and applications of nuclear energy. Its coverage extends the conventional aspects of the subject, because it has become very evident in recent years that much of the great body of knowledge of nuclei, acquired several decades ago, is highly relevant to other field such as solid state, modern spectroscopy, chemistry, biological / medical physics and technology of power production.

In a book of moderate size it is not possible to give a comprehensive treatment, in depth, of the whole of subatomic physics, for the student community at the degree level. However, I have tried to add my experience of teaching, 4 credit semester courses, on the subject a few years during my tenure (1971 - 2000) to the M.Sc. students at the Department of Physics of the University of Kerala, Kariavattom campus, Thiruvananthapuram. My experience in research in the field of Mossbauer spectroscopy has certainly an impact in the quality of the contents of the book. Further, as an author of four books, I could prepare this book in its own uniqueness for instance providing student friendly features. I have incorporated a good deal of **Worked out Examples with solutions** at appropriate places and Review Questions including their answers at the end of each Chapter. Outline solutions are deliberately avoided so as to reduce the volume of the book. I have kept the mathematics as simple as possible. I assume knowledge of the basics of special relativity and basic quantum mechanics

The many bibliographic references have been arranged in alphabetical list to enable students as well as faculty, for their academic references.

The present book is designed rather to meet the needs of the academic community who wishes to adopt the whole or parts of the book as a text for the prescribed syllabus of any course containing nuclear physics. It is hoped that the book will be of interest to those whose work lies inter-disciplinary fields, for example health physics, industrial physics, and related fields.

Remarks on this book are invited to the attention of the author by chsd1976@gmail.com.

15 January 2015   S. Devanarayanan

PREFACE of this Edition

This Revised Edition of the book: *A Text Book on Nuclear Physics- For Graduate Students,* has been brought up after getting suggestions from the readers. No change in the text was made. However, the font size is changed from 10 to 11, for easy reading for normal vision. Further, the contents s expanded for higher utility of the book, with page numbers, and each chapter begins with a better get up.

The author would like to thank Mrs. Chitra, Mr. Ajith Devan, Ms. Aparna and Mr. Sumesh Subramanian for their encouragements.

November 2016   S. Devanarayanan

CONTENTS

Preface
Contents

Pages

Chapter 1: INTRODUCTION 1
Physics the most fundamental Science – Objectives of study of Nuclear Physics - structure of matter – Turning points in development of Nuclear physics – Neutron-proton hypothesis – Review Questions

Chapter 2: NATURAL RADIOACTIVITY- Flashback 8
Introduction – Chart of nuclides – Radio-active decay –Radio-activity α–, β–, γ– Radiations – Decay Series – Radioactive equilibrium- Induced activity – Activity in Earth – Cosmic rays – Review Questions.

Chapter 3: STABLE PROPERTIES OF NUCLEI I - Charge, Mass 34
Introduction – Nuclear charge – Measurement of Nuclear charge –Rutherford α – Scattering – Nuclear mass –proton-electron hypothesis – Thomson's mass spectrograph – Principles of mass spectroscopy –Aston, Dempster, Bainbridge, Nier, Double-focussing, Time of flight spectrometers – Mass of Neutron – Review Questions.

Chapter 4: STABLE PROPERTIES OF NUCLEI II- Nuclear Magnetic Moments 53
Introduction – Intrinsic nuclear spin I – properties –proton-neutron theory – Parity – Symmetry –Nuclear magnetic dipole moment - μ_N, $\vec{\mu}_p$, g_I, -Measurement of I - NMR, NMR setup –Mossbauer method, parameters, δ, ΔE_Q $\vec{H}_n$ - experimental setup, Nuclear Zeeman splitting, Applications – Review Questions.

Chapter 5: STABLE PROPERTIES OF NUCLEI III: Electric Quadipole Moments 80
Nuclear Electric moment – Shape Q_I – Quadripole moment – measuring Q_I - NQR, Mossbauer methods – Nuclear charge density & radii –Nuclear size – Review Questions. -

Chapter 6: THE INTERNUCLEON POTENTIAL 93
Preliminaries, binding energy, mass defect, packing fraction – Stability of nuclides – Nature of nuclear force – Two nucleon interaction – Deuteron and its properties – $n - p$ Scattering – Deuteron ground state -, μ_D, Q_D, di-neutron, di-proton – Photo-disintegration –Yukawa (meson) theory & potential – Range of nuclear force Discovery of meson – Pions –OPEP, TPEP, Nucleonic structure – Review Questions

Chapter 7: STABILITY OF NUCLEI 111
Introduction – Chart of Nuclides, Segre chart'

Chapter 8: NEUTRON – ITS DISCOVERY AND APPLICATIONS 115
Discovery of neutron, Bethe-Becker & Chadwick's experiments, Types of neutrons, Neutron sources- Neutron properties, M_n, spin, μ_n, decay, life-time $\tau_{1/2}$ - Neutron detection – Interaction of neutrons with materials, neutron diffraction, NAA, & applications – Neutron interactions, elastic & inelastic scattering, Absorption, Neutron capture – Review Questions.

Chapter 9: PARTICLE DETECTORS 129

Introduction – Gas ionization curve – Gas counters, Proportional-, GM counters – Cloud & Diffusion – Nuclear emulsion detectors – Scintillation type – Solid state detectors – High energy detectors, Bubble & Spark chamber, Cerenkove – Counter electronics – Image detectors. – Review Questions-

Chapter 10: DYNAMIC PROPERTIES OF NUCLIDES I - WEAK INTERACTIONS- BETA DECACY 144

Nuclear reaction equation - β^- – , β^+ – ,decays, Ec process,- β^- - ray sp[ectrum, Sargent diagram, - Range od beta rays- Energetics of β^- decay – Fermi's QF theory – Double β^- decay, Neutrino burning, - Polarizations of β^- ,- Neutrino helicity, Neutrino detection – Beta ray spectrometer – Review Questions.

Chapter 11: DYNAMIC PRPERTIES OF NUCLIDES II – EM INTERACTIONS- GAMMA DECAY 160

Preliminaries – 1234 Rule and energetic of γ-rays, FWHM of γ-rays, Lorentzian, Voigt profiles,- Absorption of γ-rays, Compton, internal production, Pair production, annihilation, Range of, - Multi pole moments , classification of γ-rays, , Selection rules,   Parity conservation,   Weisskopf estimates - - Bremstrahlung – Energetic of γ^- - Angular correlations - γ^- ray spectrometer – Mossbauer effect - Review questions.

Chapter 12: DYNAMIC PROPERTIES OF NUCLIDES III - STRONG INTERACTIONS- ALPHA DEACY 188

Alpha decay & spectra – Geiger-Nuttal rule - En3rgerics of α- decay , disintegration equation - α- particle spectra – Theory of α- -decay, Geiger-Nuttal formula, Barrier penetration, Tunnel effect -- α- particle sources – Review questions.

Chapter 13: MODELS OF NUCLEAR STRUCRTURE 206

Introduction – Uniform particle Model, Independent particle Model – Shell Model , Magic numbers, Electric nuclear quadrupole moments, and I - Wave mechanics of shell model, Shell structure, - Relationships between nuclear I and μ , Calculation of μ, Schmidt formula – Nordheim's rules, Double scattering – Conclusions – Liquid Drop Model – Volume, surface & Coulomb energies, Weizsaecker formula, Line of stability, Mass parabolas, - Abundance of nuclides – Fermi Gas Model – Collective Model – Unified Models, Nilsson & Optic models – Review Questions.

Chapter 14:   PARTICLE ACCELERATORS   250

Introduction – ES Generators, Cockroft-Walton, Van de Graff & TANDEM – Cyclotron – Linac – Synchro-cyclotron – Betatron - Synchrotron and different types – Colliding beam type - Review Questions.

Chapter 15:   NUCLEAR REACTIONS   267

Introduction - Based of CN Model – Direct Reactions – Reaction Cross-section, dynamics of - Scattering , Q of reaction, Reaction types, - Continuum theory of - Method of mass determination from - Differential & Total scattering Cross-section, CM and L- systems- and their relationships, in Elastic & Inelastic cases, Mean free path - Reaction rate - High energy reactions – High energy and Nuclear physics - Review Questions.

Chapter 16:   FISSION OF NUCLIDE   303

Introduction – Theory of Fission , Energetic of fission fragments – Q-value of binary fission, E_B/A stability plot, - Fission Barrier , Bohr-Wheeler criterion, Fission mechanism – Spontaneous Fission – Induce fission –critical value, n-capture, other products of fission, Spectrum of neutrons, prompt & delayed neutrons, Transuranic elements - Nuclear Reactors, Chain reaction, Moderators, Four factor reactor formula, - Nuclear Power- Features of nuclear reactor – BWR, HWR (PHW), Breeder type, Breeding cycles, Uranium & Thorium – Conclusions – Review Questions. -

Chapter 17:   THERMO-NUCLEAR REACTIONS (NUCLEAR FUSION)   336

Introduction – Fusion – Reasons for thermo-nuclear fusions -, Hydrogen burning – Nucleo-synthesis , PP cycle, CN Cycle, Basic exo-thermic reactions – Fusion in Sun and Stars – Origin of elements – Review Questions.

Chapter 18:   PARTICLES AND THEIR INTERACTIONS   354

Introduction – Sub-atomic particles Chronology, History, - Families of particles Units in nuclear physics - Intrinsic particle properties, Conservations laws , Electric charge, Lepton number, Baryon number, Isospin, - Strong Interaction, - EM interaction -Weak interactions , Strangeness, Associated production, Hypercharge, Resonances, Hypernuclei, -Quark Model of Particles – Further details, Eight-fold way, Spin-$\frac{1}{2}$ Baryon Multiplet, Spin-0 Meson Multiplet, The Decuplet of Baryons, Confinement and Quark Properties, Gell-Mann-Nishijima Mass Formula - Hadron Multiplets, Baryon Multiplet, Beta Decay Revisited, e Quark Structure – Weak interaction and Unification, An Exotic Five-quark particle – Standard Model, Evolution of matter, Proton decay, $\hat{C}\hat{P}$-invariance, $\hat{C}\hat{P}\hat{T}$ Theorem - **The Higgs boson – String theory,** Extra Dimensions, M-theory, United theory – Review Questions.

Chapter 19:   NUCLEAR POWER   409

Introduction , Turning Points in Nuclear India – Nuclear battery, - Basic principle of a Reactor , Fission chain reaction, Nuclear fuel cycle, , Uranium Preparation, Moderators, n Reactor formula, , Atomic Pile, - Types of Reactors, First Reactor in India, Nuclear

Power Plant, PWR, BWR, HWR, SGHWR, Gas-cooled Reactor, Power Plants of India – Breeder Reactors, Breeding cycle, U & Th –LMFBR LWBR, Thermal breeder Reactor, IFR, World's safest ATBR – Nuclear Weapons, A-Bomb, H-Bomb, Neutron Bomb (ERW), - Chronology of fabricated devices, - Application of Plasma, Methods of confinement, - DCX Machine, Tokomak, Stellarator, Laser Fusion, - Cold Fusion – Review Questions,

Chapter 20:   NUCLEAR INDIA   442
Introduction   - Establishment of DAE, - India's First Reactor, - India's First Bomb, Missile Programme, , Missiles of India, Nuclear Weapon State, - Pockhran II – Nuclear Power Plants, ATBR, - India's Operating Reactors, - Indo-US Nuke Deal , Indo-Jap Nuke Deal – Conclusions.

Appendixes
    A: BIBLIOGRPHY   453
    B: VALUES OF PHYSICAL CONSTANTS   457
    C: LIST OF ELEMENTS   460

About the Author

ooooooo0oooooo

Chapter 1
INTRODUCTION

Chapter 1

INTRODUCTION

"The high destiny of the individual is to serve rather than to rule"   -Albert Einstein

"The great marvel is not the size of the enterprise, its secrecy or cost, but the achievement of scientific brains in putting together infinitely complex pieces of knowledge held by many men in different fields of science into a workable plan."   -US President Truman

1   PHYSICS: THE MOST FUNDAMENTAL OF ALL NATURAL SCIENCES

1.1   The Pyramid of Science

A discernible hierarchy (not of social value or of intellectual power) exists in Science. It is known (Leon Lederman, 1993) that there exists a pyramid of science. The base of the pyramid is mathematics, as it does not depend on any other disciplines. Physics (and astronomy) lies on the next layer of the pyramid, because it relies on mathematics. Next on the upper layer falls chemistry, which depends on physics, and not the vice-versa. Akin to this is physical chemistry, mathematical physics. Next comes biological sciences, (include biochemistry and biophysics), which invariably are dependent on both chemistry and physics. Further upper layers of the pyramid become increasingly blurred and less definable, as one reaches physiology, medicine, psychology, biotechnology, *etc*. In a nutshell one may accept the old saying that physicists defer only to mathematicians, whereas mathematicians defer only to GOD.

Rudolf L. Mossbauer (2002), Nobel Laureate, expressed as follows - "Physics attempts to reduce the course of natural events to comprehensive principles - the laws of nature. These principles are universally valid, nothing can escape them. Whereas everything, that is material in the world, is continually subjected to change; the laws of nature are timeless. At any time and in any place we can trust in them and build on them".

It was not until the 17th Century that the methodology of modern physics was founded by Johannes Kepler, Galileo Galilei and Isaac Newton detaching individual occurrences from their contexts, examining them quantitatively with the help of experiments, and ultimately mathematically formulating fundamental physical principles.

The electronic structure of matter was introduced at the end of the 19th Century by physicists. Joseph John Thomson discovered (1897) the *electron*, the first indivisible particle to date. At the dawn of the 20th Century Albert Einstein revolutionized our notion of space and time applying his Theory of Relativity. Jointly with Max Planck he discovered *photons* as elementary particles of light

Werner Heisenberg, Erwin Schrödinger, Paul A.M. Dirac and Wolfgang Pauli solved the problem of wave-particle dualism with the development of their Quantum Theory. These discoveries mark the beginning of Modern Physics in the 20th Century. Further new and exciting phenomena and laws of nature have been discovered by physicists. The goals of research in the field of physics are often far removed from anything that humans are actually able to experience. However, sooner or later the discoveries made have a determining influence on human lives. This is what makes *physics the most fundamental of all the natural sciences*. It is a formative part of human culture. Physics spans a wide range of themes, from the cosmological dimensions to the elementary particles of an atomic nucleus.

The decoding of deoxyribonucleic acid (DNA), the development of the MRI and the determination of the sequence of the human genome are two among the many discoveries shaped by techniques in physics. It is an accepted fact that physical research is an integral part of natural and engineering science. It forms the basis of modern technology; for example, the semiconductor transistor, discovered in 1947, by physicists John Bardeen and Walter Brattain, has fundamentally altered electronic engineering. This advancement, and the development of solid state devices, and their miniaturization in LSI (large-scale integrated circuits), VLSI, has led to develop PCs and communication technologies.

1.2 OBJECTIVES OF STUDY OF NUCLEAR PHYSICS

Back in 1900, physicists believed that they had discovered more or less all the laws governing the Universe. The Laws of Classical Mechanics (due to Isaac Newton), Electromagnetism (due to Andre Marie Ampere, Karl Friedrich Gauss, Michael Faraday and James Clerks Maxwell) and Thermodynamics (due to James Prescott Joule, Lord Kelvin and Ludwig Boltzmann) seemed to explain pretty much everything.

However, at the beginning of the 20th Century two new theories emerged that revolutionized the way physicists looked at the Universe – Albert Einstein's theory of Special Relativity and Quantum Mechanics (due to such physicists as Niels Bohr, Louis deBroglie, Werner Heisenberg, Erwin Schrödinger and Paul AM Dirac). The unification of these theories led to the idea of *antimatter*. This in turn led to the mystery of why the Universe contains more matter than antimatter. It is this mystery that, at SLAC the experiment called **BaBar** experiment using the electron-positron collider, was investigated.

Until 1932, the *elementary* particles were the electron, the proton and the neutron. A perusal of the Particle Data Book will show more than 200 particles listed. They consist of the *gauge bosons* and *material particles*. Particle physics is the study of the fundamental constituents of matter and their interactions. Modern theory, called the **Standard Model**, tries to explain all the phenomena of particle physics in terms of the properties and interactions of a small number of particles of tree distinct types, *viz.*, **leptons, quarks** and **gauge bosons**. The fundamental forces involved are the EM interaction, the Strong force the Weak interaction and Gravity.

1.2.1 Structure of Matter

Matter exists in three forms, *viz.*, solids, liquids, vapours, all made of molecules, which in turn are made up of atoms. The quest for understanding the structure of matter dates back for at least a couple of millennia. The alchemists' efforts during the middle ages resulted in the science of chemistry, which in turn led to the idea of the '**elements**', *i.e.*, the atoms. Although it was thought at first that atoms are indivisible, it is now known that they have definite structure.

In the subatomic world the laws of motion and the types of forces that govern Nature deviate from Newton's laws. At present scientists Quantum Mechanics and Relativity, and three more forces in addition to Gravity (Electro-magnetic, Strong and Weak). QM applies to microscopic objects of subatomic sizes. Relativity applies to fast moving objects. The limit of QM for macroscopic objects, and the limit of relativity for slow moving objects, is in agreement with Newton's laws of motion. Understanding of layers of matter and their uses, molecules, atoms, nuclei, protons & neutrons and other "*elementary*" particles, quarks & leptons has reached sufficiently enough.

In this book I give a detailed analysis of the nucleus.

1.2.2 Size of Atoms and Nuclei

It was Lord Ernest Rutherford in 1911 who established the fact that the atom consists of a nucleus surrounded by electrons. Their physical dimensions are (Fig 1.1):

Diameter of an atom is $\sim 10^{-8}$ cm = 0.1 nm

Diameter of a nucleus is $\sim 10^{-12}$ cm = 10 fm
Diameter (classical)) of an electron is ~ 5 fm

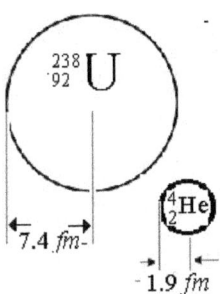

Radii of typical nuclei
Fig 1.1 Physical dimensions of nuclei.

1.2.3 Nuclear physics, more or less, as is known today, might rather arbitrarily be said to have started with the discovery of the neutron by James Chadwick in 1932. Atomic dimensions are already much too small for *direct* visual observations, but can be studied by *indirect* methods. The end of the 21st Century it was understood that no single country or a Laboratory can conduct high speed particle experiments, but only a collective funding to build a huge Laboratory facility involving co-operation from hundreds of scientists collaborating to study a problem will lead to advancement of knowledge in particle physics is possible. The principal source of study of nuclear physics is from data obtained by indirect methods. This is to understand how the nucleus is constructed, viz.,

i) What are the building blocks of the nuclei?
ii) How do they move relative to one another?
iii) What are the laws ruling the forces which hold the constituents of a nucleus together?
iv) Which are the elementary particles that combine to form other particles?
v) Formation of Grand Unification of physical theories, to understand nature.
vi) What is GOD particle?

Worked out Example 1.1
The nucleus of iron ^{56}Fe has matter density 2×10^{17} kg m^{-3}. Obtain its radius.

Solution: Step # 1 $\rho = 2 \times 10^{17}$ kg m^{-3}, A = 56, $\rho = M/V$,

Step # 2 $\rho = M/[\frac{4}{3}\pi R^3]$, $R = \left(M/[\frac{4}{3}\pi\rho]\right)^{1/3} = \left(\frac{(56)(1.66 \times 10^{-27} \text{ kg })}{(\frac{4}{3}\pi)(2 \times 10^{17} \text{ kg m}^{-3})}\right)^{1/3}$

$R = 4.8 \times 10^{-15}$ m = 4.8 fm

1.3 TURNING POINTS IN THE DEVELOPMENT OF NUCLEAR PHYSICS

The idea that matter consists of an assembly of a-toms was derived from Democritus (430 BC). Chemists had no answer to the question, say, why one gram of oxygen combines always with eight grams of hydrogen, not in any other proportion, as the resulting *water molecule* contained a fixed number of **atoms** of each kind. John Dalton (1803 - 8) concluded from his work that 2 atoms of hydrogen and one of oxygen combined to create a molecule H_2O. Michael Faraday derived in 1833 the laws of electrolysis, each atom or molecular fragment carried a **fixed electric charge**. A chronological listing of discoveries that shaped the ideas on atoms and nuclei is provided below:

1. Dimitri Ivanovich Mendeleyev (1872) was responsible for finding Periodicity of valence & developing the Periodic Table of Elements.
2. Antoine-**Henri Becquerel** (1896) discovered natural radioactivity, in Uranium salts, revealing that 'atoms are no more indivisible',
3. Joseph John Thomson (1897) discovered electrons
4. **Marie Curie** (1898) chemically separated radioactive Radium from the ores.
5. Lord Ernest Rutherford (1899) identified alpha rays and beta rays
6. Paul Ulrich Villard (1900) discovered gamma rays
7. **Albert Einstein** (1905) announced the equivalence of mass and energy equation: $E = m\,c^2$ is a cornerstone in the development of nuclear energy.
8. Frederick Soddy (1910) discovered statistical decay law in nuclear physics..
9. Victor Hess (1910) discovered cosmic rays
10. Ernest **Rutherford** (1911) formulated the Nuclear model of the atom.
11. J. J. Thomson's (1912) was responsible for the discovery of isotopes.
12. Robert A. Millikan (1913) measured experimentally the charge on an electron
13. H.G.J. Moseley (1913) introduced the concept of atomic number, from X-ray spectra.
14. Niels Hendrik David **Bohr** (1913) who gave the first successful physical model for the Hydrogen atom.
15. E. Rutherford (1919) discovered the phenomenon of nuclear transmutation by the disintegration of Nitrogen (Induced nuclear transmutation).
16. Otto Hahn & Lise Meitner (1921) discovered nuclear isomers.
17. Paul A. M. Dirac (1928) predicted the positron.
18. G. Gamow, R. Gurney & E. Condon (1928) who successfully explained Alpha-decay by using Quantum Mechanics.
19. **Ernest Orlando Lawrence** (1930) invented cyclotron
20. **Wolfgang Pauli** (1930) predicted the neutrino
21. Sir James E. Chadwick (1932) discovered the neutron.
22. **Werner Karl Heisenberg** (1932) gave *n-p* hypothesis.
23. Carl D. Anderson (1932) discovered the positron, the first antiparticle.
24. **Enrico Fermi** (1934) gave the theory of Beta-decay.
25. Otto Hahn, Lise Meitner, Fritz Strassmann (1934) split the atom for the first time; the epochal experiment
26. Hidekei Yuckawa (1935) gave theory of Meson as exchange particle between nuclear constituents.
27. Carl D. Anderson & S.H. Neddermeyer (1936) discovered mu-mesons.
28. Lise Meitner and Otto R. Frisch (1938) are responsible for Uranium Fission.
29. Hans A. Bethe (1938) showed that nuclear fusion is responsible for power in the Sun.
30. Leo Szilard and Walter Zinn (1939) demonstrated that fission reactions to be self-sustaining due to nuclear chain reactions.
31. Enrico Fermi (1942) designed the Atomic pile (first nuclear reactor), a sustained controllable nuclear chain reaction.

32. **Manhattan Project** (1942) was set up in the USA under the command of Brigadier General Leslie Groves. Scientists recruited to produce an atom bomb included J. Robert Oppenheimer (USA), David Bohm (USA), *etc*. Use of uranium and plutonium. The first three completed bombs were successfully tested at Alamogordo, New Mexico on 16th July, 1945.
33. Julius Robert Oppenheimer (1943), associated with about 200 of the best scientists, designed "**Little Boy**" $^{238}_{92}$U bomb (dropped over Hiroshima) and "**Fat Man**" plutonium bomb (dropped over Nagasaki).
34. Bombing Hiroshima (August 6, 1945) and Nagasaki (August 9 1945) in Japan.
35. Cecil Frank Powell (1946) discovered the Pi-meson.
36. Maria Goeppert-Mayer (1946) developed her "nuclear shell model".
37. Edward Teller (1952) leads a team to build the first Hydrogen bomb.
38. Clyde L. Cowan, Jr. and Frederick Reines (1955) observed the mysterious particle neutrino.
39. Tsung-Dao Lee & Chen NingYang (1956) observed the violation of conservation of parity in beta-decay.
40. Glenn Seaborg (1944 -1958) discovered 8 elements related to uranium, *viz*.,: americium, curium, berkelium, californium, einsteinium, fermium, mendelevium, and nobelium. When element 106 was discovered, it was named after him, seaborgium.
41. **Murray Gell-Mann** (1963) discovered the Quarks.
42. Murray Gell-Mann, George Zweig, Oscar Greenberg, Yoichiro Nambu and Yuval Ne'eman (1977) developed quantum chromo-dynamics (QCD) theory of strong interactions.
43. (1964) The first three quarks (up, down, and strange) are hypothesized.
44. **Steven Weinberg**, *et al*. (1970s) gave the Standard Model of nucleus as a fundamental and well-tested physics theory to describe the building blocks of the Universe.
45. (1974) Evidence for a fourth quark was found in November of 1974. Two experiments simultaneously announced the discovery of a meson with a mass of about $3.1\, GeV/c^2$, called the J meson by one group and the ψ meson by the second. It was later determined to be a combination of charm and anti-charm quarks. Since neither group had priority on the discovery, the meson is now called J/ψ Like many particles discovered in the 20th Century, it is given the name "charmonium".
46. (1977) The discovery of the bottom quark
47. (1995) Mass of the top quark was determined. The top is so massive and short lived that it does not live long enough to combine with other quarks to form a hadron. In fact the top quark is more massive than many atoms.
48. Maxim Polyakov, Dmitri Diakonov, and Victor Petrov (1997) predicted the existence of a **pentaquark** with a mass about 50 % heavier than that of a hydrogen atom. Atoms are formed of two types of elementary particles, *viz*., electron and quarks. The discovery of quarks – particles combination makes up the protons and neutrons present in the nuclei of atoms.
49. (2003) Strong evidence for the existence of the pentaquark.
50. **Higg's boson** (nick name The GOD particle) is so central to the state of physics today, so crucial to our final understanding of the structure of matter, yet it is elusive. It is the primary reason for building the SSC (Superconducting Super Collider) at the CERN, near Geneva, Switzerland. Reports of finding Higg's boson is in question (2011).

(It is learnt that Homi J. Bhabha suggested the name 'meson', now used for a class of elementary particles. India (1972) exploded its first atomic bomb).

1.4 NUCLEUS AND NEUTRON-PROTON HYPOTHESIS

Once the electron was discovered, there was development of models of the atom. In the early 1930s, it was thought that the atom was composed of just three particles, *viz.*, proton, electron and neutron, even before neutron was experimentally observed.

1.4.1 What are the Constituents of Nuclei in 1930s?

1. Proton: It is the nucleus of the Hydrogen atom
   Charge: Positive, +e
   Mass: ~1830 times the mass of an electron
2. Neutron: Postulated in theory of β- decay
   Charge: Zero
   Mass: ~ 0.1 % higher than the mass of a proton.

1.4.2 Definition of a Nucleon

Nucleon is a generic name for both proton and neutron. The various nuclei are different combinations of neutrons and protons. A nucleus with atomic number Z contains A nucleons, *i.e.*, Z number of protons and $(A - Z)$ number of neutrons.
Neutron-Proton Hypothesis for the nucleus was based on this.

1.4.3 What is a Nuclide (or Nuclear species)?
Any specific combination of protons and neutrons is called a nuclear species

1.4.4 Classification Systems and Nomenclature
A review of some of the commonly used terms and the system for classifying nuclei, based on convenience and tradition is given below..

1. Proton number (Atomic number) Z
2. Neutron number N
3. Mass number $A = (N + Z)$
4. Stable nuclei $Z = \dfrac{A}{[1.98 + 0.0155\, A^{2/3}]}$
5. Nuclide of an element $E(A, Z)$ $^{A}_{Z}El$
6. Isotopes (*iso* → equal; *topes* → place) Nuclides with identical Z but different N
7. Isobars   Nuclides with the same A;
   eg., $^{202}_{80}Hg$ & $^{202}_{82}Pb$
8. Isotones   Nuclides with constant N, but different Z.;
   eg. $^{13}_{6}C$ & $^{14}_{7}N$
9. Isomers   Two nuclides (Nuclei of the same species) in different excited states of which at least one is '*metastable*'. $^{A}_{Z}El$ & $^{Am}_{Z}El$
10. Light nuclei   Those nuclei in which $N=Z$; *i.e.*, up to $^{40}_{20}Ca$
11. Heavy nuclei   Nuclei having $N > Z$

12.	Isodiapheres	Nuclei having the same excess of neutrons over protons, $(A - 2Z)$; $^{37}_{17}Cl$ & $^{39}_{18}Ar$
13.	Atomic mass, M	Exact value of mass of a neutral atom relative to that of a neutral $^{12}_{6}C$
14	Atomic mass unit	$1\ amu \equiv 1u = \frac{1}{12}$ (mass of $^{12}_{6}C$)
15	Other physical Data	

Atomic weight $= N M_n + Z (M_p + m_e)$ - (binding energy)

$1\ u = 931.5\ MeV/c^2 = 1.66 \times 10^{-27} kg$

$m_e = 0.00054858\ u = 0.511\ MeV = 9.1094 \times 10^{-31}\ kg$

$M_p = 1.007276\ u = 938.27\ MeV = 1.67262 \times 10^{-27}\ kg$

$M_n = 1.007825\ u = 938.78\ MeV = 1.67353 \times 10^{-27}\ kg$

$^1_1H = 1.008665\ u = 939.57\ MeV = 1.67493 \times 10^{-27}\ kg$

Radius of a nucleus is $R = r_0\ A^{1/3}$ with $r_0 = 1.2 \times 10^{-15}\ m = 1.2\ fm$

For example,

$R_{He} = (1.2\ fm)(4)^{1/3} = 1.9\ fm;$
$R_{Cu} = (1.2\ fm)(64)^{1/3} = 4.8\ fm;$
$R_U = (1.2\ fm)(238)^{1/3} = 7.4\ fm.$
$R_{Am} = (1.2\ fm)(243)^{1/3} = 7.5\ fm.$

REVIEW QUESTIONS

RQ. 1.1 Find the number of neutrons, protons and electrons in a) ^3He and b) ^{206}Pb .(Answers: a) 1, 2, 2 b) 124, 82, 82).

RQ. 1.2 How many protons, neutrons and electrons are present in the atom $^{235}_{92}U$? (Answer:, 92, 143, 92).

RQ. 1.3 The binding energy per nucleon for elements near iron in the Periodic Table is about $8.90\ MeV$. Calculate the atomic mass, including electrons, of $^{56}_{26}Fe$? (Answer: M =55.9 u).

RQ. 1.4 Find the binding energy of $^{107}_{47}Ag$, which has an atomic weight of $106.905\ u$. (Answer: $915\ eV$).

RQ. 1.5 Determine the classical radius of an electron. Given: $c = 2.997925 \times 10^8\ ms^{-1}$, $m_e = 9.1094 \times 10^{-31}\ kg$, $\varepsilon_0 = 8.8542 \times 10^{-12} C^2 N^{-1}\ m^{-2}$, $e = 1.6021 \times 10^{-19}\ C$ (Answer: $r_0 = 2.818\ fm$).

@&@&@&@&@&@&

Chapter 2

NATURAL RADIOACTIVITY – A FLASHBACK

Chapter 2

NATURAL RADIOACTIVITY – A FLASHBACK

"The important thing in science is not so much to obtain new facts as to discover new ways of thinking about them". ~William Lawrence Bragg

2 INTRODUCTION

Of the 6,000 species of nuclei that can exist in the universe, about 2,700 are known, but only 270 of these are *stable*. The rest are *radioactive*, that is, they spontaneously decay. The driving force behind all **radioactive decay** is the ability to produce products of greater stability than one had initially. In other words, radioactive decay releases energy and because of the high energy density of nuclei, that energy release is substantial. The phenomenon of radioactivity has played a significant role in the development of both atomic and nuclear physics. In this Chapter the radioactive disintegration of nuclei will be discussed, for providing the graduate students as a flashback, since I consider it as an essential introduction to all those pursue further in Nuclear physics. Therefore this chapter may be treated as a prerequisite to the graduate course programme.

Nuclei can undergo a variety of processes resulting in the emission of radiation of EM (X-rays and gamma rays) or corpuscular type ($\alpha-, \beta-$, and positrons, internal conversion electrons, Auge electrons, neutrons, protons, fission fragments, among others).

2.1 Nuclear Disintegration

Antoine Henri Becquerel (1896) discovered natural radioactivity in uranium. In 1898 Marie Sklodowska and Pierre Curie discover polonium and another new radioactive element, which they name "radium" The three distinct types of accelerated particles from radioactive decay are named after the first three letters of the Greek alphabet: α-(alpha), β-(beta), and γ-(gamma) separated by a magnetic field positive alpha particles bend one direction negative beta particles bend opposite neutral gamma rays do not bend at all. Credit goes to Ernest Rutherford (1898) for identifying alpha rays and beta rays and to Paul Ulrich Villard (1900) for gamma rays. Beta rays are proved to consist of high speed electrons by Fritz Geisel, Antoine-Henri Becquerel, and Marie Curie (1898).

In these radioactive processes, nuclear radiation occurs in other forms, including the emission of protons or neutrons or spontaneous fission of a massive nucleus. Of the nuclei found on Earth, the vast majority are stable. This is so because almost all short-lived radioactive nuclei have decayed during the history of the Earth. There are approximately 270 stable isotopes and 50 naturally occurring radioisotopes (radioactive isotopes). Thousands of other radioisotopes have been made in the laboratory

Radioactive decay will change one nucleus to another if the product nucleus has a greater nuclear binding energy than the initial decaying nucleus. The difference in binding energy (comparing the before and after states) determines which decays are energetically possible and which are not. The excess binding energy appears as kinetic energy or rest mass energy of the decay products.

2.2 THE CHART OF THE NUCLIDES

The Periodic Table of elements is of limited use to the nuclear physicist, as it gives only limited information about the nuclear properties of an element. The Chart of the Nuclides is a plot of nuclei as a function of proton number, Z, and neutron number, N. as schematically shown in Fig 2.1. There is the N = Z line, a diagonal line.

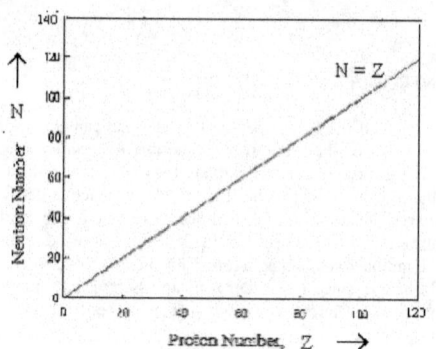

Fig 2.1 The N = Z line Nuclide chart.

2.2.1 The Radioactive Decay Law (Statistics of Disintegration)

The nuclear system does not age as does a biological system. Radioactive disintegration is a statistical one-shot process and is spontaneous. It was Schweidler in 1905 that advanced the idea that the active decay is statistical in character.

2.2.2 Why is Radioactive Decay called Spontaneous?

Energies involved in nuclear disintegrations are so large compared with interaction energies associated with thermal agitation, visible photons, chemical binding, and all other ordinary influences, that nuclear decays are independent of environment.

2.2.3 Decay Rate [λ, Decay or disintegration constant]

At what rates do nuclear disintegrations take place? Every nuclear decay involves the emission of a β- or an α- particle from the nucleus, called the *parent*, which is transformed as the *product* or *daughter* nucleus. The parent and daughter are governed by some activity (R) versus time (t) law. To quantify the representation, consider

N_t = Number of parent nuclides initially present (at time $t = t$),
ΔN = Number of decays occurring during t to $(t + \Delta t)$

$$\Delta N = - \lambda N_t \Delta t \qquad (2.2.1)$$

where the negative sign denotes decay.

Number of disintegrations per second $= \dfrac{\Delta N}{\Delta t} = - \lambda N_t$

$$\therefore \frac{dN_t}{dt} = \lim_{\Delta t \to 0} \frac{\Delta N}{\Delta t} = - \lambda N_t \qquad (2.2.2)$$

λ = *Decay or disintegration constant*, characteristic of the decay for a given nuclide.

2.2.4 Decay Law

Ernest Rutherford & Frederick Soddy experimentally established that all radioactive processes follow an exponential law. If
N_0 = Number of parent nuclides initially undecayed at $t = 0$, from equation (2.2.2) on integration

$$\int_{N_0}^{N_t} \frac{dN_t}{dN} = \int_{0}^{t} -\lambda \, dt$$

gives $\quad N_t = N_0 \, e^{-\lambda t}$ (2.2.3)

This is known as the *Radioactive decay Law*.

2.2.4.1 Why is the Radioactive Decay Law called statistical?

The expression for probability

$$P = \frac{\Delta N}{N_t} = \lambda \, \Delta t$$

refers to the decay of the parent nuclide in a short time interval Δt.
Let Q_n = probability that a given nuclide does not decay during n number of intervals Δt, then

$$Q_n = (1 - P)^n = (1 - \lambda \frac{t}{n})^n.$$

For the reason

$$\lim_{n \to \infty} Q_n = \frac{N_1}{N_0}$$

and using the binomial expansion for $(1 - \lambda \frac{t}{n})^n$, the above expression yields the statistical formula (2.2.3). This is a law of chance!

2.2.4.2 Experimental result of Rutherford & Soddy
Graphically the experimental result of Rutherford & Soddy is illustrated in Fig 2.2.

2.2.5 Source Activity (R) of a Sample

The Activity (or Strength) of a radioactive sample substance is the *number of decays per second*, defined as

$$R = -\frac{dN_t}{dt} = (\lambda \, N_0) \, e^{-\lambda t} \quad (2.2.4)$$

The activity depends only on the number of decays per second, not on the type of decay, the energy of the decay products, or the biological effects of the radiation.

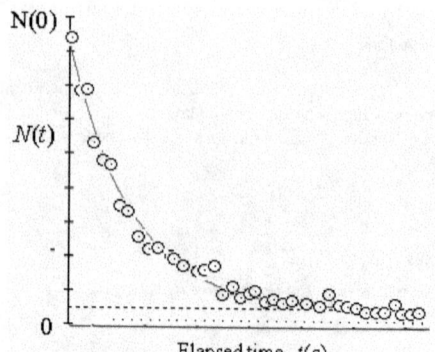

Fig 2.2 The Radioactive Law - Graphical

2.2.6 Units of Radioactivity

a) Traditionally activity is measured in Units of Curies (Ci). Originally, it is defined as the activity of 1 gm of pure radium, $^{226}_{88}Ra$.

$1\ Ci$ = The activity of 1 g of radium $^{226}_{88}Ra$

$\qquad = 3.7 \times 10^{10}$ disintegrations per second. (2.2.5)

This of course is a very large unit.

b) The number of disintegrations (decays) per second, or activity, from a sample of radioactive nuclei is measured in **Becquerel** (Bq), after Henri Becquerel. It is defined as

$\qquad 1\ Bq$ = One disintegration per second. (2.2.6)

c) Another unit of activity is the **Rutherford** (rd)

$\qquad 1\ rd = 10^6$ disintegrations-s^{-1} (2.2.7)

$1\ Ci = 3.7 \times 10^{10}\ Bq$

$1\ m\ Ci = 37\ MBq$

$100\ kBq = 2.7\ \mu\ Ci$

$1\ Roentgen = X$ – ray quantity producing ionization of 1 $esu\ cm^{-3}$

$\qquad = 2.58\ Coul\ kg^{-1}$ in air at STP

2.2.7 Mean Life (τ)

Combined ages of atoms in the group, t to ($t + \Delta t$), is $t\,dN$. Soddy showed that the average (or mean) life time τ of a single atom is

$$\tau = \frac{\sum\limits_{\substack{all\ atoms \\ 0}} t.dN}{\int\limits_{N_0} dN}$$

This becomes simplified, by using the transformation $\int\limits_{N_0}^{0} dN \rightarrow \int\limits_{0}^{\infty} dt$,

$$\tau = \frac{e^{-\lambda t}}{\lambda}\bigg|_0^\infty = \frac{1}{\lambda} \qquad (2.2.8)$$

2.2.8 Half Life ($\tau_{1/2}$)

Another calculation which is conceptually important radioactive statistics is as follows. The decay rate of a nuclide can be characterized by the term *half-life*, $\tau_{1/2}$, the time during which the number of parent nuclides reduces by a factor of 2 (Fig 2.3), *i.e.*,

$$N(\tau_{1/2}) = \frac{N_0}{2} \qquad (2.2.10)$$

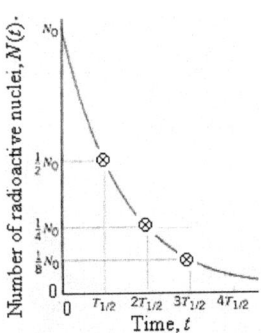

Fig 2.3 Half life of activity illustrated

It can be shown that

$$\tau_{1/2} = \frac{\ln 2}{\lambda} = \frac{0.693}{\lambda} = \tau \ln 2 \qquad (2.2.11)$$

or $\lambda = \frac{0.693}{\tau_{1/2}}$

or $\tau = 1.44 \, \tau_{1/2} = \frac{\tau_{1/2}}{\ln 2} \qquad (2.2.12)$

Some half-lives for decaying radio-nuclides are listed in Table 2.1.

In this notation the exponential form (2.2.3) of the radioactive law can be replaced by

$$N_t = N_0 \, 2^{-t/\tau_{1/2}} \qquad (2.2.13)$$

2.2.9 Determination of λ of a Substance

<u>Case 1</u>. If λ = small, $\tau_{1/2}$ = large. λ can be computed from a measurement of disintegration rate, $-\frac{dN_t}{dt}$ because $-\frac{dN_t}{dt} = (\lambda \, N_0)$.

Table 2.1   Some Half-lives for Radioactive Decay

Isotope		Half-life	Decay Mode
Polonium	$^{214}_{84}Po$	1.64×10^{-4} s	α, γ
Krypton	$^{89}_{36}Kr$	3.16 min	β^-, γ
Radon	$^{222}_{86}Rn$	3.83 da	α, γ
Strontium	$^{90}_{38}Sr$	28.5 yr	β^-
Radium	$^{222}_{88}Ra$	1.6×10^3 yr	α, γ
Carbon	$^{14}_{6}C$	5.73×10^3 yr	β^-
Uranium	$^{238}_{92}U$	4.47×10^9 yr	α, γ
Indium	$^{115}_{49}In$	4.41×10^{14} yr	β

Case 2. λ = large enough so that **activity** $R = -\dfrac{dN_t}{dt} = (\lambda \, N_0)$ gives the log R versus t plot, the slope of which will be $(-\lambda)$.

Worked out Example 2.1

It is found that ^{60}Co decays to ^{59}Co with $\tau_{1/2}$ =5.3 yrs.. i) Find out the activity of a ^{60}Co source of 0.015g, and ii) Determine the activity of the source 2 years later. (Given, $N_A = 6.0225 \times 10^{23} \, mol^{-1}$)

Solution:: $\boxed{Step\#1}$ $\tau_{1/2}$ =5.3 yrs, $N_A = 6.0225 \times 10^{23} \, mol^{-1}$, m= 0.015 g, $1 \, Ci = 3.7 \times 10^{10} \, s^{-1}$,

$\lambda = \dfrac{0.693}{\tau_{1/2}}$

$\boxed{Step\#2}$ N_0 = # of ^{60} atoms present at the start

$= \dfrac{(0.015g)}{60 g/mol}(N_A = 6.0225 \times 10^{23} \, mol^{-1}) = 1.5 \times 10^{20}$

$\boxed{Step\#3}$ 1) Activity, $\dfrac{dN}{dt}\bigg|_{t=0} = \dfrac{dN_t = N_0 \, e^{-\lambda t}}{dt}\bigg|_{t=0} = \lambda N_0$

$= \dfrac{(0.693)}{\tau_{1/2}} N_0 = \dfrac{(0.693)}{5.3yr}(1.5 \times 10^{20}) = 1.90 \times 10^{19} \, yr^{-1}$

$= \dfrac{(0.693)}{(5.3)(365)(24)(3600)s}(1.5 \times 10^{20}) = 6.2 \times 10^{11} \, s^{-1}$

$= (6.2 \times 10^{11} \, s^{-1})/(, 1 \, Ci = 3.7 \times 10^{10} \, s^{-1}) = 16.9 \, curies$

$\boxed{Step\#4}$ ii) $\dfrac{dN}{dt}\bigg|_{t=2yrs} = (\lambda \, N_0) \, e^{-\lambda t} = (16.9 \, Ci)[\exp(-(0.63)(2yr)/(5.3.yr)] = 13 \, Ci$.

2.2.10 Nuclear Level Diagram

A student of nuclear physics will be making use of Nuclear Energy Level diagrams which provide a compact and convenient way of representing changes that takes place during a nuclear transformation. For a nucleus with atomic number Z, and mass number A, the energy levels are plotted as horizontal lines on some arbitrary vertical scale. The nuclear spin I and parity of each of these energy state will also be shown. This pattern is for nucleus (Z, A). In the same plot nuclei with (Z-1, A) and Z+!, A) are also plotted. A simplest energy level diagram is depicted in

Fig 2.4. In the illustration, a β – electron decay is shown, thereby the radioactive parent nucleus $^{A}_{Z}X$ in excited state decay emitting β – ray and get transformed to a product nucleus $^{A}_{Z+1}Y$

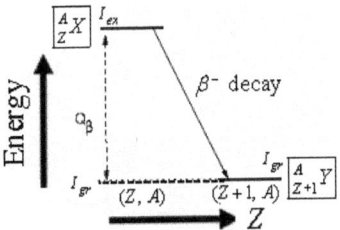

Fig 2.4 Nuclear energy level diagram

Lederer & Shirley (Ref 48) have included have tabulated several isotopes and energy level diagrams.

2.3 RADIOACTIVITY AND RADIATION

Nuclei can be stable or unstable, depending on the number of protons and neutrons. Unstable nuclei will go through one or more nuclear transformations until they reach stability.

2.3.1 Radioactivity Alpha (α-) Decay

Alpha (α-) particles are $^{4}_{2}He$ nuclei and are generally emitted by very heavy nuclei. α- particles are mono-energetic, with energy around 5 or 6 MeV. The decay is indicated by
$(Z, A) \rightarrow (Z-2, A-4) + \alpha$

The energy of the α- particle is particular to the specific nucleus. In the case of ^{241}Am disintegration, most (85%) of the α- particles emitted have a kinetic energy of 5.5 MeV, around 14% have a kinetic energy of 5.4 MeV, and around 1% have a kinetic energy of 5.3 MeV. The distribution of energies and relative intensities of these particles are collectively referred to as the α- spectrum of the ^{241}Am disintegration.

α- rays penetrate paper (Fig 2.5), but not plastics, lead and concrete.

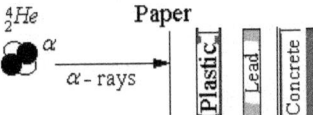

Fig 2.5 α- rays penetrability substances

An α- spectrum is representative of each specific α- decay. Therefore, it is used to identify the α- emitting radionuclide. This identification process is called alpha spectrometry.

2.3.2 Radioactivity - Beta (β-) Decay

Beta particles are fast electrons. Radio-active nuclides from light elements undergoes beta (β-) decay. Basically the emission process is the decay of neutron into proton as illustrated in Fig 2.6.

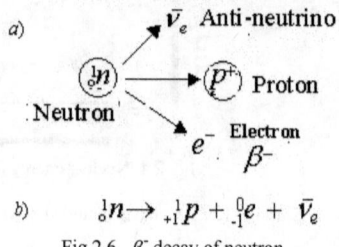

Fig 2.6 β^- decay of neutron

A basic characteristic of the β – decay is the continuous energy spectrum of the β – electrons; the reason being the sharing of the Q-value between the electron and the anti-neutrino emitted. β^- rays can penetrate substances like paper and plastics, but not lead or concrete. (Fig 2.7).

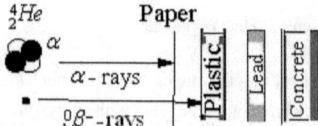

Fig 2.7 β^- rays penetrability substances

2.3.2.1 β^--decay.

An example of Negative Beta (β^-) radioactivity is the disintegration of $^{32}_{15}P$:

$$^{32}_{15}P \rightarrow\ ^{32}_{16}S + \beta^- + \bar{\nu}_e + Q_{\beta^-} \qquad (2.3.1)$$

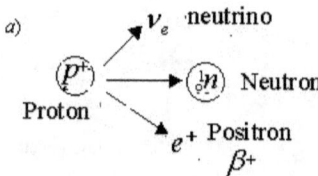

b) $^{1}_{+1}p \rightarrow\ ^{1}_{0}n + ^{0}_{+1}e + \nu_e$

Fig 2.8 Positive β- decay of proton

2.3.2.2 $_1^0e$ or β^+ - decay.

An example of $_1^0e$ or β^+ or positive β-radioactivity (Fig 2.8) is the disintegration of $_{11}^{22}Na$ (89% of the cases of disintegration):

$$_{11}^{22}Na \rightarrow\ _{10}^{22}Ne\ +\ \beta^+\ +\ v_e\ +\ Q_{\beta^+} \qquad (2.3.2)$$

The β-energy spectrum is *continuous*, which is due to sharing of the energies of β with an antineutrino $\bar{v}_e$ or neutrino v_e. It is known that this continuous energy variation of the β-particle is between zero and a maximum value, which is characteristic of the nuclide undergoing β-decay.

2.3.2.3 Electron Capture (EC)

In the case of *Electron capture* (EC) radioactivity, an electron from the closest shell (usually K-shell) of the atom of the radioactive nuclide (Z))is absorbed into the nucleus and reacts with a proton. This causes the formation of a new nucleus (Z-1) followed with emission of an electron neutrino v_e. The new nucleus if in an excited state, emits γ-rays. Further, the re-arrangement of electrons in the atom (*i.e.*, from upper shells into the empty space in the lower K-shell, leading to the emission of X-rays.

Example::EC radioactivity is $_{53}^{125}I$

$$_{53}^{125}I\ +\ e^-(K\text{-shell}) \rightarrow\ _{52}^{125}Te\ +\ v_e\ +\ Q_{EC} \qquad (2.3.3)$$

The β^- electrons and internal conversion electrons (emitted by ^{125}I) are totally different; it is in their origin (nucleus or electron shells) and energy spectra.

2.3.3 Radioactivity -Gamma (γ-) rays

Gamma (γ-) rays are produced in the disintegration of radioactive atomic nuclei. These are mionoenergetic EM radiations, with energies ranging from 10 keV to presently extend up to a few TeV. When an unstable radioactive atom decays to a more stable state, the resulting daughter nucleus may be in an excited state. Subsequently, the daughter nuclide de-excite to a lower state emitting a γ-ray photon. The energy and relative intensities of the γ-rays are characteristic to each particular nucleus. Therefore, the measure of energies and relative intensities emitted by a nucleus can be used to identify specific radio-nuclides.

Electron-positron annihilation (pair annihilation) also results in γ-ray emission.

$$_{-1}^{0}e\ +\ _{+1}^{0}e\ \rightarrow\ 2\gamma \qquad \text{(Annihilation radiation)} \qquad (2.3.4)$$

Therefore, every time a positive beta radionuclide is used, we should include γ-radiation for all problems involving shielding, dosimetry and radiation hazard evaluation.

γ-rays can also be generated in the decay of some unstable subatomic particles, such as the neutral pion (π^0) (Fig 2.9).. In the illustration below a proton () is predicted to decays (life time 8.2 x 10^{23} yrs) to a neutral pion π^0 and positron $_{+1}^{0}e$. Subsequently π^0 decays to gives two γ-rays.

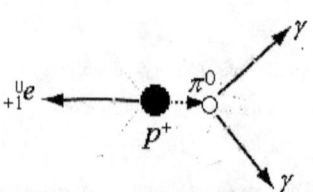

Fig 2.9 Decay of π^0 causes emission of γ-rays

Fig 2.10 is self-explanatory on the Gamma rays penetrate through matter.

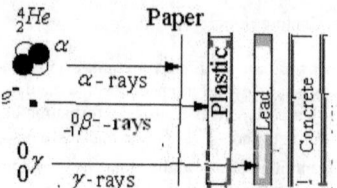

Fig 2.10 γ-rays penetrability substances

A *rad* is a unit of absorbed energy.
$$1 rem = 1\ rad = 0.1\ J/kg$$

2.3.4 Internal Conversion

Internal conversion is a mechanism of radioactive decay where a nucleus in a high metastable state interacts with one of the atom's inner orbital (K or L))electrons and causes the atom to eject its bound electron. These electrons are different from β-electrons. The internal conversion electrons have discrete energy values.

2.3.5 Neutron Emission

Neutrons $_0^1 n$ emission is a process bringing stability for an unstable atom. Actually $_0^1 n$ is emitted from the nucleus. An example is
$$_4^{13}Be \rightarrow\ _4^{12}Be + _0^1 n$$
Induced fission (Induced nuclear reactions) also produces $_0^1 n$. But spontaneous fission, such as occurs inside nat U, produces a very small number of neutrons. However, inside the nuclear reactor, the number of $_0^1 n$ can be very large. The reaction between α- particles (He nuclei) and Be is the most frequently used method to produce neutrons:
$$_2^4 He\ (\alpha) + _4^9 Be \rightarrow\ _6^{12}C + _0^1 n + Q_n \qquad (2.3.5)$$
Sources of $\alpha-$ such as Ra, Po, Am or Pu can be used, giving the neutron source is named Ra-Be, Po-Be, Am-Be or Pu-Be, respectively.

2.3.6 The Displacement Laws

Frederick Soddy and Kazimiers Fajans in 1913 described the physical nature of the α- and β- particles, by means of the Displacement Laws (also known as Fajans & Soddy Laws).. These enabled explanation for the production of various chemical elements from radioactive processes.

Law 1:
When a radioactive parent $^{A}_{Z}El$ loses an α-particle the product element $^{A-4}_{Z-2}El$ is found to be an element displaced two places to the left in the Periodic Table. and lowers the mass by four units (Fig 2.11).

$$^{A}_{Z}El \xrightarrow{\alpha} {^{A-4}_{Z-2}El} \qquad (2.3.6)$$

Example: $^{226}_{88}Ra \xrightarrow{\alpha} {^{222}_{86}Rn} \qquad (2.3.7)$

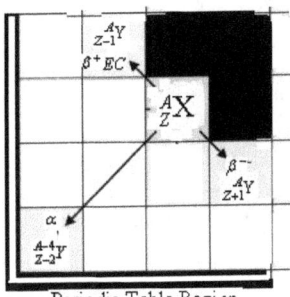

Fig 2.11 Displacement Laws depicted

Law 2:
An element $^{A}_{Z}X$ is displaced by one unit to the right $^{A}_{Z+1}Y$ in the Periodic Table as a result of the loss of a β-particle, with the atomic mass remaining the same.

$$^{A}_{Z}X \xrightarrow{\beta} {^{A}_{Z+1}Y} \qquad (2.3.8)$$

Examples: $^{14}_{6}C \xrightarrow{\beta} {^{14}_{7}N} \qquad (2.3.9)$

$^{239}_{93}Np \xrightarrow{\beta} {^{239}_{94}Pu} \qquad (2.3.10)$

2.3.7 Isotopes:

Examine the following process:

$$^{232}_{90}Th \xrightarrow{\alpha} {^{228}_{88}MsTh(I)} \xrightarrow{\beta} {^{228}_{89}MsTh(II)} \xrightarrow{\beta} {^{228}_{90}RdTh} \xrightarrow{\alpha} \qquad (2.3.11)$$

It is seen that $^{232}_{90}Th$ and $^{228}_{90}RdTh$ both have the same atomic number, but different atomic mass. They have the same chemical nature. They are said to form a pair of **isotopes** of the element Thorium. It can be said that all elements have isotopes if unstable nuclides are taken into account.

2.4 RADIOACTIVE DECAY SERIES (Genealogy of Nuclides)

Practically all the radioactive elements lie in the range of Z = 81 to Z = 92. As a result of F. Soddy's work (1910) when they are presented in a *A versus Z* plot, these are found to form members of 3 naturally occurring or radioactive series, showing successive transformations. The parents are referred to as ***ancestors***, and all other daughter nuclides are called ***members*** of the Series.

The α-, β-, and γ-decays show that the mass numbers A and atomic number Z of a given member of a radioactive Series are related to those of an ancestor by

$$A = A_0 - 4 N_\alpha$$
$$Z = Z_0 - 2 N_\alpha + N_{\beta^-} \qquad (2.4.1)$$

where N_α and N_{β^-} are respectively the number of α- and β- particles that are emitted in arriving at the given member. These two equations (2.4.1) suggest the existence of <u>4 different Series of radioactive elements</u>, clearly by a different *m* value in

$$A = 4n + m \qquad (2.4.2)$$

where *n*-value is different for the ancestor in each Series.
This means each Series is a family consisting of a succession of daughter products (members) all ultimately derived from the same ancestor (great grand parent nuclide). The end product of each Series is a <u>Stable Nuclide</u>!

2.4.1 The Thorium-232 Series (4 n – Series, m = 0)

$$A = 4n \qquad (2.4.3)$$

The longest lived ancestor of this Series is $^{232}_{90}Th$ having $\tau_{1/2} = 13.9 \times 10^9$ years (greater than 5 times the age of Earth). The end product the Series is $^{208}_{82}Pb$, which is stable.

2.4.2 Neptunium-237 Series (4 n + 1 series)

The members of the 4 n + 1 series have mass numbers specified by

$$A = 4n + 1 \qquad (2.4.4)$$

2.4.3 Uranium-238 Series

The members of this series are given by

$$A = 4n + 2 \qquad (2.4.5)$$

2.4.4 Actinium (Uranium-235) Series

$$A = 4n + 3 \qquad (2.4.6)$$

The longest lived ancestor of this Series is $^{235}_{92}U$. The sequences of the α – and β – rays that lead from parent nuclide to stable end product in this series are shown in Fig. 2.12. The salient features of the four series are listed in Table 2.2.

Table 2.2   The 4 Radioactive Series

Mass Number, A	Series	Ancestor	Half life, $\tau_{1/2}$, yrs	End product
1. $A = 4n$	Thorium	$^{232}_{90}Th$	1.39×10^{10}	$^{208}_{82}Pb$
2. $A = 4n+1$	Neptunium	$^{237}_{93}Np$	2.25×10^6	$^{209}_{83}Bi$
3. $A = 4n+2$	Uranium	$^{238}_{92}U$	4.51×10^9	$^{206}_{82}Pb$
4. $A = 4n+3$	Actinium	$^{235}_{92}U$	7.07×10^8	$^{207}_{82}Pb$

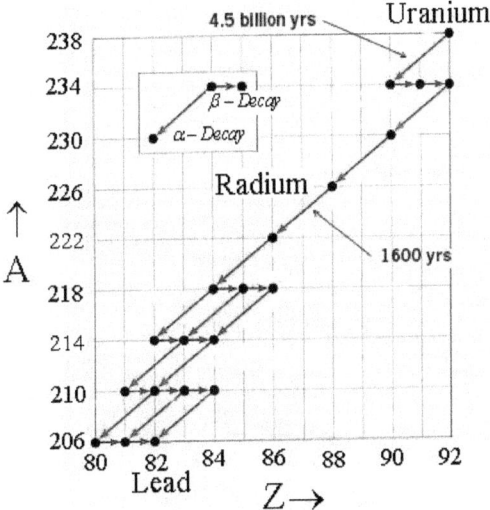

Fig 2.12 Actinium (Uranium-235) Series

2.5 RADIOACTIVE EQUILIBRIUM

When each radionuclide in decay chain decays at the same rate it is produced, it can be said that there is radioactive equilibrium for that decay chain. A knowledge of the equilibrium for a given decay series, enables to estimate the amount of radiation that will be present at various stages of the decay

2.5.1 Activities of A and B

Consider the frequently occurring case where the parent A decays to product B which in turn is radioactive, and disintegrates to C.

$$A \xrightarrow{\lambda_A} B \xrightarrow{\lambda_B} C$$

where C is stable. Application of the radioactive Decay Law gives

$$\frac{dN_A}{dt} = -\lambda_A N_A$$

$$\frac{dN_B}{dt} = \lambda_A N_A - \lambda_B N_B$$

$$\frac{dN_C}{dt} = \lambda_B N_B$$

where λ_A and λ_B are the corresponding decay constants. For longer chains the equations for the additional nuclides are derived as above.

It can be shown that

The activity of B is (initial conditions, $N_A = N_{Ao}$, $N_B = 0$ at $t = 0$)

$$N_B = N_{Ao} \frac{\lambda_A}{\lambda_B - \lambda_A} \left(e^{-\lambda_A t} - e^{-\lambda_B t} \right) \qquad (2.5.1)$$

$$N_B \lambda_B = N_A \lambda_{Ao} \frac{\lambda_B}{\lambda_B - \lambda_A} \left(e^{-\lambda_A t} - e^{-\lambda_B t} \right) \quad (2.5.2)$$

$$N_C = N_{Ao} \left[1 + \frac{\lambda_A}{\lambda_B - \lambda_A} (\lambda_A e^{-\lambda_B t} - \lambda_B e^{-\lambda_A t}) \right]$$

It will be seen that the activity of B is not given by $\frac{dN_B}{dt}$, but by $\lambda_B N_B$. The value N_B attains maximum at

$$t_{Max} = \frac{Ln \frac{\lambda_B}{\lambda_A}}{\lambda_B - \lambda_A}$$

The activity of B is maximum when

$$\boxed{\lambda_B N_B(t_{Max}) = \lambda_A N_A(t_{Max})}$$

This is known as Ideal Equilibrium.
At any other time, the ratio of the daughter to its immediate parent in any three or longer chain is

$$\boxed{\frac{\lambda_B N_B}{\lambda_A N_A} = \frac{\lambda_B}{\lambda_A N_A} \left[1 - (e^{-(\lambda_B - \lambda_A)t}) \right]}$$

Three cases arise from this equation:
i) If $\lambda_A > \lambda_B$, the ratio increases with t,
ii) If $\lambda_B > \lambda_A$, gives a state of <u>Transient equilibrium</u>, where the ratio is greater than 1,
iii) If $\lambda_B \gg \lambda_A$, one gets the <u>secular equilibrium</u>, and ratio equals unity.

Example: $^{90}_{38}Sr \xrightarrow[28\ yrs]{\beta^-} {}^{90}_{39}Y \xrightarrow[28\ yrs]{\beta^-} {}^{90}_{40}Zr$

2.5.2 Transient Equilibrium for A and B

Transient radioactive equilibrium occurs when the parent nuclide and the daughter nuclide decay at essentially the same rate. For transient equilibrium to occur, the parent must have a long half-life when compared to the daughter. An example of this type of compound decay process is $^{140}_{56}Ba$ which decays by beta emission to $^{140}_{57}La$, which in turn decays by beta emission to stable $^{140}_{58}Ce$ (Fig 2.13).

$$^{140}_{56}Ba \xrightarrow{300\ hr} {}^{140}_{57}La \xrightarrow{40\ hr} {}^{140}_{58}Ce\ (Stable). \quad (2.5.3)$$

At relatively long times the ratio of the activities of the two becomes

$$\frac{N_B \lambda_B}{N_A \lambda_A} = \frac{\tau_{1/2A}}{\tau_{1/2A} - \tau_{1/2B}} \quad (2.5.4)$$

$$N_A \lambda_A = N_B (\lambda_B - \lambda_A) \quad (2.5.5)$$

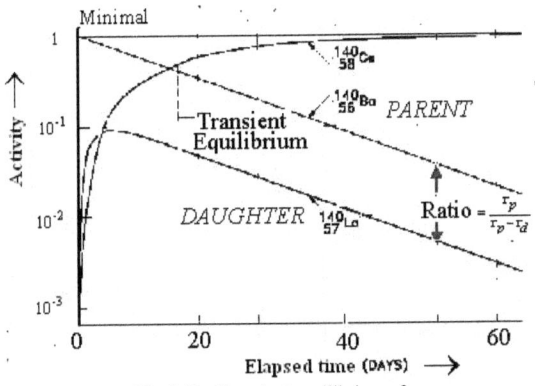

Fig 2.13 Transient equilibrium of

$$^{140}_{56}Ba \xrightarrow{300\ hr} {}^{140}_{57}La \xrightarrow{40\ hr} {}^{140}_{58}Ce\ (Stable)$$

2.5.3 Secular Equilibrium

When $\tau_{1/2A} \gg \tau_{1/2B}$, the decay product generates radiation more quickly. Within about 7 $\tau_{1/2B}$, the activities of A and B are equal, and the amount of radiation (activity) is doubled. Beyond this point, the decay product decays at the same rate it is produced--a state called "secular equilibrium." (Fig 2.14).

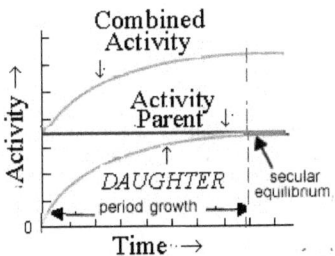

Fig 2.14 Secular Equilibrium

2.5.4 Ideal (or Permanent) Equilibrium of A and B

If A and B elements are present together, for them to be in permanent equilibrium,

$\tau_{1/2A} \gg \tau_{1/2B}$, and

$$N_A\ \lambda_A = N_B\ \lambda_B \qquad (2.5.6)$$

Example: The $^{238}_{92}U$ series, where 1 Ci of $^{238}_{92}U$ in equilibrium with its 13 daughter products have a total activity of 14 Ci.

2.5.5 Decay Product Has a Longer Half-Life

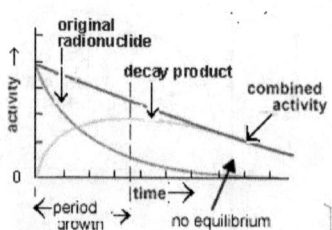

Fig 2.15 No Equilibrium of Parent, Daughter and combined

If the half-life of the decay products is much longer than that of the original radionuclide, its activity builds up to a maximum and then declines. The original radionuclide eventually decays away and no equilibrium occurs.(Fig 2.15).

2.6 INDUCED (or ARTIFICIAL) RADIOACTIVITY

2.6.1 History

In 1934, Irene Curie (the daughter of Pierre and Marie Curie) and her husband, Frederic Joliot, announced the first synthesis of an artificial radioactive isotope. They bombarded a thin piece of aluminum foil with α-particles produced by the decay of polonium and found that the aluminum target became radioactive. Chemical analysis showed that the product of this reaction was an isotope of phosphorus.

$$^{27}_{13}Al + ^{4}_{2}He \rightarrow ^{30}_{15}P + ^{1}_{0}n \qquad (2.6.1)$$

In the next 50 years, more than 2000 other artificial radio-nuclides were synthesized.

In the shorthand notation the parent (or target) and the daughter nuclides are separated by parentheses that contain the symbols for the particle that bombards the target and the particle or particles released in this reaction. For example, equation (2.6.1) becomes

$$^{27}_{13}Al\,(\alpha, ^{1}_{0}n)\,^{30}_{15}P \qquad (2.6.2)$$

The nuclear reactions used to synthesize artificial radio-nuclides are characterized by enormous activation energies (Chapter 15). Three devices are used to overcome these activation energies: linear accelerators, cyclotrons, and nuclear reactors. Linear accelerators or cyclotrons can be used to excite charged particles such as protons, electrons, α-particles, or even heavier ions, which are then focused on a stationary target. The following reaction, for example, can be induced by a cyclotron or linear accelerator.

$$^{24}_{12}Mg + ^{2}_{1}H \rightarrow ^{22}_{11}Na + ^{4}_{2}He \qquad (2.6.3)$$

Because these reactions involve the capture of a positively charged particle, they usually produce a neutron-poor nuclide.

Artificial radio-nuclides are also synthesized in nuclear reactors, which are excellent sources of slow-moving $^{1}_{0}n$, or **thermal neutrons**. The absorption of a neutron usually results in a neutron-rich nuclide. The following neutron absorption reaction occurs in the cooling systems of nuclear reactors cooled with liquid sodium metal.

$$^{23}_{11}Na + ^{1}_{0}n \rightarrow ^{24}_{11}Na + \gamma \qquad (2.6.4)$$

In 1940, absorption of thermal neutrons was used to synthesize the first elements with atomic numbers larger than the heaviest naturally occurring element, uranium. The first of these truly artificial elements were neptunium and plutonium, which were synthesized by Edwin M. McMillan and Philip H. Abelson by irradiating ^{238}U with neutrons $_0^1n$ to form ^{239}U

$$^{238}_{92}U + ^1_0n \rightarrow ^{239}_{92}U + \gamma \qquad (2.6.5)$$

which undergoes β^--decay to form ^{239}Np and then ^{239}Pu.

$$^{239}_{92}U \rightarrow ^{239}_{93}Np + \beta^- \qquad (2.6.6)$$

$$^{239}_{93}Np \rightarrow ^{239}_{94}Pu + \beta^- \qquad (2.6.7)$$

Larger bombarding particles were eventually used to produce even heavier Tran-uranium elements.

$$^{253}_{99}Es + ^4_2He \rightarrow ^{256}_{101}Md + ^1_0n \qquad (2.6.8)$$

$$^{246}_{96}Cm + ^{12}_6C \rightarrow ^{254}_{102}No + 4\, ^1_0n \qquad (2.6.9)$$

The half-lives for α-decay and spontaneous fission of a nuclide decrease as the Z of the element increases. Element 104, for example, has a half-life for spontaneous fission of 0.3 s. Elements, therefore, become harder to characterize as the atomic number increases. Recent theoretical work has predicted that a magic number of protons might exist at $Z = 114$. This work suggests that there is an island of stability in the sea of unstable nuclides, as shown in the figure below. If this theory is correct, super-heavy elements could be formed if we could find a way to cross the gap between elements $Z = 109$ through $Z = 114$.

There is some debate about the number of neutrons needed to overcome the proton-proton repulsion in a nucleus with 114 protons. The best estimates suggest that at least 184 neutrons, and perhaps as many as 196, would be needed. It is not an easy task to bring together two particles that give both the correct number of total protons and the necessary neutrons to produce a nuclide with a half-life long enough to be detected. If we start with a relatively long-lived parent nuclide, such as ^{251}Cf ($\tau_{1/2} = 800$ yrs) and bombard this nucleus with a heavy ion, such as ^{32}S, we can envision producing a daughter nuclide with the correct atomic number, but the mass number would be too small by at least 16 *amu*.

$$^{251}_{98}Cf + ^{32}_{16}S \rightarrow ^{282}_{114}X + ^1_0n \qquad (2.6.10)$$

An expanded periodic table for elements up to $Z = 168$ is shown in the figure below. Elements 104 through 112 are transition metals that fill the 6d-orbitals. Elements 113 through 120 are main-group elements in which the 7p- and 8s-orbitals are filled. The next sub-shell is the 5g atomic orbital, which can hold up to 18 electrons. There is reason to believe that the 5g- and 6f-orbitals will be filled at the same time. The next 32 elements are, therefore, grouped into a so-called super-actinide series

2.7 DISINTEGRATION and the SERGENT DIAGRAM

N. Feather *et al.*(1948) constructed Sargent diagrams constructed for negative-electron-active species having $80<Z<94$ (as also for the restricted group with $80<Z<84$), and for both negative-electron-active and positron-active species having $Z<22$, are discussed in the light of theory. The majority of first-forbidden transitions in the group of heaviest radio-elements give a Sargent line parallel to the allowed line, corresponding to disintegration constants smaller by a factor of roughly 10 only than the disintegration constants of allowed transitions of equal energy. Transitions with no parity change between states of zero spin appear to give a second-forbidden line also parallel to the allowed line.

With the lightest radio-elements "super-allowed" transitions are identified which require Wigner Selection Rules for their interpretation, but there is also a group of transitions which appears to be ordinarily allowed (i.e. subject to spin and parity selection rules only). Compared with these allowed transitions in the lightest elements, allowed transitions in the heaviest radio-elements are less probable by a factor of about three only, when comparison is made for equal energy release and account is taken of the effect of the coulomb field of the nucleus.

On the assumption that all (even, even) nuclei have even parity and zero spin in the ground state it is concluded that the β-particles of maximum energy observed with RaB, RaC and MsTh₁ are emitted in transitions to excited states of the final nuclei RaC, RaC' and MsTh₂, respectively. This conclusion makes it likely that the accepted value of the total γ-ray emission from RaB (Gray 1937) is too small, that the total disintegration energy Ra.C' is 3.78 MeV. and that MsTh₁ emits a low-energy γ-radiation which has not as yet been detected. On the same assumption it is concluded that the disintegrations $UX_1.UX_2$ and UX_2. U_{11} possess features at present unknown or insufficiently understood.

2.8 RADIOACTIVITY IN THE EARTH

It is known that when the earth was formed 4.6 billion years ago, it contained many radioactive isotopes. Since then, all the shorter lived isotopes have decayed. Only those isotopes with very long half lives (100 million years or more) remain, along with the isotopes formed from the decay of the long lived isotopes.

These naturally-occurring isotopes include uranium and thorium and their decay products, such as radon. The presence of these radio-nuclides in the ground leads to both external gamma ray exposure and internal exposure from radon and its progeny.

Worked out Example 2.2

The radioactive dating technique, is behind some archeological discoveries. In a sample of material containing ^{14}N there is an activity of $2.8 \times 10^7 \, Bq$. The half life of ^{14}C is 5730 yr. a) Determine the decay constant od ^{14}C in s^{-1}, b) Estimate the population of ^{14}C in the sample? C) What will be the activity of the sample after 1000 yr. and d) Find out the activity after four times the half life? Given: $1 \, yr = 3.15 \times 10^7 \, s$)

Solution: Step #1 Given: Activity, $R_0 = (\lambda \, N_0) = 2.8 \times 10^7 \, Bq$, $\tau_{1/2} = 5730 \, yr$,

$\lambda = \dfrac{0.693}{\tau_{1/2}}$, $1 \, yr = 3.15 \times 10^7 \, s$, $R = R_0 \, (\tfrac{1}{2})^n$

Step # 2   a) $\lambda = \dfrac{0.693}{\tau_{1/2}} = \dfrac{0.693}{(5730 y \, r)(3.15 \times 10^7 s \, yr^{-1})} = 3.84 \times 10^{-12} \, s^{-1}$

Step # 3   b) $R_0 = (\lambda \, N_0) = 2.8 \times 10^7 \, Bq$,

$N_0 = \dfrac{R_0}{\lambda} = \dfrac{2.8 \times 10^7 \, Bq}{3.84 \times 10^{-12} s^{-1}} = 7.3 \times 10^{18}$ nuclei

Step # 4   c) $R = R_0 \, e^{-\lambda t}$

$= (2.8 \times 10^7 \, Bq) \, e^{-(3.84 \times 10^{-12} s^{-1})(1000 \, yr)(3.15 \times 10^7 s \, yr^{-1})}$

$= 2.5 \times 10^7 \, Bq$

Step # 5) $R = R_0 \, (\tfrac{1}{2})^n = R_0 \, (\tfrac{1}{2})^4 = R_0 /16 = = 1.7 \times 10^6 \, Bq$

Natural Background Radiation

Everybody gets exposed to ionizing radiation from natural sources at all times. This is called natural background radiation, and its main sources are the following:
1. Radioactive substances in the earth's crust;
2. Emanation of radioactive gas radon from the earth;

3. Cosmic rays from outer space which bombard the earth;
4. Trace amounts of radioactivity in the body

2.8.1 Cosmic Radiation

Cosmic rays are extremely energetic particles, primarily protons, which originate in the Sun, other stars and from violent cataclysms in the far reaches of space. Cosmic ray particles interact with the upper atmosphere of the earth and produce showers of lower energy particles. Many of these lower energy particles are absorbed by the earth's atmosphere. At sea level, cosmic radiation is composed mainly of muons, with some gamma-rays, neutrons and electrons. Because the earth's atmosphere acts as a shield, the exposure of an individual to cosmic rays is greater at higher elevations than at sea level. For example, the annual dose from cosmic radiation in Denver is 50 *milliRem* while the annual dose at sea level is 26 *milliRem*.

2.8.2 Natural Radioactivity in the Body

Small traces of many naturally occurring radioactive materials are present in the human body. These come mainly from naturally radioactive isotopes present in the food we eat and in the air we breathe.

These isotopes include tritium (^3H), ^{14}C, and potassium (^{40}K).

2.8.3 Biological Effects of Ionizing Radiation and Mechanisms of Damage

It is well known that radiation can be hazardous to living organizms. Injury to living tissue results from the transfer of energy to atoms and molecules in the cellular structure. Ionizing radiation causes atoms and molecules to become ionized or excited. These excitations and ionizations can:
a) Produce free radicals,
b) Break chemical bonds,
c) Produce new chemical bonds and cross-linkage between macromolecules, and
d) Damage molecules that regulate vital cell processes (e.g. DNA, RNA, proteins).

The cell can repair certain levels of cell damage. At low doses, such as that received every day from background radiation, cellular damage is rapidly repaired.

At higher levels, cell death results. At extremely high doses, cells cannot be replaced quickly enough, and tissues fail to function.

2.8.4 Tissue Sensitivity

In general, the radiation sensitivity of a tissue is:
a) proportional to the rate of proliferation of its cells
b) inversely proportional to the degree of cell differentiation

For example, the following tissues and organs are listed from most radiosensitive to least radiosensitive:

Most Sensitive:	Blood-forming organs
	Reproductive organs
	Skin
	Bone and teeth
	Muscle
Least sensitive:	Nervous system

This also means that a developing embryo is most sensitive to radiation during the early stages of differentiation, and an embryo/fetus is more sensitive to radiation exposure in the first trimester than in later trimesters.

2.8.5 Prompt and Delayed Effects

Radiation effects can be categorized by when they appear.

Prompt effects: effects, including radiation sickness and radiation burns, seen immediately after large doses of radiation delivered over short periods of time.

Delayed effects: effects such as cataract formation and cancer induction that may appear months or years after a radiation exposure

2.9 COSMIC RAYS

Cosmic rays is the name given to extremely energetic radiation which continually bombard the earth's atmosphere from outer space.. Before the development of very high-energy accelerators, all that was known about high-energy reactions had been learned from studies of cosmic ray interactions. The origin and nature of these radiations have puzzled scientists since they are first discovered in the early 1900s. Cosmic rays are distinguished as 'primary' and 'secondary'. Victor Hess in 1912 detected the cosmic rays using his detector carried by a balloon at a n altitude of more than 5 kms in Earths' atmosphere. He was awarded the Nobel Prize in Physics in 1936.

2.9.1 Primary Cosmic Rays

The majority of the energetic particles that enter the earth's atmosphere from space are protons having energies in the range 10^9 to $10^{20} eV$. As soon as they enter the heavier layers of the atmosphere these primary rays interact with it and produce a greater variety of the secondary particles, including electrons, photons and mesons of all types.

Primary cosmic rays are composed primarily of protons and α-particles (99%), with a small amount of heavier nuclei (~1%) and an extremely minute proportion of positrons and anti-protons

2.9.2 Secondary Cosmic Rays

Secondary cosmic rays, caused by a decay of primary cosmic rays as they impact an atmosphere, include neutrons, pions, positrons and muons. Of these four, the latter three were first detected in cosmic rays The process of origin of secondary cosmic rays is schematically learnt from Fig 2.16.

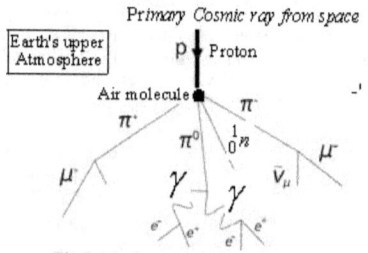

Fig 2.16. Secondary cosmic rays

REVIEW QUESTIONS

R.Q. 2.1 Objective Test

1) What type of shielding should be used when working with ^{32}P?
a) Paper   b) Plastic (plexiglass) c) Lead d) No shielding is needed

2) Consider having 0.250 *mCi* of ^{32}P, what will the activity (approx) be after 30 days?
a) 0.0625 *mCi* b) 0.250 *mCi*   c) 0.125 *mCi* d) 0.03125 *mCi*

3) Select which is not a source of natural background radiation?
a) Radioactive substances in the earth's crust   b) Cosmic rays from outer space which bombard earth
c) Trace amounts of radioactivity in the body   d) Air travel

4) An isotope has the same number of _____ as all other isotopes of that element, but contains a different number of _____.
a) neutrons, protons   b) protons, neutrons c) protons, electrons d) none of the above

5) Which among the following is not an example of ionizing radiation?
a) alpha particle radiation   b) beta particle radiation c) x rays   d) microwaves

6) The greatest hazard associated with alpha radiation is:
a) external exposure b) inhalation c) ingestion   d) b and c

7) Which activity will give you the most dose?
a) dental x-ray b) using radioactive materials in a research lab   c) smoking   d) coal burning power plant

8) Which dose will result in 100% mortality even with the best available medical treatment?
a) 800 rem b) 300 mrem c) 450 rem d) 820 mrem

9) Which lists tissues / organs from least sensitive to most sensitive?
a) nervous system, skin, muscle, blood forming organs   b) nervous system, bone and teeth, skin, blood forming organs c) blood forming organs, nervous system, bone and teeth, muscle d) muscle, skin, bone and teeth, nervous system

10) When should radiation safety be notified that a lab is moving?
a) Before the move   b) During the move   c) After the move   d) Radiation Safety does not need to be notified

11) When radioactivity is not in use:
a) No surveys need to be performed   b) No surveys need to be performed, but a survey map needs to be included with the monthly surveys indicating no radioactivity was used. c) Monthly surveys must be performed

12) Name the type of radiation causes the most biological damage for a given amount of energy deposition?
a) neutron   b) gamma   c) alpha d) beta

13) A routine survey indicates that a bench-top has removable contamination of > 200dpm / 100cm2. What action should be taken?
a) No action needs to be taken   b) Radiation Safety must be notified   c) The area needs to be decontaminated   d) Decontamination is optional

14) Which of the following is not a violation?
a) Failure to perform and document monthly radioactive contamination surveys during months in which radioactive materials were used.
b) Evidence of eating or drinking in a radiologically posted room (i.e., presence of candy wrappers, soda cans, coffee-stained cups, etc.)
c) Survey meter out of calibration or use of inoperable or inappropriate survey meter.
d) Going to the beach instead of coming to work.

15) How should sharps that have been used for radioactive materials be disposed?
a) Placed in the regular trash   b) Placed in an approved sharps container marked radioactive
c) Placed in a biohazard bag in the radioactive solid trash   d) Placed in the solid radioactive trash

16) If you double your distance from a radiation source you reduce your exposure by a factor of:
a) 4   b) 2   c) 10   d) 1

17) Which type of radiation is the least penetrating?
a) gamma   b) alpha   c) neutron   d) beta

18) Which of the following isotopes can be mixed together in solid waste containers?
a) ^{32}P, ^{33}P, ^{14}C b) ^{51}Cr, ^{35}S, ^{32}P c) ^{35}S, ^{14}C, ^{3}H d) ^{32}P, ^{33}P, ^{51}Cr

19) What is proper attire when working with radioactive materials?
a) Skirt, sandals, lab coat, gloves, and safety glasses.   b) Shorts, lab coat, sandals and safety glasses.
c) Pants, safety glasses, lab coat, gloves and closed toe shoes.   d) Skirt, safety glasses, closed toe shoes and gloves.

20) The most likely possible health effect from working with low levels of radiation is a _____ elevated risk of _____.

a) Very slightly / developing cancer   b) Significantly / developing diabetes
c) Moderately / contracting AIDS   d) Significantly / birth defects in children

21) Which of the following statements is not true?
a) An inventory of radioactive materials must be maintained by indicating use on the Isotope Use record delivered with each isotope.
b) Radioactive materials may be used in any room that is associated with the PI even if it is not on their permit.
c) Radioactive materials may only be ordered through the Radiation Safety Office.
d) You should reduce your time around radioactivity to the minimum required without interfering with your work.

R.Q. 2.2   An activity of 320 d / minute for ^{13}C was found for a piece of wood weighing 50 g. The corresponding activity for a living plant is 12 d / minute / g. What is the age of the wood. Given $\tau_{1/2}$ = 5730 yrs for ^{14}C. [Answer 5170 yrs].

RQ. 2.3 The half-life of ^{32}P is 14.3 days. If 250 μCi (= 9.25 MBq) is bought and used precisely after 43 days, find the activity. [Answer 250 $e^{-\frac{(ln\ 2)(43)}{14.3}}$ = 31.25 μCi]
This quantity is 8 times less than was initially bought. Therefore, it is better to order the radionuclide as close as possible to the date of use

RQ. 2.4 The half-life of ^{14}C is 5730 yrs. What will be the activity of 1 mCi (= 37 MBq) of ^{14}C after the interval of 43 days? [Answer (1) $e^{-\frac{(ln\ 2)(43)}{(5730)(365)}}$ = 0.999986 μCi].

R.Q. 2.5 On average, 35 eV is required to produce an ion pair. How many ion pairs are produced by a alpha particle with a 5.0 MeV kinetic energy? If the range of the alpha particle is 10 cm, what is the average ion-pair density?

R.Q. 2.6 The half-life of technetium-99 is 6 hours. What mass of $^{99}_{43}Tc$ remains from a 10.0 g sample after 18 hours? [Answer (18hrs / 6hrs = 3 half-lives; 10g/2 = 5g; 5g/2 = 2.5g; 2.5g/2 = 1.25g) 0 1. 25g]

R.Q. 2.7 A radioactive isotope decays from 8.00 g to 0.25 g in 30 days. What is the isotopes half-life? [Answer (8 g / 2 = 4 g, 4 g / 2 = 2 g, 2g / 2 = 1g, 1 g / 2 = 0.5 g, and 0.5 g / 2 = 0.25 g. The isotope underwent 5 half-lives. 30 days / 5 = 6.). Half-life is 6 days].

R.Q.2.8 Suppose you begin with 1.0×10^{-2} g of a pure radioactive substance and 4 h later determine that 0.25×10^{-2} g remain. What is the half-life of the substance? [Answer 2.0 hrs.]

R.Q. 2.9 The half-life of polonium is 3 minutes. For a sample of 32 g of polonium, prepare a table of the amount of polonium left after every three minutes. Plot the decay curve on graph paper and from it find the amount of polonium left after 13.5 minutes [Answer 1.5 g]

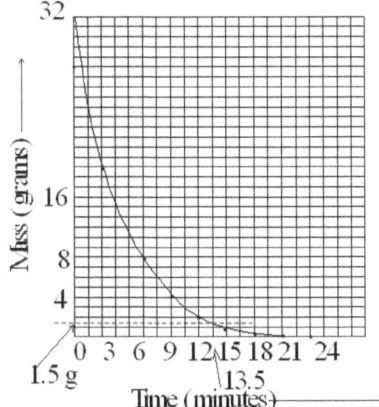

R.Q. 2.10 The ^{14}C content decreases after the death of a living system with a half-life of 5739 years. If the ^{14}C content of an old piece of wood is found to be 12.5% of that of an equivalent present-day sample, how old is the piece of wood? [Answer t = 17,217 yr.]

R.Q.2.11   A radioactive source has decayed to 1/128 of its initial activity in 50 days. What is the half life? [Answer $\tau_{1/2} = 50/7 = 7.1\ d$]

R.Q. 2.12   The half life of a specific element was calculated to be 5200 *yrs*. Calculate the decay constant (λ). [Answer 1.33×10^{-4}]

R.Q. 2.13   If a watch contains a radioactive substance with a decay rate of 1.40×10^{-2} and after 50 *yrs* only 25 mg remain, calculate the amount originally present. [Answer 50.3 mg]

R.Q. 2.14   A rock contains 0.257 mg of ^{206}Pb for every mg of $^{238}_{92}$U. The half-life decay for uranium to turn into lead is 4.5×10^9 *yrs*. How old is the rock? [Answer $\lambda = 1.5 \times 10^{-10}$, $\tau = 1.7 \times 10^9$ *yrs*].

R.Q. 2.15 $^{90}_{38}$Sr (strontium-90) has a half-life of 28.5 *yrs*. How long will it take for 98% of a sample of strontium-90 to disappear? [Answer $t = -\ln(0.02)/0.0243\ yrs^{-1} = 161\ yrs$].

R.Q. 2.16   Find an expression for defining τ as the time required on an average for the activity of a sample to decrease to $\dfrac{1}{e}$ of its initial value. [Answer $N(\tau) = \dfrac{N_0}{e}$].

R.Q. 2.17   Polonium-210 is known to have a half life of 140 days and emits alpha particles each of energy 5.3 MeV Find the mean power output per gram of the isotope during the first half life.
Given $N_A = 6 \times 10^{23}\ mole^{-1}$. [Answer $E = \dfrac{1}{2}[\dfrac{6 \times 10^{23}}{210}].[\dfrac{5.3 \times 10^6 \times 1.6 \times 10^{-19}}{140 \times 24 \times 3600}] = 100\ W$]

R.Q. 2.18   Calculate the velocity of an alpha particle which will have the same kinetic energy as the energy of a photon of gamma radiation of $\lambda = 4.5 \times 10^{-13}\ m$.
Given $h = 6.62 \times 10^{-34}\ J - s$, $c = 3 \times 10^8\ m - s^{-1}$, $m_\alpha = 6.62 \times 10^{-27}\ kg$ [Answer $v = 1.15 \times 10^7\ m - s^{-1}$]

R.Q. 2.19   Living wood takes in radioactive ^{14}C from the atmosphere during the process of photosynthesis, the proportion of ^{14}C to ^{12}C atoms being 1.25 to 10^{12}. When the wood dies ^{14}C decays, whose half life is 5600 *yrs*. 4.0 g of carbon from a piece of dead wood gave a total count rage of 20.0 disintegrations per *minute*. Estimate the age of the piece of wood.
Given $N_A = 6 \times 10^{23}\ mole^{-1}$, 1 *yr* = 3.16×10^7 s; ln 2 = 0.693. [Answer 8700 *yrs*]

R.Q. 2.20   The isotope $^{40}_{19}$K, with a half life of 1.37×10^9 years, decays to $^{40}_{18}$Ar, which is stable. Moon rocks from the Sea of Tranquility show that i) the ratio of these K atoms to Ar atoms to be 1/7, ii) Certain other rocks give a value of 1/4 for this ratio. Estimate the age of these rocks, stating clearly any assumptions you make [Answer i) All atoms were initially $^{40}_{19}$K, $\therefore$ Age = 3 x 1.37 x 10^9 yrs, ii) $^{40}_{19}$K : Ar = 1:4 $\Rightarrow$ $^{40}_{19}$K : Ar + $^{40}_{19}$K = 1:5; 5 = 2^n $\Rightarrow$ $\dfrac{\log 5}{\log 2}$ = n $\Rightarrow$ 2.32 x 1.37 x 10^9 years]

R.Q. 2.21 A small volume of a solution which contained a radioactive isotope of sodium had an activity of 12000 disintegrations / minute when it was injected into the bloodstream of a patient. After 30 hours the activity of $1.0\ cm^3$ of the blood was found to be 0.50 disintegrations / minute. If the half-life of the sodium isotope is taken as 15 hrs, estimate the volume of blood in the patient. [Answer $6000\ cm^3$]

R.Q.2.22 Uranium ores contain one radium-226 atom for every 2.8×10^6 uranium-238 atoms. Calculate the half-life of $^{238}_{92}U$ given that the half life of $^{226}_{88}Ra$ is 1600 years and $^{226}_{88}Ra$ is a decay product of $^{238}_{92}U$ [Answer $4.5 \times 10^9\ yrs$].

R.Q. 2.23 It is found that 9 U-238 atoms are found for every 8 He atoms present in a uranium bearing rock. Assuming that the decay process which eventually converts a U atom to Pb involves the emission of 8 alpha particles, calculate the age of the rock. (Half-life of U-238 = $4.5 \times 10^9\ yrs$) [Answer $6.8 \times 10^8\ yrs$].

R.Q. 2.24 In an experiment the activity of 1.2 mg of radioactive ^{40}KCl was found to be $170\ s^{-1}$. Taking molar mass of ^{40}KCl to be $0.075\ kg\text{-}mol^{-1}$, find the number of ^{40}K atoms in the sample and hence find the half-life of ^{40}K. ($N_A = 6.0 \times 10^{23}\ mole^{-1}$) [Answer $1.2 \times 10^9\ yrs$].

R.Q. 2.25 $^{210}_{83}Bi$ is a radioactive isotope of bismuth with a half-life period of 5.00 days, which emits β^- – particles. This isotope is source of β^- – for an experiment which is to continue for 300 hours. Assuming that the strength of the source must not fall below $10\ \mu Ci$, what strength of source is required at the start of the experiment? [Answer $< 57\ \mu Ci$]

R.Q. 2.26 The activity of a mass of $^{14}_{6}C$ is $5.0 \times 10^8\ bq$ and the half-life is 5570 years. Estimate the number of $^{14}_{6}C$ nuclei present. ($\ln 2 \approx 0.69$) [Answer 1.27×10^{18}].

R.Q. 2.27 The half life of $^{30}_{15}P$ is 2.5 minutes. Calculate the mass of $^{30}_{15}P$ which has an activity of $1 \times 10^{15}\ bq$. ($N_A = 6.0 \times 10^{23}\ mole^{-1}$; $\ln 2 = 0.69$) [Answer $1.1 \times 10^{-5}\ gm$].

R.Q. 2.28 The activity of a particular radioactive nuclide falls from $1.0 \times 10^{11}\ bq$ to $2.0 \times 10^{10}\ bq$ in 10 hours. Calculate the half-life of the nuclide. [Answer $4.3\ hrs$].

R.Q. 2.29 Determine the activity of a 1 g sample of $^{90}_{38}Sr$, whose $\tau_{1/2}$ against β – decay is 28 years

R.Q. 2.29 Find the energy released in the alpha-decay $^{238}_{92}U \rightarrow\ ^{234}_{90}Th +\ ^{4}_{2}He$. Given $M(^{238}_{92}U) = 238.050786\ u$, $M(^{234}_{90}Th) = 234.043583\ u$, $M(^{4}_{2}He) = 4.002603\ u$ (Answer $E = mc^2 = (0.0046\ u)(931.5\ \frac{MeV}{u\,c^2}) = 4.29\ MeV$).

RQ. 2.30 State the decay law of radioactivity. Find the expression for the number of radioactive atoms left after a certain time interval.

Write a note on the penetrating power of $\alpha-, \beta-$ and $\gamma-$ rays.

RQ. 2.31 State the general characteristics of natural radioactivity. Show that radioactivity is essentially a nuclear phenomenon.

Derive the exponential law governing radioactive changes and that it is only an approximate empirical law from probability consideration.

Write a note on 'artificial radioactivity'.

RQ. 2.32 An annual dose of 5 rem of X-rays is received by a 100 kg worker. What is the energy in J deposited in the system? Ii) (Answer: 1 J)

RQ. 2.33 If $^{60}_{27}Co$ has half-life 5.25 yr, what mass it should have to have an activity of 1.0 Ci ? (Answer: 8.8×10^{-7} kg).

R.Q.2.34. Consider 250 mCi (9.25 MBq) of ^{32}P are purchased. After 43 d what will be its activity? Given $\tau_{1/2} = 14.4\ d$. (Answer: 31.25 μCi).

R.Q. 2.35. The $^{14}_{6}C$ has $\tau_{1/2} = 5730$ yrs. After 43 d what will be its activity of a sample of 1 mCi (37 MBq) after 43 days? (Answer: 0.999986 mCi).

RQ. 2.36. In a detector 900 counts are seen for a source during 5s. What is the count rate per second and the error in the measurement? (Answer: $(900 \pm 30)/5$ $cts\ /s$).

RQ. 2.37. A source of radioactivity is found to have a mean count rate of 1 $cts\ /s$ Given a period of 4 seconds, find out the probability that a) the count rate is zero. B) One count is 4 seconds.

(Answer: $P = (4)^0 \frac{e^{-4}}{0!} = 0.0183$; $P = 0.0733$).

&%&%&%&%&

Chapter 3
STABLE PROPERTIES OF A NUCLEUS - I
Charge, Mass

Chapter 3

STABLE PROPERTIES OF A NUCLEUS - I
Charge, Mass

'The manner of giving is worth more than the gift' - Pierre Comeille

3. INTRODUCTION

The atomic nucleus is a tiny massive entity at the center of an atom. Occupying a volume whose radius is 1/100,000 the size of the atom, the nucleus contains most (99.9%) of the mass of the atom. In describing the nucleus, we shall describe its composition, size, density, and the forces that hold it together.

In this Chapter, the subject matter for description will be on nuclear charge, density and mass. The nucleus is composed of protons (charge = +1; mass = 1.007 **atomic mass units** (u) and neutrons. The number of protons in the nucleus is called the **atomic number** Z and defines which chemical element the nucleus represents. The number of neutrons in the nucleus is called the neutron number N, whereas the total number of neutrons and protons in the nucleus is referred to as the mass number A, where $A = N + Z$

3.1. Basic (Static) Properties

Certain time-independent (static) properties of nuclei influence in various ways the behaviour of atoms and molecules. The subject of Atomic Physics and Molecular Physics are highly developed. Nuclear influences on the behaviour of atoms and molecules are well understood. This gives the reasons as to why the use of certain observations within the fields of Atomic and Molecular Physics is used to determine nuclear static properties.
The BASIC properties of a nucleus are:
1. Nuclear Charge,
2. Nuclear Mass,
3. Nuclear Spin or Angular momentum
4. Nuclear Parity, and
5. Nuclear Electro-Magnetic Moments.

The tightly bound group of nucleons in a nucleus is in many ways analogous to the bound electrons in the atom. The last two properties exert considerable influence on the behaviour of nuclei. If there are no external forces / torques acting on a system, its angular momentum is conserved. In Quantum Physics conserved quantities are represented by quantum numbers. (see for example, S. Devanarayanan, 2005).

3.2. NUCLEAR CHARGE

3.2.1. The early work in Atomic Physics of Rutherford, Barkla, Moseley, Chadwick and others conclusively established the fact that
Atomic number (Z) of an element = # of extra-nuclear electrons in an atom of the element

= # of charge units present in its nucleus.
Thus, in an electrically neutral atom,
1) Nuclear charge = | (# of atomic electrons) x (electronic charge) |
= | - Z e |
2) Sign of nuclear charge = Positive

3.3. MEASUREMENT OF NUCLEAR CHARGE.

3.3.1. Rutherford Scattering of Alpha Particles

3.3.1.1. Experimental Setup:

An experimentalist of extraordinary ability, Ernest Rutherford turned his attention to the Plum-pudding model of atom by J.J. Thomson (Devanarayanan, 2005). He wanted to put the Thomson model to an experimental test. For this he used gold as target and Alpha particles as bombarding particles. Being malleable metal which can be rolled and made into thin foils, gold was selected as the target. His purpose was to measure the angular distribution of α-particles from their paths (scattered) by gold atoms. The experiment was performed by Hans Wilhelm Geiger & E. Marsden in 1909, in Rutherford's laboratory. The experimental setup (Fig 3.1) consisted of a radioactive Source (Radon) emitting energetic (E = 5.5 MeV) α-particles, a collimator arrangement, the Target (gold foil of thickness $\sim 10^{-4}$ cm = 1 μm), and a moveable α-particle Detector. The detector was a chip of ZnS mounted in front of a microscope, the whole of which could be adjusted to look at various scattering angle, θ. The entire apparatus had to be kept in a vacuum chamber (to avoid absorption of α-particles by air).

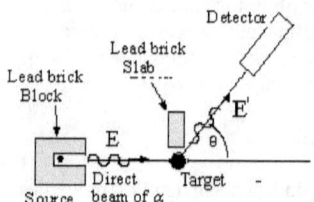

Fig 3.1 Rutherford scattering experimental set up

3.3.1.2. Results

1. Most of the α-particles were found to be scattered through very small angles (i.e. passed through the foil undeviated).

2. But slightly < 2% was scattered through angles greater than > $3°$. This distribution of α-scattering angle spectrum, denoted by $N(\theta).d\theta$. where $N(\theta)$ = # α-particles scattered at angular limits θ, $N(\theta).d\theta$ = # α-particles scattered at angular limits θ to $\theta + d\theta$. To a good approximation, $N(\theta).d\theta$ is given by the Gaussian

$$N(\theta).d\theta = \frac{2N\theta}{\overline{\theta^2}} e^{-\theta^2 / \overline{\theta^2}} = \frac{2N\theta}{\overline{\theta^2}} [e^{-9} \quad (\approx 10^{-4})] \quad (3.3.1)$$

3. A few ($\sim 10^{-2}$ %), i.e. 1 out of every $\sim$8000 α-particles were observed to be scattered through angles > $90°$, that is, it was *back-scattered*.($\theta = 180°$).

$$\frac{N(>90°)}{N} \Box e^{-90^2} \approx 10^{-3500} \tag{3.3.2}$$

4. These back-scattering of α-particles could not be understood by the Thomson model of the atom.
5. Rutherford interpreted the results of Geiger & Marsden experiment as indicating that the positive charge of the atom must be much more compact than Thomson model predicted. The atomic radius of the positive charge, as per the Thomson model had estimated to be $\sim 0.5 \ \overset{o}{A}$. This is to be revised downward. To realize the back-scattering. For this,

$$T = \frac{1}{2}MV^2 \leq \int_R^\infty \left(\frac{2k\,Ze^2}{r^2}\right) dr \tag{3.3.3}$$

i.e., $R \leq \dfrac{2kZe^2}{T} = 4.6 \times 10^{-3} \ \overset{o}{A}$. (3.3.4)

6. Geiger & Marsden had again to perform the α-ray scattering experiment, to show concretely that Rutherford's theory gave the true picture of the atom model.

3.3.2. Theory of α-scattering

There are four important assumptions:
1. Point mass particle are under impact
a. Projectiles, viz. α-particles, are point mass particles
b, Target nuclei are point mass particles
c. α-particles have positive point charges
d. Target nuclei are positive point charges.
e. Coulomb inverse square law force (i.e. electro-static repulsive and central) is the only interaction between these colliding particles, as the distances are small.
2. The nucleus is so massive compared to the α-particle that the collision is *elastic* (i.e., there is no recoil).
3. The First Born approximation is valid, i.e., the incident and outgoing particle can be considered as plane waves.
4. Both the incident and target particles are spin Zero.

m_α Mass of α-particle moving with velocity v_α
$Z_\alpha e$ Charge on the projectile (α-particle)
$Z e$ Charge on the target nucleus

$$F = \frac{(Z_\alpha Z e^2)}{4\pi\varepsilon_0 r^2} \tag{3.3.5}$$

θ Scattering angle (Fig 3.2)
p   momentum of the incident particle
x Impact parameter of the collision = minimum distance to which α-particle approach the target nucleus.
ε Eccentricity of orbit

$$\varepsilon = \left(1 + \frac{2(kE)\,p^2 x^2}{M(Z_1 Z e^2)^2}\right)^{1/2} \tag{3.3.6}$$

$\alpha = \theta$ in the limit $r \to \infty$

$$\cos\alpha = \frac{1}{\varepsilon} \tag{3.3.7}$$

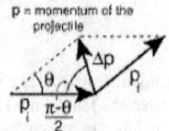

Fig 3.2 Schematics of the α-ray scattering

Since change in linear momentum

$$\Delta p = 2 m_\alpha v_\alpha \, Sin\frac{\theta}{2} \qquad (3.3.8)$$

$$\Delta p = Impulse = \int_0^\infty (Force) \, dt = \int (F \cos \varphi) \, dt \qquad (3.3.9)$$

From the equations above one gets

$$Cot\frac{\theta}{2} = \frac{4 \pi \varepsilon_0 \, r^2 \, m_\alpha v_\alpha^2}{(Z_\alpha Z \, e^2)} x \qquad (3.3.10)$$

Letting, collision radius, $\ b \ = \ \dfrac{|Z_\alpha Z| e^2}{4 \pi \varepsilon_0 \, m_\alpha v_\alpha^2}$, $\qquad (3.3.11)$

$$x = b \, Cot\frac{\theta}{2} \qquad (3.3.12)$$

b is the value of x for which $\theta = 90°$ (Fig 3.3).

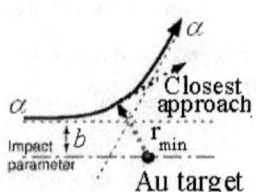

Fig 3.3 Impact parameter and closest approach

This relation between x and θ can be interpreted as: those particles with impact parameters between x and (x + dx) are scattered through an angle θ and $\theta - d\theta$.
Define *scattering cross-section* σ as

$$\sigma = \pi x^2$$

$\therefore \qquad \sigma(\geq \theta) = \pi x^2 = \pi b^2 \cot^2 \dfrac{\theta}{2}.$ $\qquad (3.3.13)$

For backscattering, this becomes

$$\sigma(\theta = 90°) = \pi b^2 \qquad (3.3.14)$$

3.3.3. Rutherford's Scattering Formula

It can be shown that the fraction f of incident α-particles scattered within a distance $b/2$,

$$f = \pi n t \left(b \cot\frac{\theta}{2}\right)^2 \tag{3.3.15}$$

where n = density of nuclei / cm^3 of the foil
t = thickness of the foil
A = Area of the foil on which the particles strike
Total sensitive area in the foil effecting scattering = $n\, t\, \sigma\, A$
N_i = # of particles incident normally upon unit area of foil

# of particles that will be scattered within a cone between θ and $\theta - d\theta$ and received in annular area dS is

$$\delta N = N_i\, df = -N_i\, \pi n t\, b^2 \cot\frac{\theta}{2} \csc^2\frac{\theta}{2}\, d\theta \tag{3.3.16}$$

The negative sign is indicative of the decrease of f with increase of θ. In another form

$$N(\theta) = N_i \left(df/dS\right)$$

$$= N(\theta) = N_i \left(df/dS\right) = \left[\frac{N_i\, n t\, b^2}{4\, a^2}\right]\left(\frac{1}{\sin^4(\frac{\theta}{2})}\right) \tag{3.3.17}$$

a = distance of the elemental volume in the annular ring from the target.
In another form, the *Rutherford scattering Law* becomes

$$N(\theta) = \left[\frac{N_i\, n t\, Z_\alpha^2\, Z^2\, e^4}{4\, a^2\, (4\pi\, \varepsilon_o)^2\, (kE)^2}\right]\left(\frac{1}{\sin^4(\frac{\theta}{2})}\right) \tag{3.3.18}$$

Thus
$N(\theta) \propto t$, *the thikness of the target*
$\propto n$, the *atomic density of the target*
$\propto (Z\,e)^2$, *the square of nuclear charge of the target* nuclei
$\propto \left(1/kE\right)^2$, *the inverse square of kE of the α- particle*
$\propto 1/\sin^4(\frac{\theta}{2})$, i.e., *falls* off very rspidly with increase of θ

The differential scattering cross section, $d\sigma$, for scattering into solid angle $d\Omega$ at mean angle θ, using Fig 3.4, is

$$d\sigma = |2\pi x\, dx| = \frac{b^2}{4}\left(\frac{1}{\sin^4(\frac{\theta}{2})}\right) d\Omega \tag{3.3.19}$$

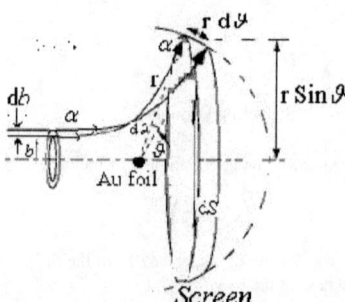

Fig 3.4 Differential scattering geometry

Worked out Example 3.1

A beam of $7.7 MeV$ energetic α-particles incident upon a gold foil $0.3 \mu m$ thick is scattered through angles greater than $45°$. Find out the fraction of the beam scattered, Given $Z = 79$, $\rho = 1.932 \times 10^4 \ g \ cm^{-3}$, atomic weight $197 \ kg.kmol^{-1}$ for gold, $N_A = 6.0225 \times 10^{23} \ mol^{-1}$).

Solution: $\boxed{Step \# 1}$ Given $Z = 79$, $\rho = 1.932 \times 10^4 \ g \ cm^{-3}$, $M = 197 \ kg.kmol^{-1}$,

$N_A = 6.0225 \times 10^{23} \ mol^{-1}$, $\theta = 45°$, kE=7.7 MeV, $f = \pi n t \left(b \cot \frac{\theta}{2}\right)^2$; $b = \frac{|Z_\alpha Z| e^2}{4 \pi \varepsilon_0 \ m_\alpha v_\alpha^2}$

$\boxed{Step \# 2}$ # of Au atoms in the foil = $n = \rho N_A / M$; $n = \frac{[1.932 \times 10^4 \ kg \ m^{-3}]\{6.0225 \times 10^{23} mol^{-1}\}}{197 \ kg \ kmol^{-1}}$

$= 5.9 \times 10^{28} \ atoms / m^3$;

$\boxed{Step \# 3}$ $\frac{Z_\alpha e^2}{4\pi \varepsilon_0 \ kE} = \frac{(79)(1.6021 \times 10^{-19} C)^2 [8.9875 \times 10^9 \ Nm^2 C^{-2}]}{(7.7 \times 10^6 \ eV)[1.6021 \times 10^{-19} \ J/eV]}$

$\boxed{Step \# 4}$ $f = \pi n t \left(b \cot \frac{\theta}{2}\right)^2$

$f = \pi(5.9 \times 10^{28} \ atoms / m^3)[0.3 \mu m] \left(\frac{(79)(1.6021 \times 10^{-19} C)^2 [8.9875 \times 10^9 \ Nm^2 C^{-2}]}{(7.7 \times 10^6 \ eV)[1.6021 \times 10^{-19} \ J/eV]}\right)^2 \cot^2 \frac{\pi}{8}$

$= 7 \times 10^{-5} = 0.007\%$

3.3.4. Nuclear Charge Determination from Optical Spectra

Careful studies of the spectra of certain atoms (Example, H, $He^=$, Li^{2+}, etc.), as interpreted by Bohr' theory, showed the effect of the nuclear charge in these atoms and ions.

3.3.4.1. Nuclear Charge Determination from X-ray Spectra

One of the most striking methods of determining charge Z was shown by Moseley. He related the frequency of the characteristic X-rays to the atomic number, Z, by using Balmer's formula

$$v_{K_\alpha} = c R (Z-1)^2 \left[\frac{1}{1^2} - \frac{1}{2^2}\right] \quad (3.3.20)$$

$$v_{K_\alpha} = \frac{3}{4} c R (Z-1)^2 \quad (3.3.21)$$

where c = velocity of light
R = Rydberg Constant

Z = Atomic number of the target element emitting X-rays
v_{K_α} = Frequency of the K_α - X-ray line.

X-ray scattering experiments by Barkla (1909) had indicated that for an element $_Z^A El$, $Z \approx \frac{A}{2}$ as a Rule of thumb before the corroboration and extension by Moseley.

3.3.4.2. Absolute Determination of Ze

James Chadwick (1920) repeated and refined Geiger-Marsden's experiments. α-particles from a source were scattered by a thin foil, in the form of annular ring, through angle θ and detected by scintillation counter. The total number N_i falling on the scintillator screen on the axis of the cone, the areas of the annular ring and the screen are known. $N(\theta)$ is measured. The results of Chadwick's experiments using platinum (Pt), silver (Ag), and copper (Cu) foils are

Element	Ze	Z
Cu	29.3±0.5 e	29
Ag	46.3±0.8 e	46
Cu	77.4±0.8 e	78

Worked out Example 3.2

A 5MeV energetic of He nucleus collides with a target of silver. Determine the distance of closest approach between the incident and target particle. (Given: Z=47 for Silver, $1\ eV = 1.6021 \times 10^{-19}\ J$).

Solution: | Step # 1 | Given: $E = T + U = (\frac{1}{2}mv^2) + (\frac{qQ}{4\pi \varepsilon_o r})$, q = +2e, Q = +47e;

E_1 = 5 MeV.

$1\ eV = 1.6021 \times 10^{-19}\ J$; $k = (1/4\pi \varepsilon_o) = 8.9875 \times 10^9\ Nm^2C^{-2}$; $e = 1.6021 \times 10^{-19}\ C$

| Step # 2 | Before collision,

$E_1 = (\frac{1}{2}mv^2) = 5\ MeV = (5 \times 10^6\ eV)(1.6021 \times 10^{-19}\ J/eV) = 8 \times 10^{-13}\ J$

| Step # 3 | At the point of closest approach, $E_2 = (\frac{qQ}{4\pi \varepsilon_o r}) = (\frac{(2e)(47e)}{4\pi \varepsilon_o r}) = (\frac{94e^2}{4\pi \varepsilon_o r})$

| Step # 4 | Energy being conserved at the collision, $E_1 = E_2$

$8 \times 10^{-13}\ J = (\frac{94e^2}{4\pi \varepsilon_o r})$, from which $r = \frac{1}{4\pi \varepsilon_o} (\frac{94\ e^2}{8 \times 10^{-13} J})$

$= (8.9875 \times 10^9\ Nm^2C^{-2})(\frac{94\ (e=1.6021 \times 10^{-19} C)^2}{8 \times 10^{-13} J}) = 2.7 \times 10^{-14}\ m$

3.4. NUCLEAR MASS

The mass is the home address of a particle. The exact value of the nuclear mass is of great importance in understanding the nucleus and its behaviour.

Nuclear Mass ≈ the neutral atomic mass or isotopic mass, M
  ≈ 99.975% of the mass of an atom

Frederick Soddy (1910) established for the first time that atoms of a single element may have *isotopes*.

J.J. Thomson (1912) confirmed the existence of isotopes.

3.4.1. Units of Nuclear mass

a) Physical Scale:

1 mass unit (M.U.) = $\frac{1}{16}$ (Mass of the Stable Isotope of Atomic Oxygen, $^{16}_{8}O$) (3.4.1)

b) Unified Scale:

IUPAP met at Ottawa in September 1960 and adopted the mass unit as the *atomic mass unit (a m u)*, denoted as $1\,u$. In this scale

$$1\,u \equiv 1\,amu = \frac{1}{12}(\text{Mass of the neutral Carbon atom, } ^{12}_{6}C) \qquad (3.4.2)$$

$$= \frac{\text{Mass of 1 mole of } ^{12}_{6}C}{12\,N_A}$$

$$= 1.660\,565 \times 10^{-27}\,kg = (1.660\,565 \times 10^{-27}\,kg) \times c^2\,J$$

$$= 931.48\,MeV/c^2\,J \qquad (3.4.3)$$

c) Conversion Factor

Mass in ^{16}C Scale ≡ (1.000 3179) (Mass in ^{12}C Scale) (3.4.4)

Worked out Example 3.3

Given the carbon $^{12}_{6}C$ nucleus has radius, about 3 *Fm*, and a mass of 12.0 *u*. Determine the average density of nuclear material. Compare this with the density of water.

Solution: Step # 1 Density, $\rho = M/V$, $V = \frac{4}{3}\pi r^3$; M= 12.0 u, r =3 Fm

Step # 2 $\rho = M/[\frac{4}{3}\pi r^3] = \frac{(12.0\,u)[1.66 \times 10^{-27}/u]}{4\pi(3 \times 10^{-15}m)} = 1.8 \times 10^{17}\,kg/m^3$

Step # 3 $\rho/\pi_{water} = \frac{1.8 \times 10^{17}\,kg/m^3}{1000\,kg/m^3} = 2 \times 10^{14}$

3.4.2. Proton-Electron Hypothesis

As soon as the Rutherford nuclear atom model was introduced there were observations suggesting the integral nature of atomic masses. This could be interpreted to mean that the nucleus itself is built up of particles having a mass ≡ 1 *MU*. Accordingly nucleus was thought to consists of {Z protons + (A-Z) extra protons + (A - Z) electrons}. This is the *proton-electron hypothesis*. This hypothesis is successful only to explain β-ray radioactivity, whereas both the *deBroglie hypothesis* and the angular momentum consideration could not support.

3.4.3. Parabolic Method for Isotopic Mass
(JJ Thomson's Mass Spectrograph)

This method, described in undergraduate books in Physics, was used by J.J. Thomson (1912) for determining $\frac{n\,e}{M}$ of the positive ions of an element formed in the form of gases, example, neon (Ne). The positive ions were analyzed by traversing them under externally applied electric (E) and magnetic (B) fields acting parallel to each other and simultaneously to form on the screen detector as parabolae. Fig 3.5 describes the apparatus.

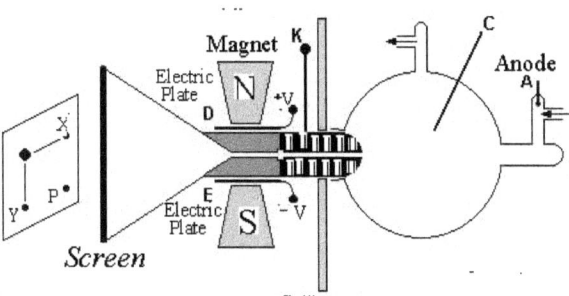

Fig 3.5 Thomson's Mass spectrograph

For a beam of positive ions traveling in the x-direction, E_y and B_y are the electric and magnetic fields both in the y-direction, it can be shown that the y and z-deflections of the beam components are related by the expression

$$z_1^2 = \left[\frac{B_y^2}{2E_y} L\left(\tfrac{1}{2}L+D\right)\right]\left(\frac{n\,e}{M}\right) y_1 \qquad (3.4.5)$$

which is a parabola. L is the length of the es deflector plates.
The first element to be analyzed was Neon of atomic weight 20.2. Two parabolae were identified as due to the two isotopes ^{20}Ne and ^{22}Ne.
The disadvantages of the parabola method are
1.   It is not accurate
2.   It is limited to gases
3.   It does not yield the abundances of the isotopes.

3.4.4. Principles of Mass Spectroscopy

An instrument capable of measuring accurately <u>atomic mass</u> is a *Mass Spectroscope*.
Why does one measure the mass of an atom? This is because *free* nuclei cannot be obtained getting rid of all the atomic electrons of the atom! Francis William Aston (1919) redesigned Thomson's spectrograph for getting more precession. The spectrometers developed by Arthur J. Dempster, Kenneth Tompkins Bainbridge, J. Mattauch, Nier, and others have different designs.

3.4.4.1. Principles

1.   Ion source
     It provides a supply of ionized atoms of the type to be measured by the rest of the instrument.
2.   Analyzer
     The ions leave the ion source with a) different degrees of ionization, e, 2e, 3e, b) different energies (or velocities, V), and c) different masses, M. it is necessary to select the ions of a narrow range of energies. Since V and M are two unknown parameters, two different operations are to be performed on the ions beam. This can be done in various ways, through the use of combination of electric and magnetic fields.
     a)   Energy Filter
It is possible to select charged particles of specific energy in one of two ways:

(i) Acceleration energy filter: Letting the initial KE of the particles to be $(\frac{1}{2}MV_i^2) \approx 0$, they emerge out with $(\frac{1}{2}MV^2)$ after getting accelerated through an electric field of potential U V, between two plane plates, (Fig 3.6) as given by

$$\frac{1}{2}M(V^2 - V_i^2) = neU \ ; \ E \parallel V_i \qquad (3.4.6)$$

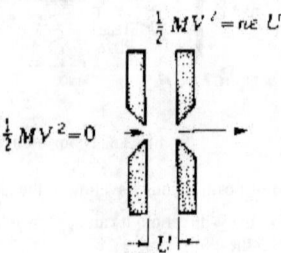

Fig 3.6 Acceleration Filter

(ii)   E.S. Filter: The ion beam has to enter at right angles to the electric field, E, maintained between two curved plates of radius r (Fig 3.7).

$$\frac{MV^2}{r} = neE \ ; \ E \perp V_i \qquad (3.4.7)$$

This yields particles of the same energy and charge the same radius of curvature. This is called the *cylindrical capacitor* energy filter.

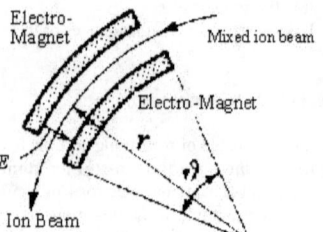

Fig 3.7 ES Filter

b)   Magnetic momentum filter

When ions pass through a magnetic field B (Fig 3.8), particles with the same momentum MV and the same charge ne will have the same curvature r, given by

$$r = \frac{c}{B}\frac{MV}{ne} \qquad (3.4.8)$$

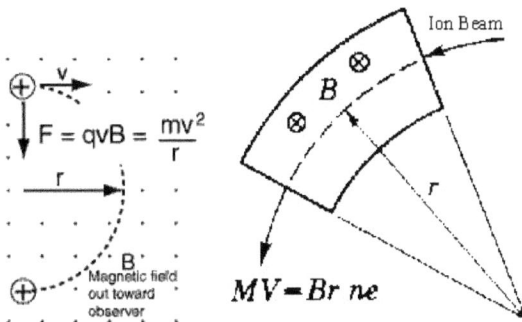

Fig 3.8 Principle of magnetic momentum filter

c) Velocity Filter:
The use of crossed electric E and magnetic fields B exert forces in opposite directions can select charged particles of equal velocity (Fig 3.9). The magnitudes of E and B must be such that

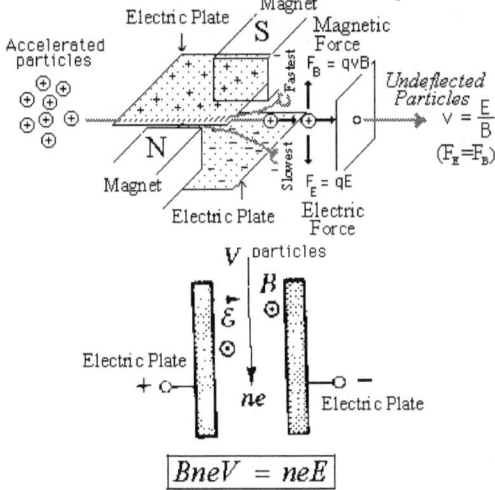

Fig 3.9 Principle of velocity filter

$$V = \frac{E}{B} \qquad (3.4.9)$$

3.4.4.2. Parts of a Mass Spectrograph
All Mass Spectrometers consist essentially of
(1) An ion source
(2) An Energy / Velocity filter
(3) Two analyzing elements, which is a critical combination of ES and Magnetic deflection applied to the beam so as to get ions of $\frac{M}{ne}$ focused at a single line in space.
(4) Recording of the image.

3.4.4.3. Aston's Mass Spectrograph

Designed in 1919, it was a *single or directional / velocity focussing* apparatus. The E and B fields are applied one after the other and in different directions (Fig 3.10). Francis W. Aston received the Nobel Prize in 1922 for nuclear mass measurement.

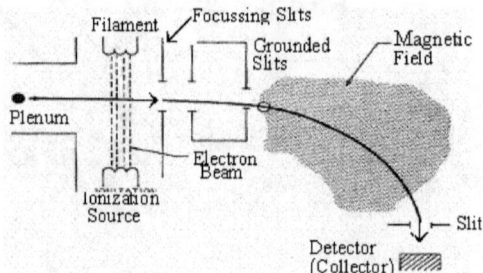

Fig 3.10 Aston's Mass Spectrograph

The detector is a photograph.

$$d_1 = \frac{1}{2} \frac{neE}{MV^2} L^2 \qquad (3.4.10)$$

is the deflection of the beam in the E-field,
L = length of the ES deflector in the B-field, having
b as length of path,

$$d_2 = \frac{1}{2} \frac{Bne}{MV^2} b^2 \qquad (3.4.11)$$

If θ_e and θ_m are the angular deflections of the ions in E- and B-fields, respectively, to get a net *velocity dispersion*,

$$\frac{d\theta_e}{dV} + \frac{d\theta_m}{dV} = 0, \text{ the condition is}$$

$$\theta_m = -2\theta_e. \qquad (3.4.12)$$

Velocity focusing results when

$$|\theta_m| > 2\theta_e \qquad (3.4.13)$$

The accuracy of the instrument is determination of M with 100 *ppm*.

3.4.4.4. Dempster Mass Spectrometer

Arthur J. Dempster in 1922 built an instrument the outline of which is as in Fig. 3.11.

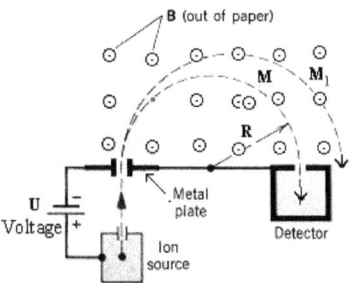

Fig 3.11 Dempster's Mass Spectrometer

$$\frac{n\,e}{M} = \frac{2\,U\,c^2}{B^2\,R^2}, \quad \theta_m = 180° \qquad (3.4.14)$$

The detector is an electrometer. This is a *single differential focusing* instrument, $\theta_m = 180°$ was used. The two isotopes of potassium, ^{39}K and ^{41}K, in relative intensity 18 : 1 were found.

3.4.4.5. Dempster's Mass Spectrograph

It is a *single or directional focusing* instrument. A cylindrical capacitor energy filter is used. The electric and magnetic sectors give

$$\frac{n\,e}{M} = \frac{E}{B^2}\,\frac{R_e}{R_m^2} \qquad (3.4.15)$$

The isotopes of Ytterbium element were studied. ^{235}U was discovered (by Dempster) with its abundance ▢ $\frac{1}{140}$ of the common isotope ^{238}U.

Worked out Example 3.4

In a magnetic spectrograph, a beam of singly charged ions of chlorine with $5.0 \times 10^4\,ms^{-1}$, determined by velocity selector, is incident at right angles into a region of magnetic field of 0.15T. Given the two isotopes of chlorine have masses 34.97 u and 36.97 u, respectively, find the radii of the paths of the two beams in the field region.($e = 1.6021 \times 10^{-19}\,C$)

Solution: Step # 1 Given: Mass of $^{35}Cl = M^{35}Cl = (34.97\,u)(1.66 \times 10^{-27}\,kg/u)$

$M^{37}Cl = (36.97\,u)(1.66 \times 10^{-27}\,kg/u)$, $R = \frac{MV}{qB}$, $B = 0.15$ T, $V = 5.0 \times 10^4\,ms^{-1}$

$e = 1.6021 \times 10^{-19}\,C$

Step # 2 $R = \frac{MV}{qB}$; $R^{37}Cl = \frac{((34.97\,u)(1.66 \times 10^{-27}\,kg/u)[5.0 \times 10^4\,ms^{-1}]}{(1.6021 \times 10^{-19}\,C)(0.15\,T)} = 0.17\,m$

$R^{37}Cl = \frac{((36.97\,u)(1.66 \times 10^{-27}\,kg/u)[5.0 \times 10^4\,ms^{-1}]}{(1.6021 \times 10^{-19}\,C)(0.15\,T)} = 0.18\,m$

3.4.4.6. Bainbridge Spectrograph

It is a *directional (single) focusing* instrument. It uses a preliminary velocity selector, containing an E-field E and B-field B_o in crossed positions (Fig 3.12). The detector is a photographic plate.

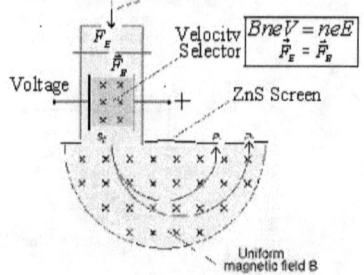

Fig 3.12 Bainbridge Spectrograph

$$V = \frac{c\,E}{B_O} \qquad (3.4.16)$$

$$M = \frac{R\,n\,e\,B\,B_O}{c^2\,E} \qquad (3.4.17)$$

Using this Kenneth T. Bainbridge (1936) studied Germanium and was found to consist of ^{70}Ge, ^{72}Ge, ^{73}Ge, ^{74}Ge, and ^{76}Ge.

3.4.4.7. **Mass Spectrograph of Bainbridge & Jordan**

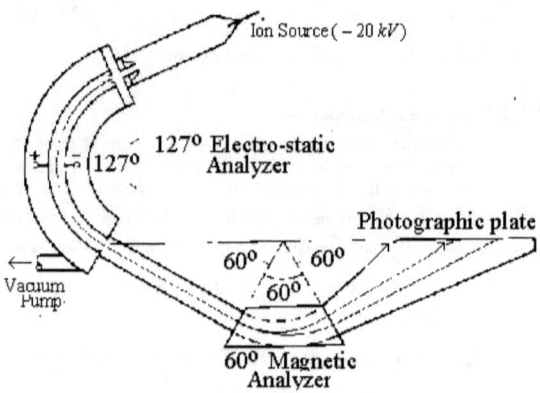

Fig 3.13 Mass Spectrograph of Bainbridge & Jordan

Employing ES and magnetic deflection with a photographic detector KT Bainbridge & E Jordan bent the ions through $\theta_e = 127°\ (=\frac{\pi}{\sqrt{2}})$ and $\theta_m = \frac{\pi}{3}$. The mass scale was linear over a large portion of the detector plate. The ten isotopes of tin (Sn) and one isotope of iodine are obtained in this spectrograph (Fig 3.13).

3.4.4.8. Nier's 60° Sector Spectrometer

Alfred O Nier designed this spectrometer, as shown in Fig 3.14.

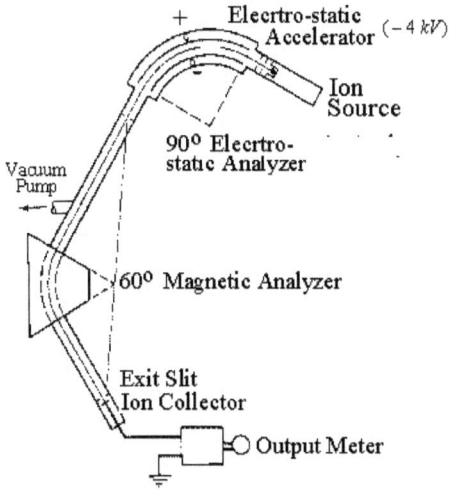

Fig 3.14 Nier's Spectrometer

3.4.4.9. Double Focusing Spectrograph of Mattauch

The primary limitation of single focusing instruments is one of resolution, of the order of 1000 mostly, and few with 10,000. (Resolution of 100 means a mass difference of 1 part in 100, i.e., $\frac{M}{ne}$ =100 & 101). Instead these low resolutions, in double focusing geometry. In 1936, J. Mattauch got high resolutions by correcting for the energy spread of the ion beam. He used cylindrical ES analyzer. Ions leaving the θ_e = 31°51' ($=\frac{\pi}{4\sqrt{2}}$) electric sector are focused at infinity. Then in the magnetic analyzer, with θ_m = 90° sector, both $\frac{M}{ne}$ separation and direction focusing are achieved.

3.4.4.10. Nier-Johnson Double focusing Spectrograph

In 1953, this instrument was designed having a θ_e = 90° electric sector and a θ_m = 90° magnetic sector satisfying double focusing geometry at a particular point. Resolutions up to 40,000 with a mass range capability of >2000 mass units were attained.

3.4.4.11. Time-of-Flight Spectrometers

Converging Annular Time-of-flight Mass Spectrometer was designed in 1955.The Kore's novel spectrometer geometry which starts with a ring shaped ion source. The ions are extracted perpendicular to the ring and deflected inwards so that as they proceed through the spectrometer each annular bunch of ions converges to a relatively small size by the time it reaches the detector. This allows a relatively large source to be used with a small detector giving the best of both worlds in terms of sensitivity, mass resolution and cost.

3.4.5. Nuclear Mass by the Doublet Method of Mass Spectroscopy

Direct method of measuring masses of atoms is not accurate (accuracy only ☐ 0.01%). Mass spectrometers of the last half of the 20th Century were particularly accurate at comparing two compounds with nearly equal $\frac{M}{ne}$ values and art measuring small differences in masses. The doublet method is a procedure which is relatively insensitive to calibrations errors and nonlinearities in the instrument.

For example, in most of the measurements, the frequently used *fundamental doublets* are as in Table 3.1.

Table 3.1   Fundamental Mass Doublets

Fundamental doublets	$\frac{M}{ne}$	Mass difference(amu)
1. $(^{12}C\ ^1_1H)^+ - (^{16}O)^+$	≅ 16	$a = (36.369 \pm 0.02) \times 10^{-3}$
2. $(^2_1H)^+ - (^{12}C)^{2+}$	≅ 6	$b = (42.230 \pm 0.019) \times 10^{-3}$
3. $(^1_1H)^+ - (^2H)^+$	≅ 2	$c = (1.539 \pm 0.002) \times 10^{-3}$

where: $(^{16}O) \equiv O$; $(^{12}C) \equiv C$; $(^2H) \equiv D$; $(^1H) \equiv H$

Then the following three simultaneous equations in O, C, D, and H are written, with the obtained doublet separations a, b, and c.

1) $M_C/1 + 4 M_H/1 - M_O/1 = a = 0.0363$ amu
2) $3 M_D/1 - M_C/2 = b = 0.0422$ amu
3) $2 M_H/1 - M_D/1 = c = 0.0015$ amu

Using $M_O = 16$ amu, gives

$$M_{(12_C)} \equiv M_C = 12 + \tfrac{1}{4}(-6c - 2b + 3a)\ amu = 12.00384\ amu$$

$$M_{(2_H)} \equiv M_D = 2 + \tfrac{1}{8}(-2c + 2b + a)\ amu = 2.0147\ amu$$

$$M_{(1_H)} \equiv M_H = 1 + \tfrac{1}{16}(6c + 2b + a)\ amu = 1.008058\ amu$$

3.5.1 MASS OF NEUTRON
The mass of the neutron is obtained from a combination of mass spectrometric data and nuclear-reaction data.

3.5.2 Methods other than Mass spectrometers
This is made from
a)   Energetics of nuclear reactions, and
b)   Radioactive processes.

REVIEW QUESTIONS

R.Q. 3.1 A beam of alpha particles having $8.8\ MeV$ is striking a gold foil of thickness $0.2\ \mu m$. Find the fraction scattered through an angle $\theta \geq 90°$. [Answer:

$(n = \rho N_A / M = 5.9 \times 10^{22}\ atoms/cm^3,\ \dfrac{(Z_\alpha\ e^2)}{4\pi\varepsilon_0\ kE} = 1.3 \times 10^{-14}\ m$,

$Z = 79, \rho = 19.32\ gm/cm^3$), $df|_{\pi/2}^{\pi} = 6.25 \times 10^{-6}$]

R.Q. 3.2 Arrive at a relationship between the impact parameter x, the distance of closest approach b and the scattering angle, θ, in the case of Coulomb elastic scattering. For what θ value will the impact parameter x becomes equal to b? [Answer $x = b\ Cot\dfrac{\theta}{2}$; $\theta = 53.2°$].

R.Q. 3.3 In a scattering experiment, a scintillation detector of area $0.1\ cm^2$ was at 5 cm apart from the gold foil target and that it made an angle of $45°$ with the direction of the direct beam of α-particles. If the incident beam consisted of 10^8 particles of energy $8.8\ MeV$, what would be the number of particles detected? [Answer $(\rho = 19.32\ gm/cm^3$

$N_i = 10^8/s$, $Z = 79$, $n = \dfrac{\rho N_A}{M} = 5.9 \times 10^{22}\ atoms/cm^3$, $\dfrac{(Z_\alpha\ e^2)}{4\pi\varepsilon_0\ kE} = 1.3 \times 10^{-14}\ m$,

$dS = 0.1\ cm^2$, $N(\theta) = 107\ cm^2/s$), $N(\theta).dS = 10.7/s$].

R.Q. 3.4 Alpha particles from Polonium are directed normally against a thin foil of gold of thickness $10^{-7}\ m$. Given $\rho = 19.32\ gm/cm^3$ for gold. Determine the fraction of incident α-particles scattered through angles $\theta \geq 90°$. [Answer: $df|_{\pi/2}^{\pi} = 8.5 \times 10^{-6} \approx 0.00085\%$].

R.Q. 3.5 A target of beryllium having $10^{19}\ atoms/cm^2$ normal to the direction of incidence is bombarded by a beam of $1.00\ MeV$ protons. What is the probability that a given proton will be scattered by the Coulomb field of beryllium nucleus through an angle $\theta = 138°$ into a solid angle, $d\Omega = 10^{-2}\ Sr$? Given Mass of beryllium to proton is 8.96. $mc^2 = 0.511\ MeV$, $e^2/4\pi\varepsilon_0\ m_o\ c^2 = 2.82 \times 10^{-15}\ m$. [Answer: 2.17×10^{-10}].

R.Q. 3.6 A beam of α-particles of $7.7\ MeV$ is bombarding a gold foil $3 \times 10^{-7}\ m$ thick. Find the fraction scattered through angles $\theta > 45°$. Given: for gold $\rho = 19.32\ gm/cm^3$, $Z = 79$, atomic weight is $197\ kg/kmole$ and $N_A = 6.02 \times 10^{26}\ atoms/kmole$. [Answer: $n = \rho N_A / M = 5.91 \times 10^{28}\ atoms/m^3$, $df| = 7 \times 10^{-5} \approx 0.007\%$].

R.Q. 3.7 A beam of α-particles of $10\ MeV$ is bombarding a Copper foil (Z = 29). Calculate the distance of closest approach of the particles to Copper nuclei. [Answer: $\dfrac{(Z_\alpha\ Z e^2)}{4\pi\varepsilon_0\ b} = 1.6 \times 10^{-12}\ J$, $b = 8.4 \times 10^{-15}\ m$].

R.Q. 3.8 In a scattering experiment, a scintillation detector of area $2.5\ mm^2$ was at 5 cm apart from the aluminium foil target of $1.0\ \mu m$ thickness and that it made an angle of $\theta = 60°$ with the direction of the direct beam of α-particles. If the fraction of α-particles of energy $9.0\ MeV$

scattered was found to be 1 in 10^8, calculate the atomic number of aluminium. (Take relative density of aluminium as 2.7).[Answer: Z = 13].

R.Q. 3.9 In a Thomson mass spectrograph, the electric plates are 50 *mm* long and the electric and magnetic fields are $E_y = 15\ kVm^{-1}$ and $B_y = 0.5\ T$. Find the equation of the parabola for the simply ionized ^{20}Ne isotope. The mass of the nucleon is 1.67×10^{-27} kg and the electronic charge is $1.6 \times 10^{-19} C$. [Answer $y = 10\ z^2$].

R.Q. 3.10 In a mass spectrograph, ions of the same charge and velocity with mass numbers (A) 35 and 37 enter a uniform magnetic field. The radius of circular path for ions of mass number 35 is found to be 0.45 *m*. Calculate the radius of the path due to the ions of A = 37. [Answer 0.48 *m*].

R.Q. 3.11 Find the binding energy per nucleon of [a] carbon-12 and [b] carbon-14 given that neutron mass = 1.00898 u, proton mass = 1.00759 u, C_{12} mass = 12.000000, C_{14} mass = 14.003242 u and 1u = 931 *MeV*. Which isotope of carbon is expected to be less abundant in nature and why? [Answer 7.71 *u* and 7.59 *u*].

R.Q. 3.12 In a Dempster Mass Spectrometer, the radius of curvature in the magnetic sector is 5 cm. $B = 0.384\ Wb\ m^{-2}$. What voltage must be applied to accelerate the following type of singly charged particles to a velocity such that they will be recorded? (a) ^6_3Li, (b) ^7_3Li, and (c) $^{41}_{19}K$? [Answers (a) 2,964 V, (b) 2,540 V, and (c) 434 V].

R.Q. 3.13 A.O. Nier measured one set of doublet data as is given below:
1. $CH_4 - O \equiv a \cong (36381.5 \pm 0.9)\ \mu$ amu
2. $C_4 - SO \equiv b \cong (33016.4 \pm 1.3)\ \mu$ amu
3. $C - O_2 - S \equiv c \cong (17754.3 \pm 0.9)\ \mu$ amu

Determine the mass of aluminium, M_{Al} using $C_2\ H_3 - {}^{27}Al \equiv d = (41.9451 \pm 23)\ m$ amu and in the Carbon scale.
[Answer: $12\ M_H - M_C = 3a - b + c$ so that
$M_H = 1 + \frac{1}{12}(3a - b + c) = 1.0078235$ *amu*
$M_{27Al} = C_2\ H_3 - (41.9451 \pm 23)\ m$ amu = 26.981525 *amu*].

R.Q. 3.14 In an investigation with a double focusing spectrometer, Nier obtained the following mass doublet measurements.

TABLE 3.2 Mass Doublets

Fundamental doublets		$\Delta M \times 10^4 u$
1. $(^{12}C)_4$	$- {}^{32}S\ {}^{16}O$	$\cong a = 331.82$
2. $^{12}C\ (^{16}O)_2$	$- {}^{12}C\ {}^{32}S$	$\cong b = 177.82$
3. $(^{12}C)_3\ (^1H)_8$	$- {}^{12}C\ {}^{16}O$	$\cong c = 729.67$
4. $(^{12}C)_6\ (^1H)_4$	$- {}^{12}C\ (^{32}S)_2$	$\cong d = 873.26$

Using the ^{16}O as the standard, determine the atomic masses of ^1H, ^{12}C, and ^{32}S
[Answer: $^1_1H = 1.008987$ *amu*; $^{12}C = 12.003844$ *amu*, and $^{32}S = 31.982274$ *amu*].

R.Q. 3.15   In a Nier's instrument the following mass doublet was detected: $(^{12}C)_{12}\,(^{1}H)_{16} - {}^{A}Gd \equiv d = (198.05 \pm 0.09) \times 10^{-3}\, u$. Determine the mass of Gd, from the knowledge of that of hydrogen on the ^{12}C scale. Given the following data:

TABLE 3.3   Mass Doublets

1. $CH_4 - O \equiv a \cong$	$(36381.5 \pm 0.9)\,\mu\,amu$
2. $C_4 - SO \equiv b \cong$	$(33016.4 \pm 1.3)\,\mu\,amu$
3. $C - O_2 - S \equiv c \cong$	$(17754.3 \pm 0.9)\,\mu\,amu$

[Answer: $12\,M_H - M_C = 3a - b + c$; $M_H = 1 + \frac{1}{12}(3a - b + c) = 1.0078235\, u$

Uncertainty $= \frac{1}{12}\sqrt{[(3\,\Delta a)^2 + (\Delta b)^2 + (\Delta c)^2]} \cong 0.3\,\mu\,u$ $M_{Gd} = 159.92713\, u$].

R.Q. 3.16   Give an account of mass doublet method of determining the atomic masses with greater accuracy.

R.Q. 3.17   Give the theory and working of a mass synchrometer.

RQ. 3.18   Give the theory of working of Aston's Mass spectrograph. Describe briefly the various methods used for the separation of isotopes.

RQ. 3.19   Give the theory and practice of Rutherford's researches on the large angle scattering of α-particles by matter. Indicate the importance of the results obtained.

RQ 3.20   Describe the construction and action of any mass spectrograph and show how isotopic masses have been accurately determined.

&%&%&%&%&%

Chapter 4
STABLE PROPERTIES OF A NUCLEUS - II:
Nuclear Magnetic Moments

Chapter 4

STABLE PROPERTIES OF A NUCLEUS - II:
Nuclear Magnetic Moments

"The saddest aspect of life right now is that science gathers knowledge faster than society gathers wisdom".   ~Isaac Asimov

4.1   INTRODUCTION

Certain time-independent (static) properties of nuclei influence in various ways the behaviour of atoms and molecules. The subject of Atomic Physics and Molecular Physics are highly developed. Nuclear influences on the behaviour of atoms and molecules are well understood. This gives the reasons as to why the use of certain observations within the fields of Atomic and Molecular Physics is used to determine nuclear static properties. In the previous Chapter 3, it was learnt that an atomic nucleus is an assembly of Z protons and N (= A-Z) neutrons confined to a region whose linear dimension is of the order of 10 fm or less. It was also seen that nuclear masses   In this Chapter let us consider a second of the general properties of nuclei, *viz.*, the intrinsic spin, in terms of which they can be characterized. It will be described how the magnetic properties exert considerable influence on the behaviour of nuclei. The nuclear electric moment, second static property of importance, will be dealt with in the next Chapter.

4.1.1   Spin of Nucleons

Protons and neutrons, which constitute nuclei, both have spin quantum number, $s_p = s_n = \frac{1}{2}$ The nucleons, by virtue of their motion within nuclei, may have both spin and orbital angular momenta. The orbital angular momentum quantum number of a nucleon is designated by $\ell = 0, 1, 2, ...$.

4.1.2   Nucleus – A Quantum System

Is the wave nature of matter relevant to the nucleus?
For this one has to examine if the deBroglie wavelength (λ) of the nucleus is of the order of the size of the nucleus (see Exercise 4.1). From the results of Example 4.1 it is seen that $\lambda = h / p = 9.0 \times 10^{-15} m$ is close to the nuclear size of $10 \times 10^{-15} m \equiv 10\, fm$. This clearly shows that the wave nature of matter is indeed relevant in the sub-atomic world. Thus there is the need to treat the bound state of nucleons within a nucleus by Quantum Mechanics rather than Classical Theory is evident.

Even before the origin of '*fine structure*' of atoms was understood on the basis of spin-orbit coupling, some optical lines were found to exhibit '*hyperfine structure*'.

4.2.   INTRINSIC NUCLEAR SPIN

4.2.1   Nuclear Angular Momentum

A number of distinct effects are responsible for the hyperfine structure of spectral lines. One of them is that the nucleus of the atom also has **intrinsic spin**, I According to Quantum Mechanics, like all angular momenta, nuclear spin angular momentum is quantized. Nuclear angular momentum quantum number is given the symbol, I, and permits both integer and half integral values.

$$I = 0, \frac{1}{2}, 1, \frac{3}{2}, 2, \frac{5}{2}, etc. \qquad (4.2.1)$$

4.2.2 Nuclear Magnetic Quantum Number

To account for the hyperfine structure, the concept of components of nuclear angular momenta, denoted by m_I, is introduced.

$$m_I = 0, \pm\frac{1}{2}, \pm 1, \pm\frac{3}{2}, \pm 2, \pm\frac{5}{2}, etc \qquad (4.2.2)$$

4.2.3 Failure of Proton-electron Hypothesis:- Invalid

Both the electron and proton are Fermi particles, having both $\frac{1}{2}$- spin values. This means each has spin angular momentum, $\frac{1}{2}\hbar$.

According to the *proton-electron* hypothesis,

Nucleus with I = integer, has {# of protons + # of electrons} = Even, and

Nucleus with $I = \frac{1}{2}$ − integer, has {# of protons + # of electrons} = Odd.

Thus the total angular momentum of the atom of this nucleus,

$$F = I + J \qquad (4.2.3)$$

where J = Total electronic orbital angular momentum
F = Hyperfine quantum number.

It is beyond the scope of this book to further extend discussion of the analysis of hyperfine structure.

The above prediction, however, is not obeyed, for example, even in the simplest case of a nuclear system like deuteron, $^2_1H \equiv D$. This means $m_I = \pm\frac{3}{2}, \pm\frac{1}{2}$, since $I = (\frac{1}{2} + \frac{1}{2} + \frac{1}{2}) = \frac{3}{2}$.

However, the observed value is $I = 1$. This shows that there is something that cannot be reconciled with the hypothesis of nuclear electron.

4.2.4 The Proton-neutron theory is Valid!

On the other hand, each nucleon, i.e. proton or neutron, is a Fermi particle having spin $s = \frac{1}{2}$. The total angular momentum of a nucleus containing a group of nucleons is obtained in either of two ways: j - j coupling or L-S coupling.

$$I \begin{cases} = \sum_i j_i, & \text{where } j_i \pm s_i \\ = L \pm S, & \text{where } L = \sum_i l_i, \ S = \sum_i s_i; \end{cases} \qquad (4.2.4)$$

depending on the choice of nucleus: light or heavy.

4.2.5 Nuclear spin and Nuclear Mass

In general, it is found experimentally that *"Nuclear spin is very closely related to the Mass number of the nucleus"*, as follows:

(i) Even-Even nuclei: (Z − even, N= even)
$A = [Even\ N + even\ Z]$ = even, $I = 0$.
This is due to the *pairing effect*.

The angular momentum of the nucleus = SUM of the spin value of the unpaired nucleons.
Examples: ^{12}C, ^{16}O, ^{22}S,
(ii) Even-Odd nuclei: (Z – even, N= odd)
$A = [Even\ Z + odd\ N] = [Odd\ Z + Even\ N] = odd;\quad I = Half-integer$
The angular momentum of the nucleus = The spin value of the last or unpaired nucleon.
Examples: ^{1}H, ^{11}B, ^{19}F, ^{31}P, ^{35}Cl, ^{79}Br,
(iii) Odd-Odd nuclei: (Z – odd, N= odd)
$A = [Odd\ Z + odd\ N] = even;\quad I = Integer$
Examples: ^{2}H, ^{14}N, .

4.2.6 Measurement of Nuclear Angular Momentum

The values of I have been measured by nuclear reactions for most nuclei., not only for their ground state but also for excited states.

4.2.7 NUCLEAR PARITY, Π

4.3.1 Conservation of Parity

Parity Π is a concept of great importance in Atomic and Nuclear Physics. It is a property of the wave function $\psi(r_1, r_2, r_3, ... r_A)$, describing a quantum mechanical system, containing two or more (A) particles, such as a nucleus. Parity operator $\hat{\Pi}$ is defined as the one having the property of reflecting each coordinate r_j through the origin ($r_j \to -r_j$), viz.,

$$\hat{\Pi}\psi(r_1, r_2, r_3, ..., r_A) = \psi(-r_1, -r_2, -r_3, ..., -r_A)$$
$$= (\pm 1)\ \psi(-r_1, -r_2, -r_3, ..., -r_A) \quad (4.3.1)$$

A state with $\hat{\Pi} = +1$ (or -1) is said to have even (or odd) parity.

The usefulness of parity in atomic physics indicates that the EM interaction is invariant under the space inversion transformation (see, for example, Devanarayanan, 2005). In addition to the conservation of total energy, momentum, and angular momentum (as in Classical Physics), Parity is also *normally* conserved in nuclear processes.

<u>Note 1</u>: In a nucleus the *strong interaction* determines the energy levels, and the parity is conserved.

<u>Note 2</u>: Parity Π is violated in nucleus where *weak interaction* is present. But it does not alter the situation because the amplitude of the impurity of weak interaction contributed to parity is only 10^{-6} to 10^{-7} of the regular amplitude of ψ.

Further, one can mean that parity is a constant of the motion; parity is **conserved** in nuclear processes, if

$$[\hat{\Pi}, \hat{H}] = 0 \quad (4.3.2)$$

where $\hat{H}$ is the nuclear Hamiltonian concerned.

4.3.2 How to Indicate the Parity Π of a Nuclear State?

Parity of a nuclear state is indicated by a superscript with + *or* – sign on the angular momentum symbol I (Table 4.1). It is a customary convention to assign both the proton and the neutron POSITIVE parity.

Examples: $1^+, 1^-, \dfrac{1}{2}^-, 0^+, 7^-, \dfrac{3}{2}^+, ..$

Table 4.1 Parity of Typical Nuclides

Nuclide	I	Parity	Nucleus
^1_1H	$\frac{1}{2}^+$	Even	Odd Z-Even N
^2_1H	1^+	Even	Odd Z-Odd N
^4_2He	0^+	Even	Even Z-Even N
^7_3Li	$\frac{3}{2}^-$	Odd	Odd Z-Even N
$^{176}_{71}Lu$	7^-	Odd	Odd Z-Odd N

4.3.3 Measurement of Parity of a Nuclear State

The parity of a nuclear state is measured by studying reactions. Consider the reaction

$$^{16}O + d \rightarrow {}^{17}O + p$$

is 'deuteron stripping'. The parity of ^{16}O and ^{17}O is given by $(-1)^\ell$
Parity of a nucleus Π = (Parity of neutrons) x (Parity of protons).   (4.3.3)

4.3.4 Classification of Particles

According to their spin quantum numbers, particles are classified as listed in Table 4.2 below:

TABLE 4.2 Classification of Interactions

Name of the Class	Interaction type	Quantum #, S	Name of Particles
1. BARYONS	Strong	$\frac{3}{2}$	Ω
		$\frac{1}{2}$	$\Xi, \Sigma, \Lambda, n, p$
2. MESONS	Strong	0	η, K, π
3. LEPTONS	Weak	$\frac{1}{2}$	μ, e, ν_μ, ν_e
4. MASSLESS BOSONS	EM	1	γ
	Gravitation	2	graviton

4.4 SYMMETRY

Another significant concept in Atomic and Nuclear Physics is that of *Symmetry*. If two identical *non-interacting* particles move in the same potential well, with coordinates 1 and 2, wave functions

$$\psi_{nk} = \psi_n(1) \cdot \psi_k(2). \quad (4.4.1)$$

Since particles cannot be labeled (being identical) this in invalid. On the other hand, this difficulty can be avoided, one can write

$$\psi_S = \frac{1}{\sqrt{2}} [\psi_n(1).\psi_k(2) + \psi_n(2).\psi_k(1)] \quad (4.4.2)$$

$$\psi_A = \frac{1}{\sqrt{2}} [\psi_n(1).\psi_k(2) - \psi_n(2).\psi_k(1)] \quad (4.4.3)$$

where ψ_S and ψ_A are *symmetric* and *anti-symmetric* functions to the exchange of the 1 and 2. If the particle has intrinsic spin, and Φ is the spin function,

$$\psi_S = \frac{1}{\sqrt{2}} [\psi_n(1).\psi_k(2) + \psi_n(2).\psi_k(1)] \cdot \Phi \quad (4.4.4)$$

$$\psi_A = \frac{1}{\sqrt{2}} [\psi_n(1).\psi_k(2) - \psi_n(2).\psi_k(1)] \cdot \Phi. \tag{4.4.5}$$

In general, *symmetry is conserved.*
ψ_S is associated with Bosons and ψ_A is related to Fermions. (4.4.6)

4.5 NUCLEAR MAGNETIC DIPOLE MOMENT

4.5.1 Why does a nucleus possess (nuclear) angular momentum?
This is because
(i)   Nucleus has a finite size,
(ii) It has nuclear mass,
(iii) It has nuclear spin
(iv) Nucleus has (positive) charge.

4.5.2 Nucleus as a bar Magnet
The static magnetic properties of a nucleus are specified in terms of its magnetic moments. Since the charge of the nucleus is distributed uniformly over its solid surface, the spinning would make the nucleus equivalent to a circular current, which in turn is equivalent to a magnetic dipole.
Let q, ϖ, S and i represent, respectively, the nuclear charge, angular velocity, area of current loop, and current

$$i = (\frac{q}{c})[\frac{\varpi}{2\pi}] \tag{4.5.1}$$

This means the nucleus is a *magnetic dipole*, called nuclear magnet, with *magnetic dipole moment*, $\vec{\mu}_I$

$$\vec{\mu}_I = i S = (\frac{q}{c})[\frac{\varpi}{2\pi}](\pi r^2) \tag{4.5.2}$$

$$= \left(\frac{q}{2Mc}\right) |\overleftarrow{L}_I, \text{Angular momentum} \tag{4.5.3}$$

While the direction of the resulting magnetic field (of the nuclear bar magnet) is dictated by the **Left Hand Rule**, the direction of the $\overleftarrow{L}_I$ vector is given by the **Right Hand Rule**.

4.5.2.1 Nuclear Magnetic-Gyro Ratio γ_I

Nuclear Magneto-Gyro Ratio

$$\gamma_I = \frac{\text{Magnetic moment}, \vec{\mu}_I}{\text{Angular Momentum}, \vec{L}_I} \tag{4.5.4}$$

gives $\gamma_I = \left(\frac{q}{2Mc}\right). \tag{4.5.5}$

4.5.2.2 Proton Magnetic Moment $\vec{\mu}_p$

$$\vec{\mu}_{Iz} = \left(\frac{Ze}{2M_p c}\right) m_I \hbar \tag{4.5.6}$$

$$\vec{\mu}_p = \left(\frac{eh}{4\pi M_p c}\right) m_I \tag{4.5.7}$$

4.5.2.3 Nuclear Magneton (1 $nm \equiv \mu_N$), the unit of nuclear magnetic moment

$$1 \text{ nm} \equiv \mu_N = \left(\frac{eh}{4\pi M_p c}\right) = 5.0504 \times 10^{-27} J\, T^{-1} \qquad (4.5.8)$$

Experimental measurements show that

1. Electron, $\mu_e = \frac{1}{2}\mu_B$
2. Proton, $\mu_p = 1\text{ nm} \equiv \mu_N$ (4.5.9)
3. Neutron, $\mu_n = 0\text{ nm} \equiv 0$

4.5.2.4 The Nuclear g-factor, g_I

The nucleus will have both the orbital and spin magnetic moment, yielding the resultant magnetic moment, $\vec{\mu}_I$ one gets

$$\vec{\mu}_I = \gamma_I \hbar I = g_I.I.\mu_N \equiv g_I.I.\text{ nm} \qquad (4.5.9)$$

$$\vec{\mu}_{I,z}\Big|_{Maximun} \text{ (is the state with } m_I = I) \equiv \mu_I \qquad (4.5.10)$$

$$\vec{\mu}_{I,z} = g_I.m_I.\mu_N \equiv g_I.I.\text{ nm} \qquad (4.5.11)$$

4.5.2.5 Nuclear magnetic moments of Typical Nuclides:

Table 4.3 lists the values of spin and magnetic moments of typical nuclides (Shirley, 1978).

TABLE 4.3 Nuclear Magnetic moments of Typical Nuclides

Nuclide	Spin, I	μ_I in units of μ_N
n	½	-1.9130418
^{1}H (p)	½	+2.7928456
^{2}H (D)	1	+0.8574376
^{4}He (α)	½	0
^{17}O	5/2	-1.89279
^{57}Fe	3/2	+0.50
^{57}Fe	½	+0.09062293
^{57}Co	7/2	+4.733
^{93}Nb	9/2	+6.1705
^{155}Gd	3/2	-0.2700

4.6 MEASUREMENT OF NUCLEAR SPIN

The magnetic properties of the nucleus have interested the physicist ever since they were first postulated to explain the hyperfine structure of spectral lines. The various methods by which the nuclear magnetic moment of free protons or nuclei can be measured are given below.

1. Optical spectroscopy
2. Radio frequency spectroscopy – Molecular beam resonance
3. Radio frequency spectroscopy – Nuclear magnetic resonance
4. Mossbauer spectroscopy

4.6.1 NUCLEAR MAGNETIC RESONANCE (NMR, PMR)

For a review of the NMR the reader may refer to any standard book on modern Spectroscopy (or "Quantum Chemistry" by S. Devanarayanan, 2008).

NMR has found wide applications in the field of diagnosis in the field of medicine. They are: i) NMR pins down Alzheimer's clue, as it has demonstrated that an enzyme previously shown to protect brain cells from the characteristic fibrous tangles associated with Alzheimer's disease also helps inhibits formation of the amyloid peptide plaques (APPs) seen in this disease. Using NMR to determine protein structure has two enormous advantages over traditional crystallographic methods. The first is that it works with proteins that cannot be crystallized, but perhaps even more importantly it can reveal structural details and dynamics of proteins in their native state, whether in solution or embedded in a cell membrane. In NMR it is possible, quite literally on the back of an envelope, to make exact predictions of the outcome of quite sophisticated experiments. The experiments chosen are likely to be encountered in the routine NMR of small to medium-sized molecules, but are also applicable to the study of large biomolecules, such as proteins and nucleic acids.

4.6.1.1 Nucleus is a tiny bar magnet (See Section 4.5.3.4). Equation (4.5.9)

$$\vec{\mu}_I = \gamma_I \hbar I = g_I . I . \mu_N \equiv g_I . I . nm \qquad (4.6.1)$$

4.6.1.2 Energy of a Nucleus in Magnetic Field:

$$E_H = -\vec{\mu}_I \cdot \vec{H}$$
$$= - g_I . m_I . H . nm \qquad (4.6.2)$$

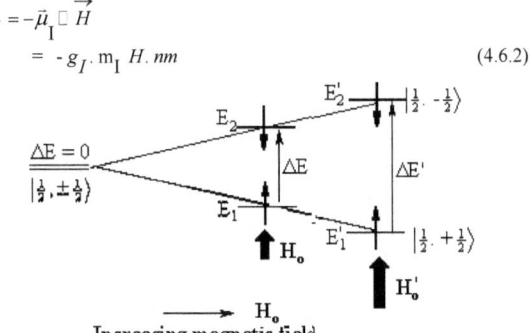

Fig 4.1 nuclear Zeeman split up of the energy level for ^1H nucleus

where m_I can have $(2I+1)$ values. This nuclear Zeeman split up of the energy level for ^1H nucleus which has $I = \frac{1}{2}$ is shown in Fig.4.1.

OBSERVE the difference in the energy level splitting in the case of atomic and nuclear systems. The upper and lower sublevels for the state $\left| I = \frac{1}{2} \right\rangle$ are

$\left| I = \frac{1}{2}, m_I = -\frac{1}{2} \right\rangle$ and $\left| I = \frac{1}{2}, m_I = +\frac{1}{2} \right\rangle$ in the **nuclear case**, whereas they are

$\left| I = \frac{1}{2}, m_I = +\frac{1}{2} \right\rangle$ and $\left| I = \frac{1}{2}, m_I = -\frac{1}{2} \right\rangle$ in the **atomic case**. The answer is the difference in the sign of charges of atomic electron and nucleus!

4.6.1.3 Larmor Precession

When bathed in a magnetic field $\vec{H}$, the nucleus of intrinsic dipole moment acquires an additional energy $E_H = -\vec{\mu}_I \cdot \vec{H}$. At the same time, the nuclear magnet experiences a torque, $\vec{T}$ given by

$$\vec{T} = [\vec{\mu}_I \wedge \vec{H}] = -\left(\frac{q}{2Mc}\right)[\vec{L}_I \wedge \vec{H}] \quad (4.6.3)$$

The torque makes the angular momentum to precess around the direction of the field $\vec{H}$ as shown in Fig 4.2. The angular frequency of precession, known as the Larmor frequency, ω_L is given by

$$\omega_L = g_I \cdot \left(\frac{e}{2Mc}\right) H = \left(\frac{g_I H \, nm}{\hbar}\right) \quad (4.6.4)$$

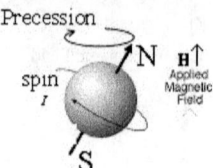

Fig 4.2 Larmor precession

4.6.1.4 Condition of Thermal Equilibrium

An assembly of nuclear magnets, having two states $\left|+\frac{1}{2}\right\rangle$ and $\left|-\frac{1}{2}\right\rangle$ is in equilibrium at temperature T, is determined by the *Boltzmann criterion*,

$$N_\uparrow = N_\downarrow \, e^{-\frac{\hbar \omega}{k_B T}} \quad (4.6.5)$$

where the *Bohr frequency condition*, viz.,

$$E_\uparrow - E_\downarrow = \hbar \omega_L \quad (4.6.6)$$

is applied.

4.6.1.5 Magnetic Resonance Absorption

If a magnetic field $\vec{H}_\omega$ of radio frequency, in the radiation is applied such that it is rotating in the same direction as the nuclear precession, and is at right angles to the field $\vec{H}$ and in the plane of $\vec{H}$ and $\vec{\mu}_I$, resonance takes place, and energy from the radiation will be absorbed such that $\Delta E = E_\uparrow - E_\downarrow$ at $\omega = \omega_L$ This is the NMR absorption.

In the case of Hydrogen nucleus, NMR is termed as PMR (Proton Magnetic Resonance) or H-NMR.

4.6.1.6 NMR Set up

The salient features of an NMR spectrometer are shown schematically in the self-explanatory diagram in Fig 4.3. and its block diagram in Fig 4.4.

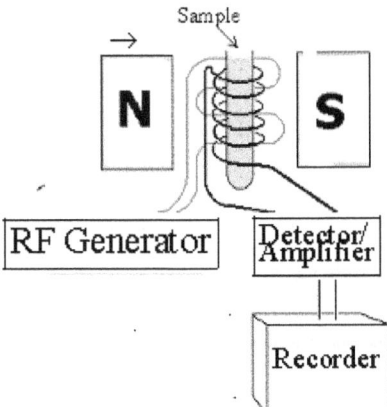

Fig 4.3 Block diagram of NMR Setup.

For spin 1/2 nuclei the energy difference between the two spin states at a given magnetic field strength will be proportional to their magnetic moments. For the four common nuclei noted above the magnetic moments are: ^1H μ = 2.7927, ^{19}F μ = 2.6273, ^{31}P μ = 1.1305 & ^{13}C μ = 0.7022. The following diagram gives the approximate frequencies that correspond to the spin state energy separations for each of these

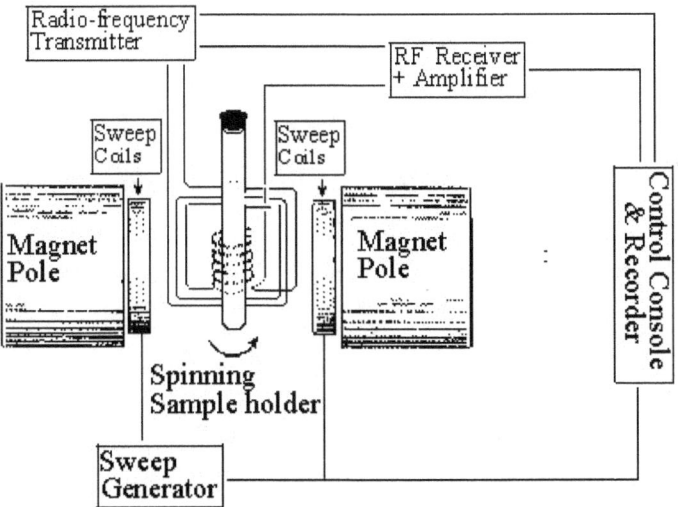

Fig 4.4 Schematic diagram of NMR spectrometer

nuclei in an external magnetic field of 2.34 T. The formula in Fig 4.5 shows the direct correlation of frequency (energy difference) with magnetic moment (h = Planck's constant).

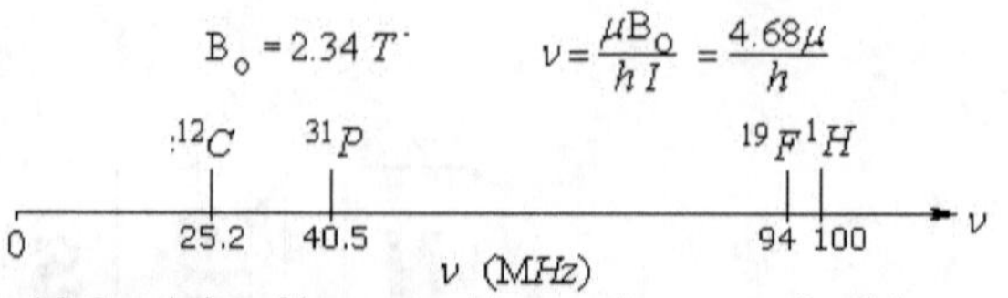

$$B_o = 2.34\ T \qquad \nu = \frac{\mu B_o}{hI} = \frac{4.68\mu}{h}$$

```
      ¹²C      ³¹P                          ¹⁹F ¹H
       |        |                             |  |
───────┼────────┼─────────────────────────────┼──┼──────► ν
0     25.2    40.5                          94 100
                        ν (MHz)
```

Fig 4.5 Correlation of frequency with magnetic moment of nuclei

4.7 MOSSBAUER TECHNIQUE (Nuclear Resonant Scattering, NGR)

4.7.1 Mössbauer spectroscopy

It is a spectroscopic technique based on the Mossbauer Effect. In its most common form, Mössbauer Absorption Spectroscopy, a solid sample is exposed to a beam of gamma radiation, and a detector measures the intensity of the beam that is transmitted through the sample, which will change depending on how many gamma rays are absorbed by the sample. The atoms in the source emitting the γ-rays are the same as the atoms in the sample absorbing them. In the Mössbauer Effect a significant fraction of the gamma rays emitted by the atoms in the source do not loose any energy due to recoil and thus have almost the right energy to be absorbed by the target atoms. The γ-ray energy is varied by accelerating the γ-ray source through a range of velocities with a linear motor. The relative motion between the emitter source and absorber sample results in an energy shift due to the Doppler Effect.

In the resulting spectra, gamma ray intensity is plotted as a function of the source velocity. At velocities corresponding to the resonant energy levels of the sample, some of the gamma-rays are absorbed, resulting in a drop in the measured intensity and a corresponding dip in the spectrum. The number, positions, and intensities of the dips (also called peaks) provide information about the chemical environment of the absorbing nuclei and can be used to characterize the sample.

4.7.2 Two major OBSTACLES to get Gamma ray Resonance (Mossbauer Effect):

a) The 'hyperfine' interactions between the nucleus and its environment are extremely small, and
b) The recoil of the nucleus as the gamma-ray is emitted or absorbed.

4.7.2.1. RECOIL OF FREE ATOMS

An energy quantum (γ-radiation) gets emitted from a free (isolated) radioactive nuclide of mass M; it acquires a recoil energy, E_R

$$E_R = \frac{E_\gamma^2}{2M c^2} \qquad (4.7.1)$$

This much energy is removed from the nuclear transition energy, E_t, of the emitting nuclide. This results in the energy E_γ of the γ-radiation

$$E_\gamma^{emitter} = (E_t - E_R) \qquad (4.7.2)$$

Similarly an absorbing similar nucleus will recoil on absorption of E_t, and effectively the γ-radiation will have energy

$$E_\gamma^{ab} = (E_t + E_R) \qquad (4.7.3)$$

4.7.2.2 THE DOPPLER EFFECT (Christian Andreas Doppler, 1845).
If a <u>moving source</u> is emitting waves of actual frequency f_S, then an <u>observer stationary</u> relative to the medium detects waves with a frequency f given by:
(if the source moving <u>away / toward</u> observer)

$$f = \frac{V}{(V \mp v_S)} f_S \qquad \text{moving source} \qquad (4.7.4)$$

$$\Delta \lambda = (V - v_S)/f_S \qquad (4.7.5)$$

where V = the speed of sound.
v_S = the speed of the source
(if the observer moving <u>away / toward</u> source)

$$f = \frac{(V \mp v_S)}{V} f_o \qquad \text{moving observer} \qquad (4.7.6)$$

$$\Delta f = \frac{V}{c} f_O$$

This is Doppler Effect.
$\Delta \lambda$ is $-ve$; (λ dectreases "*Blueshifted*")
Δf is $+ve$; (f increases)
The Doppler shift due to thermal motion of atoms. for Maxwell distribution of the speeds of atoms, the emission and absorption lines also have a shape of Maxwell distribution (Gaussian) with the Doppler width $2\delta_D$

$$2\delta_D = 2\sqrt{E_R k_B T} \; eV \qquad (4.7..7)$$

4.7.2.3. Energy profile :
As atoms move due to random <u>thermal</u> motion, the gamma-ray energy E_γ has a spread of values E_D caused by the Doppler Effect. This produces a gamma-ray <u>energy profile</u> as shown in Fig 4.6.

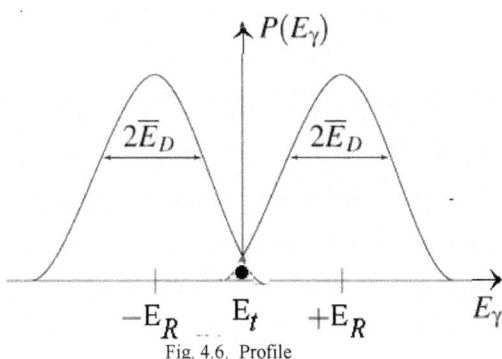

Fig. 4.6. Profile

Resonant overlap in free atoms. The overlap shown shaded is greatly exaggerated

4.7.2.5 NATURAL LINE WIDTH Γ ($=\Gamma_n$) OF GAMMA EMISSION AND ABSORPTION
A ray is an extremely monochromatic energy quantum whose line broadening (width Γ of nuclear level) is determined only by its lifetime τ ($=\tau_m$) with the Heisenberg's <u>Uncertainty Principle</u>

$$\boxed{\Gamma\tau=\hbar}. \tag{4.7.8}$$

i.e., $$\Delta\nu=\Gamma_n=\frac{1}{2\pi\,\Delta t}=\frac{1}{2\pi\,\tau_m} \tag{4.7.9}$$

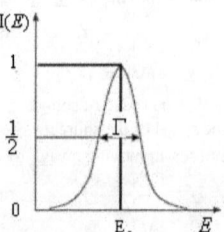

Fig. 4.7 The profile of energy distribution; Natural line width defined

The mean life τ_m of the $I=\frac{3}{2}$ state in $^{57*}Fe$ is 1.4×10^{-7} s. The energy distribution is given by a Lorentzian (i.e., Breit-Wigner) profile (Fig.4.7.2) with a FWHM (Full Width at Half Maximum) of $\Gamma_{nat} = 4.7 \times 10^{-9}$ eV. (Fig 4.7)

4.7.3.. PROBABILITY f

Probability f is an important characteristic of the Mössbauer Effect. For a single atom crystal with cubic symmetry the expression for f is

$$f = e^{-<x^2>/\lambda^2} \tag{4.7.10}$$

where $<x^2>$ = Average square displacement of atoms at the direction of emission of γ -- quantum,
λ = Wavelength.
Factor f under room temperature has a comparably high value only for restricted number of isotopes (^{57}Fe, ^{119}Sn, ^{181}Ta and some others) and depends on the temperature T of crystal.

4.7.4.1. THE REQUIRED CONDITIONS TO OBSERVE THE MOSSBAUER EFFECT
a) The atom (ion, molecule, etc.) must be in the solid state, to avoid recoil and thermal broadening.
b) The γ-ray energy $E_\gamma \approx E_t$ must be fairly low (10 to 100 keV) to obtain an appreciable number of recoil-less events

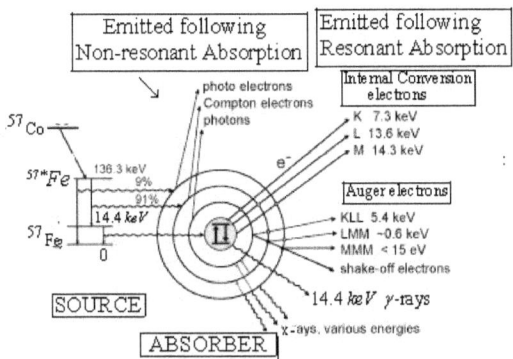

Fig. 4.8. Competing processes in a Nuclear environment

c) The life times (mean life) τ of nuclear excited states giving rise to such γ-rays must typically be in the range $\tau = 10^{-6} - 10^{-11}\ s$, since longer lived nuclear species emit lines which are extremely narrow for detection whereas the shorter lived species give lines which are broad and get lost in the counting statistics.

d) The internal conversion electron coefficients α should be as small as possible ($\alpha = 0 - 20$), and

e) The absorber sample may have the Debye temperature, Θ_D, preferably high.

These are clearly illustrated in the self-explanatory diagram (Fig 4.8).

4.7.4.2. FAILURE OF GAMMA RAY RESONANCE IN GASES

Attempts of Mössbauer to observe gamma-ray resonance in gases failed due to energy losses to recoil, preventing resonance. The energy E_γ of emitted gamma ray has $E_\gamma < E_t$, but to be resonantly absorbed it must be $E_\gamma^{ab} = (E_t + E_R)$ due to the recoil of the absorbing nucleus. To achieve resonance the loss of the energy due to this must be overcome in some way. He was able to observe resonance in solid Iodine, which raised the question of why gamma-ray resonance was possible in solids, but not in gases. For a μ^3 volume solid

$$E_R(Solid, \mu^3 volume) = E_R(Free\ atom) \qquad (47.11)$$

$$E_R(Bound\ atom) \cong 10^{-15} E_R(Free\ atom) \qquad 4.7.12)$$

Mössbauer proposed that for the case of atoms embedded into a solid, under certain circumstances, a fraction of the nuclear events could occur essentially without recoil. He attributed the observed resonance to this *recoil-free fraction* of nuclear effects.

4.7.5.1 HOW TO OBSERVE MOSSBAUER EFFECT?

When the atoms are within a solid matrix the effective mass of the nucleus is very much greater. The recoiling mass is now effectively the mass of the whole system, making E_R and E_D very small. If the gamma-ray energy is small enough the recoil of the nucleus is too low to be transmitted as a phonon (vibration in the crystal lattice) and so the whole system recoils, making the recoil energy practically zero a recoil-free event. That is why an emission (absorption) spectrum represents a superposition of shifted line with Doppler width

$$2\delta_D = D = 2\delta_D = 2\sqrt{E_R k_B T},$$

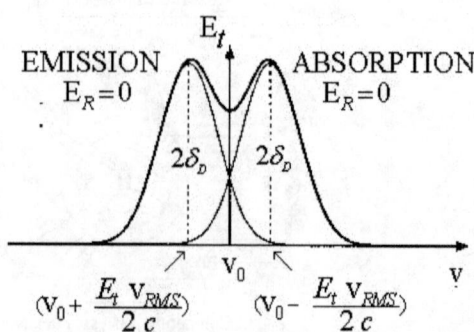

Fig.4.9. An emission (absorption) spectrum represents a superposition of shifted lines, (according to Rayleigh's criterion) with Doppler width D (at $(E_t \pm E_R)$) and non-shifted line with the width to be close to Γ_n).

At $(E_t \pm E_R)$, and non-shifted line with the width to be closed to natural (Fig 4.9). In the case of free atoms,

$$\delta_D = E_t\, V_{RMS}/c \qquad (4.7.13)$$

To use a Mössbauer source as a spectroscopic tool one must be able to vary its energy over a significant range. This is accomplished by Doppler shifting the energy of the gamma beam. If v and c are the frequency and velocity of γ–rays and V is the relative velocity of the source of the observer, frequency shift, Δv is given by

$$\Delta v = V v/c \; Hz \qquad (4.7.14)$$

$$E_\gamma = (E_t \pm V/c)\, eV \qquad (4.7.15)$$

Or the energy shift, $\Delta E_\gamma = (E_\gamma - E_t)$ of the γ–ray transition energy E_t

$$\Delta E_\gamma = (E_\gamma - E_t) = V E_t/c \; eV \qquad (4.7.16)$$

Moving the source at a velocity of 1 mm/s toward the sample will increase the energy of the photons by 14.41 $keV(v/c) = 4.8 \times 10^{-8}\, eV$ or ten natural line width. The "$mm\, s^{-1}$" is a convenient Mössbauer unit and is equal to $4.8 \times 10^{-8}\, eV$ for ^{57}Fe. A Mössbauer spectrometer consists of a source which may be moved relative to the sample and a counter to monitor the intensity of the beam after it has passed through the sample. The Mössbauer spectrum is a plot of the counting rate against the source velocity, i.e., the beam energy. If the sample nuclear levels are not split and $I = \frac{3}{2} \rightarrow I = \frac{1}{2}$ transition energy equals that of the source.

As discussed below, the hyperfine interactions will split the nuclear levels of the sample and complicate the Mössbauer spectrum. (See later Section).

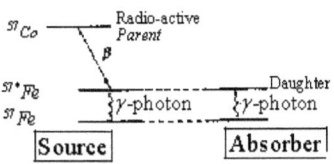

Fig 4.10 Relevance of Source and Absorber

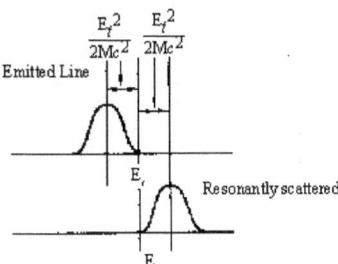

Fig 4.11. Resonant scattering principle

Doppler expression is

$$\Gamma = 10^{-8} eV = 14.41 \, keV(V/c) \qquad (4.7.17)$$

Worked out Example 4.1

A free nucleus ^{57*}Fe emits a gamma rat of frequency $3.5 \times 10^{18} \, s^{-1}$. What is the recoil velocity of the nucleus ?, Calculate the Doppler shift of the γ-ray to an outside absorber.

Solution: Step # 1 $\lambda = \frac{h}{Mv}$; $\Delta v = v\frac{V}{c} Hz$; $v = 3.5 \times 10^{18} \, s^{-1}$; $c = 2.997925 \times 10^{8} \, ms^{-1}$; $h = 6.6256 \times 10^{-34} \, J-s$; $M(Fe)-57) = 57g \, mol^{-1}$; $N_A = 6.0225 \times 10^{23} \, mol^{-1}$.

Step # 2   Recoil velocity, $V = \frac{P_\gamma}{M} = \frac{h\,v}{c\,M}$

$$= \frac{(6.6256 \times 10^{-34} J-s)(3.5 \times 10^{18} s^{-1})(6.0225 \times 10^{23} mol^{-1})}{(2.997925 \times 10^{8} ms^{-1})(57g \, mol^{-1})}$$

$= 8.2 \times 10^{3} \, m\,s^{-1}$

Step # 3   Doppler shift, $\Delta v = v\frac{V}{c} Hz = \frac{(3.5 \times 10^{18} s^{-1})(8.2 \times 10^{3} cm\,s^{-1})}{2.997925 \times 10^{8} m\,s^{-1}} = 9.6 \times 10^{11} s^{-1}$.

Worked out Example 4.2

The meta-stable state of ^{57}Fe has life time $1.5 \times 10^{-7} s$. Calculate the line width of the gamma ray emission.

Solution: Step # 1 $\tau_m = (\ln 2)(\tau_{1/2})$; $\tau_{1/2} = 1.5 \times 10^{-7} s$; $\hbar = (h/2\pi) = 1.054 \times 10^{-34} J-s$; $1 eV = 1.6021 \times 10^{-19} J$.

Step # 2 $\tau_m = (\ln 2)(\tau_{1/2}) = (1.5 \times 10^{-7} s)(\ln 2) = (1.5 \times 10^{-7} s)(0.693) = 1.0395 \times 10^{-8} s$.

Step # 3 $\quad \Delta v = \frac{1}{2\pi \Delta t} = \frac{1}{2\pi \tau_m} = \frac{1}{2\pi (1.0395 \times 10^{-8} s)} = 1.532 \times 10^7$ Hz.

In conventional instruments, the energy of the source photon is varied over a small range (tens of neV) using the Doppler Effect. The source is repetitively accelerated through a range in velocities (from a few to hundreds of mm s^{-1}) to add or subtract energy to the photons being emitted. When a match in the energy of the source photon and the absorber transition energy is achieved, resonant absorption occurs. Because subsequent emission of the absorbed photon has no directional probability, in contrast to the source photon directed at the detector, a decrease in the intensity of the background signal is observed at the energies (velocities) where resonant absorption occurs, thus giving rise to a Mössbauer spectrum (Fig. 4.11).

Although more than half the elements in the Periodic Table have isotopes exhibiting the Mössbauer Effect, the ^{57}Fe isotope is the most favorable isotope for Mössbauer spectroscopy. This is because a) the recoil energy associated with absorption of the γ-rays of 14.41 keV (I = 3/2 to I = 1/2 transition) is low, b) the half-width of the resonant line is narrow (3×10^{-13} times the energy of the γ-rays), and c) the natural abundance of ^{57}Fe a high (2.14%). The main advantage of ^{57}Fe Mössbauer spectroscopy is that it is a ^{57}Fe specific technique with greater sensitivity than X-ray diffraction (XRD). For example, Fe oxidation states and local environments are identifiable for samples with Fe contents as low as 0.5 wt.%.

4.7.5 EXPERIMENTAL SETUP

4.7.5.1 Mössbauer Source:

Mössbauer spectroscopy facilitates the study of minute energy changes and hyperfine interactions which are otherwise undetectable. Full exploitation of this high intrinsic resolution depends on the availability of sources which efficiently emit recoil-less radiation of narrow line width, satisfying all the conditions mentioned in Section 4.7.3. They are available commercially, either in one of the following matrices.

a) ^{57}Co in Rh matrix b) ^{57}Co in Pd matrix c) ^{57}Co in Cu matrix, etc., from 5 to 200 mCi activity.

4.7.5.2 Basic Apparatus

The basic apparatus consists of a source o γ-radiation, an absorber containing the appropriate nuclide, and a suitable detector system for low energy γ-radiation. The detector can be a counter, a scintillator counter (NaI (Tl) crystal) or a solid state detector like Ge(Li). In order to observe resonance the source is mounted on an electro-mechanical transducer, which works as a Doppler drive system, which is driven with a triangular wave, varying the energy of the γ-rays according to
According to

$$E_\gamma = E_t \left(1 \pm \frac{v}{c}\right) \quad (4.7.18)$$

i.e., $\quad \boxed{v = \Gamma_n {c}/{E_\gamma} \approx 1 \; mm \; s^{-1}} \quad (4.7.19)$

A velocity range of $0 - 60 \; mm \; s^{-1}$ will be adequate to cover all primarily used Mossbauer isotopes. The absorber can be studied down to liquid helium temperatures (close to the Absolute zero) by means of cryostats and at higher temperatures by means of furnaces.

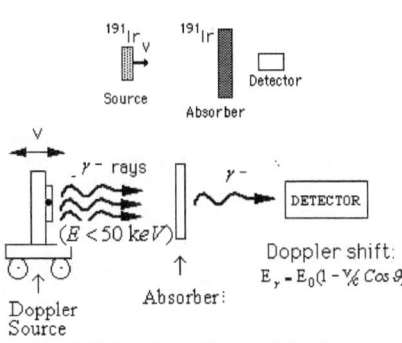

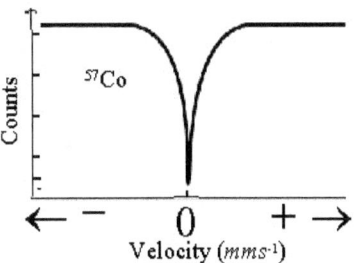

Fig. 4.12 Mössbauer Doppler Drive System

For convenience the energy scale of a Mössbauer spectrum is quoted in terms of the source velocity, v as shown in Fig 4.13.

Fig 4.13. Mossbauer spectrum of an absorber of identical material of that of the source.

In Fig 4.13 the absorption peak occurs at $0 \; mms^{-1}$, where *source and absorber are identical*.

Worked out Example 4.3

Determine the Doppler velocity in mms^{-1}, corresponding to the natural line width of the gamma ray emitted from the 23.8 keV excited state of ^{119}Sn nucleus with half-life of $1.85 \times 10^{-8} s$.

Solution: Step # 1 $\quad \tau_m = \dfrac{\tau_{1/2}}{\ln 2}; \; v = \dfrac{\Delta E}{h}; \; v = \dfrac{\Gamma_n c}{V} \equiv \dfrac{h\Gamma_n c}{E_\gamma} \equiv \dfrac{\hbar c}{E_\gamma \, \tau_m};$

$E_\gamma = 23.8 \; keV$;;; $\tau_{1/2} = 1.85 \times 10^{-8} s;$

$h = 6.6256 \times 10^{-34} J \, s; \quad c = 2.997925 \times 10^8 \, m \, s^{-1}; \; 1 \, eV = 1.6021 \times 10^{-19} J$

Step # 2 $\quad \tau_m = (\ln 2)(\tau_{1/2}) = (1.85 \times 10^{-8} s)(\ln 2) = 2.67 \times 10^{-8} s$

Step # 3 $\quad \Delta v = \Gamma_n = \dfrac{1}{2\pi \Delta t} = \dfrac{1}{2\pi \, \tau_m} = \dfrac{1}{2\pi \, (2.67 \times 10^{-8} s)} = 6 \times 10^6 \, s^{-1}$.

Step # 4 $\quad v = \dfrac{\Delta E}{h} = \dfrac{(E_\gamma = 23.8 \, keV)((1.6021 \times 10^{-19} J/eV))}{(h = 6.6256 \times 10^{-34} J \, s)(10^7 \, ergs/Js)} = 5.754 \times 10^{18} \, Hz$

Step # 5 $v = \dfrac{\Gamma_n c}{v} \equiv \dfrac{h \Gamma_n c}{E_\gamma} \equiv \dfrac{\hbar\, c}{E_\gamma\, \tau_m} = \dfrac{(6.6256 \times 10^{-34}\, J\, s)(2.998 \times 10^8\, m\, s^{-1})}{2\pi\, (23.8\, keV)(2.67 \times 10^{-8} s)(1.6021 \times 10^{-19} J)}$

$= 0.03107\ m\ s^{-1} = 3.11\ mm\ s^{-1}$

4.7.6 Mossbauer Parameters and Hyperfine Interactions

Three effects arise in Mossbauer Effect due to the hyperfine interactions between the nuclear energy levels and the atomic / molecular electrons as given in Table 4.4 along with their manifestations in a spectrum.

Table 4.4

Types of electron-nuclear (hyperfine) interaction EL or ML		Manifestation
1. E0	(Electric Monopole)	Chemical (Isomer) Shift, δ
2. E2	(Electric Quadrupole)	Quadrupole Splitting, ΔE_Q
3. M1	(Magnetic Dipole)	Hyperfine Zeeman Splitting, ΔE_H

The nuclear Hamiltonian can be expressed as

$$\hat{H} = \hat{H}_0 + E_0 + M_1 + E_2 + \ldots \qquad (4.7.20)$$

where $\hat{H}_0$ represents all of the terms of the Hamiltonian other than the hyperfine interactions, E_0 refers to the electric monopole interactions, M_1 the magnetic dipole interactions and E_2 the electric quadrupole interactions.

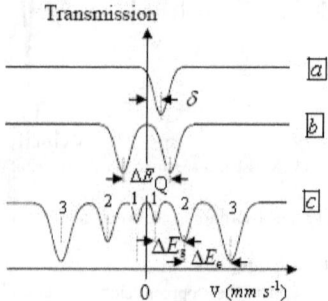

Fig 4.14 Three parameters δ, ΔE_Q and ΔE_H of a Mossbauer spectrum illustrated through cases $\boxed{a}$, $\boxed{b}$ & $\boxed{c}$.

These contribute respectively the three Mossbauer parameters called:
1. *Isomer shift*, δ corresponding to magnetic dipole interaction.
2. *Nuclear Zeeman splitting*, ΔE_H corresponding to magnetic dipole interactions.
3. *Quadrupole splitting*, ΔE_Q corresponding to electric quadrupole interactions.

These are shown in Fig 4.14.

4.7.6.1 Centre of Shift

The Center Shift (CS) of a Mössbauer spectrum, which sets the *centroid* of the spectrum, is composed of two factors: the Chemical Isomer Shift, δ, and the Second Order Doppler Effect (SODS), meaning that

$$CS = \delta + SODS \qquad (4.7.21)$$

4.7.6.2. Chemical Isomer Shift (δ)

The Isomer Shift arises due to the non-zero volume of the nucleus and the electron charge density due to s-electrons within it leading to an electric monopole (Coulomb) interaction which alters the nuclear energy levels. The volume of the nucleus in its ground and excited states are different and the s-electron densities are affected by the chemical environment. This relationship between s-electron density and nuclear radius is given by

$$\delta = \tfrac{2}{3}\pi Z e^2 \{|\psi_s(0)_E|^2 - |\psi_s(0)_A|^2\}\{\langle R_e^2\rangle - \langle R_g^2\rangle\} \qquad (4.7.22)$$

$$\delta = \tfrac{Ze^2}{5\varepsilon_o E_\gamma}\{\rho_e(0) - \rho_g(0)\} \; \square \left(\tfrac{\Delta R}{R}\right) \; mm \; s^{-1}$$

$$\delta = [\text{Nuclear factor}] \, (\text{Solid state factor}) \qquad (4.7.23)$$

where $\langle R_g^2\rangle$ and $\langle R_e^2\rangle$ are the mean square radii of the ground and excited nuclear states, $\rho(0)_E \equiv |\psi_S(0)_E|^2$ and $\rho(0)_A \equiv |\psi_S(0)_A|^2$ are the electron densities at the emitting and absorbing nuclei, e is the electronic charge, R is the effective nuclear radius, c is the velocity of light, E_γ is the energy of the Mössbauer gamma ray, $\Delta R = (R_e - R_g)$ and Z is the atomic number.

Any difference in the s-electron environment between emitter and absorber thus produces a shift in the resonance energy of the transition (Fig 4.15). This shift cannot be measured directly and so a suitable reference is necessary, such as a specific source or an absorber. Isomer shifts are quoted relative to α – Fe at room temperature (any isomer shifts quoted which use a different calibration material are quoted relative to α – Fe for consistency).

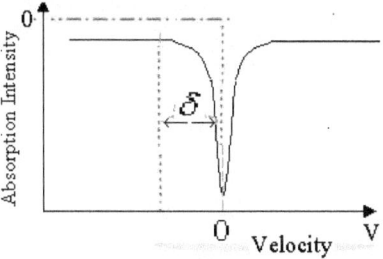

Fig, 4.15 The Mossbauer spectrum of stainless steel using ^{57}Co in Pd matrix

The Isomer Shift is useful for determining valency of states, ligand bonding states, electron shielding and the electron-drawing power of electronegative groups of the Mössbauer atom. As the wave functions of the s-electrons penetrate into outer shells changes in these shells will directly alter the s-electron charge density at the nucleus. For example, Fe^{2+} and Fe^{3+} have electron configurations of $(3d)^6$ and $(3d)^5$, respectively. The ferrous ions have less s-electron density at the nucleus due to the greater screening of the d-electrons. This produces a positive Isomer Shift greater in ferrous ion than in ferric ion.

$$\delta\,(Fe^{2+}\,ionic) > \delta\,(Fe^{3+}\,ionic) > \delta\,(Fe^{2+}\,cov) > \delta\,(Fe^{3+}\,cov) \qquad (4.7.24)$$

Table 4.5   Isomer Shift in Tin compounds; relative to SnO_2

Valence state	Electron configuration	Isomer shift δ (mm s^{-1})
Sn^{4+}	$5s^0 5p^0$	~ 0.0
Sn^{IV}	$(5 s^1 p^3)$	~ 1.3
Sn^0 (4-covalent)	$5(s^1 p^3)$	~ 2.1
Sn^{II}	$5s^2 (+p^x)$	~ 3.5
Sn^{2+}	$5s^2 5p^0$	> 3.71

Table 4.5 lists the Mossbauer δ (mm s^{-1}) of a series of tin compounds.

4.7.6.3. Quadrupole Splitting (Δ or ΔE_Q)

A nucleus that has a spin quantum number $I > \frac{1}{2}$ has a non-spherical charge distribution. The magnitude of the charge deformation, Q, is given by

$$eQ = \int \rho\, r^2 (3 \cos^2\theta - 1)\, d\tau \qquad (4.7.25)$$

where e is the charge on the proton, ρ is the charge density in a volume element $d\tau$ at a distance r from the center of the nucleus and making an angle θ to the nuclear spin quantization axis. The sign of Q indicates the shape of the deformation. Negative Q is due to the nucleus being flattened along the spin axis, an elongated nucleus giving positive Q.

An asymmetric charge distribution around the nucleus causes an asymmetric electric field at the nucleus, characterized by a tensor quantity called the Electric Field Gradient (EFG), ∇E. The electric quadrupole interaction between these two quantities gives rise to a splitting in the nuclear energy levels. The interaction between nuclear moment and EFG is expressed by the Hamiltonian

$$\hat{H}_{Eq} = -\frac{1}{6} e\, Q\, \nabla E \qquad (4.7.26)$$

where ∇E may be written as

$$\nabla E_{ij} = -\frac{\partial^2 V}{\partial x_i \partial x_j} \qquad (4.7.27)$$

$$\{x_i, x_j\} = \{x, y, z\}$$

where V is the electrostatic potential.

There are two contributions to the EFG; i) lattice contributions from charges on distant ions and ii) valence contributions due to incompletely filled electron shells. If a suitable coordinate system is chosen the EFG can be represented by three principal axes, V_{xx}, V_{yy} and V_{zz}. If an asymmetry parameter is defined using these axes as

$$\eta = \left(\frac{V_{zz} - V_{yy}}{V_{xx}} \right) \qquad (4.7.28)$$

where $V_{xx} \geq V_{yy} \geq V_{zz}$ so that $0 \leq \eta \leq 1$, the EFG can be specified by two parameters: V_{zz} and η.

The excited state of ^{57}Fe has a spin $I = \frac{3}{2}$. The EFG has no effect on the $I = \frac{1}{2}$ ground state but does remove degeneracy in the excited state, splitting it into two sub-states $m_I = \pm\frac{1}{2}$ and

$m_I = \pm \frac{3}{2}$ where the $m_I = \pm \frac{3}{2}$ states are higher in energy for positive V_{zz}. The energy eigenvalues for $I = \frac{3}{2}$ have exact solutions given by

$$E_{Eq} = \frac{e^2 q Q}{4I(2I-1)} \{3 m_I^2 - I(I+1)\} \sqrt{\left(1+\frac{\eta^2}{3}\right)} \quad (4.7.29)$$

while the energies for higher spin states require analytical methods to calculate the energies.

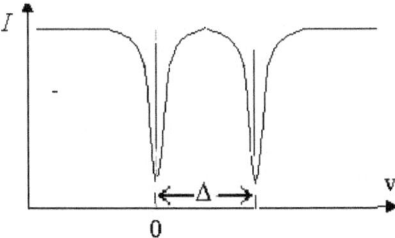

Fig. 4.16   Schematic ^{57}Fe quadrupole split Mossbauer spectrum

The now non-degenerate excited states of ^{57}Fe give rise to a doublet in the Mössbauer spectrum as illustrated in Fig 4.13. The separation between the lines, Δ, is known as the Quadrupole Splitting and is given by

$$\Delta = \frac{e^2 q Q}{2} \sqrt{\left(1+\frac{\eta^2}{3}\right)} \quad (4.7.30)$$

$$\Delta = E_Q = e q V_{zz} = \text{(Nuclear factor)} \cdot \text{[Solid state factor]} \quad (4.7.31)$$

with the line intensities being equal for polycrystalline samples. Texture or orientation effects can lead to asymmetric doublets

As the nuclear quadrupole moment is fixed the magnitude and sign of Δ give information about the sign of the EFG and magnitude of η.

An electric field gradient (EFG) acting on a nucleus depends on the local symmetry of the site in which the nucleus is situated.. Conclusions concerning the point symmetry at the site of a Mossbauer ion can be derived from the analysis of the quadrupole splitting observed in the spectrum. If several chemically or crystallographically inequivalent sites exist, in a solid compound, it is possible, in many cases to distinguish between them and to determine their local symmetries and relative abundances. The ratio of Fe^{2+}/Fe^{3+} ions in the compound can be determined. Crystal structure, coordination chemistry of large molecules and crystal phase transitions are also topics which are investigated by Mossbauer spectroscopy.

Worked out Example 4.4

The Mössbauer spectrum of ^{57}Fe of ferrocene [$Fe(C_5H_5)_2$] at $T = 20$ K consisted of two lines at -0.50 mm s^{-1} and $+1.88$ mm s^{-1}, relative to a stainless steel source. What are the isomer shift δ_{IS} and the quadupole splitting constant Δ for the excited nucleus, in mm s^{-1}?

Solution: Step # 1   Doublet location -0.50 mm s^{-1} and $+1.88$ mm s^{-1}

| Step # 2 | Quadruple separation, $= (+1.88\ mm\ /\ s - 0.50\ mm\ /\ s) = 2.38\ mm\ s^{-1}$.

$\Delta = 55\ MHz$

| Step # 3 | $\delta_{IS} = [(-0.50) + \frac{(1.88 + 0.50)}{2}]\ mm\ s^{-1} = 0.69\ mm\ s^{-1}$

$\delta_{IS} = 7.9\ MHz$

4.7.6.4 Magnetic Splitting

Magnetic hyperfine splitting is caused by the dipole interaction between the nuclear spin moment and a magnetic field, *i.e.*, Zeeman splitting. The effective magnetic field experienced by the nucleus is a combination of fields from the atom itself, from the lattice through crystal field effects and from external applied fields. This can be considered for now as a single field, $\vec{H}_n$, whose direction specifies the principal z axis. The Hamiltonian for the magnetic hyperfine dipole interaction is given as
$$\hat{H} = -\vec{\mu}_I \cdot \vec{H}_n = -g_I\ \mu_N \vec{I} \cdot \vec{H}_n$$
(4.7.32)

where $\mu_N = nm =$ the Nuclear Magneton,

$\vec{\mu}_I =$ the nuclear magnetic moment,

$\vec{I} =$ the nuclear spin and

$g_I =$ the nuclear g-factor.

$$\Delta E_H = -g_I\ \mu_N\ \vec{H}_n\ m_I = [\text{Nuclear factor}]\ (\text{Solid state factor})\quad (4.7.33)$$

where m_I is the magnetic quantum number representing the z component of I (*i.e.*, $m_I = \pm I, \pm(I-1), ...$). The magnetic field splits the nuclear level of spin I into $(2I+1)$ equi-spaced non-degenerate substrates. This and the selection rule of $\Delta m_I = 0, \pm 1$ produces splitting and a resultant spectrum as shown in Fig 4.14 for a $\left|\frac{3}{2}\right\rangle \rightarrow \left|\frac{1}{2}\right\rangle$ transition.

When placed in a magnetic field a nucleus with a magnetic moment has a dipole interaction with the field. The interaction raises the degeneracy of nuclear states with angular momentum quantum number I > 0 to (2I +1). For example, the ^{57}Fe ground state I =1/2 splits into two, and excited state I = 3/2 splits into four. The selection rule of
$$\Delta m_I = 0,\ \pm 1 \qquad\qquad (4.7.34)$$
gives six possible transitions.(six finger pattern) of ^{57}Fe.

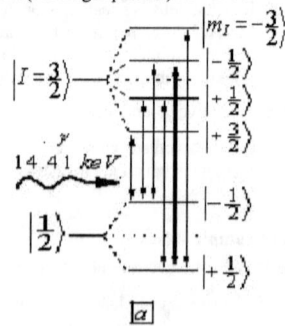

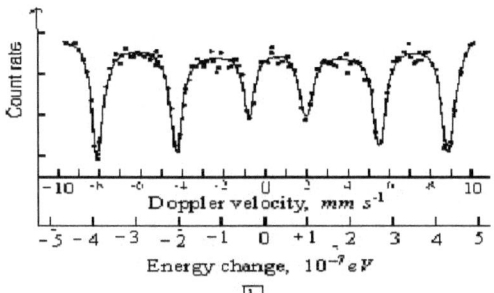

Fig 4.17 (a) The origin of Zeeman splitting (b) Zeeman split Mossbauer spectrum
[Adopted from: Kistmer & Sunyar, Phys. Rev. Lett, 4, (1960) 412].

The splitting is directly proportional to the magnetic field applied and so Mössbauer technique provides a way of measuring it. The magnetic field at the nucleus has several terms associated with it.

For ^{57}Fe spectrum of the standard material used is a thin foil of α-Fe, which has no isomer shift and almost no quadrupole splitting. But it has an ideal spectrum which is a symmetric six component nuclear hyperfine split pattern. The six peaks occur at

$5.15, \; 2.91, \; 0.67, \; -1.03, \; -3.27, \; -5.51 \; mm \; s^{-1}$

$\Delta E_H = -g \, H_n \, m_I \; nm \quad \Delta E_H = -\mu_I \, H_n \, m_I \, / \, I$

$$\Delta E_H = \frac{\delta E_t}{c} - g \, H_n \, m_I \; nm \tag{4.7.35}$$

4.7.6.4.1 Stainless steel

A standard material giving a singlet (single line) Mossbauer spectrum is that of stainless steel.

$$\delta = -0.30 \; mm \; s^{-1} \; (\equiv -1.2 \, x10^{-4} \, cm^{-1}). \tag{4.7.36}$$

4.7.6.4.2 Thin soft iron, α-Fe

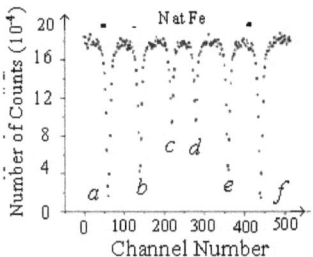

Fig 4..18 A raw ^{57}Fe spectrum of a standard material α-Fe, showing a plot of the number of Mossbauer counts versus channel number of the MCA.(having 512 channels).

α-Fe has no isomer shift and no quadrupole splitting. Bu it has the beautiful six-finger pattern due to nuclear Zeeman Effect, as seen earlier. The intensity ratio (Fig. 4.15) is

$$a:b:c:d:e:f = 3:2:1::1:2:3. \quad (4.7.37)$$

It also gives

$$H_n = 1.07 \times 10^{-7} / g_{Iex} = -33.3 \ T \text{ for } \alpha\text{-Fe}. \quad (4.7.38)$$

Thus there is only one Fe site in α-Fe and it is shown that the *internal magnetic field* is -33.3 T whose direction is opposite to that of the magnetization vector.

A spectrum of the substance under study is then taken. The final analysis of the sufficiently treated spectrum, in terms of hyperfine interactions is made by using a suitable computer program in a PC, which provides the values for the Mossbauer parameters for a complex spectrum.

Worked out Example 4.5

The ground and excited states of a particular Mossbauer niclide are $\left|\frac{3}{2}\right\rangle$ and $\left|\frac{5}{2}\right\rangle$, respectively. A) If the nucleus is under the influence of internal EFG, but no applied magnetic field, determine the gamma ray spectrum, and b) On the other hand, how would the spectrum look like if the EFG were zero, but the magnetic field at site were non-zero.?

Solution: Step # 1 $I'' = \left|\frac{5}{2}\right\rangle, I' = \left|\frac{3}{2}\right\rangle$

Step # 2 The non-degenerate nuclear states are: for $I' = \left|\frac{3}{2}\right\rangle$, the four sublevels $\left|m_{I'}\right\rangle$ are, $\pm\frac{3}{2}, \pm\frac{1}{2}$. For the excited $I'' = \left|\frac{5}{2}\right\rangle$, the $\left|m_{I''}\right\rangle$ are the six levels, $\pm\frac{5}{2}, \pm\frac{3}{2}, \pm\frac{1}{2}$ ..

Step # 3 a)When $EFG \neq 0$, and $H_n = 0$, the # allowed transitions = 5; and the # spectral lines =5.

Step # 4 b) $EFG = 0$, and $H_n \neq 0$ Selection rule is $\Delta m_I = 0, \pm 1$.
allowed transitions = 12; and the # spectral lines =12.

4.7.7 Information from a Mossbauer Spectrum

The Mössbauer technique provides information about the valence state, coordination number, crystal field strengths [e.g., low spin and high spin Fe(III)], and magnetic ordering temperatures. In contrast to XRD, it also provides information on compounds that do not exhibit long-range order (poorly crystalline or amorphous materials). Common Fe-oxide phases such as magnetite and hematite, are readily distinguished from each other and from Fe in layer silicates and predominantly Fe(II) compounds.

Worked out Example 15.6

A Mossbauer spectrum of α-Fe was recorded using a 14.41 keV gamma radiation and a MCA having 256 channels. The centroid of the spectrum occurs at the 62.32 channel. For ^{57}Fe, $E_\gamma = 14.41 \ keV$ and this gives 1 $mm \ s^{-1} \rightarrow 4.800 \ x10^{-8} \ eV \rightarrow 11.61 \ MHz$

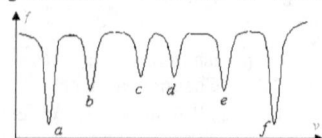

Given that the ground state has $g_{lg} = 0.180, \mu_{lg} = 0.090$ nm, by measuring the separation between the various components of the six-line spectrum, one gets for the excited state of ^{57}Fe,

$\Delta E(|\frac{1}{2},-\frac{1}{2}\rangle \to |\frac{3}{2},-\frac{3}{2}\rangle) - (|\frac{1}{2},-\frac{1}{2}\rangle \to |\frac{3}{2},-\frac{1}{2}\rangle) = 1.07 \times 10^{-7} eV = (a-b)$

$\Delta E(|\frac{1}{2},-\frac{1}{2}\rangle \to |\frac{3}{2},-\frac{1}{2}\rangle) - \Delta E(|\frac{1}{2},-\frac{1}{2}\rangle \to |\frac{3}{2},+\frac{1}{2}\rangle) = 5.04380 \times 10^{-9} eV$

$\Delta E(|\frac{1}{2},+\frac{1}{2}\rangle \to |\frac{3}{2},+\frac{1}{2}\rangle) - (|\frac{1}{2},-\frac{1}{2}\rangle \to |\frac{3}{2},+\frac{1}{2}\rangle) = 1.9 \times 10^{-7} eV = (b-d)$

Determine the internal magnetic field in the absorber.

Solution: Step # 1 $E_\gamma = 14.41$ keV; $g_{Izx} = -0.1015, \mu_{Iex} = -0.153$ nm, $\Delta E = g_I H_n$ nm.;

1 mm $s^{-1} \to 4.800 \times 10^{-8}$ eV $\to 11.61$ MHz;. $E_M = -g_I \mu_N \vec{H} m_I$

Step # 2   It gives $H_n = 1.07 \times 10^{-7} / g_{Iex} = -33.3$ T for α-Fe.

4.7.8 APPLICATIONS

This versatile, highly sensitive, and nondestructive technique has a wide range of applications in various fields including geochemistry, soil science, and materials science.

The following information briefly describes 1) performance specifications and operational overview to facilitate user planning, and 2) Mössbauer spectroscopy.

Since the gamma emission is recoil-free, it can be resonantly absorbed by stationary atoms, i.e., also in a solid. The nuclear transitions are very sensitive to the local environment of the atom and Mossbauer spectroscopy is a sensitive probe of the different environments an atom occupies in a solid material.

A perusal of the expressions for all the three Mossbauer parameters, viz., isomer shift, quadrupole splitting and the magnetic hyperfine splitting in equations (4.7.13), (4.7.21) and (4.7.23) show clearly that all the three contain a nuclear factor. Therefore Mossbauer measurements can be useful for the understanding and determination of stable properties of nuclei, described in Chapter 3, Chapter 4 and Chapter 5.

REVIEW QUESTIONS

R.Q. 4.1   Typical alpha particle kinetic energies are of order 5 MeV = $8 \times 10^{-13} J$. Given $M_\alpha = 6.7 \times 10^{-27}$ kg, Find i) velocity, ii) a momentum, and iii) de Broglie wavelength [Answer $v = 1.1 \times 10^7$ m s^{-1}, $p = mv = 7.4 \times 10^{-20}$ N s, $\lambda = h/p = 9.0 \times 10^{-15}$ m].

R.Q. 4.2   Determine i) nucleonic configuration, and ii) the parity for a nuclide having A = 6. [Answer: $1s^4 \ 1p^2$. Π = even].

R.Q. 4.3   Determine the i) nucleonic configuration and ii) parity for a nucleus with A = 7. [Answer: $1s^4 \ 1p^3$. Π = negative].

R.Q. 4.4   Determine i) nucleonic configuration, and ii) parity for the odd nuclide, A = 19. [Answer: [Closed Shell] + $(2s$ or $1d)^3$,]. Π = positive \.

R.Q. 4.5   For the ^{127}I nucleus $I = \frac{5}{2}$, $\mu_I = 2.79$ nm. Assuming $H = 10^4$ Gauss, find the frequency at which NMR absorption occurs. [Answer: $\nu = 8.52$ MHz].

R.Q. 4.6 For proton ($I = \frac{1}{2}$,) in the magnetic field mo 10,000 Gauss, resonance is observed at 42.576 *MHz*. Calculate the nuclear g_I and γ_I. [Answer: $\varpi_L = \gamma_I$ H; $\gamma_I = \frac{g_I \cdot nm}{h}$; $g_I = \frac{\lambda_I \cdot h}{nm}$].

R.Q. 4.7 For proton in the magnetic field of 1.4092 *Wb* m^{-2}, NMR resonance is observed at 60 MHz. Calculate the nuclear g_I and γ_I for the proton. (Answer: $\varpi_L = \gamma_I$ H; $\gamma_I = \frac{g_I \cdot nm}{h}$; $g_I = \frac{\lambda_I \cdot h}{nm}$].)

R.Q. 4.8 For proton in the magnetic field of 2.3490 *Wb* m^{-2}, NMR resonance is observed at 100 MHz. Calculate the nuclear g_I and γ_I for hydrogen nucleus. [Answer: $\varpi_L = \gamma_I$ H; $\gamma_I = \frac{g_I \cdot nm}{h}$; $g_I = \frac{\lambda_I \cdot h}{nm}$].

R.Q. 4.9 For ^{13}C in the magnetic field of 1.4092 *Wb* m^{-2}, NMR resonance is observed at 19.25 MHz. Calculate the nuclear g_I and γ_I for ^{13}C. [Answer: $\varpi_L = \gamma_I$ H; $\gamma_I = \frac{g_I \cdot nm}{h}$; $g_I = \frac{\lambda_I \cdot h}{nm}$].

R.Q. 4.10 For ^{19}F in the magnetic field of 1.4092 *Wb* m^{-2}, NMR resonance is observed at 56.445 MHz. Calculate the nuclear g_I and γ_I for ^{19}F. [Answer: $\varpi_L = \gamma_I$ H; $\gamma_I = \frac{g_I \cdot nm}{h}$; $g_I = \frac{\lambda_I \cdot h}{nm}$].

R.Q. 4.11 For ^{31}P in the magnetic field of 1.4092 *Wb* m^{-2}, NMR resonance is observed at 24.288 MHz. Calculate the nuclear g_I and γ_I for ^{31}P. [Answer: $\varpi_L = \gamma_I$ H; $\gamma_I = \frac{g_I \cdot nm}{h}$; $g_I = \frac{\lambda_I \cdot h}{nm}$].

R.Q.4.12. A spectral line of 400.0 *nm* is ahifted to 399.9 *nm* by relative motion of the source. What is the speed of the sou8rce along the line of sight?(Ans: $V = 7.5 \times 10^4 \, ms^{-1}$ The Doppler shift being +ve indicates 'Blue shift', caused by motion toward the observer.)

R.Q. 4.13. Obtain the energy of recoil of the ^{57}Fe nucleus Given: $c = 2.997925 \times 10^8 \, ms^{-1}$;; $M(Fe)-57) = 57g \, mol^{-1}$; $N_A = 6.0225 \times 10^{23} \, mol^{-1}$ (Ans: $E_{recoil} = 0.002 \, eV$).

R.Q. 4.14. How many Fe (iron) nuclei would have to recoil together to keep the gamma γ – within the natural linewidth, Γ Given $\Gamma_{nat} = 4.8 \times 10^{-8} \, eV$, $c = 2.997925 \times 10^8 \, ms^{-1}$;; $M(Fe)-57) = 57g \, mol^{-1}$; $N_A = 6.0225 \times 10^{23} \, mol^{-1}$.
(Ans; $N = 200,000$)

R.Q. 4.15. A free nucleus $^{57*}Fe$ emits a gamma rat of frequency $13.5 \times 10^{18} \, s^{-1}$. What is the recoil velocity of the nucleus ?, Calculate the Doppler shift of the gamma ray to an outside absorber. (Ans: $= 8.2 \times 10^3 \, m \, s^{-1}$; $= 9.6 \times 10^{11} s^{-1}$).

R.Q. 4.16. The meta-stable state of ^{57}Fe has life time $1.5 \times 10^{-7} s$. Calculate the line width of the gamma ray emission. (Ans: $= 1.532 \times 10^7$ Hz)

R.Q. 4.17. Determine the Doppler velocity in, mm/sec, corresponding to the natural line width of the gamma ray emitted from the 23.8 keV excited state of ^{119}Sn nucleus with half-life of $1.85 \times 10^{-8} s$. (Ans: , $\Delta v = \Gamma_n = \frac{1}{2\pi \Delta t} = \frac{1}{2\pi \tau_m}$, $\tau_m = (\ln 2)(\tau_{1/2})$, $v = \frac{\Delta E}{h}$,

$v = \frac{\Gamma_n c}{v} \equiv \frac{h \Gamma_n c}{E_\gamma} \equiv \frac{\hbar c}{E_\gamma \tau_m} = 3.11 \; mm \; s^{-1}$).

R.Q. 4.18 In a Mössbauer set up, ^{57}Co *in stainless steel matrix* giving 14.412 keV gamma radiation and a thin absorber is used. Find out the FWHM of the radiation (Γ_γ) and the Doppler velocity corresponding to the FWHM. Given $\tau_{1/2} = 97.7$ ns. (Answers $\Gamma_\gamma = 2(0.467 \times 10^{-8}$ eV), $v = 0.192 \; mm \; / \; s$).

R.Q. 4.19 The Mössbauer spectrum of ^{57}Fe of ferrocene $[Fe(C_5H_5)_2]$ at $T = 20 \; K$ consisted of two lines at $-0.50 \; mm \; s^{-1}$ and $+1.88 \; mm \; s^{-1}$, relative to a stainless steel source. What are the isomer shift δ_{IS} and the quadripole splitting constant Δ for the excited nucleus, in $mm \; s^{-1}$? (Ans: $\Delta = 55 \; MHz = 2.38 \; mm \; s^{-1}$; $\delta_{IS} = 7.9 \; MHz = 0.69 \; mm \; s^{-1}$).

R.Q. 4.20 A Mossbauer spectrum of α-Fe was recorded using a 14.41 keV gamma radiation and a MCA having 256 channels. The centroid of the spectrum occurs at the 62.32 channel. For ^{57}Fe, $E_\gamma = 14.41 \; eV$ and this gives $1 \; mm \; s^{-1} \rightarrow 4.800 \; x10^{-8} \; eV \rightarrow 11.61 \; MHz$

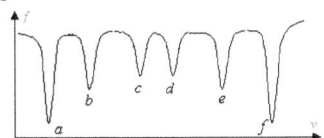

Given that the ground state has $g_{Ig} = 0.180$, $\mu_{Ig} = 0.090 \; nm$, by measuring the separation between the various components of the six-line spectrum, one gets for the excited state of ^{57}Fe,

$\Delta E(|\tfrac{1}{2},-\tfrac{1}{2}\rangle \rightarrow |\tfrac{3}{2},-\tfrac{3}{2}\rangle) - (|\tfrac{1}{2},-\tfrac{1}{2}\rangle \rightarrow |\tfrac{3}{2},-\tfrac{1}{2}\rangle) = 1.07 \; x10^{-7} eV = (a-b)$

$\Delta E(|\tfrac{1}{2},-\tfrac{1}{2}\rangle \rightarrow |\tfrac{3}{2},-\tfrac{1}{2}\rangle) - \Delta E(|\tfrac{1}{2},-\tfrac{1}{2}\rangle \rightarrow |\tfrac{3}{2},+\tfrac{1}{2}\rangle) = 5.04380 \; x10^{-9} eV$

$\Delta E(|\tfrac{1}{2},+\tfrac{1}{2}\rangle \rightarrow |\tfrac{3}{2},+\tfrac{1}{2}\rangle) - (|\tfrac{1}{2},-\tfrac{1}{2}\rangle \rightarrow |\tfrac{3}{2},+\tfrac{1}{2}\rangle) = 1.9 \; x10^{-7} eV = (b-d)$

Determine the g-factor, magnetic moment of the excited state of ^{57}Fe and the internal magnetic field in the absorber. [Answers: $\mu_{Iex} = -0.153 \; nm$ $H_n = -33.3 \; T$ for α-Fe].

R.Q. 4.21 Write a short note on nuclear moments.

&&*&*&*&*

Chapter 5
STABLE PROPERTIES A NUCLEUS - III:
Electric Quadripole Moments

Chapter 5

STABLE PROPERTIES A NUCLEUS - III:
Electric Quadripole Moments

"Let noble thoughts come to us from every side" –   Rig Veda, I-81-1

5.1 INTRODUCTION

In Chapter 4 it was seen how the magnetic moment of nuclei which gives information about the way in which magnetic properties at the nuclear site. Discussion about the nuclei so far has implicitly assumed that nucleus has spherical symmetry. But many nuclei are not spherical in shape. The second most important property is the electric quadrupole moment. These two aspects on a nucleus form the subject matter for the present Chapter.

5.2. ELECTRIC MOMENTS OF A NUCLEUS

5.2.1 ES Interaction Energy

If $\rho_N(r_N)$ = charge density of the nucleus, with electric charge Ze

$\rho_e(r_e)$ = charge density originating from charges external to the nucleus (in the atomic and molecular system), the ES interaction energy of the charge system is

$$W = \iint \frac{\rho_N(r_N) \, \rho_e(r_e) \, d\tau_N \, d\tau_e}{|r_N - r_e|} \qquad (5.2.1)$$

Introducing the ES potential V, instead of $\rho_e(r_e)$, the expression becomes

$$W = \int_\tau \rho_N(r_N) \, d\tau_N \int \frac{\rho_e(r_e) \, d\tau_e}{|r_N - r_e|}$$

i.e. $$W = \int_\tau \rho_N(r_N) \, V(r) \, d\tau \qquad (5.2.2)$$

Expanding the potential In Taylor series, about the nuclear centre,

$$V(\vec{r}) = V(0) + \left.\frac{\partial V}{\partial x_j}\right|_0 x_j + \frac{1}{2!} \sum_{jk} x_j x_k \left.\frac{\partial^2 V}{\partial x_j \partial x_k}\right|_0 + \cdots \qquad (5.2.3)$$

$$W = \left(V(0) \int_\tau \rho(r) x_j \, d\tau \right) + \left\{ \left.\frac{\partial V}{\partial x_j}\right|_0 \int_\tau \rho(r) x_j \, d\tau \right\}$$

$$+ \left\{ \frac{1}{2!} \sum_{jk} x_j x_k \left.\frac{\partial^2 V}{\partial x_j \partial x_k}\right|_0 \int_\tau \rho(r) x_j x_k \, d\tau \right\} + \cdots \qquad (5.2.4)$$

with $q_{jk} = [3 x_j x_k - \delta_{jk} r^2]$

and electric field, $\vec{E} = -\nabla V \qquad (5.2.5)$
and using that total nuclear charge,

$$q = \int \rho_N(r_N) \, d\tau = Ze \qquad (5.2.6)$$

Electric dipole moment vector, $\vec{p}$

$$\vec{p} = \int \vec{r}\, \rho_N(r)\, d\tau \qquad (5.2.7)$$

Electric Quadrupole tensor Q_{jk}

$$Q_{jk} = \int \rho_N(r)\, [3 x_j x_k - \delta_{jk}\, r^2]\, d\tau \qquad (5.2.8)$$

Using Laplace's equation, $\nabla^2 V(r) = 0$, one arrives at

$$W = (= q V(0)) + \vec{p} \cdot \vec{E}(0) - \frac{1}{6} \sum_{jk} Q_{jk} \left. \frac{dE_k}{dx_j} \right|_0 + \qquad (5.2.9)$$

i.e.. $\qquad W = E0 + E1 + E2 + \qquad (5.2.10)$

These are *Multipole Radiations*, which depends on angular momentum $L\,\hbar$, characterized by 2^L, and one gets the EL radiations differentiated depending on if $L = 0, 1, 2, 3$.
For a nucleus is in a definite parity,
$\qquad\qquad$ E1 = 0, as $\qquad\qquad\qquad\qquad\qquad (5.2.11)$

$$\left(\vec{r}\right)_i \rightarrow \left(-\vec{r}\right)_i \qquad (5.2.12)$$

It can be further shown that a nucleus can have
$\qquad$ E2 $\neq$ 0, only if its $I \geq 1 \qquad\qquad\qquad (5.2.13)$

$$Q_{jk} = \sum_{i=1}^{Z} e_i\, [3 x_j x_k - \delta_{jk}\, r^2]\, d\tau$$

$$Q_{zz} = \sum_{i=1}^{Z} e_i\, [3 z_i^2 - r_i^2]\, d\tau \qquad (5.2.14)$$

5.3 NUCLEAR SHAPE

5.3.1.1 Do all nuclei have spherical shape?

For a nucleus is in a definite parity, $\left(\vec{r}\right)_i \rightarrow \left(-\vec{r}\right)_i$, i.e. it has spherical symmetry of state. Such nuclei only have spherical shape!
Examples: ^1_1H, ^4_2He,

Nuclei such as ^2_1H, ^7_3Li, $^{14}_7N$, ^{35}Cl, have nuclear spin $I > \frac{1}{2}$, i.e. $I \geq 1$, and do not have spherical symmetry for time average of volumetric charge distribution within nuclei (Fig 5.1).

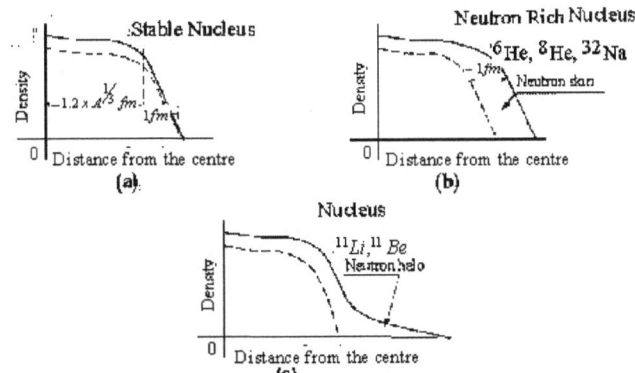

Fig 5.1 Density of charge distribution versus distance from nuclear centre of typical nuclei

5.3.1.2 Nuclei with Ellipsoidal Shape?

Most nuclei assume approximately the shape of an ellipsoid of revolution.
If $2b$ = diameter along the symmetry (i.e. spin) axis,
$2a$ = diameter at right angles to the spin axis,
Q_I = deviation from spherical symmetry (of charge distribution) of nucleus with spin I,

$$Q_I = \frac{1}{e} \int [3 z_i^2 - r_i^2] \rho \, d\tau = \frac{1}{e} \int \rho r^2 \, [3\cos^2\theta - 1] \, d\tau \qquad (5.3.1)$$

$$\cos\theta = m_I / \sqrt{I(I+1)} \qquad (5.3.2)$$

It can be shown that

$$E2 = \frac{1}{4} e Q_I V_{zz} \qquad (5.3.3)$$

$$V_{zz} = \nabla^2 V$$

5.3.1.3 Electric Quadripole Moment

The quantity $e Q_I$ is termed the electric quadripole moment of the nucleus (Fig 5.2).

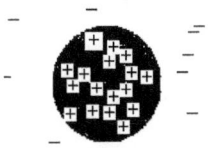

Quadrupole Moment
Fig 5.2 Nuclear Electric Quadripole Moment

Assuming uniform charge density distribution within nucleus, quantum mechanically,

$$\rho = \frac{Ze}{\int d\tau} = \frac{Ze}{[\frac{4}{3}\pi a^2 b]} \qquad (5.3.4)$$

and then one can get *the intrinsic* value

$$Q_I = -\frac{2}{5} Z (a^2 - b^2)_{m_I = I} \qquad (5.3.5)$$

5.3.1.4 Unit of eQ_I

The unit of electric quadrupole moment is *barn*.

$$1 \, barn \equiv 1 \, b = 10^{-24} \, cm^2 \qquad (5.3.6)$$

5.3.1.5 Different Shapes on Nuclei

The expression (2.9.19) for eQ_I can be put in the form

$$Q_I = -\frac{2}{5} Z R^2 \left(\frac{\Delta R}{R}\right)_{I \geq 1} \qquad (5.3.7)$$

where $R \approx \dfrac{a+b}{2}$ = average radius of the nucleus, $\qquad (5.3.8)$

$$\Delta R = \frac{b-a}{2}$$

$$\eta = \frac{b-a}{R} \qquad (5.3.9)$$

The values of Q_I for typical nuclei and their shape are described in Table 5.1.

TABLE 5.1 Value of Q_I for Typical Nuclei

	Value of Q_I	Nuclear shape	I	Typical nuclei
1.	$Q_I = 0$	Sphere	$I \leq \frac{1}{2}$	^1_1H, ^4_2He
2.	$Q_I = +ve$	Prolate Ellipsoid	$I \geq 1$	$^{119}_{50}Sn$,
3.	$Q_I = -ve$	Oblate Ellipsoid	$I \geq 1$	$^{57}_{26}Fe$,

The largest values of Q_I measured were for $^{167}_{68}Er$ and $^{176}_{71}Lu$, for which $\left(\dfrac{\Delta R}{R}\right) \approx 25$ to 30%.

5.4 MEASUREMENT OF Q_I

To measure Q_I there are two methods experimentally,
1. Nuclear Quadrupole Resonance (NQR)
2. Mossbauer Spectral Technique (NGR)

5.4.1. NUCLEAR QUADRUPOLE RESONANCE (NQR)

5.4.1.2. Principle

It is a branch of RF Spectroscopy, which is concerned with magnetic resonance absorption in crystals. In the place of the external magnetic field in NMR, an electric field gradient (EFG, q or V_{zz}) is present in the NQR. The factors involved are:

5.4.1.3. Nuclear Quadrupole Moment

Expressions (5.3.5) or (5.3.7) is relevant here.

5.4.1.4. Interaction with an Electric Field (Quadrupole Coupling)

The interaction of a nucleus with eQ_I and an axially symmetric EFG at the site of that nucleus (Fig 5.3) is given by the energy

$$E_Q = \frac{e\,Q_I\,V_{zz}}{4\,I\,(2I-1)}\left\{3m_I^2 - I(I+1)\right\} \qquad (5.4.1)$$

$e\,Q_I\,V_{zz} \equiv e\,Q_I\,q$ = Quadrupole coupling constant.

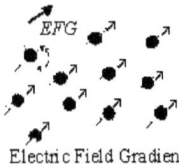

Electric Field Gradient
Fig 5.3 EFG

This axially symmetric case is only an approximation, and the actual case gives

$$E_Q = \frac{e\,Q_I\,V_{zz}}{4\,I\,(2I-1)}\left\{3m_I^2 - I(I+1)\right\}\left(1+\frac{\eta}{3}\right)^{\frac{1}{2}} \qquad (5.4.2)$$

where the asymmetry η is

$$\eta = \frac{V_{xx}-V_{yy}}{V_{zz}} = \frac{b-a}{R} \qquad (5.4.3)$$

which causes, for integral spins, *first-order splitting* only of the degenerate $m_I = \pm 1$ levels., while for half-integral spins experience only small *second-order shifts*.

5.4.1.5 Larmor Precession

This interaction results in a torque, which for appropriate sign of $e\,Q_I\,V_{zz}$, tends to align the nuclear axis and the reference z-axis./ As in the case of Zeeman Effect, the nuclear spin momentum and the μ_I will respond to this torque by precessing about the z-axis. In nuclear Zeeman Effect the precession frequency in a constant, and independent of θ.

$$\cos\theta = m_I/\sqrt{I(I+1)} \qquad (5.4.4)$$

5.4.1.6 Interaction in a Magnetic Field

If the quadrupole system is subjected to a static magnetic field, a nuclear Zeeman Effect results due to interaction with the μ_I. The Zeeman level pattern depends strongly on the relative orientation of the electric field $\vec{E}$ axis and the magnetic field $\vec{H}$.

(a)   If $\vec{H} \mathrel{\|} \vec{E}$, $\Delta E_Q = -\dfrac{\mu_I\,H\,m_I}{I}$ (5.4.5)

(b)   If $\vec{H} \perp \vec{E}$, $\Delta E_Q = \pm(1+\dfrac{1}{2})\dfrac{\mu_I\,H}{2I}$ (5.4.6)

5.4.1.7 NQR Transitions

Transitions between the quadrupole levels take place with the selection rule,

$$\Delta m_I = \pm 1 \qquad (5.4.7)$$

NQR frequencies are given by

$$\nu_Q = \frac{eQ_I\, q}{4I(2I-1)h}\{2|m_I|+1\}\left(1+\frac{\eta}{3}\right)^{\frac{1}{2}}, \qquad (5.4.8)$$

for $0 < m_I < (I-1)$.

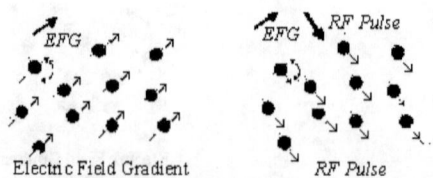

Electric Field Gradient   RF Pulse
Fig 5.3 the interaction of an RF field with EFG

5.4.1.8. MOSSBAUER SPECTROSCOPY

Nuclear Quadrupole Moments of the excited nuclide can be determined by measuring tone of the Mossbauer parameters, viz., quadrupole splitting Δ, as described in Chapter 4.

Quadrupole Splitting (Δ or ΔE_Q)

A nucleus that has a spin quantum number $I > \frac{1}{2}$ has a non-spherical charge distribution. The magnitude of the charge deformation, Q, is given by

$$eQ = \int \rho\, r^2 (3\cos^2\theta - 1)\, d\tau \qquad (4.7.25)$$

where e is the charge on the proton, ρ is the charge density in a volume element $d\tau$ at a distance r from the center of the nucleus and making an angle θ to the nuclear spin quantization axis. The sign of Q indicates the shape of the deformation. Negative Q is due to the nucleus being flattened along the spin axis, an elongated nucleus giving positive Q.

An asymmetric charge distribution around the nucleus causes an asymmetric electric field at the nucleus, characterized by a tensor quantity called the Electric Field Gradient (EFG), ∇E. The electric quadrupole interaction between these two quantities gives rise to a splitting in the nuclear energy levels. The interaction between nuclear moment and EFG is expressed by the Hamiltonian

$$\hat{H}_{Eq} = -\frac{1}{6} e\, Q\, \nabla E \qquad (4.7.26)$$

where ∇E may be written as

$$\nabla E_{ij} = -\frac{\partial^2 V}{\partial x_i \partial x_j} \qquad (4.7.27)$$

$$\{x_i, x_j\} = \{x, y, z\}$$

where V is the electrostatic potential.

There are two contributions to the EFG; i) lattice contributions from charges on distant ions and ii) valence contributions due to incompletely filled electron shells. If a suitable coordinate system is chosen the EFG can be represented by three principal axes, V_{xx}, V_{yy} and V_{zz}. If an asymmetry parameter is defined using these axes as

$$\eta = \left(\frac{V_{zz} - V_{yy}}{V_{xx}}\right) \qquad (4.7.28)$$

where $V_{xx} \geq V_{yy} \geq V_{zz}$ so that $0 \leq \eta \leq 1$, the EFG can be specified by two parameters: V_{zz} and η..

The excited state of ^{57}Fe has a spin $I = \frac{3}{2}$. The EFG has no effect on the $I = \frac{1}{2}$ ground state but does remove degeneracy in the excited state, splitting it into two sub-states $m_I = \pm \frac{1}{2}$ and $m_I = \pm \frac{3}{2}$ where the $m_I = \pm \frac{3}{2}$ states are higher in energy for positive V_{zz}. The energy eigenvalues for $I = \frac{3}{2}$ have exact solutions given by

$$E_{E_q} = \frac{e^2 q Q}{4I(2I-1)} \{3\ m_I^2 - I(I+1)\} \ \sqrt{\left(1 + \frac{\eta^2}{3}\right)} \qquad (4.7.29)$$

while the energies for higher spin states require analytical methods to calculate the energies.

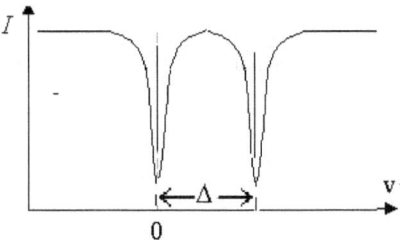

Fig. 4.16   Schematic ^{57}Fe quadrupole split Mossbauer spectrum

The now non-degenerate excited states of ^{57}Fe give rise to a doublet in the Mössbauer spectrum as illustrated in Fig 4.13. The separation between the lines, Δ, is known as the Quadrupole Splitting and is given by

$$\Delta = \frac{e^2 q Q}{2} \sqrt{\left(1 + \frac{\eta^2}{3}\right)} \qquad (4.7.30)$$

$$\Delta = E_Q = e\ q\ V_{zz} = \text{(Nuclear factor)} \ [\text{Solid state factor}] \qquad (4.7.31)$$

with the line intensities being equal for polycrystalline samples. Texture or orientation effects can lead to asymmetric doublets

As the nuclear quadrupole moment is fixed the magnitude and sign of Δ give information about the sign of the EFG and magnitude of η.

An electric field gradient (EFG) acting on a nucleus depends on the local symmetry of the site in which the nucleus is situated.. Conclusions concerning the point symmetry at the site of a Mossbauer ion can be derived from the analysis of the quadrupole splitting observed in the spectrum. If several chemically or crystallographically inequivalent sites exist, in a solid compound, it is possible, in many cases to distinguish between them and to determine their local symmetries and relative abundances. The ratio of Fe^{2+}/Fe^{3+} ions in the compound can be determined. Crystal structure, coordination chemistry of large molecules and crystal phase transitions are also topics which are investigated by Mossbauer spectroscopy.

Worked out Example 5.1

The Mössbauer spectrum of ^{57}Fe of ferrocene [$Fe(C_5H_5)_2$] at $T = 20$ K consisted of two lines at $- 0.50$ $mm\ s^{-1}$ and $+1.88\ mm\ s^{-1}$, relative to a stainless steel source. What is the quadupole splitting constant Δ for the excited nucleus, in $mm\ s^{-1}$?

Solution: Step # 1   Doublet location $- 0.50\ mm\ s^{-1}$ and $+1.88\ mm\ s^{-1}$

Step # 2   Quadruple separation, $= (+1.88\ mm\ /\ s - 0.50\ mm\ /\ s) = 2.38\ mm\ s^{-1}$.
$\Delta = 55\ MHz$

5.4.2 NUCLERAR CHARGE DENSITY AND NUCLEAR RADII

5.4.2.1 Nuclear Charge Density

Information about the distribution of nuclear charge has been obtained from the Optical Spectra. Quantum Theory predicts that some of the very penetrating atomic electrons say the orbital s-electrons, will spend a fraction of their time within the nucleus. The Coulomb potential between the nucleus and the electron over the region inside the nucleus will depend on the actual charge distribution inside the nuclear volume. Three situations arise:
(i)   Charge concentrated at the centre of the nucleus,
(ii)   Charge uniformly distributed throughout the sphere,
(iii)   The entire charge are on the surface of nucleus.
It is known that
For case (ii)

$$U(r) = \frac{-Ze}{R}\left[\frac{3}{2} - \frac{1}{2}\left(\frac{r}{R}\right)^2\right]_{r \leq R} \tag{5.4.9}$$

5.4.2.2 Expression for Nuclear Charge Density

Electron scattering is the tool of choice for determining the density distributions of nuclei. A schematic introduction is given here. This ignores all relativistic effects and spin interactions, but illustrates the importance of the form factor and its relation to the scattering potential. It is also a quick introduction to scattering.

Consider a plane wave $e^{i\mathbf{k}\cdot\mathbf{r}}$ incident on a target nucleus fixed at the coordinate origin. This wave is scattered to $e^{i\mathbf{k'}\cdot\mathbf{r}}$ by the potential $V(r)$. We use perturbation theory (first Born approximation) in order to find the amplitude for finding k'. The probability for finding a particle M is given by the square of the amplitude which is also proportional to the differential cross section, *i.e.*

$$M^2 \propto \frac{d\sigma}{d\Omega} \propto |<k'|V|k>|^2 \tag{5.4.10}$$

One can evaluate this integral (with $\mathbf{q} = \mathbf{k'} - \mathbf{k}$):

$$\tag{5.4.11}$$

Now, if the charge distribution is not point-like, the potential $V(r)$ can be written

$$V(r) = \int \rho(r') \frac{K}{|\mathbf{r}-\mathbf{r'}|} dr' = \int \rho\{|\mathbf{r}-\mathbf{r'}|\} \frac{K}{r'} dr' \tag{5.4.12}$$

The above matrix element M then becomes

$$M = \iint e^{-i\mathbf{q}\cdot\mathbf{r}} \rho(|\mathbf{r}-\mathbf{r'}|) \frac{K}{r'} dr' dr \tag{5.4.13}$$

This can be split into two independent integrals by means of the substitutions $R = r - r'$ and $dr = dR$.
Then

$$M = \iint e^{-i\,q\,r^0} e^{-i\,q\,r^R} \rho(R) \frac{K}{r'} dr' dR$$
$$= 4\pi \frac{K}{q^2} F(q) \qquad (5.4.14)$$

The function $F(q^2)$ is known as the form factor and is the Fourier transform of the charge density.

5.5 NUCLEAR SIZE

Methods which measure the extent of the nuclear force rather than the nuclear charge, utilize the determination of nuclear radius.

Radius of an atom is $\Box\, 10^{-10}\,m$, and this is within the nucleus. In the scattering experiment by Rutherford the Coulomb Law is valid when the projectile-target distance exceeds $\Box\, 10^{-14}\,m$, which considers a finite size for the target nucleus. On the other hand for4 back-scattering, the nucleus appears as point charge to the α – particles. At the distance of closest approach $r = r_c$, when the

kE of the α – particle = ES repulsion.
This condition gives

$$r_c = \frac{2\,Z_\alpha Z\,e^2}{4\pi\,\varepsilon_o\,m_\alpha\,v^2} \qquad (5.5.1)$$

$r_c = 4.3 \times 10^{-14}\,m$ for gold foil and $7.68\,MeV\,\alpha$ – particles.

5.5.1 Rutherford Scattering Fails!

In 1954, Farwell *et al.* reported, based on their scattering studies with $10 - 45\,MeV$ α – particles, that for $kE > 27.5\,MeV$ the Coulomb scattering fails! The experiments conducted by them in 1957 with different targets gave the result that the *Charge Radius* R_C is

$$R_C = r_o\,A^{1/3} \qquad (5.5.2)$$

Radius constant, $r_o = 1.2\,to\,1.3\,fm$ $\qquad (5.5.3)$
A = Mass number of the nucleus

5.5.2 Electron Scattering by Nuclei

The electron scattering experiment (R. Hofstadter ,1953) is a method to determining nuclear radii. Here *nuclear charge distribution* is determined and involves the assumption that charge distribution and the matter distribution in nuclei is essentially the same. In the region of energies $100\,MeV$ and higher, when the deBroglie $\lambda\,\Box$ a few *fm*, electrons act as EM probes, This situation of electrons is preferred because
(a) Electrons do purely interact Electro-Magnetically with nuclei,
(b) Conjoined with their small mass, and
(c) Possible high resolution because of their small inherent size.

5.5.2.1 An intense beam of > 250 *MeV* electrons produced from an electron accelerator (Hofstadter *et al.*(1957)), which strikes a target in a scattering chamber. In their experimental results of elastic scattering intensity as a function of scattering angle and they compared the differential scattering cross section for electron ^{12}C scattering at 420 *MeV* with First Born approximation calculations, using the nuclear charge distribution

$$\rho(r) = \rho_0 \, e^{-(\frac{r}{a})^2} \left(1 + \frac{\omega^2}{a^2}\right) \qquad (5.5.4)$$

Other charge distributions which gave fair agreement for spherical nuclei between Z = 20 and Z = 83 are the Fermi type, the modified Gaussian and the trapezoidal types.

5.5.2.2 With (one parameter) Gaussian

$$\rho(r) = \rho_0 \, e^{-(\frac{r}{a})^2} \qquad (5.5.5)$$

to the nuclear charge, is used for light nuclei only. Example He

5.5.2.3 With (two parameter) Fermi distribution (or Woods-Saxon formula)

$$\rho(r) = \frac{\rho_0}{\left[1 + e^{(\frac{r-c}{a})}\right]} \qquad (5.5.6)$$

c = half-density radius,

$$c = 1.1 \, A^{1/3} \, fm \qquad (5.5.7)$$
$$a = 0.545 \, fm = \text{Diffuseness.} \qquad (5.5.8)$$

This was used for Ca, V, Co, In, Sb, Au and Bi.

5.5.2.4 With Trapezoidal distribution,

$$\rho(r) = \rho_0 \, e^{-(\frac{r}{a})} \left(1 + \frac{r}{a}\right) \qquad (5.5.9)$$

used for lighter nuclei, Be, Li.

5.5.3 Nuclear radii and the Liquid Drop Model

Scattering investigations with electrons, nucleons, and α – particles conform that to a first approxn. nuclei can be regarded as spherical in shape. So the volume V of the nucleus with radius R, mass number A and charge density

$$\rho(r) \begin{cases} = 0, \text{ for } r > R, \\ = \rho_0, \text{ for } r \leq R, \end{cases} \qquad (5.5.10)$$

$$V = 4\pi \int_0^\infty \rho(r) \, r^2 \, dr = \tfrac{4}{3}\pi R^3 \propto A \qquad (5.5.11)$$

This gives $\quad R \propto r_0 \, A^{\frac{1}{3}} \qquad (5.5.12)$

5.5.4 Three Different Definitions for Nuclear Radius

5.5.4.1 Nuclear Force Radius constant (Nuclear unit radius) $\quad r_0$
In the expression (5.5.12)

$$r_0 = 1.414 \times 10^{-15} m = 1.414 \, fm \qquad (5.5.13)$$

which was obtained using nuclear particle scattering on nuclei.

5.5.4.2 Coulomb (or, Electro-magnetic) radius constant, r_o

From electron scattering on nuclei one gets

$$r_o = 1.22 \times 10^{-15} m = 1.22 \, fm \tag{5.5.14}$$

5.5.4.3 Mean square radius $\sqrt{\langle r^2 \rangle}$

$$\langle r^2 \rangle^{\frac{1}{2}} = \sqrt{\frac{\int_0^\infty \rho(r) \, r^2 d\tau}{\int_0^\infty \rho(r) \, d\tau}} = r_o \, A^{\frac{1}{3}} \tag{5.5.14}$$

where $r_o = 0.94 \, fm$ (5.5.15)

5.5.4.4 Half-density Radius, c

$$c \approx (1.18 \, A^{\frac{1}{3}} - 0.48 \, fm) \tag{5.5.16}$$

$$t = (4 \ln 3) \, a \approx 2.4 \, fm = \text{Skin thickness} \tag{5.5.17}$$

$$a = 0.545 \, fm = \text{Diffuseness}. \tag{5.5.18}$$

5.6 CONCLUSIONS

From Chapters 3, 4 and 5 the important results on the stable properties of a nucleus are:

1. In an electrically neutral atom,

    Nuclear charge = | (# of atomic electrons) x (electronic charge) |
    = | - Z e |
    Sign of nuclear charge = Positive

2. The BASIC properties of a nucleus are:

    Nuclear charge, Nuclear Mass, Nuclear Angular momentum nuclear Parity and Nuclear Magnetic Moment, Nuclear shape, Nuclear Quadrupole moment and nuclear Radius and Shape.

3. The experiment on α – *particle* scattering by Rutherford with theory was discussed to point out the origin of the 'nuclear atom model' and the important interpretation of *back-scattering*.

4. The concept of 'nuclear structure'

5. Nuclear mass and development of various mass spectrometers and identification of isotopes.

6. Concept of' intrinsic nuclear spin'

REVIEW QUESTIONS

R.Q. 5.1 What are the NQR frequencies for a nuclear state $I = \frac{3}{2}$? [

[Answer: $v_Q = \frac{1}{2} e Q_I q \left(1 + \frac{\eta^2}{3}\right)^{1/2}$].

R.Q. 5.2 Show that for a nucleus $Q_I = 2 \int \rho_N(r) r^3 P_2(\cos\theta) d\tau$.

R.Q. 5.3 Show that $Q_I = 0$, $Q_I = + ve$, $Q_I = - ve$, respectively, for a spherically symmetric, prolate spheroid, and oblate spheroid charge distributions.

R.Q. 5.4 Evaluate the maximum shift in energy in the energy level that can be observed for a nucleus having quadrupole moment eQ.

[Answer $\langle W_Q \rangle = \frac{1}{8} e Q_I \frac{\partial^2 V}{\partial z \partial z} \left(\frac{2I-1}{I+1}\right)$].

R.Q. 5.5 A beam of α – particles having 7.68 MeV of Radium is striking a gold foil of thickness 0.2 μm. Find the r_c, the distance of closest approach. [Answer:

($n = \rho N_A / M = 5.9 \times 10^{22}$ atoms/cm^3,

$\frac{(Z_\alpha e^2)}{4\pi \varepsilon_0 kE} = 1.3 \times 10^{-14} m$, $Z = 79$, $\rho = 19.32$ gm/cm^3), $r_c = 4.3 \times 10^{-14} m$].

R.Q. 5.6 A beam of α – particles having 7.68 MeV of Radium is striking an aluminium foil. Find the r_c, the distance of closest approach [Answer:

($\frac{(Z_\alpha e^2)}{4\pi \varepsilon_0 kE} = 1.3 \times 10^{-14} m$, $r_c = 0.49 \times 10^{-14} m = 4.9$ fm].

R.Q. 5.7 A beam of α – particles having 25 MeV is bombarding a uranium target. Find the r_c, the distance of closest approach. [Answer: $r_c = 1 \times 10^{-14} m = 10$ fm].

R.Q. 5.8 Consider the nucleus ^{27}Al. Given the value of the nuclear force radius constant, find the nuclear matter density. [Answer: $V = \frac{4}{3}\pi \{1.414 \times 10^{-15} m \times 27^{\frac{1}{3}}\}^3 = 3.1 \times 10^{-37} cm^3$;

nuclei $= \frac{2.7 \text{ gm cm}^{-3} \times 6 \times 10^{23} \text{ mole}^{-1}}{27 \text{ gm mole}^{-1}}$; $\rho_N \approx 10^{17}$ kg $m^{-3} \approx 0.17$ nucleon m^{-3}].

R.Q. 5.9 Calculate approximately the ES force that exists between two protons in a typical nucleus, say iron, using knowledge of the nuclear radius constant.

[Answer R = $\{1.414 \times 10^{-15} m \times 56^{\frac{1}{3}} = 5.4$ fm

$F_{ES} = + (4.80 \times 10^{-10} St.C)^2 / (5.4 fm)^2 \approx 1 \times 10^6$ dynes].

R.Q 5.10 What are the NQR frequencies for a nuclear state I = 1?

[Answer: $v_Q^+ = \frac{3}{4} e Q_I q \left(1 + \frac{\eta}{3}\right)$, $v_Q^- = \frac{3}{4} e Q_I q \left(1 - \frac{\eta}{3}\right)$, and $v_Q^0 = \frac{1}{2} e Q_I q \eta$].

R.Q. 5.11 A beam of α – particles having 7.68 MeV of Radium is striking an gold and aluminium foils. Find the r_c, the distance of closest approach. [Answer: $\frac{(Z_\alpha \, e^2)}{4\pi\varepsilon_0 \, kE} = 1.3 \times 10^{-14}$ m, $r_c(Au) = 2.98 \times 10^{-14}$ m $= 29.8$ fm $r_c(Al) = 0.49 \times 10^{-14}$ m $= 4.9$ fm].

R.Q. 5.12 Find the nuclear radius of ^{16}O, ^{120}Sn, ^{208}Pb, and ^{238}U .taking the required data to two significant digits.
[Answer $R(^{16}O) = 3.62$ fm; $R(^{120}Sn) = 6.90$ fm; $R(^{208}Pb) = 8.28$ fm; $R(^{238}U) = 9.00$ fm].

R.Q. 5.13 Given the plot density of charge versus nuclear radii of 3 nuclides find out the nucleus which has the radius equal to one-half the radius of ^{236}U ?

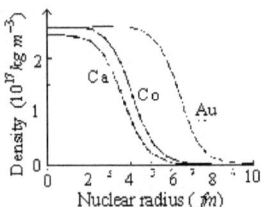

&*(&*&*&*&*&*&*&

Chapter 6
THE INTERNUCLEON POTENTIAL

Chapter 6

THE INTERNUCLEON POTENTIAL

"The plurality that we perceive is only an appearance. it is not real "
- Erwin Schrodinger

6.1   INTRODUCTION

This Chapter presents an account on the binding energy of nuclei, nuclear forces and the deuteron. The results of measurement of atomic masses of the isotopes can be presented in a variety of ways, each instructive and useful in its own manner. One method is in terms of the binding energies of the different nucleons, whereas the other method is packing fraction. The atomic masses of elements are expressed in standard Tables on Physical Data (for example, Clarke's Table), for common purposes. In nuclear work the atomic masses are used on the physical scale.

It was assumed in all the previous chapters that there is an essentially attractive force –the nuclear force - between any two nucleons, responsible for them to be held together in a nucleus. The simplest system to study the nuclear force is that of two nucleons. Usually force $\vec{F}(r)$ is defined as $\vec{F}(r) = -\vec{\nabla} V(\text{r})$, where $V(\text{r})$ is the potential. Therefore, the implications of nuclear properties for the form of $V(\text{r})$ are first studied in this chapter. To a limited extent, but in sufficient depth, a detailed analysis of the experimental data on the two nucleon system, especially the deuteron, will be presented here.

6.2   PRELIMINARIES

6.2.1.   Nuclear Binding Energy, E_B

The atomic mass of a nuclide can be understood in terms of the masses of it constituent nucleons. The *binding energy* E_B of a nucleus is defined as the energy required separating the nucleons composing a nucleus.

If $A (= Z + N)$ = Mass number or # of nucleons (each of mass m_i) in the nucleus and
M = Atomic mass (measured mass) of the nucleus, it is found that

$$M < \sum_{i=1}^{A} m_i \qquad (6.2.1)$$

The explanation to this loss of mass lies in the binding energy E_B

$$E_B = \sum_{i=1}^{A} m_i - M$$

E_B = [P.E. of nucleus + K.E. of nucleus] - (Resit mass energy of nucleons)

This gives

$$E_B = \{ Z \sqcup M_p + (A - Z) \sqcup M_n \} - {}_Z^A M \qquad (6.2.2)$$

where M_p = Mass of a proton = 1.0072765 u ,

M_n = Mass of a neutron = 1.008665 u ,

$^A_Z M \equiv M(Z,A)$ = Atomic mass of isotope $^A_Z El$
To a first approximation,
$$E_B \propto A \qquad (6.2.3)$$

Worked out Example 6.1

What is the mass of the tritium nucleus, given $_1^3$H)=3.01647 u, $m_e = 5.4858 \times 10^{-4} u$

Solution: Step #1   Given, $_1^3$H)=3.01647 u, $m_e = 5.4858 \times 10^{-4} u$

Step # 2 $_1^3$H)=[3.01647 u – 1(0.00054858u)] = 3.014932 u

Worked out Example 6.2

What is the binding energy of ^{12}C ? ($M_p = 1.007276\ u$, $M_n = 1.008665\ u$ $M(^{12}C)=12.000\ u$ $m_e = 5.4858 \times 10^{-4} u$)

Solution: Step #1 : Given, Z=, # of electrons 6, N=6, # of protons =6, $m_e = 5.4858 \times 10^{-4} u$
$M_p = 1.007276\ u$, $M_n = 1.008665\ u$ $M(^{12}C)=12.000\ u$, $1u = \frac{1}{12}$ (mass of $^{12}_6C$) $= 931\ MeV/c^2$

Step # 2 : Mass formation of Component particles= (NxM_n + ZxM_p) 12.0990 u.=

Step # 3   Loss in forming   Step # 3 $^{12}C = 12.0990\ u - 12.0000u) = 0.0990\ u$

Step # 4   Binding energy, E_B = = (0.0990 u)(= 931 MeV/c^2) = 92 MeV

6.2.2 Mass Defect, Δ

Mass defect Δ of an individual nucleus is defined as
$$\Delta = M - A \qquad (6.2.4)$$

Table 6.1

Nuclide	A	Sign of Δ
1. $^4_2$He	< 16	+ ve
2. ^{40}Ar	> 16	- ve

6.2.3 Packing Fraction, F

The divergences of observed mass of a nuclide from the integer value is expressed as a quantity known as *Packing Fraction, F*,

$$F = \frac{M-A}{A} = \frac{\Delta}{A} \quad u\ \text{nucleon}^{-1} \qquad (6.2.5)$$
$$M(Z,A) = A(1+F)\ u \qquad (6.2.6)$$

Table 6.2 Packing Fraction versus Mass scale

	Nuclide	F	Mass Scale
1.	$^{12}_{6}C$	0	^{12}C
2.	$^{16}_{8}O$	0	^{16}O

6.2.4 Packing Fraction *versus* Mass Number curve

The variation of *F versus A* curve for a large number of nuclides is illustrated in Fig 6.1.
For $20 < A < 180$, $F = $ -ve, with minimum at $A \approx 60$,
For $A > 180$, F - + ve, and increases with increasing A.

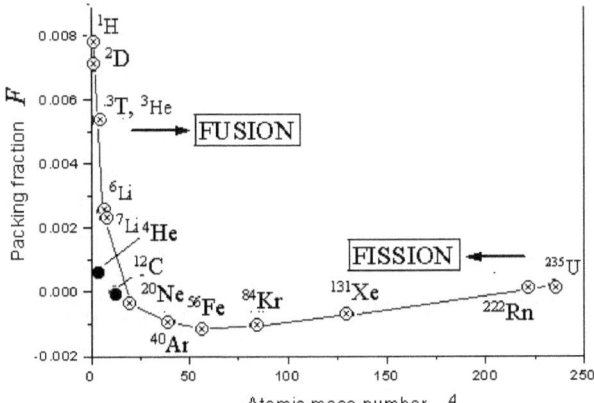

Fig 6.1 Packing fraction F versus Mass Number A plot

6.3 STABILITY OF NUCLIDES (Binding Energy per Nucleon)

Einstein's mass-energy relation, derived from the Special Theory of Relativity, is used to account for the $\Delta = M - A$, i.e. $A < M$.
The A number of free nucleons, when get bound together as a nucleus, causes a decrease in nuclear mass, and this is equivalent to an energy, ΔE

$$\Delta E = \Delta \cdot c^2 = \Delta(\text{in } u) \cdot (931.49432 \, MeV) \quad (6.3.1)$$

Thus from equation (6.2.2)

$$E_B = \left(\{ Z \cdot M_p + (A - Z) \cdot M_n \} - {^A_Z}M \right) \cdot c^2 \, ergs \quad (6.3.2)$$

$$E_B = \left(\{ Z \cdot M_p + (A - Z) \cdot M_n \} - {^A_Z}M \right) u \cdot (931.49432) \, MeV \quad (6.3.3)$$

$$\frac{E_B}{A} = \frac{Z}{A}\left(\{ M_p - M_n \} + M_n (1 + F) \right) u \cdot (931.49432) \, MeV / \text{nucleon} \quad (6.3.4)$$

Equation (6.3.4) is plotted as a $\frac{E_B}{A}$ *versus* A curve, known as the stability curve, in Fig .6.2

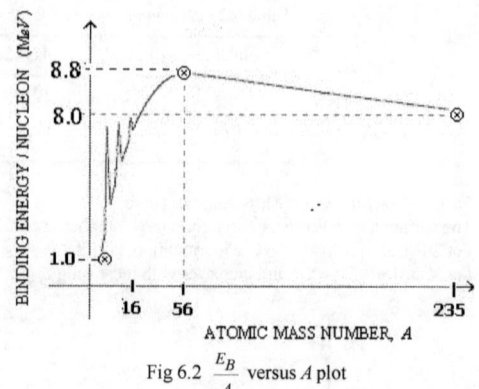

Fig 6.2 $\frac{E_B}{A}$ versus A plot

The diagram gives the results listed in Table 6.3.

Table 6.3

	A	E_B / A
1.	$1 < A < 20$	variation is considerable
2.	$20 < A < 60$	8.0 - 8.5 MeV
3.	$A \approx 60$	8.5 MeV
4.	$A > 60$	decreases slowly to 7.5 MeV

The variation in E_B / A and E_B for different values of A may be very significant in any study of nuclear forces, nuclear stability and other nuclear properties. Fig 6.2 shows the diagram E_B / A versus Z.

i) <u>Nuclear Fusion region</u>:
For $1 < A < 60$ region, energy will be liberated if light nuclei are combined together o form a new compound nucleus.

ii) <u>Nuclear Fission</u>:
For $A > 60$ region, Energy will be liberated upon a heavy nucleus is split into two medium mass fragments.

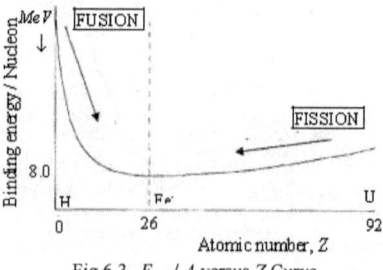

Fig 6.3 E_B / A versus Z Curve

iii) Nuclear Stability
Following facts can be seen From Fig 6.2.:
a)   The vast majority of nuclides have a binding energy E_B/A of 8 MeV, approximately.
b)   He has a particularly high value of E_B/A, much higher than the light isotopes of hydrogen.
c)   The trend for Nuclides of nucleon numbers in multiples of 4 to be particularly stable (*i.e.* have a high binding energy).
d)   Fe is the most stable nuclide.
f)   The nuclides having higher values of Z tend to be less stable, with slightly lower E_B/A.

Iron has the highest E_B/A, i Iron is the most stable nucleus, as it has the highest E_B/A,. If one looks at heavy nuclei (heavier than iron), one can find that the further to the right (greater nucleon number) from the origin of the curve (Fig 6.3) the less stable are the nuclei. This is because the E_B/A is getting lesser and less. This observation requires the explanation that the strong nuclear force that binds the nucleus together has a very limited range, and there is a limit to the number of nucleons that can be crammed into a particular space.

Worked out Example 6.3

Determine the mass defect, the total binding energy and the binding energy per nucleon for ^{14}N. (Given: $M_p = 1.007276\ u$, $M_n = 1.008665\ u$ $A(^{14}N) = 14.00307\ u$, $m_e = 5.4858 \times 10^{-4} u$ = 931 MeV .

Solution:: | Step # 1 | Given, $M_p = 1.007276\ u$, $M_n = 1.008665\ u$ $A(^{14}N) = 14.00307\ u$, A =14 $m_e = 5.4858 \times 10^{-4} u$ = 931 MeV

| Step # 2 | Exact $M(^{14}N) = 14.00307\ u - 7(0.00054858\ u) = 13.99983\ u$

| Step # 3 | Total mass of the component particles of N nucleus, $A = 7 M_p + 7 M_n$
= 7 (1.007276 u) + 7 (1.008665 u) = 14.11159 u

| Step # 4 | Mass defect, $\Delta = M - A = 0.124\ u$.

| Step # 5 | $E_B = (0.124\ u)(= 931\ MeV/u) = 104.7\ MeV$

| Step # 6 | $E_B/A = (104.7\ MeV)/14 = 7.5\ MeV/$ nucleon.

6.4   THE NUCLEAR FORCE

In order that the many properties of nuclei can be ascertained and to study problems such as nuclear structure and binding energies, the knowledge of the nature of the forces present between nucleons in the nucleus is to be understood..

6.4.1   General Nature of Force between Nucleons in a Nucleus

1) Nature:
Nuclear force is predominantly *attractive*; otherwise stable nuclei could not exist.
At distances, within nucleus, $r > 1\ fm$ the nuclear force is negligible; this is an experimental observation while determining nuclear radii. At distances $r > 10\ fm$ the interaction regulating the scattering of nucleons and the grouping of atoms into molecules is *electro-magnetic*.
2) Range & Strength:

The nuclear force is appreciable only when the nucleons are very close, $r \leq 1\,fm$; hence the nuclear force *short-range force*. The range of it can be determined by performing proton-nuclei scattering experiments. Only protons having sufficient kE to overcome the Coulomb repulsion and pass close to the scattering nucleus are affected by the nuclear force, and their scattering is different from pure Coulomb scattering, as is confirmed by experiments.

3) Charge independence:
It is confirmed by analysis of proton-proton and neutron-proton scattering experiments that the nuclear force is charge independent. Nuclear interactions F_{n-n} between any two nucleons are basically the same.

$$F_{n-p} \equiv F_{n-n} \equiv F_{p-p} \tag{6.4.1}$$

The Coulomb energy alone can account for the following:
(i)   Z = N, for light nuclei,   (6.4.2)
(ii) $E_B / A \approx$ constant, and   (6.4.3)

(iii)   The mass difference between *mirror nuclei*. Consider ^3_1H and ^3_2He which have a mass difference of $\Delta M = 0.000840\,u$.

4)   Spin dependence:
The nuclear force depends on the relative orientation of the spins of the interacting nucleons. This confirmed by
a) Scattering experiments, and
b) Analysis of nuclear energy levels.
In the two nucleon system like the (n-p) system, has a bound state called deuteron, which has the spins parallel and so $I = 1$, whereas if the two nucleons have spin anti-parallel then there is no such system exists (with $I = 0$).

5)   Non-central, but *tensorial* feature:
Nuclear force has a tensorial feature, meaning it depends on the orientation of the spins relative to the line joining the two nucleons. This is evident that to explain the properties of the ground state of deuteron, such as the magnetic dipole moment μ_D and the electric quadrupole moment eQ_D, requires LCAO of the s- and d- states. Nuclear force has two parts:

a) Strong spin-orbit interaction, and
b) Tensor force.

6) Repulsive Core:
To account for the facts that the nuclear volume
$V_N \propto A (= \#nucleons)$, i.e. there is constancy in the separation of nucleons, and certain features of nucleon-nucleon scattering.

6) Saturation (Exchange force):
Because for heavy nuclei, $E_B / A \approx$ constant, i.e. $E_B \propto A$. This requires the nucleons, having substructure, to exchange particles between them. H. Yukawa, (1936) discovered such a particle called **meson**.

6.5   THE INTERACTION BETWEEN TWO NUCLEONS

The basic problems in "Sub-atomic Physics" include following information:
1. On strength and range of proton-proton (p-p) force by (p-p) scattering,
2. On strength and range of neutron-proton (n-p) forces from n-p scattering, and
3. About the 'n – n' forces from a study of complex nuclei.

6.5.1 Three Possible Systems

The successful application of Wave Mechanics to α – *radioactivity* and in describing the energy states of the H-atom (a two body system) showed that a study of a nucleus composed of two nucleons by Quantum Mechanics was appropriate. There are three possible states of a two-nucleon system. These are:

(i) The *di-neutron* (n-n) which does not exist in nature,
(ii) The *di-proton* (p-p) ($_2^2He$) which is unstable and does not exist in nature, and
(iii) The *deuteron* (n-p) ($_1^2H$) which is stable and exists in nature.

It is obvious, therefore, that of these possible systems only the deuteron can be analyzed to know the nuclear force law F_{n-p} or nuclear potential between the neutron and proton.

6.6 THE DEUTERON

6.6.1 Experimental Data on Deuteron ($_1^2H$)

Experimental data on the ground state on the deuteron ($_1^2H$) are described below:

1. The deuteron has only one bound level.
2. The ground state energy of the deuteron,
$$E_D = -(2.226 \pm 0.003) \, MeV \, . \qquad (6.4.4)$$
3. The Spin quantum number, $I = 1$
4. Magnetic moment $\mu_D = +(0.85735 \pm 0.00003) \, nm \qquad (6.4.5)$
5. The electric quadrupole moment
$$eQ_D, \text{ where } Q_D = +0.00282 \, b \qquad (6.4.6)$$
6. Because n is electrically neutral, the nuclear force binding up the n-p system can not be electrical.
7. Because $M_p \approx M_n \approx 1.67 \times 10^{-27} \, kg$, the gravitational forces are too weak to give
$$E_D = -(2.226 \pm 0.003) \, MeV$$

The force between n and p, *i.e.*, F_{n-p} is of different origin.

6.6.2 The Ground State of the Deuteron ($_1^2H$)

6.6.2.1 Pure Central potential V_o existing between neutron and proton:

The ground state wave function of the deuteron can be written from the knowledge of the several properties given above.
Since $I = 1$,

$$I = 1 = L \pm S; \qquad (6.6.1)$$
$$S = (s_n + s_p) \qquad (6.6.2)$$
$$s_p = \frac{1}{2} \hbar \, \sigma_p, \, s_n = \frac{1}{2} \hbar \, \sigma_n \qquad (6.6.3)$$

s_n & s_p are independent. So

$$S = [\hat{s}_n, \hat{s}_p] = 0 \qquad (6.6.4)$$
$$|S|^2 = |s_n|^2 + |s_p|^2 + 2|s_n||s_p| \qquad (6.6.5)$$
$$\hat{S}^2 = \hbar^2 \, s(s+1) \qquad (6.6.6)$$

S = 0   for the *Singlet* state
S = 1   for the *Triplet* state

$$|s_n|^2 = |s_p|^2 = \frac{1}{2}\hbar^2(\frac{1}{2}+1) = \frac{3}{4}\hbar^2 \qquad (6.6.7)$$

$$s_n \cdot s_p = -\frac{3}{4}\hbar^2 \text{ for the } Singlet \text{ state}$$
$$s_n \cdot s_p = \frac{1}{4}\hbar^2 \text{ for the } Triplet \text{ state} \qquad (6.6.8)$$

$I = 1 = L \pm S$,
means one gets the 4 states listed in Table 6.4.

Table 6.4   The Deuteron

$I = 1 = L \pm S$		$^{2S+1}L_I$ states
1. L = 0, S = 1	gives	$^3S_1$ state
2. L = 1, S = 0	gives	$^1P_1$ state
3. L = 2, S = 1	gives	$^3D_1$ state
3. L = 1, S = 1	gives	$^3P_1$ state

This calculation based on the relative motion of the n and p around their centre of mass, is L = 0, and $s_p \cdot s_n$, i.e. S = 1, gives $^3S_1$ state

In $^3S_1$ state, L = 0, $\mu_D = 0$

6.6.2.2 Magnetic Moment of the Deuteron.
$$\mu_D = \mu_p + \mu_n = \{2.792810 - 1.913148\} \text{ nm } = +0.879862 \text{ nm}$$
This value of μ_D is very close to the experimentally measured value $\mu_D = +0.85735$ nm.

This confirms that the ground state of the deuteron is $^3S_1$ state.

6.6.2.3 To find the Energy of $^3S_1$ state
Applying the TISE (Time independent Schrödinger equation) to the deuteron

$$\nabla^2 \psi_D + \frac{2\mu}{\hbar^2}[E - V(r)]\psi_D = 0, \qquad (6.6.9)$$

where $\quad \mu = \frac{M_p M_n}{M_p + M_n} \cong \frac{M_p}{2} \qquad (6.6.10)$

and $\quad E = E_D = -(2.226 \pm 0.003) \text{ MeV} \qquad (6.6.11)$

where $\quad V(r) = \begin{vmatrix} -V_o, & (r \leq b), \\ 0, & (r > b). \end{vmatrix} \qquad (6.6.12)$

$$k^2 = -\frac{2\mu}{\hbar^2}\{E_D - V(r)\} = \frac{M_p}{\hbar^2}(V(r) - E_D), \qquad (6.6.13)$$

and $\quad \gamma^2 = \frac{2\mu}{\hbar^2} E_D = \frac{M_p}{\hbar^2} E_D \qquad (6.6.14)$

See Fig 6.4.

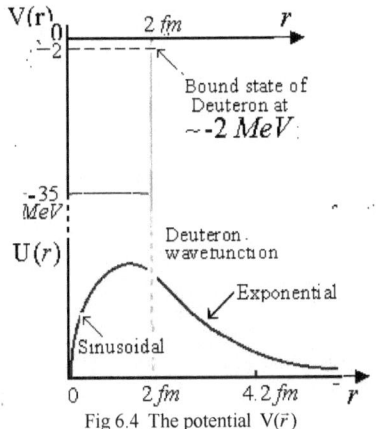

Fig 6.4 The potential $V(\vec{r})$

Solving the TISE above, one gets

$$\tan k b = -\frac{k}{\gamma} \qquad (6.6.15)$$

This is the Transcendental Equation relating b to V_o. Solving it graphically or numerically,

$$V_o b^2 \approx \frac{\hbar^2}{M_p} \frac{\pi^2}{4} = 10^{-28} \text{ MeV m}^2 = 143 \text{ MeV fm}^2 \qquad (6.6.16)$$

Choosing various values for b and V_o results in Table 6.5

Table 6.5 The Deuteron: Range b and Depth V_o

b fm		V_o (MeV)
1.	1.0 fm gives	143
2.	1.4 fm gives	52
3.	1.7 fm gives	46
4.	2.0 fm gives	37
5.	2.5 fm gives	25
6.	∞ fm gives	2.83

But beyond the limit $V_o b^2 \geq 143$ MeV fm^2 no bound state for the deuteron can exist.
Equation (6.6.16) relates range b with the (square well) potential energy V_o of the nuclear force. It is, therefore, clear from the above analysis that by using a more correct $V(\vec{r})$, the correct values of the range b and V_o will be obtained.

6.7   EXPERIMENTS: Scattering of Neutrons by Protons

The principle of the experiment accept the fact that Coulomb scattering is absent for neutrons as projectile or bombarding particles, n-p nuclear interaction can be studied. a) Two Sources: a nuclear reactor producing neutron beams at sub-thermal energies and proton accelerators giving protons as secondary particles in the *MeV* region to the highest, b) Targets are hydrogen, c) Simple attenuation methods are used to observe total cross sections, d) Partial wave analysis is employed to analyze the measured total cross sections and angular distribution. For more details, in interested reader is recommended to read advance books listed in the Bibliography.

6.7.1 Spin Dependence on $n-p$ Scattering

The experimental value of the S- wave scattering cross section $\sigma_{\ell=0}$ is much higher than the theoretical value. To explain this disagreement, the spin-dependence of nuclear forces was introduced by Eugene P. Wigner (in 1935). In the deuteron, depending on the spin directions of the neutron and proton, it may be in either in the triplet $^3S_1$ state or in the singlet $^1P_1$ state (Table 6.6). So to examine if the nuclear force is or not spin dependent can be studied by examining the cross section in each case. Experimental results on n-p scattering show that the experimental and theoretical values of cross section are different, confirming the spin dependence of the nuclear forces.

TABLE 6.6 Deuteron Potential Parameters

Deuteron State	Range, b	Potential Depth V_0 (MeV)
1. $^3S_1$ state	2.25 fm	26.2
2. $^1P_1$ state	2.51 fm	17.8

6.7.2 $\sigma(^1P_1)$ contributes the most at low energy to σ, Why?

The spins are said to be correlated. Since, in general, the spins of the neutron bream may be oriented at random, the n-p system the triplet and singlet states will be in proportion to the weight factors of the two. It has been found that

$$\sigma = \frac{3}{4}\sigma(^3S_1) + \frac{1}{4}\sigma(^1P_1) \qquad (6.7.1)$$

Experimentally, $\sigma = 20.4\ b$, (6.7.2)
and $\sigma(^3S_1) = 2.35$ to $5.0\ b$, (6.7.3)
This means $\sigma(^1P_1) = 66.6$ to $74.6\ b$. (6.7.4)

6.7.3 Electromagnetic Properties of the Deuteron (Effects of Tensor Forces)

The deuteron was treated based on a purely central potential existing between neutron and proton, in Section 6.7.2.

6.7.3.1 The Nature of the Ground State of the Deuteron

As was discussed earlier look at Table 6.7, there are 4 possible combinations compatible with $I = 1$. It is known that states having the same parity only can mix. The wave function of the deuteron is a mixed state of either

a) $^3S_1$ and $^3D_1$ states or
b) $^1P_1$ and $^3P_1$ states,

so that one, of these as the ground state of the deuteron, has a well defined parity.

Table 6.7 Deuteron State and Parity

State	Parity, $(-1)^\ell$
1. $^3S_1$	Even
2. $^1P_1$	Odd
3. $^3P_1$	Odd
4. $^3D_1$	Even

6.7.3.2 Angular Momentum Wave function of the Deuteron

To find out which one of the two mixed states seen above represents correctly the ground state of the deuteron, one has to examine the addition of two angular momenta.

Denoting the angular momentum state by $|j, m\rangle$, where $j = \ell \pm s$ and $m = \pm \ell, .., 0$ or $\pm \frac{1}{2}$.

Using *spin up* and *spin down* as $\alpha = \begin{pmatrix} 1 \\ 0 \end{pmatrix}$ and $\beta = \begin{pmatrix} 0 \\ 1 \end{pmatrix}$, one can then get the following 4 states of Table 6.8.

Table 6.8 Angular Momentum States of Deuteron

Pure State	I,M,ℓ,S	Mixed state
1. $^3S_1$	1,1,0,1	$\langle 0101 \mid 11 \rangle Y_0^0\, \alpha(1)\,\alpha(2) \equiv Y_0^0\, \alpha(1)\,\alpha(2)$
2. $^3D_1$	1,1,2,1	$\sqrt{\frac{3}{5}}\, Y_2^2\, \beta(1)\beta(2) - \sqrt{\frac{3}{10}}\, Y_2^1\, [\alpha(1)\,\alpha(2) + \beta(1)\beta(2)] + \sqrt{\frac{1}{10}}\, Y_2^0\, [\alpha(1)\,\alpha(2)]$
3. $^1P_1$	1,1,1,0	$\frac{1}{\sqrt{2}}\, Y_1^1\, [\alpha(1)\beta(2) - \beta(1)\alpha(2)]$
4. $^3P_1$	1,1,1,1	$\frac{1}{\sqrt{2}}\, Y_1^1\, \frac{1}{\sqrt{2}}[\alpha(1)\beta(2) + \beta(1)\alpha(2)] - \frac{1}{\sqrt{2}}\, Y_1^0\, \alpha(1)\,\alpha(2)$

6.7.3.3 Magnetic Moment of the Deuteron, μ_D

Accurate experimental values of the magnetic moments for the deuteron are:

$$\mu_D = +0.857351\ nm \pm 0.000019\ nm \qquad (6.7.5)$$

$$\mu_n = -1.91315\ nm \pm 0.00007\ nm, \qquad (6.7.6)$$

$$\mu_p = +2.79280\ nm \pm 0.00002\ nm \qquad (6.7.7)$$

$$\mu_n + \mu_p = +0.87965\ nm \pm 0.00007\ nm \qquad (6.7.8)$$

So there is seen a small discrepancy between the values of μ_D and $\mu_n + \mu_p$, and this $\mu_D - (\mu_n + \mu_p) = +0.0223\ nm$ must be accounted for.

Any way the bound (or, ground) state of the deuteron is either

$$\Psi(SD) = a\,\psi(^3S_1) + b\,\psi(^3D_1) \text{ or}$$
$$\Psi(PP) = a\,\psi(^3P_1) + b\,\psi(^1P_1)$$

It can be shown that

$$\langle \mu_z \rangle_{2S+1L_I} = [\tfrac{1}{2}\ 0.37960 \ \square\ \frac{I(I+1) + S(S+1) - L(L+1)}{2I(I+1)}]\ nm \qquad (6.7.10)$$

Using this formula,
$$\langle \mu_z \rangle_{^3S_1} = 0.87960\ nm$$
$$\langle \mu_z \rangle_{^3D_1} = 0.31020\ nm$$
$$\langle \mu_z \rangle_{^1P_1} = 0.5000\ nm$$
$$\langle \mu_z \rangle_{^3P_1} = 0.68980\ nm$$

But the experimental value of $\mu_D = +\ 0.857351\ nm \pm 0.000019\ nm$ which is not equal to any one of the above 4 values of $\langle \mu_z \rangle_{2S+1L_I}$.

6.7.3.4 Mixture of States to represent the deuteron
Using
$$\Psi(SD) = a\,\psi(^3S_1) + b\,\psi(^3D_1) = 0.96\,\psi(^3S_1) + 0.04\,\psi(^3D_1) \qquad (6.7.11)$$

it will be easy to calculate to get $\mu_D = +\ 0.857351\ nm$, the experimental value. So the deuteron ground state angular momentum function should be a mixture of 96% $^3S_1$ and 4% $^3D_1$, and the parity should be even.

6.7.3.5 The Quadripole Moment of the Deuteron Q_D

It was seen that the deuteron ground state angular momentum function should be predominantly (96%) the $^3S_1$ with a small (4%) admixture of $^3D_1$, and the parity should be even. This conclusion is supplemented by the fact that as otherwise a $^3S_1$ state is spherically symmetric and does not have an electric quadrupole moment. The value of Q_D can be estimated in a straightforward manner using relation (2.8.15), viz.,

$$Q_D = \int \psi_D^* \ [3\cos^2\theta - 1]\ \psi_D\ r^2 d\tau \qquad (6.7.12)$$

In terms of spherical harmonics,

$$Q_D = \frac{1}{4}\sqrt{\frac{16\pi}{5}} \int \psi_D^* \ [r^2\ Y_2^0]\ \psi_D\ d\tau \qquad (6.7.13)$$

which will yield using expression $\psi_D \equiv \psi_{SD}$ (3.2.10).

$$Q_D \neq 0. \qquad (6.7.14)$$

The result that $Q_D \neq 0$ along with the fact that $\mu_D \neq (\mu_n + \mu_p)$, confirms that the nature of the nuclear force is partially *non-central*, i.e. it has partly a *tensor component*.

6.7.3.6 Both the Di-neutron (n-n) and the Di-proton (p-p) Systems are Unstable, Why?

It has been seen that the nucleon-nucleon force must depend on the spins or orbital contributions or both. This also shows that both the di-neutron (n-n) and the di-proton (p-p) systems are not stable systems, for the nucleons (2 protons and 2 neutrons) being Fermions cannot exist with parallel spins, in the lowest state of $\vec{L} = 0$. Therefore there is no ground state for either di-neutron or for $_2^2 He$ system.

6.7.3.7 How to Represent the Component of the Tensor Force

There is no ground state for either di-neutron or for $_2^2 He$ system. In order to obtain a representation of the spherical effects, one can a a small factor of a non-central feature to the central force. The resulting total potential, $V(\vec{r})$, between two nucleons that depends on the spin, has the form

$$V(\vec{r}) = \begin{cases} V_C(r) & (\rightarrow \text{A spin independent Central potential}) \\ + V_{CS}(r) \frac{\vec{s}_1 \cdot \vec{s}_2}{\hbar^2} & (\rightarrow \text{A spin dependent Central potential}) \\ + V_T(r) \frac{\hat{S}_{12}}{\hbar^2} & (\rightarrow \text{A Tensor potential}) \\ + V_{LS} & (\rightarrow \text{A Two-body spin-orbit potential}) \end{cases} \quad (6.7.15)$$

where $V_i(r)$ are ordinary functions of r, $\vec{s} = \frac{1}{2} \hat{\sigma} \hbar$, and $\hat{\sigma}$ is Pauli spin matrix.

The non-central character of the interaction is contained in the tensor operator $\hat{S}_{12}$.

$$\hat{S}_{12} = \left\{ \frac{[3(\hat{\sigma}_1 \cdot \hat{r}_1)(\hat{\sigma}_2 \cdot \hat{r}_2)]}{|\hat{r}_2|^2} \right\} - [\hat{\sigma}_1 \cdot \hat{\sigma}_2] \quad (6.7.16)$$

This gives

$\hat{S}_{12} = 0$, for all the singlet ^1S, ^1P, ^1D, states, and

$(\hat{S}_{12} - 2)(\hat{S}_{12} - 4) = 0$, for all the triplet, ^3S, ^3P, ^3D, states.

6.7.3.8 Scattering of Neutron by Proton and Photo-disintegration

This is neutron-proton capture involves exchange of energy between the n-p system and an EM Field, as in the reactions given below:

$n + p \rightarrow d + \gamma$, or $\gamma + d \rightarrow n + p$

The minimum exchange energy = $E_B = 2.2245 \, MeV$.

The spin of the particles, viz. n, p, d allow each of the two processes mentioned above to occur by an magnetic dipole interaction, example, $^1S \rightarrow {}^3S$, for capture, $^3S \rightarrow {}^1S$ for photo-disintegration.

The observed value of the capture cross section confirms that the ^1S state of the deuteron is not a bound state. The photo-disintegration process may also take place through the electric dipole transition $^3S \rightarrow {}^3P$, which is responsible for the major part of the total cross section.

6.7.3.8 Saturation of Nuclear Force

For $A > 16$ (heavy nuclei), $E_B / A \approx $ constant, i.e.,

$$E_B \propto A. \quad (6.7.17)$$

If the interactions between all the pairs of nucleons are the same,
$$E_B \propto \frac{A(A-1)}{2} \qquad (6.7.18)$$
in the nucleus. That is a nucleon interacts with only a few neighbouring nucleons within he nucleus. This means the nuclear force is saturated!

6.7.3.9 What kind of Nuclear Force can produce Saturation?
Werner Karl Heisenberg suggested that the nuclear force must be some kind of exchange force of attraction. Both the Wigner and Majorana types of potential may be considered.

6.8 YUKAWA THEORY (or Meson Theory) OF NUCLEAR FORCES

The concept of exchange forces introduced by Heisenberg and Majorana has the property of being attractive or repulsive depending on the states of the nucleons involved. After the failure of this theory, Hidekei Yukawa (1935) predicted the existence of a particle, now called a meson, and proposed that the nuclear forces arise from the constant exchange of these particles back and forth between nearby nucleons.

6.8.1 Yukawa's arguments
Yukawa's arguments, greatly simplified, are approximately as follows:

1. EM Fields obey a wave equation that can be derived from the *Maxwell's EM Equations*. That is the scalar electric potential ($U = E_x$ or H_y) in free space obeys the differential form of Maxwell's equation
$$\{\nabla^2 - \frac{1}{c^2}\frac{\partial^2}{\partial t^2}\} U(r, t) = 0 \qquad (6.8.1)$$

2. According to Quantum Electrodynamics (QED), the EM Field which is quantized, and these quanta known as photons are the virtual carriers of the EM field (now referred to as the photon field), which are continuously emitted by an electrical charge and absorbed by the other charged particles.

3. Yukawa suggested that the meson field $\phi(r)$, by which the force between two nucleons is propagated, obeys the equation
$$\{\nabla^2 - \mu^2\} \phi(r, t) = -4\pi \rho_N. \qquad (6.8.2)$$

6.8.2 To find the potential function representing the meson field, Postulate in *Wave Mechanics*
$$\vec{P_q} \rightarrow -i\hbar \hat{\nabla}_q, \text{ and } E \rightarrow i\hbar \hat{\nabla}_t; \text{ so}$$
$$\{\nabla^2 - \frac{1}{c^2}\frac{\partial^2}{\partial t^2}\} U(r, t) = 0 \qquad (6.8.3)$$
can be written as
$$E^2 - p^2 c^2 = 0, \qquad (6.8.4)$$
relativistically,
$$E^2 = p^2 c^4 + m_0^2 c^4 \qquad (6.8.5)$$
This gives the Klein-Gordon Equation, for a spin-less relativistic free particle,
$$\{\nabla^2 - \frac{m_0^2 c^2}{\hbar^2} - \frac{1}{c^2}\frac{\partial^2}{\partial t^2}\} \phi(r, t) = 0 \qquad (6.8.6)$$
The time independent form is
$$\{\nabla^2 - \frac{m_0^2 c^2}{\hbar^2}\} \phi(r) = 0, \qquad (6.8.7)$$

whose solution is
$$\phi(r) = g \frac{e^{-\mu r}}{r} \tag{6.8.8}$$
where $\mu = \frac{m_0 c}{\hbar}$, (6.8.9)

and g = undetermined constant which may be evaluated by means of *Gauss theorem* with the result that g = 1.

This is called the *Yukawa potential*, applicable within a nucleus, just like the role of the ES potential between any two electrically charged particles.

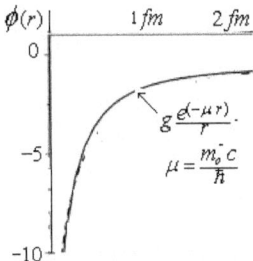

Fig 6.5 The Yukawa potential

The mechanical potential between two charged particles q and Q has interaction is

$$q V_C(r) = \frac{Q q}{4 \pi \varepsilon_0 r} \tag{6.8.10}$$

Here the charge q is the source of the photon field. $E_i = E_1 + E_2$; $E_f = E_1 + E_2 + E_\gamma$. ES interaction results from the continuous transmission of virtual photon (i.e. EM field quanta) between two charges, whether it is in a collision or bound state (eg., *positronium*) (Fig 6.6).

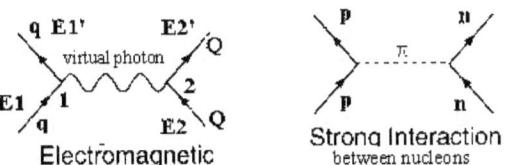

Fig 6.6 Feynman diagrams for EM and Strong interactions

By analogy in nuclear forces, the mechanical potential resulting from the interaction between nucleons (each nucleon is the source of meson field) each of strength g, one at the origin and the other at position r is

$$V_N(r) = g^2 \frac{e^{-\mu r}}{r} \tag{6.8.11}$$

One may therefore consider this nuclear interaction as arising from the continuous transfer, of virtual mesons of rest mass m_0, between the two nucleons. The term 'virtual' means that the meson cannot be released from the nucleus unless an energy of at least $m_0 c^2$ is supplied, which is the violation of conservation of energy!!

Total energy before collision $E_i = E_1 + E_2$

Total energy after collision $E_f = E_1 + E_2 + m_0 c^2$

The emission of virtual meson is followed by its re-capture immediately, within a time Δt, over which the *Law of conservation of energy* is **violated**.

6.8.3 Range of Nuclear Force

Δt = time duration of an excursion of the meson between two nucleons,
v = speed of the meson.

$$\Delta E \cdot \Delta t = \hbar, \qquad (6.8.12)$$

Consider $v = c$, and the range R of the strong nuclear force, i.e. the maximum distance that the meson can travel in Δt,

$$R_{max} = c \cdot \Delta t = c \cdot \frac{\hbar}{m_0 c^2} = \frac{1}{\mu} \qquad (6.8.13)$$

Compton Wavelength of the meson

$$\lambda_C = \frac{\hbar}{m_0 c} \qquad (6.8.14)$$

$$\frac{m_0}{m_e} = \frac{\lambda_C}{2 \pi R} \qquad (6.8.15)$$

This gives the mesonic mass $m_0 = 275\, m_e$

6.8.4 The Discovery of the Meson, the π-mesns (Pion)

Mesons were discovered experimentally in cosmic rays. The pions (π-mesns), discovered in 1947 by Cecil Frank Powell and co-workers in nuclear emulsion, have mass closest to that of the meson estimated above. $m_0 \equiv 273\, m_e$.

$\therefore \quad m_0 \equiv m_\pi$

So pion is regarded as mainly responsible for the occurrence of the nuclear forces, between the nucleons..

6.8.4.1 The Three Types of Pions

The pions are of three kinds:-
1) π^+, intrinsic spin $s_\pi = 0$, Parity odd, mass $m_\pi \equiv 273\, m_e$,
2) π^0, intrinsic spin $s_\pi = 0$, parity odd, mass $m_\pi \equiv 273\, m_e$, and
3) π^-, intrinsic spin $s_\pi = 0$, parity odd, mass $m_\pi \equiv 273\, m_e$.

The properties of the pions give them the name pseudo-scalar.

6.8.4.2 OPEP (One Pion Exchange Potential)

In this brief outline of the meson theory of nuclear forces there are several important factors not yet discussed between a n and p, the force is

$$n \rightarrow p + \pi^- \qquad (6.8.16)$$
$$p + \pi^- \rightarrow n \qquad (6.8.17)$$

i.e., that part of the force field ϕ which relates to charged pions transfers the charge from one nucleon to another.

In the reverse process, p → n + π^+ (6.8.18)

and   n + π^+ → p   (6.8.19)

Examination of parity and angular momentum conservation laws, when a pion is created and transferred, the force field between the two nucleons in the triplet state is not spherically symmetric, as suggested by the Yukawa potential, but has an additional term which is just a tensor force. This is the OPEP.

6.8.4.3 TPEP (Two Pion Exchange Potential)

Instead of the OPEP, simultaneous transfer of two pions is also possible. In an analysis of the Breit phase parameters, Feshbach and coworkers (1961) have given convincing evidence that the nuclear forces are adequately described by the OPEP and TPEP for distances greater than 0.7 *fm* between the nucleons.

6.9 The NUCLEONIC STRUCTURE

The physically observed neutron and proton are regarded as consisting of a basic *nucleonic core* surrounded by a *meson cloud*. Littauer et al. (1961), based on their electron scattering experiments, describe that a proton and a neutron in common have a core, of positive charge ≈ + 0.35 e, and probable radius 0.2 *fm*, surrounded by two distinct clouds of mesons. The inner cloud of + 0.5 e charge has radius 0.8 *fm*, in the case of proton and - 0.5 e for a neutron. The outer cloud of proton as well as neutron has charge + 0.15 e and radius ≈ 1.4 *fm*,

In 1966 the p-p scattering experiments at 90° in the range of energies 5.0 to 13.4 *BeV* indicate an internal structure to the proton along the lines of *onion shells* with an outer π cloud at a radius of 0.92 *fm*, an inner heavy cloud of radius 0.50 *fm* and a core of radius 0.32 *fm*.

REVIEW QUESTIONS

R.Q. 6.1   What is the nuclear mass of helium-3 (^{3}He) of which the atomic mass is 3.016030 *u*?
[Answer   3.016030 *u* − (2 × 0.000549 *u*) = 3.014932 *u*].

R.Q. 6.2   The nuclear radius is represented precisely as $1.2 \times A^{1/3}$ *fm*. Roughly approximate the nucleus as a sphere and calculate its mass density. Given: The mass of the nucleon 1.66×10^{-27} kg. [Answer $\rho_N \approx 2 \times 10^5$ *tonnes / mm*^{3} and is independent of *A*].

R.Q. 6.3 Calculate the binding energy of a deuteron in *MeV*. Given: $^2_1H \equiv M_D = 2.014102$ u,
$M_p = 1.0072765$ *u*, $M_n = 1.008665$ *u*, $m_e = 9.1093897 \times 10^{-31}$ kg, c = 2.9979 × 10^8 m s^{-1},
1 *u* = 1.66054×10^{-27} kg, 1 *u* = 931.4943 *MeV* c^{-2}.

R.Q. 6.4   Calculate the binding energy of argon nucleus. Given: $^{40}_{18}Ar \equiv 39.962384$ *u*,
$M_p = 1.0072765$ *u*, $M_n = 1.008665$ *u*, $m_e = 9.1093897 \times 10^{-31}$ kg, c = 2.9979 × 10^8 m s^{-1},
1 *u* = 1.66054×10^{-27} kg, 1 *u* = 931.4943 *MeV* c^{-2}.

R.Q. 6.5 Calculate the binding energy of Helium. Given: $^4_2He \equiv 4.002603\ u$,

$M_p = 1.0072765\ u$, $M_n = 1.008665\ u$, $m_e = 9.1093897 \times 10^{-31}\ kg$, $c = 2.9979 \times 10^8\ m\ s^{-1}$,

$1\ u = 1.66054 \times 10^{-27}\ kg$, $1\ u = 931.4943\ MeV\ c^{-2}$.

R.Q. 6.6 What is the binding energy of the helium atom whose mass defect is 0.030377 u?
[Answer: $m = 0.030377\ u \times 1.661 \times 10^{-27}\ kg = 5.046 \times 10^{-27}\ kg$; $E = mc^2 = 4.541 \times 10^{-12}\ J$].

R.Q. 6.7 What is the mass defect in i) atomic mass units (u), ii) kg for the lithium nucleus which is composed of 7 nucleons, and a proton number of 3? Iii) What is the binding energy in J, iv) eV? v) What is the binding energy per nucleon in eV? Given, the nuclear mass = 7.014353 u; $c = 3 \times 10^8$ m/s; $1\ J = 1.6 \times 10^{-19}\ eV/J$ and $1\ u = 1.661 \times 10^{-27}$ kg. [Answer i) mass deficit = (7.056488 u - 7.014353 u = 0.042135 u; ii) $6.9986235 \times 10^{-29}$ kg; iii) $E_B = 6.3 \times 10^{-12}$ J; iv) $E_B = 3.9 \times 10^7\ eV$ = 39 MeV.;
v) $E_B / A = 3.9 \times 10^7\ eV \div 7 = 5.6 \times 10^6\ eV$].

R.Q. 6.8 The copper nucleus has 63 nucleons and proton number 29. What is the binding energy per nucleon in eV? The nuclear mass = 62.91367 u. [Answer: $E_B / A = 28.38\ MeV \div 4 = 7.1\ MeV$].

R.Q. 6.9 Determine the time duration of excursion of the meson between two nucleons. What is the value of mesonic energy? Estimate the mass of the meson. [Answer

$\Delta t = \dfrac{R}{c} = 5.0 \times 10^{-24}$ s, $\Delta E \approx \dfrac{\hbar}{\Delta t} = 131$ MeV, $m_0 = 275\ m_e$].

R.Q. 6.10 Show that for a two-body system with central force potential the ground state is a zero angular momentum state.

R.Q. 6.11 Why is it that both ^2_2He and di-proton are unstable and do not exist in nature?

R.Q. 6.12 Describe how the binding energy, a characteristic of the deuteron can be described?

R.Q. 6.13 Describe the following characteristics of the deuteron can be described?
a) The spin, b) The isospin, c) The magnetic moment, e) The quadripole moment.

R.Q. 6.15 State the nuclear properties of Deuteron. Give an account of the theoretical description of the ground state of deuteron and hence account for the knowledge of the nuclear forces.

R.Q. 6.16 Write a short note on Exchange forces.

R.Q. 6.17 Outline the simple theory for the ground state of deuteron. If the well depth is 73 MeV, determine the well width corresponding to the ground state of deuteron.

RQ. 6.18 Discuss the nature of nuclear forces. Give an account of Yukawa's theory of nuclear forces.

RQ. 6.19 Estimate the binding energy per nucleon $^{238}_{92}U$ nucleus. The atomic masses are: 238.050 79 u for $^{238}_{92}U$, $M_p = 1.007276\ u$, $M_n = 1.008665\ u$.(Answer: 7.57 MeV)

&&*&*&*&

Chapter 7
STABILITY OF NUCLEI

Chapter 7

STABILITY OF NUCLEI

"I am become Death, the chatterer of worlds."--Shiva in "The Bhagavad Gita".

7.1   INTRODUCTION

The basic (static) properties of nuclei, considered as stable (time independent) assemblies of nucleons have been discussed in the previous Chapters 3, 4 and 5. A source of information about nuclear structure is the analysis of time dependent properties (processes), such as I. Nuclear disintegrations or decays and II. Nuclear reactions, in which there is a rearrangement of the energy or configuration of nucleons.

7.1.1   Nuclear Decay

Antoine Henri Becquerel (1896) and Pierre and Marie Sklodowska Curie discovered 3 distinct types of accelerated particles from radioactive decay named after the first three letters of the Greek alphabet: α- (alpha), β- (beta), and γ- (gamma) separated by a magnetic field positive alpha particles bend one direction negative beta particles bend opposite neutral gamma rays do not bend at all.

In these radioactive processes, nuclear radiation occurs in other forms, including the emission of protons or neutrons or spontaneous fission of a massive nucleus. Of the nuclei found on Earth, the vast majority are stable. This is so because almost all short-lived radioactive nuclei have decayed during the history of the Earth. There are approximately 270 stable isotopes and 50 naturally occurring radioisotopes (radioactive isotopes). Thousands of other radioisotopes have been made in the laboratory

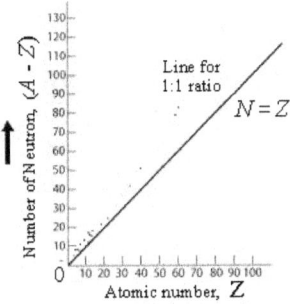

Fig 7.1   The N = Z line Nuclide chart.

Radioactive decay will change one nucleus to another if the product nucleus has a greater nuclear binding energy than the initial decaying nucleus. The difference in binding energy (comparing the before and after states) determines which decays are energetically possible and which are not. The excess binding energy appears as kinetic energy or rest mass energy of the decay products.

7.2 CHART OF THE NUCLIDES

The Chart of the Nuclides, part of which is shown above is a plot of nuclei as a function of proton number, Z, and neutron number, N. as schematically shown in the diagram in Fig 7.1. This is the N = Z line, a diagonal line.

7.2.1 Nuclear Stability and the Segre Chart

A complete list of nuclides of the elements up to Z = 102 shows that there are about 1050 different nuclides now known, but only about 25% of these are stable. It was seen in Chapter 6 that some nuclei have a combination of neutrons and protons that does not lead to a stable configuration. Thus stable configurations of nucleons are the exception rather than the rule. A plot of the neutron number N (ordinate) versus proton number Z (abscissa), of stable nuclei (of naturally occurring nuclides for $Z \leq 83$) is reproduced in Fig. 7.2 This is known as the *Segre chart* (after Emilio Gino Segre).

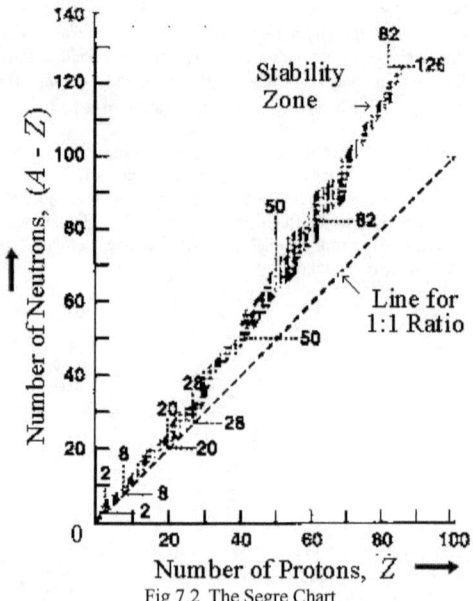

Fig 7.2 The Segre Chart

N would all fall in an area known as the *band of stability*. The band of stability also includes radio-nuclides because smooth lines cannot be drawn to exclude them. This is called the *Segre Chart* and is reproduced in Fig 7.3. The band of stability also stops at element 83 because there are no known stable isotopes above it. Elements lying outside the band of stability would be too unstable to justify the time and money for an attempt to make it. Another thing that is noticed about the band of stability is that as the number of protons increases, the ratio of neutrons to protons increases. This is because more neutrons are needed to compensate for the increasing proton-proton repulsions. Isotopes occurring above and to the left of the band tend to be β^-- emitters because they want to lose a neutron and gain a proton. Those lying below and to the right of the band tend to be β^+- (positron) emitters because they want to lose a proton and gain a neutron. Isotopes above element 83 tend to be α- emitters because they have too many nucleons

Table 7.1 Features of the Chart of the Nuclides

1. For low Mass # A nuclides, $\frac{N}{Z} = 1$,
2. For large Mass # A nuclides, $\frac{N}{Z} \rightarrow 1.59$.
3. Only 4 Lines of constant A nuclides (Isobars):are at 135° with the Z-axis. These occur at A = 96, 124, 130, and 136.
4. Two regions of instability, one above and the other below, the stability line occur.
5. A Nuclide in the region below the stability line disintegrates by:
   a) Positron β^+ emission, by low A nuclides,
   b) Positron β^+ emission or EC, by middle A nuclides,
   c) Positron β^+ emission, or EC or α-emission by larger A nuclides.
6. A Nuclide (with excess neutron) in the region above (left of) the stability line disintegrates by β^--decay.

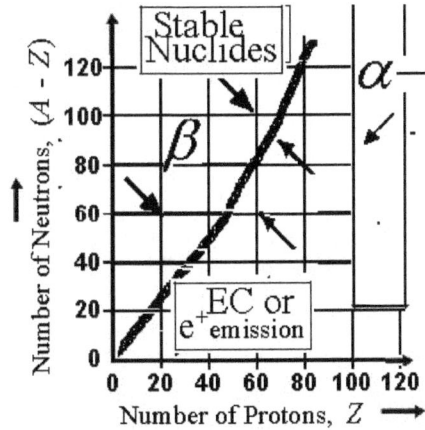

Fig 7.3 α-, β-, and γ-decays

All stable nuclei and known *radioactive nuclei*, both naturally occurring and manmade, are shown on this chart, along with their decay properties. Nuclei with an excess of protons or neutrons in comparison with the stable nuclei will decay toward the stable nuclei by changing protons into neutrons or neutrons into protons, or else by shedding neutrons or protons either singly or in combination. Nuclei are also unstable if they are excited, that is, not in their lowest energy states. In this case the nucleus can decay by getting rid of its excess energy without changing Z or N by emitting a γ-ray (Fig 7.3).

It shows the features listed in Table 7.1.as well as in Fig 7.3.:

A necessary condition for nuclear stability: E_B = +ve quantity, *i.e.*,

M (Z, A) < (Total mass of the constituents when arranged in any other combination)
M (Z, A) < (Total mass of the constituents when arranged in any other combination).
Unstable nuclides are called *radioactive nuclides*.

Radioactivity is a term used to describe any process in which the nucleus spontaneously undergoes a transformation to approach a stable configuration.

Worked out Example 7.1

Suppose the nuclide $^{226}_{88}Ra$ which can be stable or unstable. Check if this nuclide is unstable α- decay. giving daughter $^{222}_{86}Rn$. Given, $M(^{226}_{88}Ra) = 226.025406\ u$, $M(^{222}_{86}Rn) = 222.017574\ u$, $M_\alpha = 4..002603\ u$.

Solution: Step # 1   For α- decay, $Q_\alpha = (M_P - M_D - M_{He})\ c^2$,
$M_P = M(^{226}_{88}Ra) = 226.025406\ u$, $M_D = M(^{222}_{86}Rn) = 222.017574\ u$, $M(^4_2He) = 4..002603\ u$

Step # 3 $Q_\alpha = (226.025406\ u - 222.017574\ u - 4..002603\ u)(931\ MeV/u)c^2/c^2$
= +4.87 *MeV*

Sice $Q_\alpha > 0$, the parent nuclide $^{226}_{88}Ra$ is unstable to α- decay..

7.3.   NUCLEAR MASS AND STABILITY

It is considered to treat the subject of nuclear mass and stability in a more appropriate Chapter 13. Weizsacker nuclear mass formula, for example, is dealt with.

REVIEW QUESTIONS

R.Q. 7.1   What are mirror nuclei? What is its importance in nuclear physics?
R.Q. 7.2.   What is the Segre Chart ? Describe its features and explain nuclear stability..

%%*%*%*%*

Chapter 8

THE NEUTRON, ITS DISCOVERY & APPLICATIONS

Chapter 8

THE NEUTRON, ITS DISCOVERY & APPLICATIONS

"Whatever you see as duality is unreal" - Adi Shankara

8.1 INTRODUCTION

By now from all the previous chapters one has been able to regard the atomic nucleus solely as a point mass possessing positive charge. Actually, nuclei are quite complex entities, so complex that serious problems still remain in understanding their properties and behaviour. In this chapter some fundamental information about nuclei together

Until 1932, the atom was known to consist of a positively charged *nucleus* surrounded by enough negatively charged *electrons* to make the atom electrically neutral. Most of the atomic space was empty, with its mass concentrated in a tiny nucleus. The nucleus was thought to contain both *protons* and electrons because the proton (otherwise known as the hydrogen ion, H$^+$) was the lightest known nucleus and because electrons were emitted by the nucleus in *beta decay*. In addition to the β- particles, certain radioactive nuclei emitted positively charged α- particles and neutral γ- *radiation*.

The artificial transmutation of one element in the Periodic Table into another, the dream of alchemists for centuries, was first definitely accomplished by Lord Ernest Rutherford (1919) in a very simple type of experiment with nitrogen.

Twelve years earlier, he had postulated the existence of a neutral particle, with the approximate mass of a proton that could result from the capture of an electron by a proton. This postulation stimulated a search for the particle. However, its electrical neutrality complicated the search because almost all experimental techniques of this period measured charged particles.

8.2 DISCOVERY OF NEUTRON

In 1928, Walter Bothe and Herbert Becker took the initial step in the search for the neutral particle. They bombarded beryllium with α- particles emitted from a sample of RaC' ($^{214}_{84}$Po) and found that it gave off a highly penetrating radiation (Fig 8.1). There are two types of conclusion for an explanation of this phenomenon

8.2.1 γ-ray explanation:

Bothe & Becker ascertained that the radiation was electrically neutral and naturally interpreted to be high-energy γ- photons. The energy measured was □ 7 *MeV*.

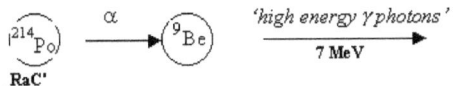

Fig 8.1 Bothe & Becker experiment

$$^9_4Be + {}^4_2He \text{ (of RaC')} \rightarrow \left({}^{13}_6C\right)^* \rightarrow {}^{13}_6C + \gamma\text{-}(\square\ 7\ MeV) \quad (8.2.1)$$

In 1932, Irene Curie and her husband Frederic Joliot (Mme. Curie-Joliot) found that this radiation ejected protons from a paraffin target. This discovery was amazing because photons have no mass. However, they interpreted the results as the action of photons on the hydrogen atoms in paraffin (and other hydrogenous materials).

i) They used the analogy of the (γ-, p) nuclear **Compton Effect**. It was possible to calculate the minimum γ-ray energy ($E_\gamma = h\nu$) needed to transfer the KE (T) to a proton, head-on-Compton scattering ($\varphi = 180°$, for maximum energy transfer) gives

$$2 M_p c^2 (h\nu - h\nu') = 2 ((h\nu)(h\nu'))(1 - \cos\varphi)$$

From this $M_p c^2 T = 2 E_\gamma (E_\gamma - T)$ \hfill (8.2.2)

With $M_p c^2 = 938$ MeV for protons, T = 5.3 MeV. This means $E_\gamma = 53$ MeV.

Thus the minimum γ-ray energy is $E_\gamma = 53$ MeV !

This result appears to be peculiar, since no nuclear radiation known at that time had such high values as $E_\gamma = 53$ MeV. Mme. Curie-Joliot showed that the radiation excited in beryllium possessed a penetrating power distinctly greater than that of any $\square$-radiation yet found from the radioactive elements.

ii) The peculiarity became even more striking if the presumed reaction

$$^9_4Be + {}^4_2He \text{ (of RaC')} \rightarrow \left({}^{13}_6C\right)^* \rightarrow {}^{13}_6C + Q \quad (8.2.3)$$

where 5 MeV α-particle is used. Using the masses of the ^9_4Be, $^{13}_6C$ and ^4_2He nuclei,

$$Q = [9.015030 + 4.003879 - 13.007505]\ u$$
$$= (0.011404\ u)(931.2\ MeV\ c^{-2})$$
$$= 10.4\ MeV$$

Therefore the total energy used is 15.4 MeV, which about 33% of $E_\gamma = 53$ MeV estimated earlier and is needed by the γ-ray if it is to knock out 5.3 MeV protons out of paraffin. Now one knows that γ-photons do not have enough energy to eject protons from paraffin. Thus no consistent results is obtained, for E_γ to be compatible with energy and momentum conservation.

8.2.2 Neutron explanation

Being dissatisfied with the γ-ray explanation, seen above, Sir James Chadwick (1932) performed a series of experiments (Fig 8.2) on the recoil of nuclei which underwent impact from radiations coming

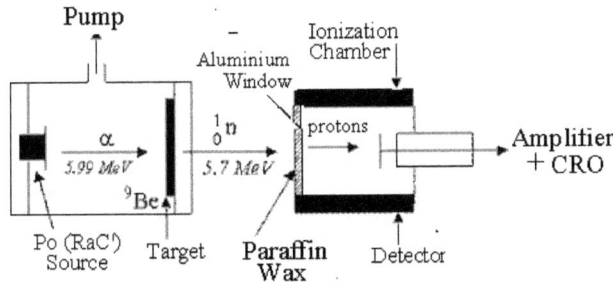

Fig 8.2 Chadwick's experimental set up

from $^{9}_{4}Be$ bombarded by α- particles. He assumed that neutral particles, called **_neutrons_**, represented symbolically $^{1}_{0}n$, are formed as result of the reaction (8.2.4):

$$^{9}_{4}Be + ^{4}_{2}He \; (+ E_1 = 5.99 \; MeV) \rightarrow \left(^{13}_{6}C\right)^* \rightarrow ^{12}_{6}C + ^{1}_{0}n + E_2 \qquad (8.2.4)$$

The results of these experiments are:

a) When neutrons $^{1}_{0}n$ from Be sample travel directly into the detector, a few counts per minute are recorded. (due to secondary ionization by N atoms in air colliding with $^{1}_{0}n$). For a distance of 3 cm between the beryllium and the detector the number of counts was nearly 4 per minute

b) A thin lead (Pb) sheet, even as much as 2 cm thick was interposed between the source vessel and the window of detector, and in front of the detector does not reduce the counting rate appreciably.

c) However, a thin sheet (about 2 mm. thick) of paraffin wax was interposed in the path of the radiation just in front of the window of the detector, the number of counts recorded increased markedly.

By placing absorbing screens of aluminium between the wax and the counter one can plot an absorption curve. From this curve it appears that the particles have a maximum range of just over 40 cm. of air, assuming that an Al foil of $1.64 \; mg \; cm^{-2}$ is equivalent to 1 cm. of air. By comparing the sizes of the number of counts (proportional to the number of ions produced in the chamber) due to these particles with those due to protons of about the same range it was obvious that the particles were protons. From the range-velocity curve for protons one can deduce therefore that the maximum velocity u_{pmax} imparted to a proton by the beryllium radiation is about $3 \cdot 3 \times 10^9 \; cm \; s^{-1}$, corresponding to an energy of about $5 \cdot 7 \; MeV$.

Using the momentum and energy conservation laws, it is known that the velocity u_p of the recoiling nucleus (proton or nitrogen) M_p is given by equation (7.1.5)

$$\begin{pmatrix} \text{Recoil velocity} \\ \text{of target nucleus, } u_p \end{pmatrix} = 2 \begin{pmatrix} \text{Initial velocity} \\ \text{of ray particle} \end{pmatrix} \left(\frac{\text{At. wt. of ray particle}}{\text{At. wt. of Nucleus + At. wt. of ray particle}} \right) \quad (82.5)$$

i.e, $\quad u_p = 2 V_n \cos\theta \; \dfrac{M_n}{M_n + M_p} = u_{pmax} \cos\theta \qquad (8.2.6)$

Maximum velocity of recoil nucleus u_{pmax}

$$u_{p\max} = \frac{2M_n}{M_n + M_p} V_n \qquad (8.2.7)$$

Chadwick then used the Wilson Cloud Chamber (see Chapter 9) and filled it with nitrogen gas and performed the experiment above to measure the range and then

$$u_{N\max} = 0.47 \times 10^7 \, m\, s^{-1}$$

Now $M_N \approx 14.1 \, M_p$, and so

$$\frac{u_p}{u_N} = \frac{M_n + \text{At. wt. of ray particle}(M_N)}{M_n + \text{At. wt. of ray particle}(M_p)} = \frac{3.3 \times 10^7 m\, s^{-1}}{4.7 \times 10^6 m\, s^{-1}}. \qquad (8.2.8)$$

yields $M_n = 1.16 M_p$. $\qquad (8.2.9)$

The nuclear reaction can be written as

$$^9_4\text{Be} + {}^4_2\text{He} \, (+E_1 = 5.99 \, MeV) \rightarrow \left({}^{13}_{6}\text{C}\right)^* \rightarrow {}^{12}_{6}\text{C} + {}^1_0 n \, (+E_2 = 5.7 \, \text{MeV})$$
$$(8.2.10)$$

J. Chadwick and Maurice Goldhaber (1934) using photo-disintegration of deuterium

$$^2_1\text{H} + \gamma \rightarrow {}^1_1\text{H} + {}^1_0 n \qquad (8.2.11)$$

could determine
$$M_n = 1.008665 \, u$$

8.2.3. Different types of Neutrons:
1) Epi-thermal neutrons have energy 1 eV,
2) Slow neutrons have □ 1 keV and
3) Fast neutrons have 100 keV - 10 MeV).

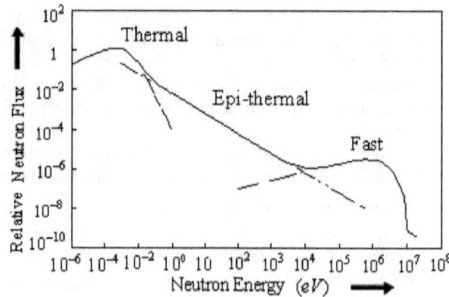

Fig. 8.3. A typical reactor neutron energy spectrum showing the various energy regions.

The energy spectrum of thermal neutrons at room temperatures, is best described by a Maxwell-Boltzmann distribution with a mean energy of 0.025 eV and a most probable velocity of 2200 ms^{-1}. Neutrons (energies from 0.5 eV to about 0.5 MeV) which have been only partially moderated, and are referred to as the epithermal neutron component. A cadmium foil 1 mm thick absorbs all thermal neutrons but will allow epithermal and fast neutrons above 0.5 eV in energy to pass through (Fig 8.7). $\qquad (8.2.12)$

8.3 SOURCES OF NEUTRONS

There is no naturally occurring radio-isotope that emit neutrons.

8.3.1 Radio-active neutron Source: α-Be Source

A mixture of beryllium (or Boron) with α – emitting ^{212}Po (RaC')(or ^{228}Ra

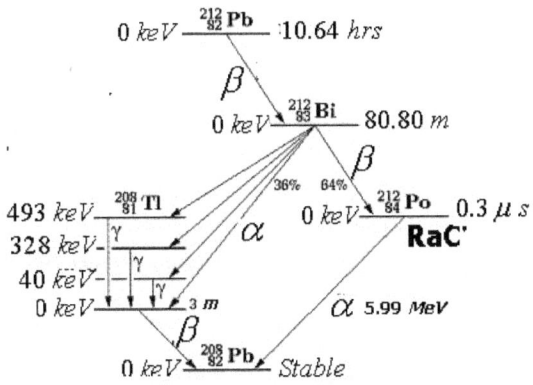

Energy Level Scheme for RaC'

Fig 8.4 Energy level scheme for RaC'

or ^{239}Pu) is a common source of neutron beam. Fig 8.4 shows the energy level scheme and decay of Po. Fig 8.5 displays the continuous spectrum of the emitted neutrons.

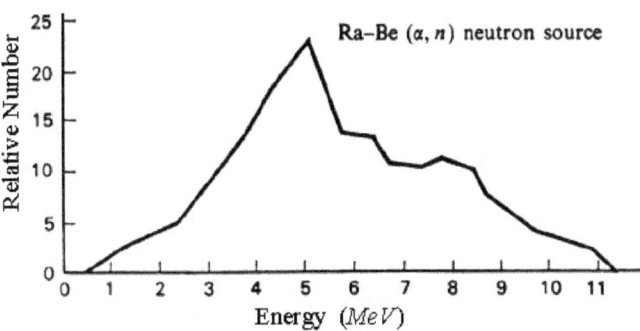

Fig 8.5. The continuous spectrum of neutrons

The reaction that follow are:
$^9_4$Be (α, n) $^{12}_6$C
$^{11}_5$B (α, n) $^{14}_7$N

8.3.2 Reactor Source

A nuclear reactor is a source of neutrons.

8.3.3 Neutron Booster
In the neutron booster, certain features of a particle accelerator and a nuclear reactor are combined.

8.3.4 Photo-neutron Source
(γ, n) reaction (8.3.1) gives neutrons.

$$\gamma + {}^{9}_{4}Be \rightarrow {}^{8}_{4}Be + {}^{1}_{0}n;$$

i.e., ${}^{9}_{4}Be\,(\gamma, n)\,{}^{8}_{4}Be$ \hfill (8.3.1)

Here radio-isotopes with single γ-ray gives mono-energetic neutron source.

8.3.5 Spallation Neutron Source (SNS)

A high-energy particle (say, proton),(green sphere)) when bombards a heavy atomic nucleus, some ${}^{1}_{0}n$ (red spheres) are knocked out, in a nuclear reaction called *spallation*. This is shown in Fig 8.6. The temperature of the bombarded nucleus becomes high and so other ${}^{1}_{0}n$ are "boiled off." For each event, *i.e.*, proton striking the nucleus, 20 to 30 ${}^{1}_{0}n$s are expelled. It is known that a beam of 500 – 1000 MeV protons (a current of 65 mA is formed) when strike a stream of a liquid alloy of Pb and Bi, a continuous beam of 10^{16} thermal (0.025 eV) neutrons per *sec* is being produced.

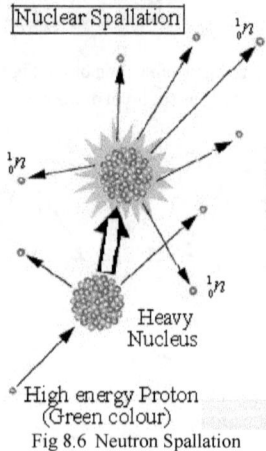

Fig 8.6 Neutron Spallation

For more details the reader may refer to the article "Pulsed Neutron Scattering for the 21st Century" by Thomas Mason (Physics Today, p44 -49, May 2006.).

8.3.6. Particle accelerator source
Neutrons are generated from a particle accelerator by means of nuclear reactions such as D-T, D-N, P-N.

Eg., ${}^{3}_{1}H\,(d, n)\,{}^{4}_{2}He$- (Q-value = 17.6 MeV) =14.1 MeV neutrons

Neutrons having the thermal energies can not be distinguished from gas molecules at the same temperature. They follow the Maxwell-Boltzmann distribution. The fraction of neutrons in the Energy E is

$$f(E) = \frac{2\pi}{(\pi k_B T)^{3/2}} e^{-E/k_B T} \sqrt{E} \qquad (8.3.2)$$

k_B = Boltzman constant,
T = Temp in K,
So the average energy = $E = \frac{3}{2} k_B T$

8.4 PROPERTIES OF NEUTRONS

8.4.1 Mass of neutron, M_n :

$$M_n = 1.00866490 \, u = 1.67493 \times 10^{-27} \, kg = 939.5656 \, MeV \, c^{-2} \qquad (8.4.1)$$

One very exact experiment for determining the mass of a neutron carried out by R.E. Bell & L.G. Elliott, depends on the precise knowledge of the masses of $_1^1H$ and $_1^2H$ precisely and studying the γ-ray production from

$$_0^1n + {_1^1}H \rightarrow {_1^2}H + h\nu. \qquad (8.4.2)$$

8.4.2 Charge of neutron:
Neutron is electrically neutral, and so its charge is zero.

8.4.3 Intrinsic spin of neutron: $s = \frac{1}{2}\hbar$

Neutron spins about an axis through its own centre of mass, and has intrinsic spin angular momentum $s = \frac{1}{2}\hbar$.

8.4.4 Magnetic moment of Neutron μ_n

Spinning hypothesis of charged particles accounts for their magnetic moments, since they are treated as tiny magnetic dipoles. It is very surprising that neutron, though carries no net electric charge, has a magnetic moment! It is usually considered that neutron spends part of its existence dissociated into a proton and a negatively charged π^--meson, according to the equation

$$_0^1n \; \square \; p^+ \, (_1^1H) + \pi^- \qquad (8.4.3)$$

During the dissociation period there could be a relatively + ve charge surrounded at $\square \, 10^{-15} \, m$ by a cloud of equal negative charge. Measurement has lead to the accepted value of the neutron magnetic moment,

$$\mu_n = -1.9130427 \, nm \qquad (8.4.5)$$

8.4.5 The Decay of Neutron

Enrico Fermi (1934) elaborated the theory of β-decay by visualizing it as the break-up of a neutron into three components; *viz.*, proton, electron and neutrino. This is spontaneous decay of neutron. J.M Robson (1950) determined the half-life of neutron.

$$\boxed{\tau_{1/2} = 12.8 \pm 2.5 \text{ minutes}} \qquad (8.4.6)$$

8.4.6 Wave nature of Neutron
The de Broglie wavelength

$$\lambda = \frac{h}{m v} \qquad (8.4.7)$$

Slow and thermal (cold) neutrons have a Maxwellian velocity distribution with average value corresponding to 0.025 eV, the Equi-partition energy of cold neutron

$$\boxed{\lambda \ (cold \ \text{neutron}) \approx 2 \ \overset{o}{A} \ (\text{or } 0.2 \ nm)}. \quad (8.4.8)$$

$$\boxed{\lambda \ (fast \ \text{neutron}) \approx 0.0003 \ \overset{o}{A}} \quad (8.4.9)$$

8.4.7 Electric Dipole Moment of Neutron

Certain theories of nucleus suggest that weak electric dipole moment should exist for a neutron. Experimentally not yet measured, it can be expressed as

Dipole moment $= (q_n) \Box (x < 2 \times 10^{-24} \ m)$ \quad (8.4.10)

The charge on a neutron is deduced to be

$$\boxed{q_n < \tfrac{1}{700} \ e}. \quad (8.4.11)$$

8.4.8. Neutron Scattering Cross section.

8.5 DETECTION OF NEUTRONS

Neutron does not produce ionization when it passes through matter. But it takes part in some nuclear reaction with an atomic nucleus and it is the products of this reaction which can be detected. The basis of several different kinds of neutron detector is the reaction

$$^{10}_{5}B + \ ^{1}_{0}n \ \rightarrow \ ^{7}_{3}Li + \ ^{4}_{2}He + 2.7 \ MeV \quad (8.5.1)$$

8.5.1 Boron trifluoride Counter

It contains a cylindrical chamber filled with gas of BF_3, enriched 90% $^{10}_{5}B$, and an electric potential of $\Box 2.5 \ kV$ between the cathode (axial wire) and anode (chamber body). The α-particles produced by n-capture causes intense ionization, in the gas, on which the action of the counter depends.

8.5.2 The $^{3}_{2}He$ Detector

This detector depends on the reaction

$$^{3}_{2}He + \ ^{1}_{0}n \ \rightarrow \ ^{3}_{1}H + \ ^{1}_{1}H \quad (8.5.2)$$

where $^{3}_{2}He$-gas is obtained fro irradiation of $^{6}_{3}Li$ in a reactor. In this detector ionization is caused by the protons.

8.5.3 There are other neutron detectors, involving 3 other different basic principles.

8.6 INTERACTION OF NEUTRONS WITH MATERIALS

These fall under three categories:
1. Neutron cross sections,
2. Macroscopic cross section
3. Neutron diffraction,

Neutron Scattering Cross section:

8.6.1 NEUTRON DIFFRACTION

Like X-rays, electrons and neutrons are also used in diffraction methods to characterize materials Quantum mechanics has taught us that a particle (say neutron) in motion is associated with it a wave. This means the energy of a neutron is related to the deBroglie wavelength by

$$\lambda (\overset{o}{A}) \cong \frac{0.28}{\sqrt{E(eV)}} \qquad (8.6.1)$$

where $E = \frac{\hbar^2}{2 M_n \lambda^2}$. It is known so that λ ($E \cong 0.08\ eV$) $\approx 1\ \overset{o}{A}$.

This wavelength of the neutron is comparable with the lattice spacing. Further because neutron has magnetic moment it can interact with magnetic electrons in a solid. Neutrons are therefore diffracted by crystal lattice in directions quite different for the incident direction. Clifford Shull and Bertram Brockhouse (1940s – 1950s) performed their pioneering elastic and inelastic diffraction experiments. Brockhouse, for example, investigated the dispersion, *i.e.*, phonon frequency versus wave vector, in the case of diamond.

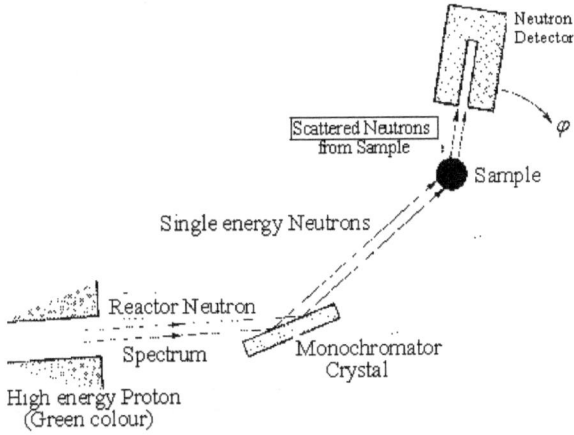

Fig 8.7 Neutron Diffractometer

One can count scattered neutrons, measure their energies and the angles at which they scatter, and map their final position (shown as a diffraction pattern of dots with varying intensities) using detectors.. From this one extracts details about the nature of materials ranging from liquid crystals to superconducting ceramics, from proteins to plastics, and from metals to micelles to metallic glass magnets.

The importance of neutron scattering (Fig 8.7) was recognized by the awarding of the 1994 Nobel Prize for Physics to Clifford Shull and Bertram N. Brockhouse. Shull pioneered the use of neutron scattering to decipher the structure of materials, and Brockhouse found ways to use it to learn about the motions of atoms in materials. The block diagram of a neutron diffracto meter is shown in Fig 8.7.

8.6.2 Neutron Activation Analysis (NAA)

Neutron Activation Analysis (NAA) is an analytical nuclear process useful for determining concentration of major and minor trace elements both qualitatively and quantitatively in samples any field of scientific or technical interest.

George de Hevesy and Hilde Levi (1936) found that samples containing certain rare earth elements became highly radioactive after exposure to a source of neutrons. They came to the conclusion that the potential of employing nuclear reactions on samples and subsequent measurement of the induced radioactivity in them (giving isotopes) would facilitate both qualitatively and quantitatively characterize the elements present in the samples.

The basic essentials required for NAA are a source of neutrons, instrumentation suitable for detecting γ-rays, and a detailed knowledge of the reactions that occur when neutrons interact with target nuclei (Fig 8.8).

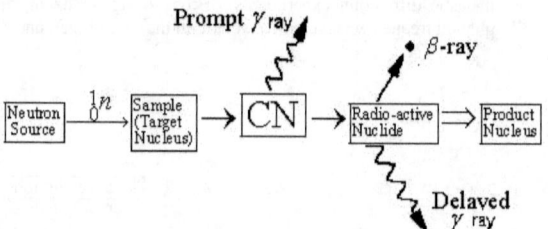

Fig. 8.8. The NAA method: illustrating the process of (n, γ) reaction followed by the emission of γ-rays.

The sequence of events are, 1) the neutron capture or (n, γ) reaction, 2) As a result of a neutron interacting with the target nucleus by non-elastic collision, a compound nucleus (CN) forms in an excited state. Its energy of excitation is due to the binding energy of the neutron with the target nucleus. 3) The compound nucleus instantaneously de-excite into a more stable configuration through emission of one or more characteristic prompt γ-rays, giving often a radioactive nuclide. 4) The radio-active nucleus decays by emission of one or more characteristic delayed γ-rays, but at a much slower rate decided by the half-life (fractions of a second to several years) of the radioactive nuclide

Therefore, based on the time of measurement, there are two types of NAA:

(i) PGNAA: *Prompt* γ-ray neutron activation analysis, where measurements are performed during irradiation, or

(ii) DGNAA: *Delayed* γ-ray neutron activation analysis, in which the measurements follow radioactive decay.

The latter operational mode is more common. About 70% of the elements have properties suitable for measurement by NAA.

An NAA technique that employs only epithermal neutrons to induce (n, γ) reactions by irradiating the samples being analyzed inside either cadmium or boron shields is called *Epithermal Neutron Activation Analysis* (ENAA).

8.6.2.1 Reactor Neutron Energy Spectrum

Of the several types of neutron sources (reactors, accelerators, and radio-isotopic neutron emitters) one can use for NAA, nuclear reactors with their high fluxes of neutrons from uranium fission offer the highest available sensitivities for most elements.

γ- rays from radioactive samples can be measured using the instrumentation consists of a semiconductor detector (see Chapter 9) and associated electronics

8.7 APPLICATIONS OF NAA

The Neutron Activation analysis is a very important tool of then century as it has found application in most area of science and technology.
1. Archaeology Characterization of archaeological specimens (*e.g.,* pottery, basalt and limestone) and to relate the artifacts to sources through their chemical signatures is a well-established application of the NAA. Data base contains more than 50,000 analyses.
2. Investigation of the redistribution of Uranium and Thorium due to Ore Processing
3. Study the Fate of hazardous elements in waste material/coal-char admixtures under Gasification--an Emerging Waste Management Technology, employing radio-tracers
4. Aquatic Species containing selenium in Selenium-contaminated Fresh-water Impoundments
5. *In-situ* Radiotracers for Dosage-Form Testing
6. Nutritional Epidemiology [Nutritional and Biochemical/Genetic Markers of Cancer]
7. Nutritional Epidemiology [A Cohort Study of the Relationship Between Diet],
8. Nutritional Epidemiology [Thyroid Cancer Study]
9. Nutritional Epidemiology {Non-Melanoma Skin cancer Study}
10. Nutritional Epidemiology (Molecular epidemiology of prostate cancer)
11. Knock-out Gene Mouse Model for Cystic fibrosis
12. Study of Calcium Metabolism
13. In Geological science
14. Study of Semiconductor materials and other high-purity materials
15. In the field of Soil Science

8.8 NEUTRON INTERACTIONS

It has been known that neutrons have no charge and have a mass slightly higher than that of a proton. Neutrons essentially interact only with the atomic nucleus (elastically and inelastically), and do not directly with electrons, because the nucleus of an atom is 10,000 times smaller than the electron cloud surrounding it. Therefore, the chance of neutrons interacting with a nucleus is very small, allowing neutrons to travel long distance through matter before interacting. (Fig 8.9).

All neutrons are initially fast neutrons which lose kinetic energy in collision with atoms in their environment until they become thermal neutrons which are captured by nuclei in matter

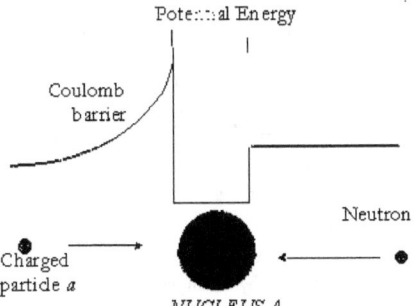

Fig 8.9 A particle approaching a nuclear potential

Free neutrons are unstable and will disintegrate in about 10.6 minutes by beta minus decay to a proton and electron if they do not interact with matter.

A threshold exists for neutrons to interact. This value is infinity for Hydrogen, -6 MeV for Oxygen, and < 1 MeV for Uranium.

8.8.1   Elastic Scattering of Neutrons

Conservation of momentum is a particularly useful physical concept when examining collisions between two or more objects taking place in a limited time and space. They interact one another in that region. That is these are collisions involving loss of momentum and energy to the colliding objects. In atomic physics theses collisions are referred to as scattering. The process scattering can be either elastic or inelastic.

In the case of scattering between impinging neutrons and target nuclei (which recoil) elastic scattering is known to lower the kinetic energy of (slow down) the outgoing neutrons ($_0^1 n$). The recoil nuclei quickly become *ion pairs* and loose energy through excitation and ionization, if the material that they pass through is biological, and is a primary mechanism by which living tissues are damaged by the neutrons. The amount of the energy transfers to a nucleus when struck by a neutron increases as the mass of the target approaches a neutron.

Consideration of the Laws of Conservation of Energy and Momentum lead to result expressed mathematically.

$$E = E_o \left\{ \frac{M - m}{M + m} \right\}^2 \qquad (8.8.1)$$

where E = Energy of scattered neutron
E_o = Initial energy of neutron
M = Mass of the scattered nucleus
m = Mass of neutron

The neutron, being slightly heavier than a proton, hydrogen is the element whose mass closely approximates that of a neutron. In neutron-hydrogen collisions the average energy transfer to the hydrogen nucleus is about ½ the energy originally possessed by the neutron. On the other hand, if the target was oxygen (mass 16 u) then the scattered neutron would retain 77% of its initial energy.

The role of hydrogenous materials or paraffin is very important for slowing down neutrons (See the role of Wax in Fig 8.2).

8.8.2   Inelastic Scattering

In may happen that a neutron may be captured or absorbed by the target nucleus. The resulting nucleus will then emit a neutron $_0^1 n$ of lower kinetic energy along with a γ – photon or a particle. With increasing energy of the neutron as well as of the size of the target nucleus, the probability of this type of interaction increases. Mathematically, the energy transferred to the target nucleus and emitted energy are described by the equation (8.8.2):

$$E = E_o - E_\gamma \qquad (8.8.2)$$

where E = Energy of the neutron after collision
E_o = Initial energy of the neutron. ($_0^1 n$)
E_γ = energy of the emitted γ – particle
An illustrative example is

$$_4^9 Be - {}_2^4 He \rightarrow {}_6^{12}C + {}_0^1 n \qquad (8.8.3)$$

Consider neutron with E_o = Initial energy of neutron
M = Mass of the scattered nucleus
m = Mass of neutron

V = Incident velocity of neutron before collision,
Applying the Laws of conservation of total kinetic energy and momentum,

$$\tfrac{1}{2}MV^2 = \tfrac{1}{2}MV_1^2 - \tfrac{1}{2}mv_1^2 \tag{8.8.4}$$

$$MV = MV_1 + mv_1 \tag{8.8.5}$$

whence $V_1 = \dfrac{M-m}{M+m} V$ (8.8.6)

$$E_o = \tfrac{1}{2}MV^2, \tag{8.8.7}$$

$$E = \tfrac{1}{2}MV_1^2 \tag{8.8.8}$$

which yields $E = E_o \left[\dfrac{M-m}{M+m} \right]^2$. (8.8.9)

Energy transferred to the target is $[E_o - E]$,

$$[E_o - E] = E_o \left\{ 1 - \left[\dfrac{M-m}{M+m} \right]^2 \right\} \tag{8.8.10}$$

Capture cross section

If σ_0 = capture cross section of neutron having velocity v_0 and energy E_o, then the capture cross section σ of the neutron at velocity v and energy E is given by

$$\dfrac{\sigma}{\sigma_0} = \dfrac{v_0}{v} = \sqrt{\dfrac{E_0}{E}} \tag{8.8.11}$$

8.8.4. Absorption of Neutrons by Matter

$$I = I_0 \, e^{-\sigma N \, t} \tag{8.8.12}$$

N = Number of atoms/cm^{-3} in the absorber
t = thickness of the absorber
σ = Microscopic cross section of neutrons
Σ = Macroscopic cross section
$\Sigma = \sigma N$ (8.8.13)

8.8.3   How does Neutron scattering affect our lives?
The fruits of neutron-scattering research are production of better materials by improving the range and quality of products used in everyday lives. For example
i) Neutrons have been used to study how bones mineralize during development and how they decay during osteoporosis, and they make it possible for us to devise and to test remedies for de-mineralizing diseases.
ii)   One in three (or about 100 million each year) of the patients benefit from isotopes produced by neutrons.
iii)   Neutrons enable production of improved polymers for the plastics used in many products, such as compact discs.

8.8.3   Neutron Capture

When a neutron approaches a nucleus (in the absorbing material) close enough for nuclear forces to be effective, the process called neutron capture may occur. The neutron is captured and forms a heavier isotope of the capturing element. In the case of an unstable isotope the neutron undergoes β – decay to give a proton as follows:

$$_0^1 n \rightarrow p^+ \left(_1^1 H\right) + e^- + \nu$$

If the energy of the neutron is known, the probability of capture by a specific nucleus can be defined by a term called "Capture Cross Section" which is expressed in Barns ($1b = 10^{-24}\ cm^2$). The capture cross section is different not only for each target nucleus, but also for each isotope of the target nucleus and for each energy level of neutron. Probability of capture being inversely proportional to the energy level of the neutron, the thermal neutron has the highest probability capture. The cross section of nuclei varies greatly from almost 0 for helium to $2.5 \times 10^6\ b$ for a Xenon nucleus.

REVIEW QUESTIONS

RQ. 8.1 Describe the experiments that led to the discovery of the neutron, indicating how its mass has been determined..

RQ. 8.2 Describe the experimental results that led to the discovery of neutrons. What are the various sources of neutrons?

RQ. 8.3. It was seen that a 10 MeV neutron beam incident on a 1 cm thick lead slab perpendicularly the out coming beam had 84.5% of its initial value. What is the total cross section of neutrons? Given, lead has A = 207.21 and density 11.3 $gm\ cm^{-3}$. $N_0 = 6.03 \times 10^{23}\ gm\ mole^{-1}$
(Ans: $\sigma = 5.1\ b$; $\Sigma = 0.168\ cm^{-1}$).

RQ. 8.4: How do neutrons get diffracted from a crystal? With a schematic diagram, describe a neutron diffractometer.

RQ. 8.5: Describe briefly elastic neutron scattering.

RQ. 8.6: Describe the various properties of neutrons.

RQ. 8.7: Write short notes on
  a) Types of neutrons.
  b) Spallation
  c) Neutron diffraction
  d) Applications of neutrons
  e) Elastic scattering of neutrons
  f) Neutron capture
  g) Neutron Detectors
  h) Neutron moderator.

RQ. 8.8. Given the cross-section of for the $^{10}B(n, \alpha)^7Li$ reaction is 753 barns for thermal (0.025 eV) neutrons. What is the cross-section at 50 eV? (Ans 16.8 b)

RQ. 8.9. Calculate the most probable energy and the average energy and the velocity of a neutron at 293 °K (Ans: 0.025 eV, 0.037 eV; 2.2 kms^{-1}).

R.Q. 8.10. A beam of thermal neutrons traversing through a foil of $_{79}^{197}$Au having thickness 2 mm emergent with about 70% of the neutrons. Given the radius of gold to be $r = 6.50\ fm$ and $\rho = 19300\ kg\ m^{-3}$, $N_A = 6.0225 \times 10^{23}\ mol^{-1}$, determine the total thermal neutron cross section for $_{79}^{197}$Au. (Answer: $\sigma_T = 102\ b$).

%%*%*%*%*%*%*

Chapter 9
PARTICLE DETECTORS

Chapter 9

PARTICLE DETECTORS

"For the wise all 'things are wiped away" - Buddha

"Get the glass eyes; / And like a scurvy politician seem to see those things thou dost not" - Shakespeare in King Lear

9.1 INTRODUCTION

Sub-atomic particles are extremely small and can not be observed directly by normal means. In the history of Nuclear and Particle physics there has been development of many different types of detectors. All these are based on the fundamental principle of the transfer of part or all of the energy of radiation to the detector mass, where it is converted into some form accessible to the physicist working in the laboratory. Modern detectors are essentially electrical in nature, and before the detector the information is transformed into electrical impulses that can be treated electronically.

A somewhat brief description of the various detectors used for different types of radiation will be presented. The interaction of radiation (to be detected) with matter (viz., ionization / excitation) is made use of in the working principle of variety of detectors. These detectors fall under different groups given below:

a) Gas counter:- i) Simple Ionization Chamber, ii) Proportional Counter, iii) Geiger-Muller (GM) Counter.
b) Conventional electronic Detectors,
c) Visual detectors:- i) Wilson's Cloud Chamber, ii) Diffusion Chamber, iii) Nuclear Emulsion Detector.
d) Scintillation Detector.
e) Solid State (Semi-conductor) Detectors.
f) High energy Detectors:- i) Bubble Chamber, ii) Spark Chamber, iii) Cerenkov Detector.
g) Semi-conductor Detectors, and
h) Charge Coupled Detectors (CCD).

9.2 GAS IONIZATION CURVE

A characteristic curve showing the number of ion pairs collected per event against applied voltage (V) between the two electrodes in a gas filled chamber is illustrated in Fig. 9.1. Consider that ß –radiation passes through the chamber. When the potential (V) between anode and cathode is relatively low, it is insufficient to have much effect on the *primary* electrons and positive ions produced and most *recombine*; the current flow is extremely low, till point A. As V is increased, at first some, there is no appreciable amount of recombination. The charges collected, but have no further effect on any gas molecules that they encounter in their passage, as the electrons and the positive ions migrate to the anode and cathode, respectively, with positive ions moving more slowly than the electrons as they are larger and heavier.

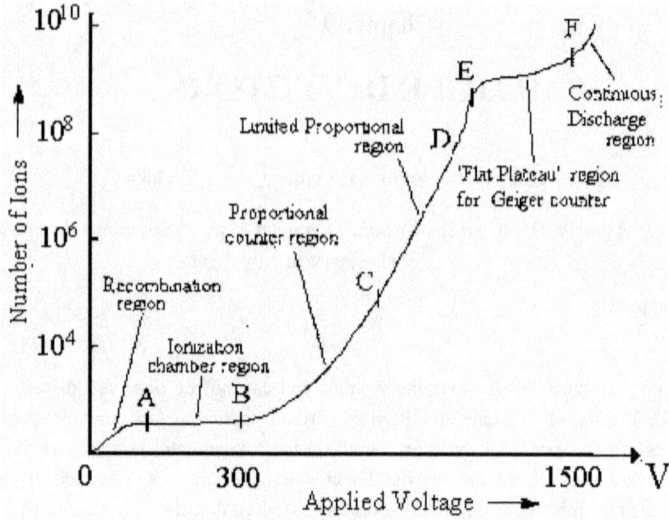

Fig 9.1 Characteristics of a Radiation Counter Tube showing the relative number of ions *versus* Voltage applied between the electrodes

Even with appropriate electronics (like a vibrating reed electrometer), the very weak current at the anode, is difficult to measure till point A in the curve (Fig 9.4). The initial region of the curve is termed "**recombination**". The next region, *viz.*, region between A and B, is known as the "*ionization chamber region*". The third region, *i.e.* ,relatively *flat plateau* begins at the potential where the field strength is able to prevent recombination, and it gets terminated when the potential V is so increased as to further affect the counting gas.

That effect is to create a field with sufficient strength to impart additional energy to the primary electrons. As they migrate to the anode they are able to ionize other gas molecules. These *secondary* electrons produce still more ionization. Each primary electron cause several secondary one, and "gas amplification" results. Thus one gains access to more substantial pulses from each decay event.

This region B to C is known as the "*proportional region*" because the pulse size is, in fact, proportional to the decay energy. It is also proportional to the potential V and leads to ultimately a larger pulse. The greater the energy of ß -particles, the more primary ionizations while the greater the potential, the more secondary ionizations.

As the potential V is increased further to point E, the entire contents of the counting tube are ionized with each event. This "*Geiger region*" is E to F. The pulse developed at the anode wire is thus the largest we can expect from a normally operating system. It is essentially the same size for all decay events no matter what the energy of the event. A plateau is seen and it may extend over 100-150 V or more since small voltage changes above the knee cannot cause more ionization. The transition from the purely proportional region to the *GM region*, *viz.* D to E, is ill-defined and is known as the "*limited proportional region*".

9.3 GAS COUNTERS

Gas filled detectors (or counters) can be designed to detect any type of ionizing radiation (α-, β-, γ-, and n). They may be filled with air and open to the atmosphere or they may be filled with a specific gas [like Boron trifluoride (BF_3) gas for neutron detection] and sealed These consist of a gas filled metal chamber (typically a cylinder) with a wire passing through the center of the chamber

9.3.1 SIMPLE IONIZATION CHAMBER

The principle used in this detector is that the radiation to be detected causes ionization in the gas, and the free charges are then colleted and measured. Relatively low voltage is applied between the electrodes, as indicated in Fig 9.1. The chamber contains a small volume of air or hydrogen at atmospheric pressure. The dimensions of the chamber are usually determined by the range in the Counter gas of the particle to be detected.

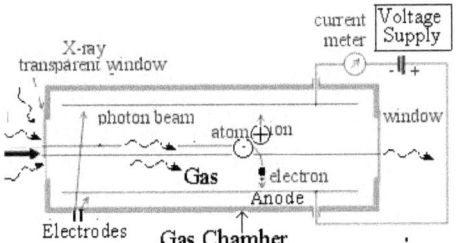

Fig 9.2 Ionization Chamber

They may be filled with air and open to the atmosphere or they may be filled with a specific gas (like Boron Trifluoride (BF_3) gas for neutron detection) and sealed. These consist of a gas filled metal chamber (typically a cylinder) with a wire passing through the center of the chamber (Fig 9.2).

9.3.2 PROPORTIONAL COUNTER

Detectors operating in the proportional region shown in the counts versus voltage curve of Fig. 9.1 are called proportional counter. It consists of a wire, commonly tungsten, with a diameter in the region of 0.0025 cm – 0.005 cm mounted along the axis of a metal cylinder (dural, brass, stainless steel) of diameter 5 cm and length 10 – 15 cm, and filled with an inert gas (argon, krypton or xenon) to a pressure of 1 atm (Fig 9.3). A positive potential of 1 – 3 kV is applied to the central wire, via an anode resistance of 10 – 100 $M\Omega$. The wire must be smooth and accurately circular in section so that the electric field is accurately uniform. To the counter gas is added 5 – 10 % of a 'quenching' gas, normally methane or carbon dioxide. This serves two distinct functions: it absorbs UV photons produced in the avalanches and absorbs similar photons released when the positive ions reach the cathode. The system is schematically shown in Fig 9.3., and the associated electronic system in Fig 9.3.

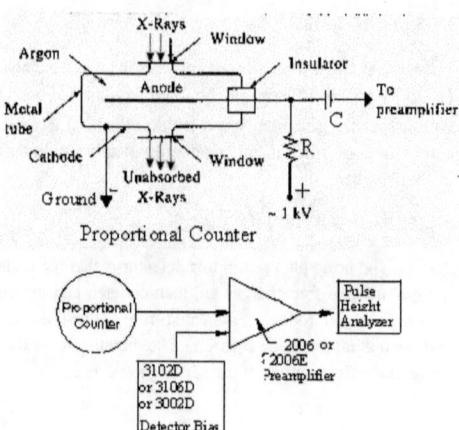

Fig 9.3 Proportional Counter

If a and b denote the radii of the fine wire (anode) and the counter, respectively, the potential difference across the tube is

$$V = (2.3) k \log (b / a) \qquad (9.3.1)$$

k = constant.

Table 9.1 Radial field E versus a values in Proportional counter
$b = 0.01\ m,\ V = 1\ kV$

a (m)	$k = V/[(2.3) \log (b/a)]$	$E = k/a\ (V\ m^{-1} \times 10^{-5})$
10^{-3}	435	4.35
10^{-4}	217	21.7
10^{-5}	145	145
10^{-6}	109	1090
10^{-7}	87	8700

Table 9.1 indicates that the increase in field E is apparent foe the thinner wire.

The counter itself is a cylinder filled with a gas. The gas is separated from the vacuum of the instrument by a thin window. The X-ray photon enters through this window where it knocks out electrons from the electron shells of the gas molecules. These electrons are attracted to a wire running down the center of the detector, which has a positive 1-3 kV potential on it. The electrons hitting the wire produce a current which is picked up by the preamp and then turned into a signal. The amount of current is proportional to the number of X-rays that are entering the detector.

9.3.3 GEIGER MULLER (GM) COUNTER

GM counter is similar to the proportional counter, having a hollow metallic cylinder electrode and an anode in the form a thin wire, enclosed in a thin glass tube; the voltage applied, however, in this tube operates only in the Geiger region. A mixture of 90% Argon and 10% some organic vapour (ethyl alcohol) or some halogen (Cl_2, or Br_2) is filled at 10 cm of Hg. Because the tube operates in the "flat region" called the *plateau* of Fig 9.1, it eliminates the need for a highly stabilized power supply (V), which is the biggest *advantage* of the GM counter. X-rays, low energy γ-rays, and β-rays are commonly detected with a GM counter. The self-explanatory schematic diagram of a GM counter is shown in Fig 9.4.

Ionization takes place along the whole length of the GM counter, and so it is not directional (In the proportional counter the process of ionization is directional and localized). The GM counter can not resolve events closer than $10^4 s^{-1}$, because of the 'dead-time' during which the counter is insensitive. There is also a 'threshold voltage' (of the order of a few hundred volts) below which the tube does not work.

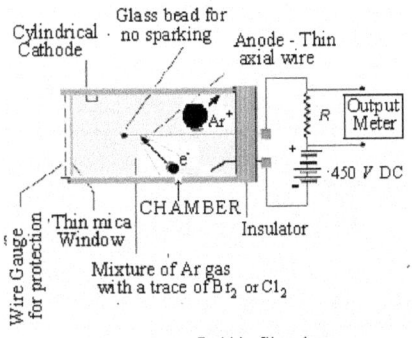

Fig 9.4 GM Counter

9.4 VISUAL DETECTORS

9.4.1 CLOUD CHAMBER

Invented by Charles Thomson Rees Wilson (1911), this was responsible for a number of important discoveries, especially prior to 1950. The working principle of the device is that a *supersaturated vapour* condenses preferentially on charged particles. Dust-free air or nitrogen and saturated water vapour at room temperature fill the Cloud Chamber (Fig 9.5). The system is put in a supersaturated state, by means of adiabatic expansion by the movement of a piston, so

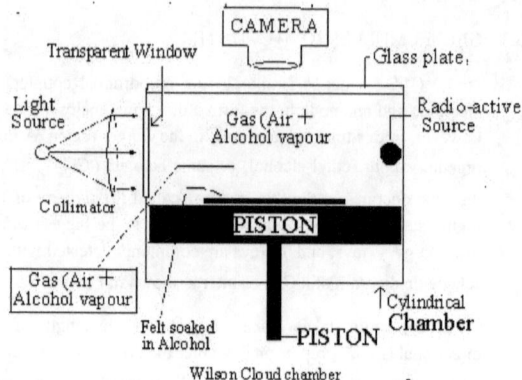

Fig 9.5 Wilson Cloud Chamber

that when charged fast particles passes through the chamber, condensation rapidly begins along the track of particles. The switching on of the illumination and the photography (stereographically) takes place immediately after the expansion before the droplets forming the tracks have dispersed (at < 0.5 s). This kind of chamber is also called a pulsed chamber because the conditions for operation are not continuously maintained.

These tracks have distinctive shapes (for example an alpha particle's track is broad and straight, while an electron's is thinner and shows more evidence of deflection. This track can be measured for range and bent in a magnetic field for energy, particle sign and momentum determination.

9.4.2   DIFFUSION CLOUD CHAMBER

This is a variation on the Cloud Chamber, but is continuously sensitive. This was designed by Alexander Langsdorf in 1936. This chamber differs from the expansion cloud chamber in that it is continuously sensitized to radiation and that the bottom must be cooled to a rather low temperature, generally as cold as or colder than dry ice. Alcohol vapour is also often used due to its different phase transition temperatures. It has a chamber having two compartments. The bottom compartment contains 'dry ice' (solid CO_2), at $-78^{\circ}C$' The upper compartment is such that its top layer is at room temperature and the bottom layer is at $-78^{\circ}C$, causing a temperature gradient in the air medium. The alcohol diffuses downward in the upper chamber as due to cooling. If a charged particle passes through such a layer, it will ionize the air molecules, causing condensation of alcohol vapour on to the air ions.. This shows up as a track.

α – particles give straight tracks, beta particles give heavy torturous tracks and gamma rays display straggly and wispy tracks.

A refinement of the design of Cloud Chamber is the Bubble Chamber.

9.4.3 NUCLEAR EMULSION METHOD

Photographic films and plates are coated with emulsions that contain silver grains. In 'normal' use, they react on exposure to light. But they also react with ionization of their by the passage of charged particles. On development the film then shows dark lines of the passage of particles. In 1930 special *nuclear emulsion* was invented containing higher densities of grains hat are sensitive to all kinds of charged particles. For example, a cosmic ray consisting of a magnesium nucleus collides with a bromine nucleus in photographic emulsion, produces a star of various particles in cosmic ray event. Since the tracks are often very short due to high density of the grain in the emulsion particles come to a quick halt. So a microscope is necessary to view them.

9.5 SCINTILLATION DETECTOR

The range of photon energies, from the negligibly small quanta of radio waves to the overwhelmingly large quanta of cosmic $\gamma-$ rays, is reflected in the wide variety of methods of detecting radiation.

The scintillation counter depends for its operation on the luminescence of certain crystals, when excited by ionizing radiation. Photons are produced in the visible region, which then allowed impinging on the photocathode of a photomultiplier. Multiplication takes place at successive dynodes by secondary emission, causing a negative charge to appear at the anode. This yields a voltage pulse which may be differentiated, amplified, and analyzed in a manner closely similar to that associated with the proportional counter. For fast counting, both organic and polymerized plastic materials are used for the detection of $\beta-$rays. The scintillator, which is thallium activated NaI crystal [NaI(Tl)], is normally coated by an optically reflecting substance (titanium dioxide or magnesium oxide) to ensure maximum channeling of light to the photomultiplier (Fig 9.6). Sodium iodide (NaI) is deliquescent and these crystals are therefore encapsulated in aluminium cans to which a beryllium window is fixed. One face of the crystal is in contact with a glass window in optical contact with the end face of the photomultiplier, using a few drops of oil or transparent grease. The whole assembly is then mounted in a light-tight container. T.

Scintillator is material which will emit photons when struck by high energy charged particles or high energy photons. Photon strikes metal plate, ejecting electrons which are pulled toward 100 V.

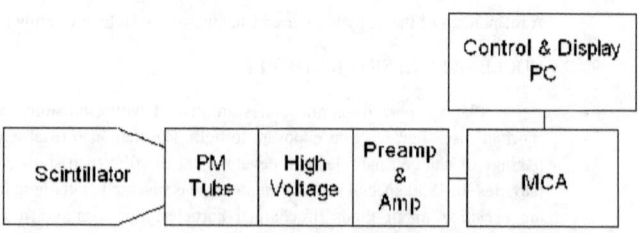

Fig 9.6 Block diagram of a Scintillation Counter

Thallium Activated Sodium Iodide [NaI(Tl)] Detectors

NaI(Tl) detectors incorporate a NaI(Tl) crystal mounted on a photomultiplier tube (PMT). In a simple PMT the electrons from the emitting surface (the photocathode) in a vacuum tube may be collected by a electrode (the anode); the current is then proportional to the rate of incidence of photons. The photoelectrons from the photocathode are accelerated to the dynode which emits several secondary electrons each time a primary electron strikes it. An accelerating potential of typically $1 - 2\ kV$ is required between the anode and the photocathode. Usually a series of from 6 to 14 such successive dynodes are used to multiply the charge by a factor of 10^5 to 10^8 PMTs are widely used for their high sensitivity and time resolution, and for operation at room temperatures.

The NaI(Tl) crystal emits a flash of light proportional to the energy of the gamma ray that interacts with the crystal. The PMT detects this light and amplifies the detected light to yield a proportional quantity of charge. This charge is converted into a voltage pulse by a charge sensitive preamplifier. The pulse can be discriminated by a window type (upper and lower level) Single Channel Analyzer (SCA) and then counted, or these pulses can be displayed as a histogram of pulse heights representing a spectrum of energy

In order to detect the gamma photon we use scintillation detectors. A Thallium-activated Sodium Iodide [NaI(Tl)] detector crystal is generally used in Gamma cameras. This is due to this crystal's optimal detection efficiency for the gamma ray energies of radionuclide emission common to Nuclear Medicine. A detector crystal may be circular or rectangular. It is typically 3/8" thick and has dimensions of 30-50 cm. A gamma ray photon interacts with the detector by means of the Photoelectric Effect or Compton Scattering with the iodide ions of the crystal. This interaction causes the release of electrons which in turn interact with the crystal lattice to produce light, in a process known as scintillation.

9.6 SOLID STATE DETECTORS

9.6.1 SEMI-CONDUCTOR DETECTOR

Both electrons and holes moving within a semiconductor may be detected by applying an external voltage, as in photoconduction, or as a current in the internal electric field of a semiconductor photodiode.

The electron energies within the crystal lattice of a pure semiconductor are almost all constrained to lie within the valence band (VB), where they occupy almost all energy levels $E = \hbar^2 k^2 / 2m^*$, where m^* is the *effective mass* of the electron and k the wave vector Above the valence band is the band gap (BG), and above this gap is the conduction band (CB). The excitation of an electron into the conduction band leaves a hole. In an applied electric field, the electron and the hole move in opposite directions, the electron moving faster than the hole. This is the action of a photoconductor. The quantum efficiency η and the responsivity S (= ratio of output to input current power) vary with the wavelength. In a direct gap semiconductor, for example GaAs, an electron in the CB can make a transition to an empty level in the VB and emit a photon of energy equal to band gap energy. For an indirect band gap case, for example Si or Ge, the transition of the electron from the CB to the VB must involve a change in k, *i.e.*, a phonon. For this reason only the direct band gap materials are efficient light emitters.

Semi-conducting radiation detectors were originally developed for use as detectors of charged particles in the study of nuclear reactions, but have been popular as photon detectors down to energies in the region of 1 *keV*. The most widely used of these detectors is the *surface-barrier detector*. This consists essentially of a p-type Si or Ge wafer (Lithium -drifted) Ge(Li) with a thin gold film evaporated on to the front surface, and operated under reverse bias conditions. In this device the acceptors in the p-type semiconductor are compensated by donor impurity atoms which are drifted into the material by means of a raised temperature and an applied electric field. This causes high resistivity so that noise can be held at the lowest level. Good detection efficiency for γ – radiation up to 1 *MeV* and above is ensured. Theses detectors have to be maintained at continuously at liquid nitrogen temperatures to ensure the lithium atoms to occupy correct positions in the germanium lattice. Si(Li) detectors are well suited for work with low energy γ – radiation.

Fractional energy resolution

$$\frac{\sigma_P}{P} = \left(\frac{FW}{E_\gamma}\right)^{1/2} \tag{9.6.1}$$

P = output pulse height, σ_P is standard deviation, FW =full width.

9.6.2 HIGH-PURITY GERMANIUM DETECTORS (HPGe)

Germanium detectors are extremely large reverse biased diodes. They include charge sensitive preamplifiers that convert the charge deposited by interacting gamma radiation into a voltage pulse. This is typically filtered and amplified for display in a pulse height histogram to exhibit the energy spectrum of the γ-rays collected. A full line of HPGe detectors is available from ORTEC including planar and coaxial configurations. HPGe detectors need to be cooled to near liquid nitrogen temperatures to operate. HPGe detectors far exceed any other detector type for gamma-ray resolution. HPGe technology has grown extensively over the past ten years. Today both laboratory and field units are available with full analysis capability with cryogenic (either liquid nitrogen or with refrigeration cooling) systems.

The instrumentation used to measure γ-rays from radioactive samples generally consists of a semiconductor detector, associated electronics, and a computer-based, multi-channel analyzer (MCA/computer). Most laboratories operate one or more hyper pure or intrinsic Ge (HPGe) detectors which operate at liquid nitrogen temperatures (77 K) by mounting the Ge crystal in a vacuum cryostat, thermally connected to a copper rod or "cold finger". Although HPGe detectors come in many different designs and sizes, the most common type of detector is the coaxial detector which is useful for measurement of γ-rays with energies over the range from about 60 *keV* to 3.0 *MeV*.

The two most important performance characteristics requiring consideration when purchasing a new HPGe detector are resolution and efficiency. Other characteristics to consider are peak shape, peak-to-Compton ratio, crystal dimensions or shape, and price.

The detector's resolution is a measure of its ability to separate closely spaced peaks in a spectrum. In general, detector resolution is specified in terms of the full width at half maximum (FWHM) of the 122-*keV* photopeak of Co-57 and the 1332-*keV* photo peak of Co-60. For most NAA applications, a detector with 1.0-*keV* resolution or below at 122 *keV* and 1.8 *keV* or below at 1332 *keV* is sufficient.

Detector efficiency depends on the energy of the measured radiation, the solid angle between sample and detector crystal, and the active volume of the crystal. A larger volume detector will have a higher efficiency. In general, detector efficiency is measured relative to a 3 x 3-inch sodium iodide detector using a Co-60 source (1332-*keV* gamma ray) at a distance of 25 cm from the crystal face. A general rule of thumb for germanium detectors is 1 percent efficiency per each 5 cc of active volume. As detector volume increases, the detector resolution gradually decreases. For most NAA applications, an HPGe detector of 15-30 % efficiency is adequate.

9.7 HIGH ENERGY DETECTORS

9.7.1 BUBBLE CHAMBER

Donald Arthur Glaser invented (1952) the Bubble Chamber. A particle detector of major importance during the initial years is of high-energy physics. The bubble chamber has produced a wealth of physics from about 1955 well into the 1970s. Because liquids have much greater density than gases, particle can interact better. It is based on the principle of bubble formation in a liquid heated above its boiling point, which is then suddenly expanded, starting boiling where passing charged particles have ionized the atoms of the liquid. The technique was perfected to work with high precision in large volumes of different liquids, (liquid hydrogen, propane, freons Dupont's trade mark for fluor compounds, *e.g.*, CF_2Cl_2 or CF_3Br) embedded in a magnetic field. The liquid in a bubble chamber served simultaneously as target and as detector with a 4π solid angle coverage; stereo cameras recorded data on film

A **bubble chamber** is a vessel filled with a superheated transparent liquid used to detect electrically charged particles moving through it (Fig 9.7). The charged particle deposits sufficient energy in the liquid that it begins to boil along its path, forming a string of bubbles. Bubble chambers are similar to cloud chambers in application and basic principle.

Bubble chamber photographs provide the aesthetically most appealing visualization of subnuclear collisions particles (K^- at 4.2 GeV/c) which are seen entering the chamber, interacts with a proton, giving rise to the reaction

$$K^- + p \longrightarrow \Omega^- + K^+ + K^0 \text{ followed by the decays}$$

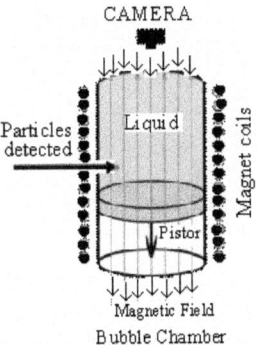

Fig.9.7   Bubble Chamber

As large-volume high-precision detectors with electronic data recording became available, and physics required ever more complex triggers, and as colliders became the high-energy accelerators of choice, retirement time arrived for bubble chambers.

A bubble chamber is normally made by filling a large cylinder with a liquid just below its boiling point; at the top of the chamber a camera looks in. The whole chamber is subject to a constant magnetic field. As the particles enter the chamber, a piston suddenly decreases the pressure in the chamber. This brings the liquid to a superheated state, in which a tiny effect, such as the passing of a charged particle near an atom, is sufficient to nucleate a bubble of vaporized liquid. At this moment, the camera records the picture. The magnetic field causes charged particles to travel in helical paths whose radius is determined by the ratio of carge-to-mass of the particle. In this way charged particles can be observed and their mass measured. However, there is no way to effectively measure their velocity (and KE).

The bubble chamber at CERN is called Gargamelle.

Bubble chambers have largely been replaced by wire chambers which allow particle energies to be measured at the same time. Another alternative technique is the spark chamber.

9.7.2 Spark Chamber

In 1960s the Spark chamber was developed at CERN and elsewhere. This consists of a series of parallel metal plates separated by a few mm and immersed in medium of inert gas.. A charged particle passing will ionize the medium, leaving a trail of ions. If high voltage is applied to every other plate, immediately after the ion trail is formed, then spark like miniature lightning bolts form along the trail, revealing the path of the particle which are photographed

A typical spark chamber consists of 6 to 128 thin metal plates which are parallel to one another and 2 -20 mm apart, a few m^2 in area set in a chamber having inert gas such as neon, at atm. pressure. A high voltage $10 - 15\ kV$ is applied to alternate plates in short bursts of $\sim 10^{-6}\ s$. It may be enough just to cause a spark to occur between the plates, when a charged particle passes through the chamber. This gives a trail of sparks behind it, which makes the tracks of the particle visible, as they spark, so that they may be photographed from the side.

9.7.3 Cerenkov Detector

The principle is the phenomenon observed by P.A. Cerenkov (1934) in which a particle moving through a dielectric medium with velocity v > c, gives out a very weak visible light. In water, glass and clear plastic whose index of refraction $n \;\square\; 1.5$, particles traveling with $v > \frac{2}{3} c$, can be expected to produce Cerenkov radiation as they pass through the substance. The direction of propagation bears a definite relation to the direction of travel of the particle, and its velocity relative to c. It is propagated on a cone of half angle φ,

$$\cos\varphi = \frac{(c/n)\ t}{v\ t}$$

where t = distance that light travels.

In the Cerenkov counter, the angle of the emitted cone is measured, and hence v of the particle is determined.

This can be used to detect the occurrence of certain nuclear interactions. Such interactions can release large amounts of energy and eject particles at highly relativistic speeds. If these interactions take place in water or another clear substance, then the Cerenkov radiation emitted as the reaction products travel through the water can be detected by photomultiplier tubes. This kind of detection is to be used in the Sudbury Neutrino Observatory to detect neutrino interactions.

Cerenkov cone

$$\cos\theta = \frac{c}{vn}$$

v = particle velocity
n = index of refraction of the medium

For water with n=1.33, the limiting angle for high speed particles is given by:

$$\theta = \cos^{-1}\frac{1}{1.33} = 41.2°$$

The threshold particle speed for Cerenkov radiation is v = c/n, which for an electron in water gives a threshold particle kinetic energy of 0.26 MeV.

$$\beta = 0.752, \quad E_{electron} = \gamma m_e c^2 = \frac{1}{\sqrt{1-\beta^2}} m_e c^2 = (1.52)(0.511 \text{ MeV}) = .775 \text{ MeV}$$

Kinetic energy = 0.775 MeV - 0.511 MeV = 0.26 MeV

The Super-Kamiokande neutrino detector facility in Japan has 11,000 photomultiplier tubes in place to detect Cerenkov radiation and is able to detect and distinguish electron and muon neutrinos.

Measurements of particle speeds can be made by measuring the angle of the Cerenkov cone, like photographing ship wakes to measure ship speeds. A portion of the light emitted by the decelerated particle is coherent and is emitted at a characteristic angle

The total amount of energy appearing in Cerenkov radiation is small compared to the total energy loss by ionization as the particle enters the medium. According to Rohlf, a relativistic particle near the speed of light will lose energy at the rate of about 200 $MeVm^{-1}$, and of that loss only about 40 $keVm^{-1}$ will be in Cerenkov radiation, about 1/5000 of the total energy.

9.8 COUNTER ELECTRONICS

Electronic components are indispensable part of nearly all modern detectors. For example, consider a Scintillating counter. Scintillation pulse is shaped and amplified in the *analog* part. The height V of the final pulse is, $V \propto$ Height of the original pulse, then the A/D Converter, the information (analog) is transformed into *digital* form. Consider then simplest case. Only pulses are accepted that have a height between V_o and $V_o + \Delta V$. If a pulse is in this window, output of the ADC is a standard pulse; if the input pulse lies outside the window, no output pulse appears. The digital part works with the standard pulse. It can be, for instance, a scaler; for every ten (or 10^n, n = integer) input pulses, one output pulse appears. The output is then a number which gives, in units of 10^n, the number of input pulses within a certain energy interval.

If many scalers are used then an MCA is used. The output of the MCA is read out into in an on-line computer for data processing.

9.9 IMAGE DETECTORS {CHARGE COUPLED DEVICES (CCDs)}

Photoelectric detectors have been described earlier in this Chapter are single detectors. An image may be recorded by assembling a 2D array of detectors, for example, an array of photodiodes in on a silicon chip. An arrangement which is widely used is the *charge-coupled device* (CCD). The three principles involved in the CCD are: photo detection, charge storage and charge transfer. In 1969, George Smith and Willard Boyle invented the first CCDs

CCD image detectors are remarkable tools for light measurement. These have been used in astronomy, X-ray diffraction, analytical spectroscopy, and consumer hand-held video and still cameras. CCDs are MOS integrated circuits (ICs) built like computer memory A CCD is often considered as a digital version of silver based photographic film. It converts the light that falls on it to digital data. Thus a CCD is an electronic memory that can be charged by light. CCDs can hold a charge corresponding to variable shades of light, which makes them useful as imaging devices for cameras, scanners, and fax machines. Because of its superior sensitivity, the CCD has revolutionized the field of astronomy and is found on many scientific space vehicles such as the Hubble Telescope.

They are used for analogue-to-digital conversion of signals, and quantum efficiency & wavelength coverage in the range 300 – 1000 *nm*. CCD sizes are specified, for example, 512 X 512, designating the number of pixels in the two-Dimensional arrays in the CCD chip in X and Y directions. A picture element is called a "pixel," the smallest region whose intensity is reported) that store and deliver photon-induced charge. These pixels, formed into one- or two-dimensional arrays in the CCD chip, are each square or rectangular, and typically range from 6 to 30 microns along an edge. Scientific-grade CCDs are particularly attractive for use as image detectors when properly cooled (typically to between -20 ^oC and -120 ^oC) and slowly read out (on the order of 20 microseconds/pixel, hence several seconds per full frame, as compared with 30 frames per second for broadcast television), because they offer small dark current, high quantum efficiency, low readout noise, and wide dynamic range. In a CCD Name (say, Tek2048) specifications are the manufacturer, CCD size, instrument specific. CCDs are operated at cryogenic temperatures (say -85 ^oC to -100 ^oC).

The useful CCD stuff is that the signal-to-noise (S?N) ratio, a quantitative measurement of data quality.,

Image generation with a CCD camera can be divided into four primary stages or functions: charge generation through photon interaction with the device's photosensitive region, collection and storage of the liberated charge, charge transfer, and charge measurement.

REVIEW QUESTIONS

R.Q. 9.1   Hydrogen ions are formed along the track of a bubble chamber using liquid hydrogen. If a proton of 1000 *MeV* passes through a bubble chamber 2 *m* long, estimate how many hydrogen ions are produced, given that the energy loss of proton is 0.4 MeV cm^{-1}. [Answer: $\left(\dfrac{1000 MeV}{(0.4\ MeV\ cm^{-1})(2\ m)}\right) = 13$].

R.Q. 9.2 Discuss about the principle and working of Cloud Chambers.

RQ. 9.3. Write a short note on the CCDs.

&%&%&%&%&%&

Chapter 10

DYNAMIC PROPERTIES OF NUCLEUS I: BETA DECAY

Chapter 10

DYNAMIC PROPERTIES OF NUCLEUS - I

BETA DECAY

"The concept of substances has disappeared from fundamental physics"

- Sir Arthur Eddington

10.1   INTRODUCTION

The radioactivity of heavy nuclides, embodied in the three processes of $\alpha-$, $\beta-$ and $\gamma-$ decay, were early indicators of nuclear activity. Although these three processes are treated together in chapter 2, the underlying interactions are quite different. However, all these nuclear decay processes are characterized by the lifetime of the decaying nuclide. The meaning of nuclear *lifetime* is formalized to some detail in chapter 2.

To the best of our knowledge, an isolated proton, a hydrogen nucleus with or without an electron, does not decay. However within a nucleus, the beta decay process can change a proton to a neutron. An isolated neutron is unstable and will decay with a half-life of 10.5 minutes. A neutron in a nucleus will decay if a more stable nucleus results; the half-life of the decay depends on the isotope. If it leads to a more stable nucleus, a proton in a nucleus may capture an electron from the atom (Electron Capture), and change into a neutron and a neutrino. The nuclear process of $\beta-$ decay will be presented in some details in this Chapter. The value of Z for beta stability (Eqn 13.7.13.) is at the bottom of the isobar given by

$$Z_{\beta\ stability} = \frac{2A}{4 + (a_c/a_a)A^{2/3}}.$$

10.2   NUCLEAR REACTION EQUATION ($\beta-$ Activity)

A nucleus can alter its N/Z ratio to achieve better stability by means of *beta* ($\beta-$) *decay*, like $\alpha-$ decay. $\beta-$ decay usually means emission of *fast* nuclear electrons.

10.2.1   Equation of β^- $-$ (or $_{-1}^{0}e$) decay

$$_{Z}^{A}X \rightarrow\ _{Z+1}^{A}Y +\ _{-1}^{0}e \tag{10.2.1}$$

Example $\quad ^{14}_{6}C \rightarrow ^{14}_{7}N + e^- + \nu \quad$ (10.2.2)

Symbolically this is shown in Fig 10.1.

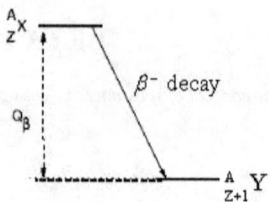

Fig 10.1 Symbolic diagram of nuclear β^- – decay.

10.2.2 β^+ – (or $_{+1}^{0}e$, positron) decay

Nuclides lying below the stability region of the N-Z chart suffer β^+ – (or $_{+1}^{0}e$, positron) decay.

$$^{A}_{Z}X \rightarrow ^{A}_{Z-1}Y + ^{0}_{+1}e \quad (10.2.3)$$

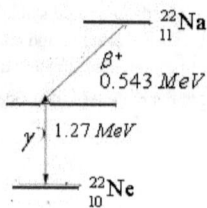

Fig 10.2 Nuclear β^+ – decay

Typically $\quad ^{22}_{11}Na \longmapsto ^{22}_{10}Ne + e^+ + \nu \quad$ (10.2.4)

Symbolically this is shown in Fig 10.2

10.2.3 EC Process

In some instances it has been observed that the parent nuclide, instead of emitting a β^+ - (or e^+ positron), captures an orbital electron from one of the innermost atomic orbitals (K, L, ..). This is called the *EC process*.

EC process is represented by

$$^A_Z X + e^- (EC) \rightarrow ^A_{Z-1}Y \qquad (10.2.5)$$

A typical example is seen in (Fig 10.3), i.e.,

$$^7_4 Be + e^- (EC) \rightarrow ^7_3 Li + \nu \qquad (10.2.6)$$

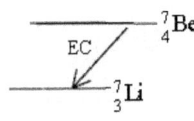

Fig 10.3

10.3 BETA RAY SPECTRUM

10.3.1 Outstanding Puzzle with the Beta-ray Spectra

To measure the energies of β^--rays Robinson's magnetic spectrometer was used.

The β^--particle having momentum p, in a magnetic field B, will move in a trajectory of radius r, dictated by

$$p = Ber \qquad (10.3.1)$$

where the KE of the particle is given by

$$E_{\beta^-} = \sqrt{m_0^2 c^4 + p^2 c^2} \qquad (10.3.2)$$

10.3.2 End-point Energy ($E_{\beta^- Max}$)

Measurements have shown that the energy of β^--particles

a) The continuous β^--spectrum was discovered in the decay of RaB ($^{214}_{82}$Pb) by Sir James Chadwick (1914) using a magnetic spectrometer. The spectrum has range from a few keV to > 15 MeV, (measured in 1927). and
b) A line spectrum having apparently two peaks (or sharp lines).
Fig 10.4 shows the continuous β^--ray spectrum of $^{210}_{83}$Bi (historical name RaE).

It is noted that the curve has

i) a definite upper limit, $E_{\beta\text{-Max}}$, known as the *end point energy*, for the energy of the disintegration electrons, and
ii) A maximum toward the passage to the low energy part of the spectrum.

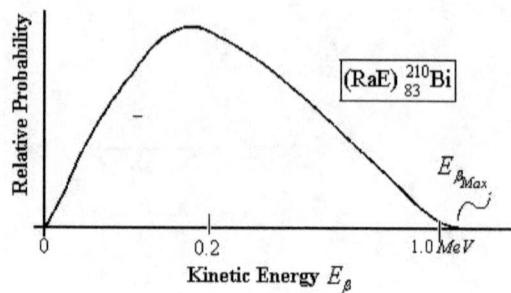

Fig 10.4 Continuous β^- - spectrum of isotope RaE ($^{210}_{83}\text{Bi}$)

10.3.3 Continuous Spectrum

The continuous β- spectrum, discovered in the decay of RaB ($^{214}_{82}\text{Pb}$) by James Chadwick (1914), using a magnetic spectrometer, had *monoenergetic* lines *and* a *continuous* component (Fig 10.4), in strong contrast with alpha and γ – ray spectra, which were known to consist of mono-energetic lines only. Interpretation of the continuous electron spectrum was the subject of considerable debate. Ernest Rutherford (1914) was of the opinion that the β -electrons were all emitted from the nucleus with the same energy, but lost different fractions of this energy to the surrounding atoms, depending on the thickness of the source traveled. This was supported by Von Bayer, Paul Harteck Otto Hahn, and Lise Meitner (1915). The point was made by Meitner (1922) that a quantized nucleus should not emit electrons of continuously varying energy. The known features of α- and γ- spectra were correctly interpreted as due to transitions of nuclei from one quantum state to another. Thus the continuous electron spectrum was a unique feature of β-decay.

C.D. Ellis and W.A. Wooster (1927) performed an experiment it was observed that the total energy released in the disintegration of a RaE ($^{210}_{83}\text{Bi}$) source inside a calorimeter thick was enough to stop all the emitted electrons. $E_{\beta Max}$ = 1.05 *MeV*, and the mean energy, E_{Mean} = 390 *keV* for.

Beta sources	Max. kinetic energy (MeV)
^{14}C	0.155
^{99}Tc	0.29
^{36}Cl	0.715
^{204}Tl	0.763
^{210}Bi	1.17
^{234}Pa	2.32

the β-electrons. The calorimeter should have measured a total energy of 1.05 MeV if the above picture was correct. In fact they observed E_{Mean} = 344 ± 34 keV, which corresponded very well with the mean energy of the emitted electrons

These results were very conclusive, but difficult to interpret at the time. Niels Bohr believed that energy conservation was violated in individual decays, although perhaps not statistically. Not only was energy apparently not conserved, but neither were momentum and angular momentum. In order to save the situation, Wolfgang Pauli (1930) proposed the idea of a very penetrating neutral particle of small mass and spin ½, emitted with the electron in β-decay. This proposal was made before Chadwick's discovery (1932) of the neutron. Pauli openly proposed his hypothesis at the Solvay Congress in Brussels in 1933. Fermi was present, and proposed the name "neutrino" to distinguish it from the neutron. Soon afterwards, Fermi developed his famous theory of beta decay.

10.3.4 Sargent Diagram

The distinction between "allowed" and "forbidden" decays was first made by Sargent in 1933 on an empirical basis. The so-called *Sargent diagram* was a log plot of the decay rate vs. the endpoint energy. The decays would line up in distinct bands, depending on their degree of forbiddenness.

Table 10.1 End point energies $E_{\beta Max}$ of continuous β^{-}-spectra

Element	Isotope	$E_{\beta Max}$ (MeV)	$\tau_{1/2}$
$^{214}_{82}$Pb	RaB	0.72	26.8 m
$^{210}_{83}$Bi	RaE	1.17	4.85 d
$^{223}_{87}$Fr	AcK	1.20	21. m
$^{228}_{89}$Ac	MsTh$_2$	1.55	6.13 h
$^{231}_{90}$Bi	UY	0.21	25.65 h

The end point energy $E_{\beta^{-}Max}$ is characteristic of the nuclide undergoing β-decay.

$$E_{\beta\text{-Total}} = m_o c^2 + E_{\beta\text{-Max}} = (M_p - M_d)c^2 \qquad (10.3.3)$$

Table 10.1 lists the $E_{\beta\text{-Max}}$ of a few nuclides.

10.4 RANGE (R_{β^-}) OF β-RAYS

Range R_{β^-} is a fairly well defined distance beyond which β-ray traveling through a material medium has no ionizing effect (Fig 10.5). A simple absorption experiment provides the value of R_{β^-} [from a plot of ln (Activity) *versus* area density ($mg\ cm^{-2}$) (of the absorber)] for many metal foils are roughly the same.

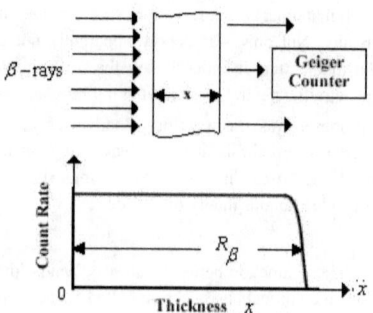

Fig 10.5 Range of β^--rays

$$R_{\beta^-} \approx 500\ mg\ cm^{-2}$$

Norman Feather has given a formula:

$$R_{\beta^-} \approx 543\ E_{\beta\text{-Max}} - 160 \qquad (10.4.1)$$

where R_{β^-} is in $mg\ cm^{-2}$ and $E_{\beta\text{-Max}}$ is in MeV.

10.5 ENERGETICS OF β-DECAY (Mass - Energy Balance)

10.5.1 β^--Decay

Applying the conservation of energy,

$$M_p c^2 = M_d c^2 + E_d + m c^2 + E_{\beta^-}$$
$$= (M_d + m) c^2 + Q_{\beta^-} \qquad (10.5.1)$$

$$\therefore \quad Q_{\beta^-} = [M(Z, A) - M(Z+1, A) - m] c^2 \qquad (10.5.2)$$

As in α- decay, the expression for E_d is given by

$$E_d = \frac{m}{M_d} E_{\beta^-} = \text{negligible.} \qquad (10.5.3)$$

$$Q_{\beta^-} = E_{\beta^-} + E_\nu \approx E_{\beta^-} \qquad (10.5.4)$$

10.5.2 β^+ – decay

$$Q_{\beta^-} = [M(Z, A) - M(Z-1, A)] c^2 \qquad (10.5.5)$$

$$Q_{\beta^+} = E_{\beta^+} + E_\nu \qquad (10.5.6)$$

10.5.3 EC Process

$$Q_{EC} = [M(Z, A) - M(Z-1, A) + m] c^2 \qquad (10.5.7)$$

10.5.4 Features of β- decay and Discrepancies

1. Puzzle in β- decay: Since the spectrum is continuous and no electron is ever observed to have energy more than is permitted by the conservation of energy.

2. The nucleons (neutron and proton) have spin-$\frac{1}{2}$. β- decay involves replacement of a neutron by a proton or vice-versa, shows that the β- particle can have only an integral relative orbital angular momentum quantum number, which violates the law of combination of quantized angular momentum.

3. In β- decay of certain nuclides measurements show that the β^- seldom moves exactly opposite the nuclide, violating the principle of conservation of linear momentum.

Worked out Example 10.1

A Tritium nucleus decays in the following way: $_1^3 H \rightarrow\ _2^3 He + \delta^- + \nu_e$. What energy is released as kE? (, given $A(_1^3 H)$=3.01647 u, $m_e = 5.4858 \times 10^{-4} u$)

Solution: $\boxed{Step\ \#1}$ Given, $A(_1^3 H)$=3.01647 u, $m_e = 5.4858 \times 10^{-4} u$,

| Step # 2 | $A({}_1^3H) = [3.01647\ u - 1(0.00054858u)] = 3.014932\ u$ |

| Step # 3 | (From Worked out Example 6.1 and
$A({}_2^3H) = [3.014932\ u + 1(0.00054858u)] = 3.015481\ u$

| Step # 4 | $KE(\delta^- + \nu_e) = 3.01592\ u = 3.015481\ u = 0.0044\ u$

$= (0.0044\ u)(931\ MeV/u) = 0.41\ MeV$

10.6 FERMI'S QUANTUM FIELD THEORY OF β- DCAY

10.6.1 Discoveries and Hypotheses

Wolfgang Pauli (1930) postulated the simultaneous emission of a third particle (charge = 0, mass = 0 and spin = $\frac{1}{2}$) (called later the *neutrino*) in β- decay to explain the continuous distribution of the β^- - particles (electrons). Since a two-body kinematics could not be reconciled with the experimental data, only with the emission of a third particle could the momentum, energy and angular momentum all could be conserved. Soon after (1931) this third particle was christened by Enrico Fermi as the *neutrino* (ν). Fermi had developed a theory of β – decay to include the neutrino.

With the discovery of the neutron by James Chadwick (1932) one could write

$${}_4^9Be + \alpha \rightarrow n + {}_6^{12}C \qquad (10.6.1)$$

and in 1932 *positron* β^+ (i.e. e^+) was discovered by Carl D. Anderson. [The origin of positrons in the *cosmic rays* was not fully understood till 1937 when Homi Jahangir Bhabha and Walter Heitler pointed out that electrons deriving from the primary cosmic rays could give rise to *bremstraahlung* quanta as a result of deflections in the electric field of nucleus, and subsequent production of e^- - e^+ pairs, as $\gamma \rightarrow e^- - e^+$..

Neutrino hypothesis was made by W. Pauli ((1933), whereas E. Fermi (1933) incorporated neutrino into his Quantum Field Theory of β- decay. Accordingly

1. Accountability of the continuous distribution of energy and the conservation of energy

In a nuclear process of β- decay, two particles are emitted by the nuclide and the total energy of these two particles (electron β^- and neutrino ν)

$$E_{\beta\text{-}Total} = \text{a constant}, E_{\beta\text{-}Max} \qquad (10.6.2)$$

i.e., $\qquad Q_{\beta^-} = E_{\beta\text{-}Max} = E_{\beta^-} + E_\nu . \qquad (10.6.3)$

2. The "*4- fermion* process" and conservation of angular momentum:

The most elementary β- decay was supposed to be $n \to p + \beta^-$ (or equivalently, $n \to p + e^-$).

Table 10.2   Conservations Laws and $n \to p + e^- + \nu$

	n	$\to$	$p + e^-$	$+ \nu$	
1. Conservation of charge	0	=	(+1) + (-1)	+ 0	ok
2. Conservation of mass or linear momentum	$(1+\Delta m)u$	=	$(1\,u) + (\Delta m\,u)$	+ 0	ok
3. Conservation of spin angular momentum	$\frac{1}{2}\hbar$	=	$\frac{1}{2}\hbar \pm \frac{1}{2}\hbar$	+ spin of ν	ok

To account for the missing energy during the disintegration of a free neutron into a proton, an invisible particle neutrino is also emitted along with an electron

$$n \to p + e^- + \nu \qquad (10.6.4)$$

Thus Table 10.2 concludes from the last row that the spin of $\nu = \frac{1}{2}$ .integral number

Subsequently it was observed that there are two kinds of neutrinos:

a) Neutrino, strictly β- neutrino, denoted as ν_e, and
b) Anti-neutrino, denoted as $\bar{\nu}_e$.

Thus β^-- decay: $\qquad n \to p^+ + \beta^- + \bar{\nu}_e \qquad (10.6.5)$

Say, $\quad {}^{14}_{6}C \to {}^{14}_{7}N + \beta^- + \bar{\nu}_e$

β^+ – decay: $\qquad p^+ \to n + \beta^+ + \nu_e \qquad (10.6.6)$

say as in $\qquad {}^{11}_{6}C \to {}^{11}_{5}B + \beta^+ + \nu_e$

EC process: $\qquad p^+ + \beta^- \to n + \nu_e \qquad (10.6.7)$

Fermi took this result to be the prototype for the weak interactions, which is the described as the **4 fermions interacting at a point**.

10.6.2 Double beta decay

Double β – decay is one process found useful with the elucidation of neutrino properties. A nucleus may occasionally energetically stable against ordinary β – decay may emit two β – particles. For example, an even-even nucleus, one bound tightly, ${}^{82}_{34}Se$ can double β – decay according to

$$ {}^{82}_{34}Se \to {}^{82}_{36}Kr + 2\beta^- + 2\bar{\nu}_e \qquad (10.6.8)$$

or as $\quad ^{82}_{34}Se \rightarrow ^{82}_{35}Br + 2\beta^-$ (10.6.9)

The second being the result of

$$^{82}_{34}Se \rightarrow ^{82}_{35}Br + \beta^- + \bar{\nu}_e$$
$$\bar{\nu}_e + ^{82}_{35}Br \rightarrow ^{82}_{36}Kr + \beta^-$$

The first stage is a virtual process, not being allowed energetically, and the second one is contrary to the finding about neutrino absorption, because it requires non-zero mass. The neutrinos in the equation (10.6.8) are the Dirac neutrinos, whereas the ones in the equation (10.6.9) are called Majorana neutrinos. The neutrino less double β – decay is important, but has not been detected. However, the double β – decay of $^{76}Ge \rightarrow \ ^{76}Se$ has established that the half-life of neutrino-less double β – decay, if it takes place, must be $\tau_{1/2} > 2.3 \times 10^{24}$ yrs.

10.6.3 Inverse β – decay of a Neutron (Neutrino Hunting)

It is certainly enormously difficult to think about trying to detect the spin of an invisible, mass-less particle, but which has been irrefutably proved for its existence. Two very beautiful but very difficult experiments were performed to

i) Measure the recoil momentum imparted by a neutrino to both the β – particle and the residual nuclide at the time of the emission of a highly energetic neutrino., and
ii) Demonstrate the '*inverse β – decay* 'of the neutron, viz.,

$$p^+ + \bar{\nu}_e \rightarrow n + \beta^+ \qquad (10.6.10)$$

where a neutrino is absorbed by a proton.

Since absorption of anti-neutrino is equivalent to the emission of a neutrino, so that one has the *neutron decay*:

$$n + \nu_e \rightarrow p^+ + \beta^-. \qquad (10.6.11)$$

The two decays given by equations (10.6.8) and (10.6.9) are equivalent in terms of Dirac's theory of *holes*!

The first recoil momentum measurement experiment was performed in 1950s by Allen and coworkers in Illinois, whereas in the inverse β – decay experiment and observation of the anti-neutrino were made by Clyde L. Cowan, Jr. and Frederick Reines (1956) in Los Alamos. The '*neutron decay*' and '*inverse β – decay* (equation (10.6.8)) are the two processes involved, respectively, to detect ν_e and $\bar{\nu}_e$.

10.6.4 Theory of β – decay

The decay of a *free* neutron, given by equation (10.6.5), viz., the "*4- fermion* process"

$$n \rightarrow p^+ + \beta^- + \bar{\nu}_e,$$

is the simplest manifestation of β-decay. Fermi described this prototype for the *weak interactions*, as 4 fermions interacting at a single point. In space-time this can be mathematically stated as

$$\psi(n) \rightarrow \psi(p) \quad (10.6.12) \quad \text{and}$$

$$\psi(\text{incoming } \nu) \;\square\; \psi(\text{outgoing } \bar{\nu}) \rightarrow \psi(e^-). \quad (10.6.13)$$

To describe this reaction multiply the ψ-s by unknown factors Γ which effect the transformations, and also multiply by a second factor, G, Fermi-coupling constant. This is the quantity which governs the intrinsic strength of the weak interactions, and so the rate of β-decay. Amplitude of β-decay, denoted by M is

$$M = G_F \left(\bar{\psi}(p) \Gamma \psi(n) \right) \left(\bar{\psi}(e^-) \Gamma \psi(\bar{\nu}) \right) \quad (10.6.14)$$

The factors Γ contain the essence of the weak interaction effects which give rise to the transformation of the particles. Treating the beta decay as a transition that depended upon the strength of coupling between the initial and final states, according to as *Fermi's Golden Rule*:

$$\lambda_{if} = \frac{2\pi}{\hbar} |M_{if}|^2 \rho_f \quad \boxed{\text{Fermi's Golden Rule}}$$

Transition probability   *Matrix element for the interaction*   *Density of final states*

(10.6.15)

The challenge was to discover the nature of these particles, as to whether they are just scalars, vectors or tensors. The choice can be better narrowed down by examining the angles of direction of emission between the emitted products of β-decay and their various energies. But the nature of the interaction which led to beta decay was unknown in Fermi's time (the weak interactions). It took some 20 years of work to work out a detailed model which fit the observations. The confirmation had to await several years till the **violation of parity in weak interactions** was detected.

The nature of the model in terms of the distribution of electron momentum p is summarized in the relationship (10.6.13) below (Fig 10.6):

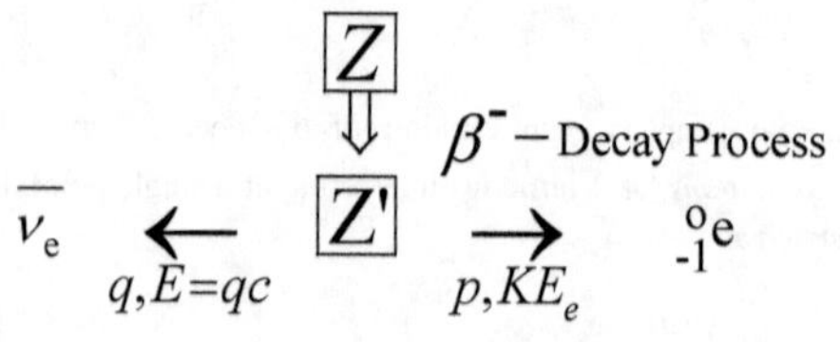

```
         Z
         ↓       β⁻ – Decay Process
    ν̄ₑ  ←   Z'  →        ⁰₋₁e
       q,E=qc     p,KEₑ
```

Fig 10.6 β^- – Decay Process

The distribution of electron momentum N(p) is given by

$$N(p) = Cp^2 (Q - KE_e)^2 \, F(Z', p) \, |M_{fi}|^2 \, S(p, q) \qquad (10.6.16)$$

where $Cp^2 (Q - KE_e)^2$ = Statistical factor derived from the density of final states

$F(Z', p)$ = Fermi function for the nuclear Coulomb interaction with the emitted particle,

$|M_{fi}|$ = Matrix element for allowed transitions gives the strength of the interaction between initial and final states,

$S(p, q)$ = Shape factor to correct the matrix element for various types of 'forbidden' decay paths.

Richard P. Feynman and Murray Gell-Mann (1956) proposed that the interaction factors Γ be a mixture of vector and axial vector quantities, to account for the effects of violation of parity in weak interactions.

Under a Lorentz transformation a vector quantity has the following well defined properties:

a) A change of sign takes place under a rotation through 180°,
b) It will appear identical under a rotation through 360°,
c) An axial vector will transform like a vector under rotations, and
d) An axial vector will change its sign opposite to that of a vector under the operation of parity (an improper operation).

10.7 POLARIZATION OF β – ELECTRONS

When parity is not violated as many left handed ones as right handed, β – electrons should be emitted in β – decay. Polarization P of the β – electrons is defined to quantify the difference in spins by

$$P = \frac{N_R - N_L}{N_R + N_L} \qquad (10.7.1)$$

where N = # specific spinning electrons in the sample. When P = 1, all the spins are right handed and, when P = -1 all the electrons are left spinning. P = 0, when electrons are emitted slowly. But when electrons have relativistically very high speeds, they are all left handed. Hans Frauenfelder and others (1957) observed the P of electrons from the β – decay of $^{60}_{27}Co$ by scattering them through a foil of heavy elements. The net polarization observed was P = - 0.4 for β – electrons traveling at 0.49 c, satisfying the prediction.

10.8 NEUTRINO HELICITY

10.8.1 Chirality

Look at your right hand. It is an asymmetric object. Its mirror image is your left hand. If you try to superimpose your two hands turned to you, you find no way to do that. That's chirality! Something is *chiral* when it is not superposable to its mirror image. Many things in nature are chiral.

10.8.2 Helicity:

One of the main properties of a particle is its spin. It figures the self-rotation of the particle. If the projection of the spin on the particle velocity direction is opposite to it, the particle is said to be of *left helicity*. On the contrary, it is said to be of *right helicity*. Fig 10.7 is a self-explanatory diagram on helicity of the particle.

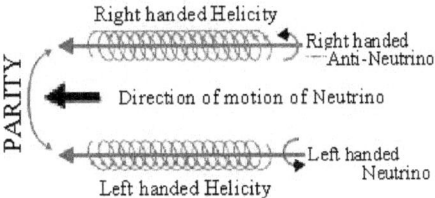

Fig 10.7 Neutrino helicity

Theoretically, a zero mass neutrino can only have a left helicity. It is very peculiar to the weak interaction, which produces only left **helicity** neutrinos. A neutrino could be of left or right **chirality**. But a zero mass neutrino having a left **helicity** would be always of left **chirality**... The terms "left" or "right" (often subscripted L and R) refer to the left or right **chirality**. In the limit of zero mass they correspond to the left or right **helicity**.

The "*4-fermion* process",

$$n \rightarrow p + \beta^- + \bar{v}_e,$$

contains very tightly the spins which can couple through the weak interactions. Neutrino spins are allowed to couple through Γ (which contains the essence of weak interaction effects), one finds that only left handed *helicity* neutrinos and right handed helicity neutrinos can take part in the weak interactions. To measure the neutrino-*helicity* H one must look for a particularly simple example of β– decay and deduce it from the helicities of other decay products, using the *conservation principle of angular momentum*.

Maurice Goldhaber and others (1958) used the Spin-0 nuclide $^{152}_{63}Eu$, which undergoes hybrid β- decay, in which the electron is captured and a neutrino is emitted, leaving an excited state of $^{152}_{62}Sm$ which is Spin-1.

$$^{152}_{62}Sm^* \rightarrow {}^{152}_{63}Eu + e^- \rightarrow {}^{152}_{62}Sm^* + \nu \atop {}^{152}_{62}Sm + \gamma \qquad (10.8.1)$$

The latter then disintegrates to its Spin-0 ground state by the emission of a photon γ.

10.9.1 Neutrno Detection

Neutrino, like e^- and e^+, cannot exist within the nucleus. It is created at the time of β – decay.

The recoil energy of a nucleus as it emits a neutrino on EC process is, non-relativistic,

$$E_R = \frac{p_\nu^2}{2 A M_n} = \frac{E_0^2}{2 A M_n c^2}$$

10.9.2 Coulomb Dissociation and Neutrno production

When one nucleus approaches another the electric field becomes gradually intensified due to the Coulomb force acting between them, and the vice-versa. This phenomenon occurs in a very short time. Since this is a very large change in intensity, it is equivalent to the emission of a virtual photon (γ – ray). Then unstable nuclei absorb the virtual photons and disintegrate. This is the phenomenon of Coulomb dissociation. This can be used to study nuclear reactions occurring in astronomical objects. The generation of solar neutrinos involves this, and is shown in Fig 10.8.

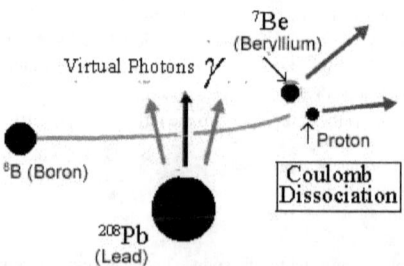

Fig 10.8 Coulomb dissociation of a nuclide and solar neutrino.

10.9.3 BETA RAY SPECTROMETER

The schematic diagram of a beta ray spectrometer is shown Fig 10.9.

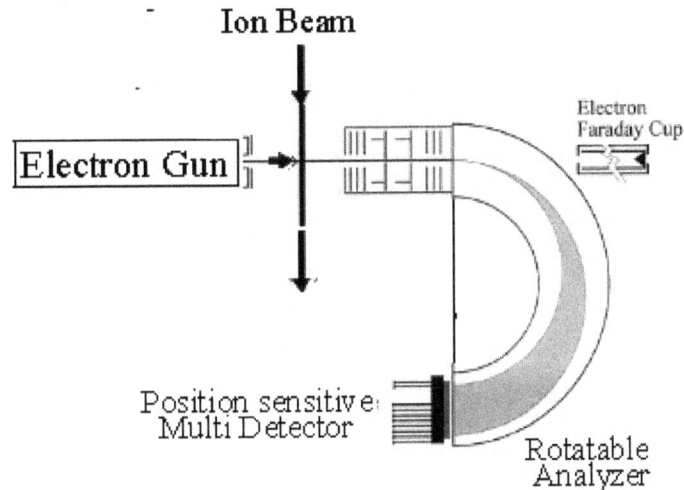

Fig 10.9 Beta ray spectrometer

Description of the instrument is beyond the scope of this book.

REVIEW QUESTIONS

R.Q. 10.1 A source of $^{32}_{15}P$ undergoes β-decay, and $R_{\beta^-} = 7.7$ mg mm^{-2} in aluminium. Find the end point energy. [Answer $E_{\beta\text{-Max}} = 1.71$ MeV].

R.Q. 10.2 State "parity law". Discuss its acceptance or violation in weak interactions?

R.Q. 10.3 Develop Fermi's theory of β-decay. What is meant by Kurie plot? Write a note on the parity violation in β-decay.

R.Q. 10.4 Give an account on Fermi's Theory of β-decay. Find the maximum kinrtic energy of the excited electron in the decay: $^{14}_{6}C \rightarrow ^{14}_{7}N + \beta^- + \bar{\nu}_e$. The atomic masses of $^{14}_{6}C$ and $^{14}_{7}N$ are 14.007685 *amu* and 14.007517*amu*, respectively. One *amu* is equivalent to 931.5 MeV.

RQ. 10.5 Outline the experimental methods in use for the study of β-ray spectra. Give a critical account of the theories proposed to explain the continuous nature of β-ray spectra.

RQ. 10.6 Give the quantum mechanical theory of Fermi which explained the observed β-spectrum. What is a Kurie plot? How is the end point of a β-spectrum related to the mass of a neutrino?

&&*&*&*&*&*

Chapter 11
DYNAMIC PROPERTIES OF NUCLEI II:
GAMMA DECAY
(RADIATIVE TRANSITIONS IN NUCLEI)

Chapter 11

DYNAMIC PROPERTIES OF NUCLEI - II
GAMMA DECAY
(RADIATIVE TRANSITIONS IN NUCLEI)

"There are not many but only One. Who sees variety and not the unity wanders from death to death"   The Upanishads

11.1   INTRODUCTION

With the production of many nuclear species in their ground or excited states through nuclear reactions, studies of such decays covers the Periodic Table. All the three types of nuclear decay processes are characterized by the lifetime of the decaying nuclide. The meaning of nuclear *lifetime* is formalized to some detail in chapter 2.

In Chapter 10 it the nuclear process of β – decay was presented. A nucleus can alter its N/Z ratio to achieve better stability by means of *beta (β –) or /and α –* decay. β – decay usually means emission of *fast* nuclear electrons. In the theory of Enrico Fermi β – decay is the *4-fermion* process, viz., $n \rightarrow p + \beta^- + \bar{\nu}_e$, which contains very tightly the spins which can couple through the *weak interactions*. Neutrino, like e^- and e^+, cannot exist within the nucleus. It is created at the time of β – decay. The γ-ray decay, presented in some details in the present Chapter, gives valuable information about the structure of nuclear states as well as the interactions responsible for them.

11.2   PRELIMINARIES

11.2.1   Origin of γ-rays

In addition to ground state configuration, nuclei may have several excited states. Both α – and β – decays can leave the parent as well as the product nuclei in excited and in many nuclear reactions excited nuclei are produced. This falls in two groups: a) Particle excitation and b) Collective excitation. Energy for the most stable value of A is at the valley of the parabola, is, Eqn (13.7.6), *viz.*,

$$Z_S = \frac{\left(\frac{A}{2} + (M_n - M_p) c^2 \frac{A}{8 a_A} + a_C \frac{A^{2/3}}{8 a_A}\right)}{\left(1 + \frac{1}{4} \frac{a_C}{a_A} A^{2/3}\right)}$$

An excited nucleus is denoted by an asterisk mark after its usual symbol:

$${}^{87}_{38}Sr^{*}$$, or $(Z, A)^{*}$

Thus ${}^{60}_{28}Ni^{*}$ returns to its ground state, i.e., *de-excited*, by emitting photons (EM radiations), dictated by the Bohr criterion (Fig 11.1). These photons are called for historical reasons gamma (γ-) rays.

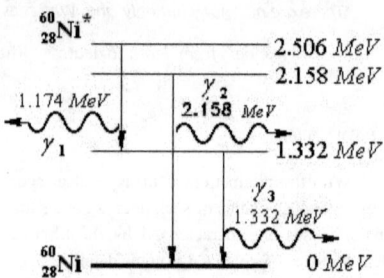

Fig 11.1 γ-decays of Ni-60

Nuclear excited states are not single particle excited states, so that a complicated rearrangement of nucleons occurs during γ-decay. Long-lived excited nucleus is called an *isomer* of the same nucleus in its ground state. An *isomeric state* is denoted by the symbol 'm' placed as a superscript after the symbol of the nuclide, as in ${}^{57}_{26}Fe^{m}$.

11.2.2   1234 Rule and Energy of γ-rays

Energy of a γ-ray is given by

$$E_\gamma = (931.5)\, m\,(at.\,wt.)\ MeV \qquad (11.2.1)$$

$$E_\gamma = h\nu = hc/\lambda = \frac{1239.8}{\lambda\,(fm)}\ MeV \qquad (11.2.2)$$

$$\lambda\,(fm) = \frac{1.2398 \times 10^{-10}}{E_\gamma\,(MeV)}\ cm \qquad (11.2.3)$$

i.e., $\lambda\,(nm) = 1234\ eV\text{-}nm / E_\gamma(eV)$ \qquad (11.2.4)

is '1234-Rule' giving the wavelength λ of γ-ray of a given energy E_γ.

11.2.3 Spectral line broadening (FWHM) of γ-rays

11.2.3.1 Line shape

A spectral line extends over a range of energies, and not a single energy. In addition its centre may be shifted from its nominal central energy / wavelength. There are several reasons for this broadening. FWHM (Full Width at Half Maximum, Γ) is an expression of the extent of a function $f(E)$. It is defined by $|E_2 - E_1|$ where E_1, E_2 are points to the left and right of the mode E_o (defined by $f(E_o)$ = max), with $f(E_1) = f(E_2) = f(E_o) / 2$. the difference between the two extreme values of the independent variable at which the dependent variable is equal to half of its maximum value. The known causes and types of spectrum-line breadths are due to: a) the Doppler effect, b) Natural breadth, and c) External effects (Collision damping, asymmetry and pressure shift and Stark effect).

11.2.3.2 Thermal Doppler (Gaussian) profile

Thermal Doppler broadening is due to the temperature T of the system. It is a form of inhomogeneous broadening and has a *Gaussian lineshape* (known as normal distribution). The line shape of Gaussian profile $D(E\,;\,\sigma)$ is given mathematically by

$$D(E\,;\,\sigma) = \frac{1}{\sigma\sqrt{2\pi}}\, e^{(-E^2/2\sigma^2)} \qquad (11.2.5)$$

It graphically is as shown in Fig 11.2. For a Gaussian (normal) distribution one has the relation

$$\text{FWHM}, \Gamma_G = 2\sigma\sqrt{-2\log(\tfrac{1}{2})} = 2.3548\,\sigma \qquad (11.2.6)$$

between Γ_G and σ, standard deviation. Also, for an atomic gaseous system

$$\Gamma_G = \sqrt{(8\ln 2)\, k_B T / Mc^2}\,.$$

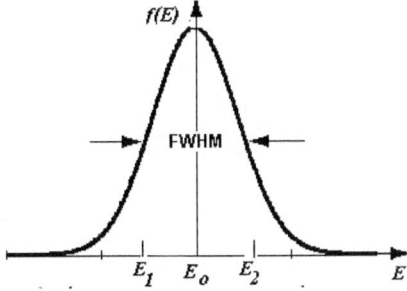

Fig 11.2 FWHM for Gaussian profile $D(E\,;\,\sigma)$

11.2.3.3 Lorentzian (or, Breit-Wigner) profile

The line shapes of transitions depend on the broadening mechanisms of the initial and final states. The Uncertainty Principle relates the life of an excited state with the precision of the energy. This is *natural broadening*. It is a homogeneous mechanism and is described by a *Lorentzian profile* (Cauchy distribution) $L(E - E_o; \Gamma_n)$, and there is no associated shift. Lifetimes (τ) of excited nuclear states and their natural width, FWHM, *i.e.*, Γ_n, can be measured very precisely by means of the Mossbauer Effect.

In the *Uncertainty Relation*

$$\Delta E \Delta t \geq \hbar \tag{11.2.7}$$

width ΔE of an energy level,

$$\Delta E = \Gamma_n \tag{11.2.8}$$

i.e., $$\Gamma_n = \frac{\hbar}{\tau} \tag{11.2.9}$$

Lorentzian profile has the form

$$L(E - E_o; \Gamma_n) = \frac{(\Gamma_n/2)}{\pi [(E-E_o)^2 + (\Gamma_n/2)^2]} \tag{11.2.10}$$

where E_o is the energy of the transition in a spin-0 nuclide, and it is the location parameter. This is the Fourier transformation of the equation of a harmonic oscillator charged particle with damping, and a periodic force. The FWHM of the Lorentzian profile is just Γ_n (the scale parameter)..

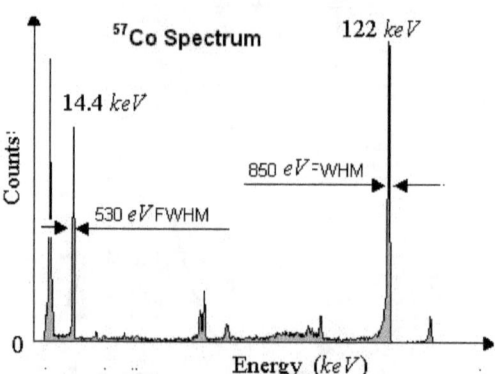

Fig 11. Typical $L(E - E_o; \Gamma_n)$ of $^{57}_{27}$Co gamma rays

11.2.3.4 Pressure broadening:

The presence of nearby particles will affect the radiation emitted by an individual particle. There may be an associated shift. This effect depends on both density and temperature of the medium, say gas.

11.2.3.5 Voigt profile

These above mechanisms can act in isolation or in combination. Assuming each effect is independent of the other, the combined line profile will be the convolution of the line profiles of each mechanism. For example, a combination of thermal Doppler broadening and impact pressure broadening will yield a Voigt profile (after Woldemar Voigt) $V(E_o; \sigma, \Gamma)$.

$$V(E_o; \sigma, \Gamma) = \int_{-\infty}^{+\infty} D(E'\ ;\ \sigma) L(E' - E_o;\ \Gamma)\, dE' \qquad (11.2.11)$$

$$\text{FWHM}, \Gamma_V \approx \Gamma_G \left(1 - c_o c_1 + \sqrt{\phi^2 + 2c_1\phi + 2c_o^2 c_1^2}\right) \qquad (11.2.12)$$

with $\phi = \Gamma_n / \Gamma_G$; $c_0 = 2.0056$ and $c_1 = 1.0593$.

11.3 ABSORPTION OF γ-RAYS BY MATTER

The processes responsible for absorption of γ-rays in matter are three types:

11.3.1 The Compton Scattering

This is the sole mechanism when low energy γ-rays are involved. A self-explanatory diagram is shown in Fig 11.3.. What can one understand from a Feynman graph is illustrated in the self-explanatory diagram of Fig 11.4. The Feynman diagram of Compton scattering is reproduced in Fig 11.5.

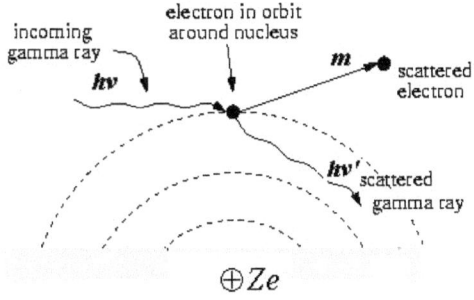

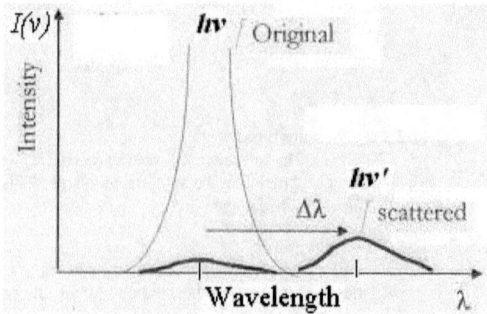

Fig 11.3 Compton Scattering by γ – rays

KE of Compton electron,

$$\frac{h\nu}{[1 + m_o c^2 / h\nu (1 - \cos\theta)]} \qquad (11.3.1)$$

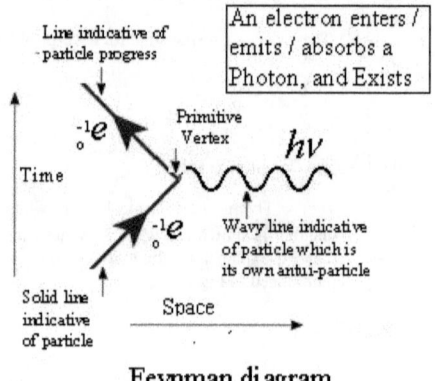

Feynman diagram

Fig 11.4. Self explanatory graph of Feynman

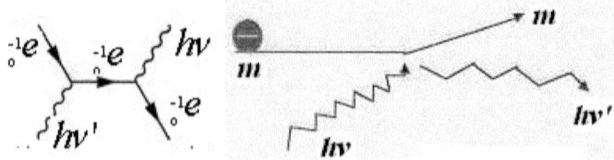

Fig 11.5 Feynman diagram for the Compton scattering

11.3.2 Internal Conversion (Conversion Electrons)

A competing electromagnetic process to γ--ray emission is that of *internal conversion*. In this process, the excitation energy of the nucleus is transferred to one of the atomic electrons, causing it to be emitted from the nucleus. The kinetic energy of the emitted *conversion* electron (different from Auger electron) depends upon the electron binding energy, B_e and the transition energy $(B_i - E_f)$

$$T_e = (B_i - E_f) - B_e \qquad (11.3.2)$$

Principally, through the photoelectric process, γ-rays interact with materials of high Z value, at low energies ($\square$ 100 keV range). β-ray emitting nuclides leave the daughter nuclides in excited states, and the daughter nuclides disintegrate further by means of γ-emission. A fraction of these γ-rays is internally converted into photoelectrons giving rise to sharp-line β-ray spectrum. Thus internal conversion competes directly with γ-emission. γ-ray interacting photo-electrically with the atom as it exits, is most readily proved in the transition between nuclear states with $I = 0$- the so-called '0 $\leftrightarrow$ 0' transitions. Here the emission of γ-rays is completely forbidden, yet internal conversion readily takes place!! This internal conversion is a '**single process**' (Fig 11.6 and 11.7).

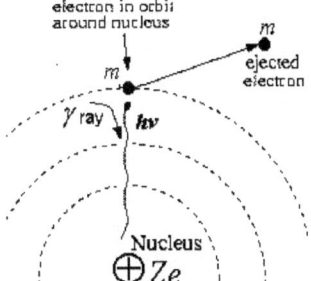

Fig 11.6 Internal conversion

The transition probability, T, for the γ-ray decay of an excited state has the form

$$T = T_\gamma + T_{i.c} \qquad (11.3.3)$$

Defining the *internal conversion coefficient*, α,

$$\alpha = \frac{T_{i.c}}{T_\gamma} = \frac{N_{i.c}}{N_\gamma} = \frac{\lambda_{i.c}}{\lambda_\gamma} \qquad (11.3.4)$$

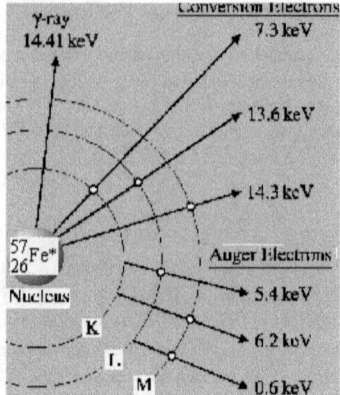

Fig 11.7 Decay scheme of Fe-57 following absorption of 14.41 *keV* gamma

where $N_{i.c}$ and N_γ are, respectively, the number of conversion electrons and γ-photons emitted per unit time in the decay, and $\lambda_{i.c}$ and λ_γ, the corresponding decay probabilities.. The values of α are dependent on the multi-polarity of a transition, particularly at low energies, and independent of the details of nuclei. It can be shown that

$$\alpha \propto \frac{Z^3}{n^3 \, E_\gamma^{5/2}} \tag{11.3.5}$$

The total decay probability is then

$$\lambda_{total} = \lambda_\gamma \, (1 + \alpha) \tag{11.3.6}$$

11.3.3 Pair Production

In this process the incident γ-ray interacts in the field of the nucleus and is completely absorbed; replacing itself with the creation of a positron-electron (e^+ - e^-) pair, as shown in Fig 11.8.

$$\gamma + (\text{Coulomb field of nucleus}) \rightarrow e^+ - e^- \tag{11.3.7}$$

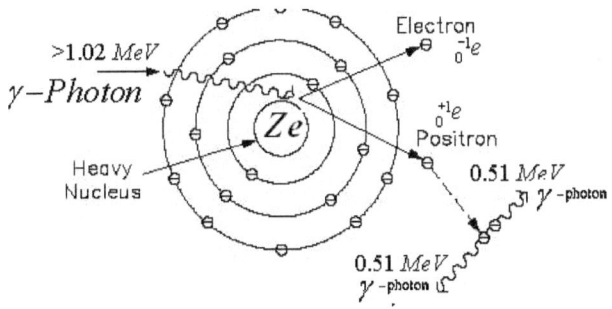

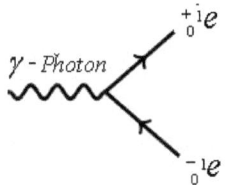

Fig 11.8 Pair production and Feynman graph

Energetics of this process is

$$h\nu = 2m_0c^2 + KE\ of\ e^+ + KE\ of\ e^- \qquad (11.3.8)$$

$$h\nu = 2(0.511\ MeV) + KE\ of\ e^+ + KE\ of\ e^- \qquad (11.3.9)$$

This brings the criterion that pair production is energetically allowed only if $E_\gamma > 1.02\ MeV$, which is called the <u>threshold energy</u> for pair production. In the process, the γ-ray is converted to an electron positron pair as shown schematically in Fig 11.8.

11.3.4 Pair annihilation

When a positron e^+ keeps colliding with the surrounding atoms until its KE tends to zero, then it captures an electron e^- from the medium and forms a *positronium atom*. Positronium is unstable and disappears in $\sim 10^{-10}\ s$, and in its place two photons are created. Applying the momentum and energy conservation laws

$$\frac{h\nu_1}{c} = \frac{h\nu_2}{c},\ and\ 2m_0c^2 = h\nu_1 + h\nu_2 \qquad (11.3.10)$$

implies that $m_0c^2 = h\nu_1 = h\nu_2 = 0.511\ MeV$.

$$e^+ + e^- \rightarrow 2\gamma \qquad (11.3.11)$$

This is the process of *pair annihilation* or the positron-electron annihilation (Fig 11.9).

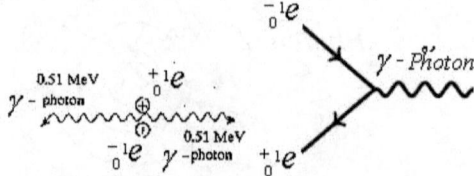

Fig 11.9 Positron-electron annihilation and Feynman diagram

11.3.5 The Competing Processes in γ-ray decay

There are a number of topics which arise in considering γ-ray decay processes. It is known from earlier subsections of this chapter that internal conversion is an additional mode of nuclear de-excitation. The processes by which absorbing materials can reduce the intensity of γ-radiation are: 1) the Photoelectric Effect, 2) the Compton Effect and 3) Pair Production. Only the first two effects occur in the present experiment as pair production requires γ-ray energies of at least 1.02 MeV. These three compete among themselves while γ-ray decay occurs for nuclei. Fig 11.10 is indicative of the dominance of these processes.

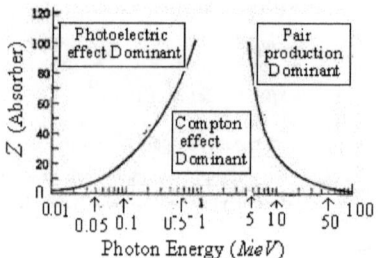

Fig 11.10 The three competing processes in γ-decay

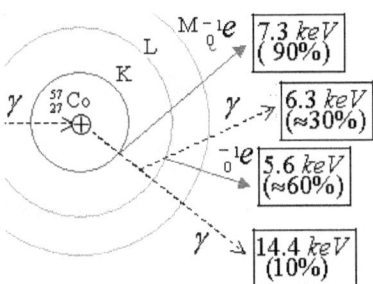

Fig 11.11 The percentage or relative intensity of the γ-emission, which competes with internal conversion IC)

11.3.6 Range (Absorption) of Gamma-Rays by a Metal

High energy γ-rays are the most penetrating of all EM radiation and are not appreciably absorbed by even several centimeters of materials such as concrete or wood. Dense metals, however, particularly lead, are effective γ-ray absorbers.

Let I be the intensity of γ-radiation detected at some distance from the source. Interposing an absorbing material of thickness x between source and detector will reduce the intensity by a certain amount. For a small increase Δx in absorber thickness, the fractional change in intensity, $\Delta I / I \propto \Delta x$ (Fig 11.12). On integrating the equation $dI / I = -dx$ yields the result:

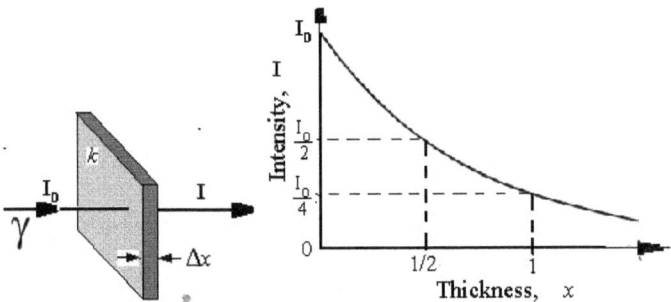

Fig 11.12 γ-radiation passing through a slab

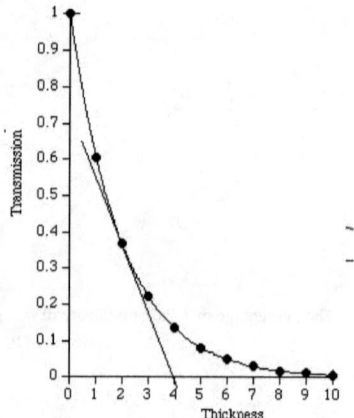

Fig 11.13 A plot of transmission versus x for a metal

$$I = I_0 \, e^{-k\,x} \tag{11.3.12}$$

where I_o is the incident intensity, that is, $I = I_0$ when $x = 0$. I is the intensity left after the radiation traverses a thickness x of material, k is the absorption (attenuation) coefficient

$$k = k_{i.c} + k_{Comp} + k_{Pair}$$

and $e = 2.718$ is the base of the system of natural logarithms

Fig 11.13 displays how transmission in and thickness of an absorber are related.

11.3.6.1   Determination of the Absorption Coefficient and Half-Value Thickness of a Material

In this experiment you will measure the absorption of the 0.662 Me γ-rays emitted by ^{137}Cs first by aluminum, then by lead. The set up is shown in Fig 11.14.

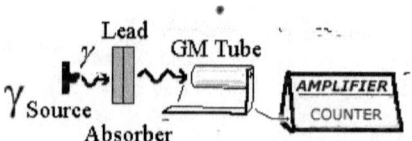

Fig 11.14 γ-ray absorption setup

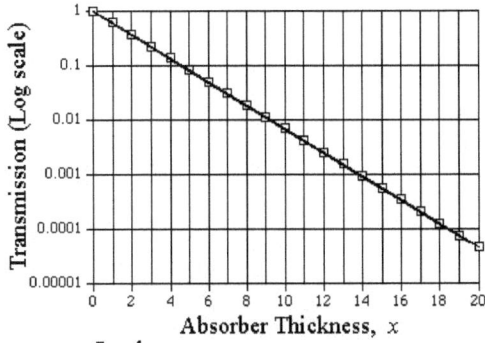

Fig 11.15 A plot of *ln* I versus *x* for a metal

Data acquisition involves adjusting the equipment with baseline at 64.2% and window at 4%. Place the source in the bottom slot of the holder and take 9 readings of the number of counts in a 10 s interval when no absorber is present. Place 0.25 inches of aluminum absorber over the source and take another nine readings. Repeat for absorber thickness of 0.50, 0.75, 1.25 and 1.75 inches. Find the average number of counts for each absorber thickness and their absolute uncertainties, i.e., N ± N /3. A plot of *ln* I versus *x* for the data yields a straight line (Fig 11.15).

11.4   MULTIPOLE CHARACTER (ML) OF γ-RADIATION (Multipole Moments)

In general, electric (charge) radiation or magnetic (current, magnetic moment radiation) can be classified into multipoles (ML), e.g., Electric dipole, quadrupole or Magnetic dipole. The radiation field will be a sum of the multipole contributions; however, usually one or two multipoles dominate.

11.4.1 (R/λ) for the Nuclear System

Atomic physics has given the facts that atomic energy levels are separated by few eV, the visible photons emitted have wavelengths $\lambda_{atomic} \sim 10^{-7} m$ and size of an atom $R_{atomic} \sim 10^{-10} m$. On the other hand, the nuclear energy levels are separated by MeV and γ-photons have wavelengths $\lambda_{nuclear} \sim 10^{-13} m$ and nuclear size $R_{nuclear} \sim 10^{-14} m$ (Table 11.1). The difference in these ratios has profound implications. For example, in case of atoms, one need consider *electric dipole radiation* only, whereas in the subatomic case, the various multipole radiations play important roles.

Table 11.1 Multipolarity of Nuclear γ-Radiation

System	Size, R	Wavelength λ	(R/λ)
Atom	$[n^2 (0.53 \text{ Å})/Z]$ $= \square 10^{-10} m$	$[1/\lambda = R_H[1/2^2 - 1/n^2]]$ $= 10^{-7} m$	10^{-3}
Nucleus	$[r_0 A^{1/3} fm]$ $= \square 10^{-14} m$	$[197/E_\gamma(MeV) fm]$ $= \square 10^{-13} m$	$10^{-1} - 10^{-2}$

The interaction (potential or force) between the nucleus and EM radiation is better useful if expressed in terms of an infinite series of powers of the factor $\dfrac{R}{\lambda}$, where

$$R = r_0 A^{1/3} fm, \qquad (11.4.1)$$

$$\lambda = \frac{\lambda}{2\pi} = \frac{\hbar c}{E_\gamma} = \frac{197}{E_\gamma(MeV)} fm \text{ of the gamma photon} \qquad (11.4.2)$$

A = Mass number of the nuclide.

$$\left.\frac{R}{\lambda}\right|_{nuclear} = \frac{r_0 A^{1/3} E_\gamma(MeV)}{197} \quad \square \quad 10^{-1} - 10^{-2} \qquad (11.4.3)$$

According to theory, for allowed γ-transitions, the decay constant of the nuclide,

λ_γ = an infinite series in powers of $(R/\lambda)^2$. $\approx$ The first non-vanishing term of the Series.

The forbidden atomic transitions are relatively more forbidden than the nuclear ones, because

$$\left.\frac{R}{\lambda}\right|_{Atoms} \bigg/ \left.\frac{R}{\lambda}\right|_{nuclear} \quad \square \quad 10^{-1} \qquad (11.4.4)$$

This has profound implications.

11.4.2 Classification of γ-rays

The expansion procedure of λ_γ separates the radiation from a nucleus into distinct types, known as *Multipole radiations*. The elementary emission / absorption process can be either *kinematical* or *dynamical* ones.

For static distribution of nuclear charge e_i at (x_i, y_i, z_i) the multipole moments have a dimensionality $\sum_i e_i x_i^L$, for a moment of order L. (i.e., for example, $L = 0$, means charge $Q = \sum_i e_i$).

Kinematical process deals with the energy and momentum of a photon emitted at a certain angle. This can be solved with the application of the laws of energy and momentum conservation.

Dynamical process deals with the decay or polarization of the photons. This can be dealt with knowledge of the form of the interaction. In quantum mechanical calculations, the angular distribution of γ- intensity in an appropriate coordinate system is represented in terms of spherical harmonics $Y_L^M(\vartheta,\varphi)$, of rank $L = 0, 1, 2, 3, \ldots$,

Electric Dipole radiation in nuclei is the quantum equivalent of the radiation produced classically by an oscillating electric dipole. The amplitude for an electric

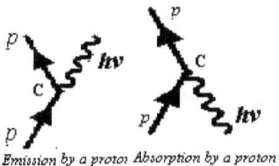
Emission by a proton   Absorption by a proton

Fig 11.16 Feynman diagrams of emission and absorption of a photon by a proton.

dipole transition between two states $|i>$ and $|f>$ is proportional to the matrix element $M_{if} = <f| \sum q\, r_i |i>$, where $\sum q\, r$ sum of the dipole moments of all the protons in a nucleus. Feynman diagram of the transition (emission / absorption) can be represented by Fig 11.16. C at the vertex is the 'em coupling' constant.

Multipole Moment

Transition Probability,

$$\lambda_\gamma(E1) \propto \frac{4 e^2 (|x_{fi}|^2 + |y_{fi}|^2 + |z_{fi}|^2) E_\gamma^3}{3 \hbar^4 c^3} \qquad (11.4.5)$$

i.e.,
$$\lambda_\gamma(E1) \propto \frac{1}{4\pi\lambda} \left(\frac{E_\gamma}{\lambda c}\right)^3 |M_{fi}|^2 \propto \frac{R^2}{\lambda^3} \qquad (11.4.6)$$

Similarly, $\lambda_\gamma(E2) \propto \dfrac{e^2}{4\pi \varepsilon_o \lambda} \left(\dfrac{E_\gamma}{\lambda c}\right)^5 R^4 \propto \dfrac{R^4}{\lambda^5}$ (11.4.7)

In general, $\lambda_\gamma(EL) \propto \dfrac{1}{\lambda}\left(\dfrac{R}{\lambda}\right)^{2L}$ (11.4.8)

The multipole radiation is characterized by its order given by 2^L, where $\sqrt{L(L+1)}\,\hbar$ is the angular momentum carried off by the photons, with respect to the source of the radiation field. Since angular momentum is quantized, the change in angular momentum is L units. For each value of L, there are two classes of radiation, called *Electric (EL) and Magnetic (ML)* multipole radiations. The two differ in parity associated with the EM radiation.

11.4.3 Parity in γ-transitions

In addition to the angular momentum selection rule, there is the parity selection rule, which determines the electric or magnetic nature of the emitted radiation. Thus even-multipole electric and odd-multipole magnetic transitions have even parity, and occur when $\Pi_i = \Pi_f$, whilst odd-multipole electric and even-multipole magnetic transitions have odd parity and occur when $\Pi_i = -\Pi_f$.

The parity of the wave function of the entire system, nucleus and the EM radiation, is *conserved*. For the combination of parities of the initial and final states, $|i>$ and $|f>$, as in

EL transitions

$$\Pi_i\,\Pi_f(EL) = (-1)^L.$$ (11.4.9)

and ML transitions

$$\Pi_i\,\Pi_f(ML) = (-1)^{L+1}$$ (11.4.10)

11.4.4 Selection Rules

11.4.4.1 Angular Momentum Conservation

A multipole of order L carries $L\,\hbar$ units of orbital angular momentum.

For a given transition the total angular momentum must be conserved. This means that the initial angular momentum,

$$I_i = I_f + L,$$ (11.4.11)

and the three terms form a closed vector triangle. The angular momentum selection rule is thus

$$|I_i - I_f| \le L \le (I_i + I_f),\ \&\ L \ne 0$$ (11.4.12)

Since the photon has an intrinsic spin of 1, a transition in which $I_i = I_f$, $L = 0$ cannot occur. A *stretched transition* is one in which the photon carries the difference between the angular momentum of the initial and final states. Thus there are no E0 and M0 radiation fields (the latter is also forbidden by the lack of magnetic monopoles). Note that even classically the EM field carries angular momentum.

In addition to the angular momentum selection rule, there is the parity selection rule, which determines the electric or magnetic nature of the emitted radiation.

11.4.4.2 Parity Conservation

It can be seen that for an E1 transition to occur between two states, the change in angular momentum should be 1, corresponding to emission of one photon. If the change in angular momentum between two states is very large, then the lower order (more probable) transitions such as electric dipole E1 or electric quadrupole E2 or magnetic dipole M1 cannot occur. In this case the difference in angular momentum is so large that the state is very long lived (M2 decay is less probable, but it finally does happen).

11.4.4.3 Selection Rules

Conservation of both angular momentum and parity lead to a set of Selection Rules for γ-ray transition, which determine which transitions are allowed.

$E1$ transition: $\Delta J = 0, \pm 1$, $\boxed{0 \to 0}$ not allowed. \hspace{1cm} (11.4.8)
$\Delta \Pi$ = 'yes'. since $L = 1$, Π = odd

$M1$ transition: $\Delta J = 0, \pm 1$, $\boxed{0 \to 0}$ not allowed. \hspace{1cm} (11.4.9)
$\Delta \Pi$ = 'no'.

$E1$ transition: $\Delta J = 0, \pm 1, \pm 2$ $\boxed{0 \to 0;\ 0 \to 1;\ 1 \to 0;\ \frac{1}{2} \to \frac{1}{2}}$ are not allowed.
$\Delta \Pi$ = 'no'. since $L = 2$, Π = even

(11.4.10)

11.4.4.4 Parity Rules

Consider energy levels E_i, with I_i, π_i and E_f with I_f, π_f and a transition $E_i \to E_f$ emitting a photon of energy E_γ.

EL	Parity change?	Example	ML	Parity change?	Example
E1	Yes	$1^- \to 0^-$	M1	No	$1^+ \to 0^-$
E2	No	$2^+ \to 0^+$	M2	Yes	$2^- \to 0^+$
E3	Yes	$3^- \to 0^+$	M3	No	$3^+ \to 0^+$
E4	No	$4^+ \to 0^+$	M4	Yes	$4^- \to 0^+$

Table 11.3   Multipole Selection Rules

Multipole	L^Π	ΔJ	$\Delta\Pi$	Multipole	L^Π	ΔJ	$\Delta\Pi$
$E1$ Electric Dipole	1^-	$0, \pm 1$	yes	$M1$ Magnetic Dipole	1^+	$0, \pm 1$	no
$E2$ Electric Quadrupole	2^+	$0, \pm 1, \pm 2$	no	$M2$ Magnetic Quadrupole	2^-	$0, \pm 1, \pm 2$	yes
$E3$ Electric Octupole	3^-	$0, \pm 1, \pm 2, \pm 3$	yes	$M3$ Magnetic Octupole	3^+	$0, \pm 1, \pm 2, \pm 3$	no
$E4$ Electric hexa-decupole	4^+	$0, \pm 1, \pm 2, \pm 3, \pm 4$	no	$M4$ Magnetic hexa-decupole	4^-	$0, \pm 1, \pm 2, \pm 3, \pm 4$	yes

All these information are summarized and listed in Table 11.3.

Fig 11.17 illustrates a few typical nuclides which emit the multipole radiations.

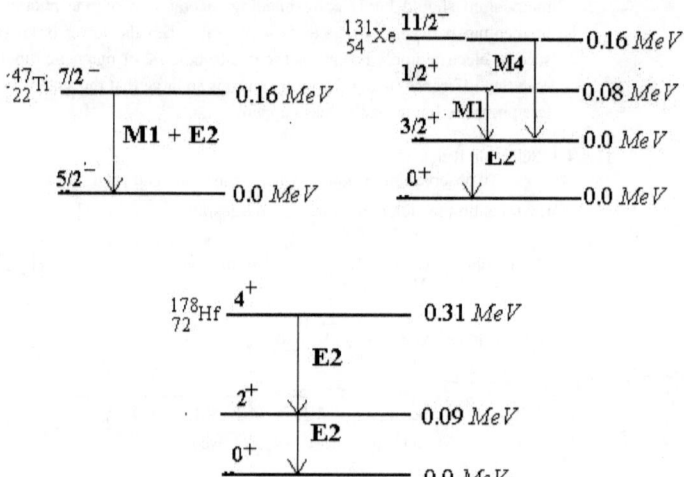

Fig 11.17 Typical γ-ray transitions to be expected

11.4.4.6 Further discussions

Examples

Let there is a nucleus in vibrating state. It has the states 0^+ (ground), 2^+ (1 phonon), and $4^+, 2^+$, and 0^+ (2 phonon states at around twice the energy of the first).

$2^+ \rightarrow 0^+$ (1 phonon to ground) will have to take away only $L = 2$. This gives $E2$.

$0^+ \rightarrow 0^+$ (1 phonon to ground) this is not permitted.

$0^+ \rightarrow 2^+$ (2 phonon to 1 phonon); $L = 2$. This gives $E2$.

$2^+ \to 0^+$ (2 phonon to ground) will have to take away only $L = 2$. This gives $E2$.

$2^+ \to 2^+$ (2 phonon to 1 phonon); $L = 1, 2, 3, 4$; $\sigma L = M1, E2, \underline{M3, E4}$.

The last two are negligible, because impossible.

$4^+ \to 2^+, 0^+$ (2 – 2 phonon) are weak transitions since their energy differs little and not detected.

$4^+ \to 0^+$ (2 - ground) $L= 4$, $E4$ is delayed and hence not detected.

$4^+ \to 2^+$ (2 – 1 phonon), $L = 2, 3, 4, 5, 6$; no change in parity occurs.

$\sigma L = \underset{\text{very hindered, and not detected}}{E2, M3, E4, M5, E6}$

11.4.5 Probability of Multipole Transitions. Transition Rates

These are determined by the Fermi's Golden Rule.(Eqn. 10.6.15)

11.4.6 The reason for NO γ – ray emission for the $\boxed{0^+ \to 0^+}$ transitions

$0^+ \to 0^+$ Transitions.

γ–transitions are not permitted, since $L = 0$.; but though they can proceed by usual internal conversion process; other possibilities are

1) Emission of 2-photon, and
2) Pair production ($^{+1}_{0}e + ^{-1}_{0}e$), provided the transition energy is $> 2\, m_e c^2$..

11.4.7 Further Discussions

From Shell Model' of nucleus for *Electric (EL)* transitions of order 2, 'the decay probability ($\propto |M_{if}|^2$) is seen to increase with L.

$$\frac{\lambda_\gamma (EL)}{\lambda_\gamma (ML)} = \frac{1}{10}\left(\frac{R}{\hbar / M_p\, c}\right)^2 = 4.4\, A^{2/3} \qquad (11.4.11)$$

whence $\lambda_\gamma (EL)$

$$\tau_{EL} \cong \left(\frac{137}{E_\gamma / m_0 c^2}\right)^{2L+1} \left(\frac{e^2 / m_0 c^2}{r_0\, A^{1/3}}\right)^{2L} \left(\frac{h}{m_0 c^2}\right)\frac{1}{S}$$

$$\cong \left(\frac{0.645 \times 10^{-21}}{S}\right) \left(\frac{140}{E_\gamma (MeV)}\right)^{2L+1} A^{-(2/3)L} \quad s \quad (11.4.12)$$

the statistical factor S is $S = \dfrac{2(L+1)}{L\cdot[(2L+1)!]^2} \left(\dfrac{3}{L+3}\right)^2$

From this, generally for a EL transition one gets

$$\tau(EL) \propto \frac{1}{\lambda}\left(\frac{R}{\lambda}\right)^{2L} \quad (11.4.13)$$

In the same way, for a ML transition

$$\tau_{ML} \cong (2\pi\nu)\left(\frac{e^2}{\hbar c}\right)(10\ S)\left(\frac{\hbar}{M c R}\right)^2 \left(\frac{R}{\lambda}\right)^{2L}$$

$$\cong \left(\frac{290 \times 10^{-21}}{S}\right)\left(\frac{140}{E_\gamma (MeV)}\right)^{2L+1} A^{-(2/3)(L-1)} \quad s \quad (11.4.14)$$

In terms of τ_{EL},

$$\tau(ML) \square\ 10 \left(\frac{\hbar}{M_p R c}\right)^2 \tau(EL) \quad (11.4.15)$$

$$\therefore \quad \frac{\tau(ML)}{\tau(EL)} \approx 4.4\ A^{2/3} \quad (11.4.16)$$

Successive multipoles have decay constants

$$\frac{\lambda(EL+1)}{\lambda(EL)} \approx \frac{\lambda(ML+1)}{\lambda(ML)} \approx \left(\frac{R}{\lambda}\right)^2 \square\ 10^{-3} \quad (11.4.17)$$

It is thus seen that for successive multipoles differ in decay constant by three orders of magnitude. Higher multipoles only become important when E1 transitions are forbidden by selection rules.

11.4.8 Weisskopf Estimates

The transition rates, under the assumption that the transition is due to a single proton moving from one shell model state to another, can be estimated from

$$T(EL) = \frac{8\pi(L+1)}{\hbar L\left((2L+1)!!\right)^2} \frac{e^2}{4\pi\epsilon_0 \hbar c}\left(\frac{E_\gamma}{\hbar c}\right)^{2L+1}\left(\frac{3}{L+3}\right)^2 cR^{2L}$$

and

$$T(ML) = \frac{8\pi (L+1)}{\hbar L((2L+1)!!)^2} \left(\mu_p - \frac{1}{L+1}\right)^2 \left(\frac{\hbar}{m_p c}\right)^2 \left(\frac{se^2}{4\pi\epsilon_0 \hbar c}\right) \left(\frac{E_\gamma}{\hbar c}\right)^{2L+1} \left(\frac{3}{L+2}\right)^2 cR^{2L-2}.$$

In the above, μ_p is the magnetic moment of the proton and M_p is the proton mass. The wave functions of the states are those obtained using a square well potential. By taking $R = r_0 A^{1/3}$ fm, and by setting the $\left(\mu_p - \frac{1}{L+1}\right)^2 = 10$, estimates can be made for the lower multipole orders. These are known as the *Weisskopf Estimates*, and are given in Table 11.4.

TABLE 11.4. Given the X-ray energy in MeV, the units are in s. Weisskopf estimates

$T(E1) = 1.0 \times 10^{14} A^{2/3} E^3$	$T(M1) = 5.6 \times 10^{13} E^3$
$T(E2) = 7.3 \times 10^{7} A^{4/3} E^5$	$T(M2) = 3.5 \times 10^{7} A^{2/3} E^5$
$T(E3) = 3.4 \times 10^{1} A^{2} E^7$	$T(M3) = 1.6 \times 10^{1} A^{4/3} E^7$
$T(E4) = 1.1 \times 10^{-5} A^{8/3} E^9$	$T(M4) = 4.5 \times 10^{-6} A^{2} E^9$

Actually, though these are only estimates are, calculations of the transition rates, yet provide values that can be compared to those experimentally measured ones..

11.5   Bremsstrahlung Process

When fast electrons are incident on heavy nuclei, it will get scattered either elastically or inelastically. In the case of inelastic case, the electron will be scattered and as slow ones and *bremsstrahlung* takes place (Fig 11.18).

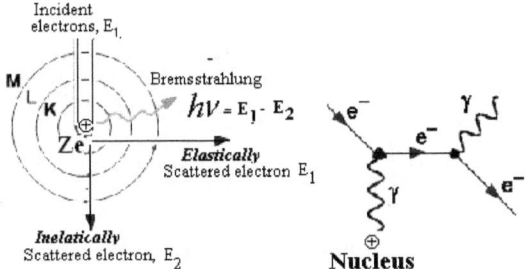

Fig 11.18   Bremstraahlung Process and its Feynman diagram

11.6 ENERGETICS OF GAMMA DECAY

Let M_o^* and M_o refer to the rest masses of the initial excited nucleus and the final state, respectively. Conservations of energy and momentum require

$$M_o^* c^2 = M_o c^2 + E_\gamma + T_f \qquad (11.6.1)$$

$$0 = p_f + T_f \qquad (11.6.2)$$

where T_f and p_f denote the recoil KE and momentum of final nucleus.

$$T_f = \frac{p_f^2}{2 M_o} = \frac{p_\gamma^2}{2 M_o} = \frac{E_\gamma^2}{2 M_o c^2} \qquad (11.6.3)$$

Typically $E_\gamma = 2$ MeV, and $A = 50$, so $T_f \approx 40$ eV is negligible.

$$E_\gamma \cong (M_o^* - M_o) c^2 \qquad (11.6.4)$$

11.7 Decay constant (λ_γ) of γ-Decay

γ-decay of an excited nucleus has $\tau_{1/2}(\gamma\text{-decay}) = 10^{-16}$ s to 100 y. This may be compared with $\tau_{1/2}$ (atomic valence electrons) = 10^{-8} s and 10^{-15} s for hole states formed after ejection of an inner electron shell. James Clerk Maxwell's equations show that an accelerated point charge e radiates EM radiation at a rate of

$$\frac{dE}{dt} = \frac{2}{3} \frac{e^2 a^2}{c^3} \quad esu \qquad (11.7.1)$$

$a = \sqrt{(a_x^2 + a_y^2 + a_z^2)}$ is the acceleration of the charge.

If a proton in nucleus is the radiating charge, then considering it to execute SHM at frequency ω,

$$\left.\frac{dE}{dt}\right|_{average} \cong \frac{e^2 R^2 \omega^4}{3 c^3} = \frac{\hbar \omega}{\tau} \qquad (11.7.2)$$

Associating the mean life for a γ-ray, the decay constant

$$\lambda_\gamma = \frac{1}{\tau} \Box \frac{e^2 R^2 E_\gamma^3}{3 \hbar^4 c^3} \qquad (11.7.3)$$

11.9 ANGULAR CORRELATIONS

The direction in which a γ-ray is emitted in a transition between two nuclear states whose spins and z-components are J_i, M_i, and J_f, M_f depends on the values of M_i and M_f. In general all magnetic substates of a decaying nuclear state are occupied with equal probability and as a result there is no preferred direction of the emission of γ-ray. For unequal populations of substates methods can be arranged, however. A widely used method is to study the directions of emission of two successive γ-rays, γ_1 and γ_2 (Fig 11.19).

It is known from detailed theory that the probability $I(\theta)$ of γ_2, detected at angle θ to the direction of γ_1-ray is

$$I(\theta) = A + B\cos^2\theta + C\cos^4\theta + \ldots$$

where the values of A, B, and the maximum power in $\cos^n\theta$ are dependent on the spins and parities of the nuclear states involved and on the multipolarities.

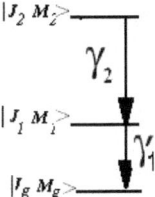

Fig 11.19 Two successive γ-rays with angular correlations

Using a radioactive source which has two γ-rays that are emitted with an anisotropic decay, the angular correlation between the two rays will be measured.

In a PAC (Perturbed Angular Correlation)-experiment one measure the characteristic frequencies ω_L, ω_0 associated with the hyperfine interaction. For this use is made of the ^{111}In (I=5/2) probe. The probe decays to ^{111}Cd through some intermediate level with a lifetime of about 122 ns. Activity is introduced into the sample by diffusion or implantation. The detector set-up consists of 4 BaF$_2$-scintillation detectors. One essentially measures the nuclear decay of the intermediate I=5/2 level (Q=0.8 b, $\mu = -0.766$ nm). In the presence of hyperfine interactions oscillations appear superimposed on the exponential.

11.10 GAMMA RAY SPECTROMETER

The energy and count rate of γ-rays emitted by radio-nuclides in substances are measured by a method called Gamma **spectroscopy**, and is an extremely important measurement. A detailed analysis of the γ-ray energy spectrum is used to determine the identity and quantity of gamma emitters present in a material. The equipment used in gamma spectroscopy includes a particle detector, a pulse sorter (multichannel analyzer, MCA), and associated amplifiers and data readout devices. The detector is often a NaI(Tl) scintillator (Fig 11.20). High resolution gamma spectroscopy often utilizes Compton suppression.

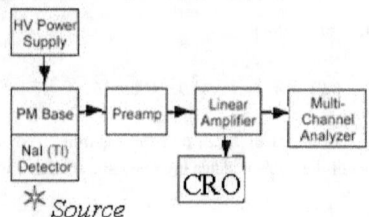

Fig 11.20 Block diagram of a Gamma ray spectrometer

11.10.1. The Gamma-ray Laser

Lev A. Rivlin (1961) first introduced the concept. Delayed induced Gamma ray emission (IGE) ^{179}Hf exposed from a 4 *MeV* high Intense Linac. There are over 800 known nuclear isomers. The five nuclear isomers that could produce IGE are ^{178}Hf, ^{180}Ta, ^{187}Os, ^{186}Pt, ^{66}Zn.

11.10 THE MOSSBAUER EFFECT

11.11.1 Introduction

If a radio-active nucleus, say ^{57}Fe, emits a γ-ray of 14.4 *keV*, the energy of this γ-photon is not really 14.4 *keV*, because the nucleus recoils back while emitting the photon, much like a gun will recoil if a bullet is fired. This recoil changes the energy of the gamma photon. The energy change is large enough so that at normal temperatures the gamma particle cannot be absorbed by another ^{57}Fe nucleus.

Nuclei and solids are quantum mechanical systems. Using these ideas a spectroscopic technique called Mossbauer technique has been developed. Key to the success of the technique is the discovery of *recoilless gamma ray emission and absorption*, now referred to as the 'Mössbauer Effect', after its discoverer Rudolph L. Mössbauer (1957) who received the Nobel

Prize in Physics in 1961 for his work. He used γ-rays of energy 0.129 MeV, corresponding to the transition from ground to the first excited state of ^{191}Ir. The Mössbauer effect has also been used to verify the prediction of gravitational redshift (the frequency of electromagnetic radiation is dependent on the strength of the gravitational field) (Pound and Rebka 1959).

11.11.2 The Phenomenon of Mössbauer Effect

Nuclei in atoms undergo a variety of energy level transitions, often associated with the emission or absorption of a γ-ray. These energy levels are influenced by their surrounding environment, both electronic and magnetic, which can change or split these energy levels. These changes in the energy levels can provide information about the atom's local environment within a system and ought to be observed using resonance-fluorescence. There are, however, two major obstacles in obtaining this information: the 'hyperfine' interactions between the nucleus and its environment are extremely small, and the recoil of the nucleus as the gamma-ray is emitted or absorbed prevents resonance

The Mössbauer Effect, otherwise known as Nuclear Gamma Ray Resonance (NGR), is essentially a form of high resolution spectroscopy. It is the phenomenon in which radioactive nuclei (say, ^{57}Fe) bound in crystal lattices emit gamma rays, having an energy equal to that of the nuclear transition, which are absorbed by similar nuclei (i.e., ^{57}Fe) in the ground state. It is very popularly referred to as "Recoil-less emission and absorption of gamma rays by solids" (Fig 11.21). In the emission of γ-ray photons from an atom corresponding to a transition to nuclear ground state, momentum must be conserved, so the atom must have small recoil. Energy balance then implies γ-rays are emitted with a spread of energies. When an atom is part of a crystal lattice, however, the entire lattice may recoil resulting in a quantized vibrational energy termed a *phonon*. If no phonon is emitted or absorbed, (A Mössbauer spectral line is an extremely monochromatic energy quanta, whose broadening Γ_n (FWHM peak-height) is theoretically determined solely by the life time from the relation) .the emitted γ-rays have a very small spread of energies, given by

$$\Gamma_n \tau = \hbar, \qquad (11.11.1)$$

where $\tau = 10^{-8} - 10^{-15} s$ is the life time of the nuclear excited level.

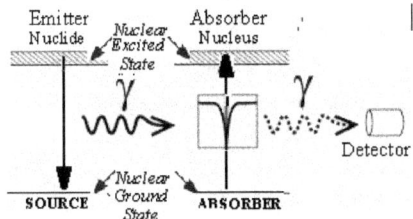

Fig 11.21 The Mössbauer effect principle

Since in Chapter 4 (Section 4.7) a detailed description of Mossbauer Technique has been given the same will not be repeated her.

11.11.3 Conversion Electron Mössbauer Spectroscopy (CEMS)

Conversion Electron Mössbauer Spectroscopy (CEMS) utilizes the emission of conversion electrons from the decay of the 14.4 keV state in the absorber to record the spectrum. This is useful for samples with thick substrates which would block transmission of gamma-rays or for studies of the surfaces of samples rather than the bulk. As the ratio α of conversion electrons to γ – rays emitted by the 14.4 keV Mössbauer event in ^{57}Fe is 8.21 the counting efficiency of CEMS is much greater than the transmission method.

REVIEW QUESTIONS

R.Q. 11.1   What is the wavelength of a 1 MeV gamma ray? (Answer: 1.23 fm).

R.Q 11.2   A sample of Uranium emits photoelectrons from the K-shell as result of interaction with γ – rays from $^{137}_{55}$Cs. A magnetic beta ray spectrometer gave the value 3083 *Gauss-cm* as momentum. Determine a) the KE of the photoelectrons, and b) E_γ of the gamma ray. Given: E_K(U) = - 115.59 keV .[Answers: a) 545.35 keV, b) 661.94 keV.].

R.Q. 11.3   Explain the phenomenon of nuclear fluorescence.

R.Q 11.4   Show that there are no E0 and M0 gamma radiations.

R.Q 11.5   Examine if E2 radiation changes parity or not.

R.Q. 11.6 Find out if the selection rules permit emission of M1 radiation by nuclei.

R.Q. 11.7 Find out the multipolarity of the radiations emitted by $^{57}_{26}$Fe nuclide having three states having specifications $|\frac{5}{2}>$, $|\frac{3}{2}>$ and $|\frac{1}{2}>_g$. [Answers: $|\frac{5}{2}> \xrightarrow{\gamma M2} |\frac{1}{2}>_g$, $|\frac{5}{2}> \xrightarrow{\gamma} |\frac{3}{2}>$ and $|\frac{3}{2}> \xrightarrow{\gamma M1} |\frac{1}{2}>_g$].

R.Q. 11.8 What is the multipolarity of the radiation emitted by $^{124}_{54}$Xe in the given energy level diagram?

```
3.58 ――――― 1⁻

1.60 ――――― 4⁺
0.52 ――――― 2⁺
²⁴₅₄Xe 0.0 ――――― 0⁺
```

R.Q. 11.9 Find out the multipolarity of the radiations emitted by $^{47}_{22}$Ti nuclide having three states having specifications $|\frac{7}{2}> \longrightarrow |\frac{5}{2}>$ [Answers: $\gamma M1 + \gamma E2$; i.e., $|\frac{7}{2}> \xrightarrow{\gamma M1} |\frac{5}{2}>$ and $|\frac{7}{2}> \xrightarrow{\gamma E2} |\frac{5}{2}>$].

R.Q. 11.10 Find the type of multipole radiation that occurs in the given diagram

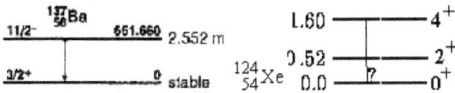

R.Q. 11.11 A radio-nuclide with radius $\approx$ 5 fm undergoes γ-decay with E_γ = 1.6 MeV. What is the decay constant? [Answer: $\lambda_\gamma \approx 2 \times 10^{15}$ s^{-1}].

R.Q.11.12 Find the Mossbauer recoil a hydrogen atom (A=1) with energy in the visible light frequencies, $E_t \cong 1eV$. [Answer $E_R = 1.93 \times 10^{-3}$ eV]

R.Q.11.13 Calculate the recoil energy E_R, the natural line width Γ_n and $E_\gamma = E_t - E_R + 2\sqrt{E_k \cdot E_R} \cos\varphi$ for a 14.4 keV transition from the first excited state to the ground state in ^{57}Fe atom, calculating the recoil energy yields: [Answer $E_R = 0.002 eV$, $\Gamma_n = 5 \times 10^{-9}$ eV].

R.Q. 11.14 In a Mössbauer set up, ^{57}Co *in stainless steel matrix* giving 14.412 keV gamma radiations and a thin absorber is used. Find out the FWHM of the radiation (Γ_γ) and the Doppler velocity corresponding to the FWHM. Given, $\tau_{1/2} = 97.7$ ns. [Answers $\Gamma_\gamma = 2(0.467 \times 10^{-8}$ eV), $v = 0.192$ mm / s].

R.Q. 11.15 The Mössbauer spectrum of ^{57}Fe of ferrocene [Fe(C$_5$H$_5$)$_2$] at $T = 20$ K consisted of two lines at -0.50 mm / s and $+1.88$ mm / s, relative to a stainless steel source. What are the isomer shift δ_{IS} and the quadrupole splitting constant Δ for the excited nucleus? [Ans: $\delta_{IS} = 7.9$ MHz, $\Delta = 55$ MHz].

R.Q. 11.16 A Mossbauer spectrum of α-Fe was recorded using a 14.41 keV gamma radiation and a MCA having 256 channels. The centroid of the spectrum occurs at the 62.32 channel. For ^{57}Fe, $E_\gamma = 14.41$ eV and this gives 1 mm s^{-1} → 4.800 x10^{-8} eV → 11.61 MHz

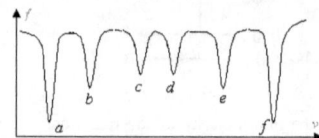

Given that the ground state has $g_{Ig} = 0.180$, $\mu_{Ig} = 0.090$ nm, by measuring the separation between the various components of the six-line spectrum, one gets for the excited state of ^{57}Fe,

$\Delta E(|\frac{1}{2},-\frac{1}{2}\rangle \to |\frac{3}{2},-\frac{3}{2}\rangle) - (|\frac{1}{2},-\frac{1}{2}\rangle \to |\frac{3}{2},-\frac{1}{2}\rangle) = 1.07 \times 10^{-7}$ eV $= (a-b)$.

$\Delta E(|\frac{1}{2},-\frac{1}{2}\rangle \to |\frac{3}{2},-\frac{1}{2}\rangle) - \Delta E(|\frac{1}{2},-\frac{1}{2}\rangle \to |\frac{3}{2},+\frac{1}{2}\rangle) = 5.04380 \times 10^{-9}$ eV

$\Delta E(|\frac{1}{2},+\frac{1}{2}\rangle \to |\frac{3}{2},+\frac{1}{2}\rangle) - (|\frac{1}{2},-\frac{1}{2}\rangle \to |\frac{3}{2},+\frac{1}{2}\rangle) = 1.9 \times 10^{-7}$ eV $= (b-d)$

R.Q. 11.17 Determine the g-factor, magnetic moment of the excited state of ^{57}Fe and the internal magnetic field in the absorber. [Answers: $g_{Izx} = -0.1015$, $\mu_{Iex} = -0.153$ nm, where the relation $\Delta E = g_I H_n$ nm is used. It also gives $H_n = \frac{1.07 \times 10^{-7}}{g_{Izx}} = -33.3$ T for α-Fe].

%%$*%*%*%&%*%*%*

Chapter 12

DYNAMIC PROPERTIES OF NUCLEI III:

STRONG INTERACTIONS

ALPHA DECAY

Chapter 12

DYNAMIC PROPERTIES OF NUCLEI - III:
STRONG INTERACTIONS
ALPHA DECAY

"I am become Death, the shatterer of worlds." — Lord Shiva in "The Bhagavad Gita".

12.1   INTRODUCTION

It was seen in Chapters 10 and 11 that in both the dynamic properties of $\beta-$ and $\gamma-$ decay processes there involve production of particles (electrons, neutrinos and photons) not present permanently in the nuclide, and Q is negative. But $\alpha-$ decay differs fundamentally from these two in that there occurs a rearrangement of nucleons resulting in the emission of an $\alpha-$ particle, as $Q = +$ve. $\alpha-$ particles with only discrete energies are observed. The lifetimes of $\alpha-$ emitters have very wide variation, for example, 1.4×10^{10} yrs for ^{232}Th and 10^{-7} s for ^{212}Po, both releasing energies of magnitude MeV.

12.2   ALPHA DECAY AND SPECTRA

12.2.1 Process of α- decay

Since the slope of the E_B / A versus A curve is negative for $A > 100$, it appears that any process which decreases A for $A > 100$ would lead to an increased E_B and hence energetically permitted. Although such a nuclide may break up in a variety of ways, the most likely is the emission of a helium nucleus, as it is the smallest particle with a high E_B / A value. It got the name α- decay, since this was the first type of emission observed in natural radioactivity, identified after a number of years as $^4_2$He.

In the usual notation the process of α- decay is expressed in general as

$$^A_Z El \rightarrow ^{A-4}_{Z-2} El + ^4_2 He \qquad (12.2.1)$$

and for $^{226}_{88}$Ra by (Fig 12.1)

$$^{226}_{88}Ra \rightarrow ^{222}_{86}Rn + ^4_2 He$$

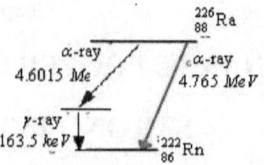

Fig 12.1 Alpha decay of $^{226}_{88}Ra$

12.2.2 α- Particle Properties

1. The α- particle is $^{4}_{2}He$ and is doubly magic nucleus. It consists of 2 protons and 2 neutrons, all in a $s_{1/2}$ shell,

2. The total nuclear spin $I = 0$,

3. The parity of α- particle is even; this means α- particle is extra-ordinarily stable,

4. α- decay interaction obeys parity conservation law and is a strong interaction.

12.2.3 Velocity of α- particles and Magnetic spectrograph

To determine the velocity v_α of α- particle emerging out from a sample of radio-nuclide a magnetic spectrograph was used. In a magnetic field B, the α- particles, from a source, of one particular velocity, are found focused on a photographic plate.

$$B\, q\, v_\alpha = \frac{m_\alpha\, v_\alpha^2}{r} \qquad (12.2.2)$$

in which putting $\frac{q}{m_\alpha} = 4.787 \times 10^7$ C.kg^{-1} yields the $v_\alpha = \square\, 10^9\ cm.s^{-1}$.

12.2.4 Range of α- particles (R_α)

Another characteristic property of α- particles is its *range* R_α in a gas such as H_2, N_2, or air. For this measurement of the ionization produced in a gas at different distances from the source of α- particles is made, say in a Wilson's Cloud Chamber. R_α is a sharply defined ionization path length.

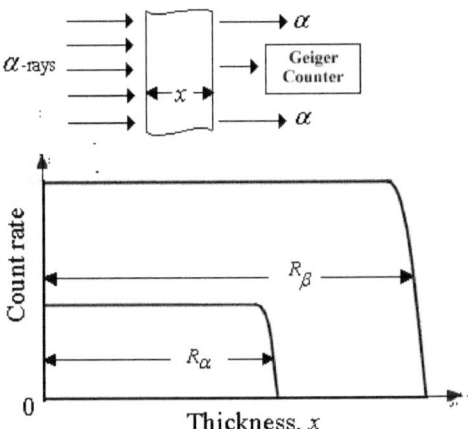

Fig 12.2 Range R_α compared with that of β – rays

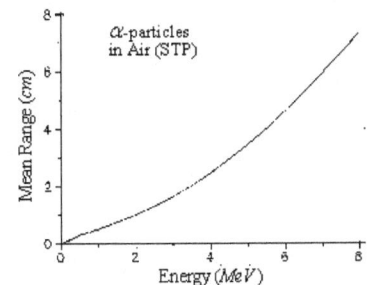

Fig 12.3 Energy versus Range of α – rays

Mono-energetic α- rays have a range given by the empirical relationship, known as Geiger's Rule,

$$R_\alpha = c\, v_\alpha^3 \qquad (12.2.3)$$

where the constant $c = 9.6 \times 10^{-28}$

It is known that, if x = distance the particle traversed from the source, Geiger's Rule can be rewritten as

$$v_\alpha^3 = b\,(R_\alpha - x) \qquad (12.2.4)$$

where v_α is in $cm\ s^{-1}$, R_α is in cm and $b = 1.03 \times 10^{27}$.

For air at 15°C and 76 cm Hg pressure, $R_\alpha[^{210}_{84}Po]$ - 38 mm.

12.2.5 Geiger-Nuttal Law

The values of R_α of the common natural α- emitters form a reciprocal relationship with their respective $\tau_{1/2}$, known as the *Geiger-Nuttal Law*, due to H. Geiger and Nuttal (1911) stated *as increase in probability of decay to increase in energy E*. The law was empirically observed as

$$\tau_{1/2} \, R_\alpha^m = \text{constant, A} \tag{12.2.5}$$

In another form

$$\ln \lambda = m \ln R_\alpha + B \tag{12.2.6}$$

with $B = \ln A$.

The converse of the law, a rapid decrease in penetration probability as the available energy is decreased, explains as to why α- decay is a rare phenomenon in the region $A = 150$ to $A = 210$.

12.2.5.1 Radioactive Series Decay

Since all the high-A nuclides are energetically unstable against α- decay one finds a whole series of nuclides where $A \xrightarrow{\alpha} B \xrightarrow{\alpha} C \xrightarrow{\alpha} D \square\square\square$. Such Series decay chains are observed in nature. It is found that the value of m is the same for all the 3 radioactive Series, whereas B is different. The relation (12.2.6) is represented as a plot in Fig 12.4.

Example of this relation Ii listed in Table 12.1.

Table 12.1 For $^{232}_{90}Th$ - Series, $\tau_{1/2} \, R_\alpha^m$ = constant, A

m	B	A	$\tau_{1/2} \, R_\alpha^m$
$\cong 60$	-44.2	10^{-84}	$\tau_{1/2} \, R_\alpha^{60} = \ln 2 \times 10^{84}$ s.mm

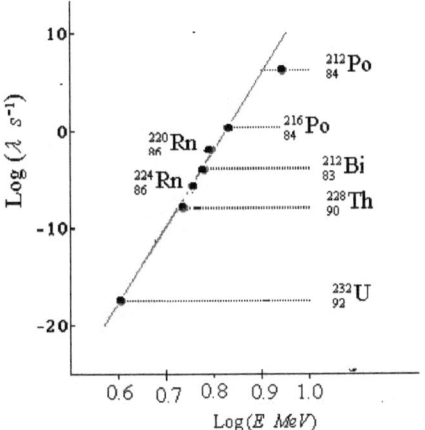

Fig 12.4 Geiger-Nuttal relation

Equation (12.2.6) can be rewritten in the form

$$\ln \tau_{1/2} = m_1 \ln E_\alpha + B_1 \tag{12.2.7}$$

Typical $\tau_{1/2}$ for α-particle emitters are $\tau_{1/2} = 10^{-6}$ to 10^{17}, compared to the nuclear life time of 10^{-21} s. $A = 4n + 2$ is the Uranium Series and starts with $^{238}_{92}U$. Historically this Series first led to the discovery of natural radioactivity by Henri Becquerel. Radium, identifies as the first radioelement by Madame Curie, is a member of this Series.

12.3 ENERGETICS OF α- DECAY

α- particles are known to have velocity of emission $v_\alpha \simeq 10^9\ cm\ s^{-1}$, which means their kinetic energies in the range $\simeq 4 \simeq 10\ MeV$.

12.3.1 Nuclear Reaction Equation (α- disintegration equation)

α- disintegration (12.2.1) can be better represented by the so-called *nuclear reaction equation*

$$^A_Z El \rightarrow ^{A-4}_{Z-2} El + ^4_2 He + Q_\alpha \tag{12.3.1}$$

This equation has to satisfy a number of conditions to be described below.

12.3.1.1 Disintegration Energy (Q_α)

Q_α in equation (12.3.1) is popularly called the *disintegration energy*.

12.3.1.2 Conservation of Charge

It is clearly seen that the equation has its left hand right hand sides do have the same number of total charge units.

$$\sum_{LHS} \text{charge} = \sum_{RHS} \text{charge} \qquad (12.3.2)$$

12.3.1.3 Conservation of Momentum

If m_α, M_p and M_d denote, respectively, the masses of the α-, parent and daughter nuclei, and the corresponding velocities by v_α, v_p and v_d, then the said equation (12.3.1) will have to obey the momentum conservation law, viz.,

$$M_p \cdot v_p = M_d \cdot v_d + m_\alpha \cdot v_\alpha \qquad (12.3.3)$$

12.3.1.4 Conservation of Energy

$$Q_\alpha = (\text{KE of } \alpha\text{-particle, } E_\alpha) + (\text{KE of daughter, } E_d) \qquad (12.3.4)$$

Eliminating v_d

$$Q_\alpha = E_\alpha \left[\frac{m_\alpha}{M_d} + 1 \right] \qquad (12.3.5)$$

In terms of mass numbers

$$Q_\alpha = E_\alpha \frac{A}{A-4} \qquad (12.3.6)$$

Another form is $Q_\alpha = E_d \dfrac{A-3}{4}$ \qquad (12.3.7)

$$Q_\alpha = (M_p - M_d - m_\alpha) \; amu \qquad (12.3.8)$$

From the E_B/A versus A curve for Helium-4, $E_B/A = 7$ MeV. For the generalized nuclide (A, Z), use of the semi-empirical mass formula (Chapter 13) gives

$$M(A,Z)\,c^2 = \left\{ \begin{array}{l} Z\,M_H + (A-Z)\,M_n \\ - \left[\begin{array}{l} -a_V\,A + a_s\,A^{2/3} + a_C\,Z(Z-1)\,A^{-1/3} \\ + a_A\,(Z-A/2)^2/A \pm a_p\,A^{-3/4} \end{array} \right] \end{array} \right\} u \qquad (12.3.9)$$

$$Q_\alpha = 28\;MeV + B(A-4, Z-2) - B(A,Z) \qquad (12.3.10)$$

Using simple calculus, $B(A-4, Z-2) - B(A,Z)$ can be expressed as

$$\Delta B = \frac{\partial B}{\partial A} \Delta A + \frac{\partial B}{\partial Z} \Delta Z \qquad (12.3.11)$$

One gets therefore

$$Q_\alpha = 28 \text{ MeV} + \left\{ \left[-4 a_V + \tfrac{8}{3} a_s A^{-1/3} + 4 a_C \left(1 - \tfrac{Z}{3A}\right) Z A^{-1/3} - 4 a_A \left(1 - \tfrac{2Z}{A}\right)^2 \right] \right\} \qquad (12.3.12)$$

Where δ is neglected. A and Z are average values between parent and product nuclides. Substitution of appropriate values shows that all nuclides with $A > 150$ are energetically unstable against α-decay. It is seen that most of the nuclides with $A > 200$ are α-emitters.

12.4 ALPHA PARTICLE SPECTRA

Energies of α-particles are found to vary slightly or these may consist of normal groups of α-particles followed by a very small fraction of α-particles of higher energy. It follows that there are two types of α-particle spectra. It is known that

$$Ln\ \tau \square C + 4 Z E_\alpha^{-\tfrac{1}{2}},$$

and so one can compare the mean lives of

$^{232}_{90}\text{Th}$ [Z(daughter) = 88,

$E_\alpha = 4.08\ MeV]$ and

$^{212}_{84}\text{Po}$ [Z(daughter) = 84, $E_\alpha = 8.95\ MeV$].

Using these yields

$Ln\ \tau(\text{Th}) \square A + 175$

$Ln\ \tau(\text{Po}) \square A + 112$

From these two one gets

$Ln\ \left[\frac{\tau(\text{Th})}{\tau(\text{Po})}\right] \square\ 63$

or $\left[\frac{\tau(\text{Th})}{\tau(\text{Po})}\right] \square\ e^{63} \square\ 23 \times 10^{27}$

12.4.1 Discrete Energy Spectrum of α-particles

The fact that the α-particles emitted by a given nuclide have a well-defined range R_α means their energy E_α is well-defined too. This means the energy spectrum (Fig 12.5) of α-particles is *discrete*.

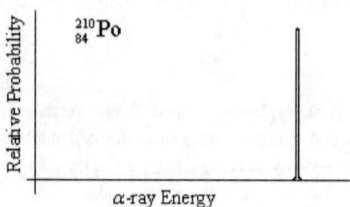

Fig 12.5 Discrete energy spectrum of alpha

It is also called *line* spectra, $^{210}_{84}$Po, $^{218}_{84}$Po, and $^{226}_{88}$Ra are examples of radio-isotopes which emit α-particles of discrete energy only.

12.4.2 Fine Structure Energy Spectrum of α-particles

Radio-isotopes like $^{212}_{83}$Bi (Th C), $^{232}_{90}$Th, $^{224}_{88}$Ra. $^{227}_{91}$Pa, etc. are found to have *fine structure* spectra. The energies are found to lie between 5.661 to 6.030 MeV consisting of two or more discrete energy values separated by a small energy difference.

A typical energy level diagram depicting the fine structure spectrum is the decay by the parent $^{212}_{83}$Bi (Th C), according to

$$^{212}_{83}\text{Bi (Th C)} \xrightarrow{\alpha} {}^{208}_{81}\text{Ta (Th C'')}.$$

with 6 groups of α-particles having E_α in the region 5.584 to 6.023 MeV. This is followed with 8 groups of γ-rays.

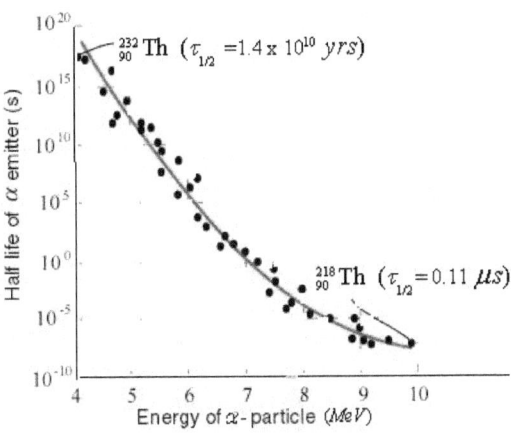

Fig 12. 6 Half-lives of alpha decay

12.4.3 Long Range Energy Spectrum of α- particles

$^{212}_{84}Po$ is found to emit a *line* spectrum at 7.68 MeV as well as a small fraction of α- rays with a much larger energy at 10.506 MeV, called *long range* spectra.

12.5 THEORY OF ALPHA DECAY

In treating the emission of α- ray through the Coulomb barrier surrounding the atomic nucleus, the strong attractive forces holding the nucleons together can be represented by an effective, strongly attractive potential. How could a particle escape through the high Coulomb barrier is a paradox, The treatment given below is just the same from the author's book **'Quantum Mechanics'** [*S. Devanarayanan, 2005*].

12.5.1 Paradox of Alpha Decay:

$^{214}_{84}Po$, a radio-active nuclide, emits α-particles with kinetic energy 7.68 MeV; and it has been found that no absorption of α-particles by $^{238}_{92}U$ foil occurs, though it may scatter them. But $^{238}_{92}U$ nuclide emits α-particles of 4.20 MeV, though this energy is less than the depth of the well of 40 MeV. This paradox was inexplicable on the arguments base on Classical Physics. Quantum Mechanics provides a straightforward explanation.

According to the one-body model of α – decay, which assumes that the α – particle is preformed in the nucleus, and confined to the nuclear interior by the Coulomb potential barrier. In the classical picture, if the kinetic energy of the α-particle is less than the potential energy represented by the barrier height, the α-particle cannot leave the nucleus. In the quantum-mechanical picture, however, there is a finite probability that the α-particle will tunnel through the barrier and leave the nucleus. This treatment of the α-decay process was one of the first successful applications of quantum theory to nuclear physics, and was presented independently by G. Gamow (1928) and R. Gurney & E. Condon (1928).

12.5.2 GEIGER-NUTTAL FORMULA: General Potential barrier - A sophisticated approach to the problem

12.5.2.1 Potential Barrier Penetration

It is necessary at first to compute the transmission and reflection coefficients of an α-particle incident on a 1-D potential barrier (Fig 12.7), mathematically expressed as

$$V(x) = \begin{cases} 0, & \text{for } x < a, \quad \text{(Region I)} \\ V_1, & \text{for } a < x < b, \quad \text{(Region II)} \\ V_2, & \text{for } x > b. \quad \text{(Region III)} \end{cases} \quad (12.5.1)$$

where $V_2 < E < V_1$. \hfill (12.5.2)

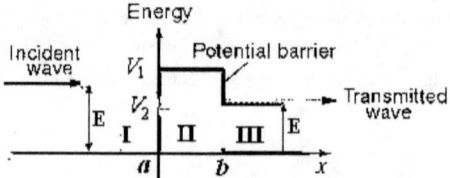

The 1-D Schrödinger Equations for the three regions are:

$$\nabla^2 u_1(x) + k_1^2 u_1(x) = 0, \qquad x < a. \qquad (12.5.3)$$

$$\nabla^2 u_2(x) + k_2^2 u_2(x) = 0, \qquad a < x < b. \qquad (12.5.4)$$

$$\nabla^2 u_3(x) + k_3^2 u_3(x) = 0, \qquad x > b.. \qquad (12.5.5)$$

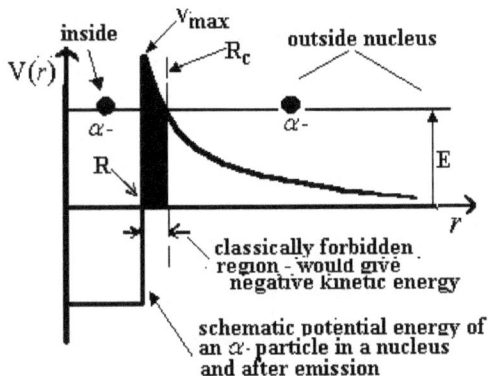

Fig. 12.7 Finite Potential Barrier and Tunnel Effect

Putting $k_1^2 = 2mE/\hbar^2$ (12.5.6)

$k_2^2 = 2m(V_1 - E)/\hbar^2$. (12.5.7)

$k_3^2 = 2m(E - V_2)/\hbar^2$ (12.5.8)

The solutions of equations (12.4.3), (12.4.4) and (12.4.5) are:

$$u_1(x) = A e^{ik_1 x} + B e^{-ik_1 x}$$ (12.5.9)

where A and B are arbitrary constants.

$$u_2(x) = D e^{-k_2 x} + C e^{k_2 x}.$$ (12.5.10)

where C and D are arbitrary constants.

$$u_3(x) = 0 + F e^{ik_3 x}.$$ (12.5.11)

where F is an arbitrary constant.

Solving and applying boundary conditions, one gets

$$\begin{pmatrix} A \\ B \end{pmatrix} = \frac{1}{2} \begin{pmatrix} (1 + k_2/ik_1) e^{k_2 a + ik_1 a} & (1 - k_2/ik_1) e^{-k_2 a + ik_1 a} \\ (1 - ik_2/k_1) e^{k_2 b - ik_3 b} & (1 + k_2/ik_1) e^{+k_2 a + ik_1 a} \end{pmatrix} \begin{pmatrix} C \\ D \end{pmatrix}$$ (12.5.12)

$$\begin{pmatrix} C \\ D \end{pmatrix} = \frac{1}{2} \begin{pmatrix} (1+ik_3/k_2) e^{-k_2 b + ik_3 b} \\ (1-ik_3/k_2) e^{+k_2 b + ik_3 b} \end{pmatrix} \begin{pmatrix} F \\ 0 \end{pmatrix} \quad (12.5.13)$$

Eliminating B, C, & D, and writing *barrier width*, c as

$$c = b - a \quad (12.5.14)$$

$$\frac{F}{A} = \frac{4 i k_1 k_2 e^{+i(k_1 a - k_3 b)}}{(k_2+ik_1)(k_2+ik_3) e^{-k_2 c} - (k_2-ik_1)(k_2-ik_3) e^{k_2 c}} \neq 0 \quad (12.5.15)$$

12.4.2.2 Barrier Penetration

The transparency, or, *transmission coefficient* τ, is defined as

$$\tau = \varpi_3 / \varpi_1 = |F/A|^2 (k_3/k_1) \quad (12.5.16)$$

ϖ_1 and ϖ_3 are Probability of finding the particle in region I and region III, respectively, at any time, t.

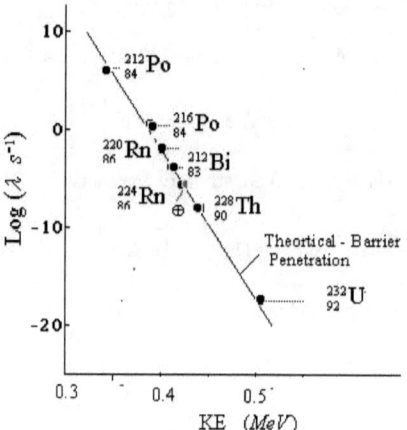

Fig 12.8 Geiger-Nuttal relation

Equation (12.4.13) can be written as

$$\tau = |F/A|^2 \frac{k_3}{k_1} \cdot G(k_1, k_2, k_3) e^{-2 k_2 c} . \qquad (12.4.17)$$

This is the Tunnel Effect.

12.5.2.3 Nuclear Potential & Tunnel Effect

Fig 17.9 is the nuclear potential field of the general form, and let it be simplified to the flat barrier form of Fig 12.7 with square well potential model.

V(r) = Coulomb potential

$$V(r) = \frac{2Ze^2}{r}, \text{ for } r_1 \leq r \leq r_2 \qquad (12.5.18)$$

where nuclear charge = + Z e, Charge on an α-particle = + 2e.

The Tunnel Effect discussed in Section 12.5.2.1 suggests that, a particle of energy E is capable of escaping from the closed region of the nucleus, because of the probability that the wave function associated with the α-particle can break through the barrier as if there were a hole in it. If ϖ_1, ϖ_2 and ϖ_3 are probabilities of finding the particle in regions I, II and III, respectively, Writing

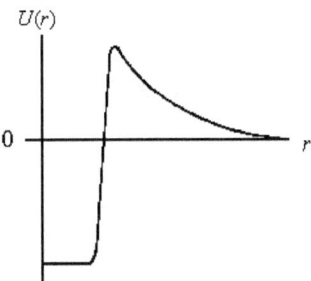

Fig.12.9 Nuclear Potential

$$k_2^2 = 2m(V(r) - E)/\hbar^2,$$

where V(r) = $-V_0$, for r < r_1

V(r) ≠ constant, for r > r_1 .

i.e., the nuclear radius $R_0 \equiv r_1 \leq r \leq r_2$ (≡ R, a turning point)

One can, therefore, replace k_2 c by

$$k_2 \, c \rightarrow \int_{r_1}^{r_2} k_2(r) \, dr$$

$$\varpi_5 \sqcup \varpi_1 \, G(k_1, k_2, k_3) \, e^{-2 k_2 c} \qquad (12.5.19)$$

i.e., $\qquad \propto e^{-\frac{1}{\hbar} \int_{r_1}^{r_2} \sqrt{2m(V(r)-E)} \, dr} \qquad (12.5.20)$

is the probability per collision of the potential barrier. If ν = number of encounters to get through the barrier per unit time,

$$\nu = 1/\varpi_{\text{o}},$$

$$\nu \propto e^{2 k_2 c}. \qquad (12.5.21)$$

If t_0 = time between two encounters,

$$t_0 \approx 2 r_1 / v, \qquad (12.5.22)$$

where v = velocity with which α-particle is leaving the nucleus

$$v = \sqrt{(2 E / m)}.$$

12.5.2.4 To evaluate the integral:

$$E = V(r_2) + KE \approx \frac{2 Z e^2}{r_2} \qquad (12.5.23)$$

assuming KE as negligibly small compared to

$$\int_{r_1}^{r_2} k_2(r) \, dr = -\frac{2}{\hbar} \int_{r_1}^{r_2} \left(\sqrt{2 m (V_1 - E)} \right) dr$$

Note: Substitute for

$$(V(r) - E) = E \left(\frac{1}{y} - 1 \right),$$

where $y = \dfrac{r E}{2 Z e^2} = \dfrac{E}{V(r)}$

$$dr = \frac{2 Z e^2}{E} \, dy, \quad \int_{r_1}^{r_2} dr \rightarrow \frac{2 Z e^2}{E} \int_{E}^{1} dy$$

$$E = \frac{r_1 E}{2 Z e^2},$$

$$\int_{r_1}^{r_2} k_2(r)\, dr \rightarrow \frac{1}{\hbar} \sqrt{2 m E}\, \frac{2 Z e^2}{E} \int_{\dot E}^{1} \sqrt{\frac{1}{y} - 1}\, dy \qquad (12.5.24)$$

This is a standard integral, and it can be evaluated to give

$$\int_{\dot E}^{1} \sqrt{\frac{1}{y} - 1}\, dy \simeq \left\{ \left[\frac{\pi}{2} r_2 - 2\sqrt{r_1 r_2} \right] \bigg/ r_2 = \frac{\pi Z e^2}{\hbar} \sqrt{\frac{2m}{E}} \right\} \qquad (12.5.25)$$

Since $\sqrt{E} \propto v$

$$\therefore k_2 c = \left(\frac{2 e^2}{\hbar} \sqrt{m} \right) \frac{Z}{v} \qquad (12.5.26)$$

$$\therefore 2 k_2 c = A \frac{Z}{v}$$

$$\frac{1}{\tau} = \frac{v}{2 R_0}\, \varpi_3$$

Substituting this equation (12.5.26) in equation (12.5.19), results in the equation:

$$\ln \tau_{1/2} = 2 k_2 c + B \; (= \text{a constant}) \qquad (12.5.27)$$

$$\ln \tau_{1/2} = \frac{AZ}{v} + B \qquad (12.5.28)$$

where both A and B are constants.

Putting $r_2 \equiv R$, $r_1 \equiv R_0$, and then eliminating $R \equiv r_2$,

$$2 \int_{r_1}^{r_2} k_2(r)\, dr$$

$$= 2^{3/2} \sqrt{\left[\frac{(0.98)(3727\, MeV)}{(197.3\, MeV\, fm)^2} \right]}\, E^{1/2} \sqrt{\left[\frac{(3.14)(2.88\, MeV\, f) Z E^{-1}}{2} - 2\left(\frac{R_0 (2.88\, MeV\, f)}{Z E} \right) \right]}$$

$$(12.5.29)$$

giving equation (12.5.29) as

$$\ln \lambda = \ln \frac{v}{2 R_0} + 2.97\, Z^{1/2}\, R_0^{1/2} - 3.95\, Z E^{-1/2} \qquad (12.5.30)$$

where E is expressed in MeV.

$$\log \lambda = \log \frac{vN}{2R_0} + 1.28\, Z^{1/2}\, R_0^{1/2} - 1.71\, ZE^{-1/2} \qquad (12.5.31)$$

where $N = \#$ of α-decay ($=1$) within the nucleus at any given instant, t.

Plot Lc λ *versus* $ZE^{-1/2}$ for a number of representative α-particle nuclides from each of the 4 decay series, *viz*,. Thorium, Neptunium, Uranium and Actinium Series having mass number, 4n, 4n + 1, 4n + 2, and 4n + 3, respectively. It will be seen that

(1) The straight line drawn through the points has a slope of (–1.71) as required by the theory. The position of the line can be used to determine R_0, the radius of the nuclide.

(2) The tremendous range of λ that is represented (corresponding to $\tau_{1/2} \approx 10^{-6}$ s to 10^{10} yrs) by the slope of the line is not subject to adjustment (This is a span of a range of 23 orders of magnitude). These two points show that the agreement of the barrier penetration theory with observations must be considered as the most satisfactory or remarkable. The position of the line gives

$$Log\,(vN/2R_0) + 1.28\, Z^{1/2}\, R_0^{1/2} \approx 55.5 \qquad (12.5.32)$$

Dependence of $Log\,(vN/2R_0)$ on $(v/2R_0)$ is quite weak. This means one gets quite accurate values of R_0. Thus, if $N = 1$ $v \approx 2 \times 10^9$ cm/s,

$R_0 \approx 10^{-12}$ cm (from Rutherford experiment) $Z \approx 85$, for radio-active α-active nuclide

$$\approx \{(55.5 - 21.0)/(1.28 \times 9.2)\}^2 = 8.5 \times 10^{-13}\ cm.$$

An additional screening correction E_{scr} to Q_α may be added to account for influence of the electron cloud on the emitted α – particle

$$E_{scr} = \left(65.3\, Z^{7/5} - 80\, Z^{2/5}\right) eV \qquad (12.4.33)$$

where Z is the proton number of the daughter nucleus. This correction is at most around 47 keV ($Z=110$) and is often neglected. The α – decay process can occur spontaneously if $Q_\alpha > 0$, though the effect of the Coulomb barrier means that the rate of α – emission does not become appreciable until the value of Q_α rises to several MeV.

A particular isotope may emit α – particles of more than one energy, if excited states in the daughter nucleus are populated by the decay. Each decay to a particular state has a different decay constant, known as the partial decay constant λ_i. The ratio of the partial to total decay constant gives the α – -decay branching ratio, b_α, and the partial half-life of a particular decay is given by

$$\tau_{1/2} = \frac{\ln 2}{\lambda_i} = \frac{T_{1/2}}{b_\alpha} \qquad (12.5.34)$$

Geiger and Nuttall showed that a reasonable straight line is obtained if the logarithm of the half-life is plotted against the logarithm of the range of α – particles in air. The range is related to the energy, and thus the α – decay Q-value. The relationship is then given by

$$\tau_{1/2} = a + \frac{b}{\sqrt{Q_\alpha}} \qquad (12.5.35)$$

where a and b are empirically determined constants that are different for each element.

12.6   ALPHA PARTICLE SOURCES

$^{244}_{96}Cm$ alpha radioactive sources with a source strength of about 5 mCi, covered with 3-μm aluminum foils that reduce the energy of emitted alpha particles from the initial value of 5.8 MeV to about 5.2 MeV. The FWHM for the alpha mode of a $^{244}_{96}Cm$ peak at 5.8 MeV is less than 100 keV.

Alpha signals from charge-sensitive preamplifiers –and similarly-x ray signals from a customized voltage-sensitive preamplifiers in the sensor head –are further amplified and filtered (semi-Gaussian pulse shapes) and then routed to peak detectors, a multiplexer, and into a 16-bit A/D converter for digitization. Signals from comparators that trigger if signals exceed a preset level initiate a sequence of logic signals necessary for peak detection (sample gate and signal hold) and the conversion process (program interrupt, alpha/x-ray flags). A microcontroller selects the appropriate input to the multiplexer and controls analog-to-digital conversion. The analyzed events are stored in the microprocessor buffer memory, building up alpha-spectrum

REVIEW QUESTIONS

R.Q. 12.1 Find an expression for the disintegration energy Q_α of an α-emission, a) in terms of its energy and mass, and b) in terms of the mass number. [Answer a) $Q_\alpha = E_\alpha \left(\frac{m_\alpha}{M_d} + 1\right)$; b) $Q_\alpha = E_\alpha \frac{A}{A-4}$]

R.Q. 12.2 Estimate the percentage value of Q_α with respect to E_α of an α-emission, for $M_p \cong 200$. [Answer: $(Q_\alpha / E_\alpha) > \square\ 2\%$.]

R.Q. 12.3 Find an expression for the disintegration energy of daughter nucleus E_d of an α-emission, in terms of the Q_α and mass number. [Answer $E_d = Q_\alpha \left(\frac{A}{A-3}\right)$]

R.Q. 12.4 Find out the kinetic energy of emission of an α-particle in the reaction; $^{232}_{92}U \rightarrow\ ^{228}_{90}Th + ^{4}_{2}He$. Given: $M(^{232}_{92}U) = 232.1095\ amu$, $M(^{228}_{90}Th) = 228.0998\ amu$, $m(^{4}_{2}He) = 4.0039\ amu$. [Answer $Q_\alpha = +5.40\ MeV$, $Q_\alpha = E_\alpha \left(\frac{M_d}{M_d + m_\alpha} + 1\right) = 5.30\ MeV$]

R.Q. 12.5 In the reaction $^{222}_{86}Rn \rightarrow\ ^{218}_{84}Po + ^{4}_{2}He + Q_\alpha$, compute the kinetic energy of the α-particle. Take the values of the required masses from standard Tables. [Answer $Q_\alpha = +5.587\ MeV$, $Q_\alpha = E_\alpha \left(\frac{M_d}{M_d + m_\alpha} + 1\right) = 5.486\ MeV$]

R.Q. 12.6 The Q value for α-particle decay of $^{214}_{84}Po$ (RaC') is 7.83 MeV. At what energy will the α-particle be emitted? [Answer $E_\alpha = 7.68\ MeV$]

R.Q. 12.7 Two nuclides of the same mass number, but one with Z = 84 and the other with Z = 82, have the same decay constant, undergo α-decay. If the nuclide with Z = 84 emits α- with energy 5.3 MeV, estimate the energy of the particle from the Z = 82 element. [Answer 5.05 MeV.]

R.Q. 12.8 Consider the alpha decay: $^{228}_{90}Th \rightarrow\ ^{224}_{88}Ra + ^{4}_{2}He$. Find the energy given out by this decay. Given: Mass of the thorium nucleus = 227.97929 u, Mass of the radium nucleus = 223.97189 u, Mass of the alpha particle (helium nucleus) = 4.00151 u [Answer 0.00589 u or E = 931.5 × 0.00589 = 5.49 MeV].

RQ. 12.9 What is the significance of the fine structure of α-ray spectra?

RQ. 12.10 What is Geiger-Nuttal law?

&&*&*&*&*&*

Chapter 13

MODELS OF NUCLEAR STRUCTURE

Chapter 13

MODELS OF NUCLEAR STRUCTURE

"Eventhough the realms of religion and science in themselves are clearly marked off from each other, nevertheless there exists between the strong reciprocal relationships and dependencies. The situation may be expressed by an image. Science without religion is lame, religion without science is blind " -
Albert Einstein

13.1 INTRODUCTION

The exact theoretical model of nuclear structure is not known. It is a practice to use specific nuclear structure models combined with simplified nuclear forces. In this Chapter the most successful theoretical models of the nucleus will be discussed. Nuclear models (each capable of explaining only a portion or aspect of the experimental results about nuclei) currently in use to explain nuclear phenomena are:

a) Independent Particle Model (IPM), and

b) Strong Interaction Model (SIM).

Though each model has its emphasis; they agree in some way, but each rests upon unique assumptions, and makes its own important predictions.

Scientists working for developing a nuclear structure model had to wait for the discovery of the neutron by James Chadwick (1932). Initially the IPM model was developed, but during 1936 – 48 interest was centered on the SIM, and then in 1958, Maria Goeppert Mayer started on the IPM.

1. The Uniform Particle Model

2. Fermi-Gas Model,

3. Alpha particle Model,

4. Liquid Drop Model (Strong Interaction Model, SIM),

5. Shell Model,

6. Complex Potential Model.

13.2.1 UNIFORM PARTICLE MODEL

Eugene P. Wigner in 1937 proposed this SIM model of nucleus. This model is moderately successful in explaining the binding energies of nuclei, but failed to agree with predicted results of the inform model.

13.2.2 Alpha particle Model

It is assumed that α- particles form subgroups inside a nucleus. These α- particles need not have separate existence, but may exchange particles one another. Low mass number nuclei can be represented by A = 4 n, where n = an integer. For example, $^{12}_{6}C$ nucleus contains 3 α-particles with small interaction between them. Since n = integer, such nuclei are stable. A = 4n +1 nuclei will have n closed structures with an additional heavy particle or one missing from this closed structure. However, the nuclei with 4n + 2 fail completely.

13.2.3 Independent Particle Model (IPM)

If the nucleons have very little interaction the nuclei in ground level and in the lowest excited levels, then the IPMs emerge. The Shell Model is the most prominent among the IPMs. Drawing upon analogies with the extra-nuclear electronic structure of atoms, Bartlett, Guggenheimer, Elsasser, and others developed early individual particle models involving closed shells of $2(2\ell+1)$ neutrons / protons, where ℓ is the angular momentum quantum number of the nucleons.

13.3 THE SHELL MODEL   [Mayer-Jensen's Shell Model]

Both the liquid drop and Fermi-gas models represent the nucleus in very crude forms. They could account for *gross* properties of nuclei. But *specific* properties nuclear excited states can not be explained by using these models.

13.3.1 Summary of experimental evidence that require accounting

It happens that a considerable body of experimental evidence (analogous to the chemical evidence responsible for the Periodic System of Elements) strongly suggests shell structure characterizes all nuclei.

13.3.2 Magic Numbers

One of the main nuclear features which led to the development of the shell structure is the existence of what are usually called the magic numbers. It is found that nuclei with even numbers of protons and neutrons are more stable than those with odd numbers. In particular, there are "*magic numbers*" of neutrons and protons which seem to be particularly favored in terms of nuclear stability:

$$\boxed{2,\ 8,\ 20,\ 28,\ 50,\ 82,\ 126} \qquad (13.3.1)$$

That such numbers exist was first remarked by Bartlett and also Elsasser in 1933. Magic number nuclei, which are nuclei with either

Z = 2, 8, 20, (28, 40), 50, 82, 114, (126), or 164, or

N = 2, 8, 20, (28, 50), 82, 126, 184, 196 (272), 318

exhibit certain properties including anomalously low masses, high natural abundances and high energy first excited states. A nucleus is said to be doubly magic if both Z and N are equal to one of the magic numbers. Example, $_2^4$He, $_{82}^{208}$Pb.

13.3.3 The main nuclear properties which show periodic variation with Z or N are:

1. Nuclear stability,
2. The absolute abundances of the various nuclides (i.e., total abundances of isotopes in the Earth),
3. The distribution of the lowest lying nuclear excited states,
4. Consideration of the binding energy of the last neutron,
5. The number of β- unstable isotopes or isotones,
6. The excitation energies of the first excited states of nuclei,
7. Small departures of the nuclear radii from formula,
8. Elemental abundance,
9. α- particle emission energies,
10. Cross-section of neutron capture,
11. Nuclear electric quadrupole moment.

13.3.4 Simplest Evidence for Nuclear Structure

Empirically,for stable nuclei

$$Z = A / [1.98 + 0.0155 \, A^{2/3}], \qquad (13.3.2)$$

Table 13.1 lists the various features.

Table 13.1 Stability and Magic number nuclides

Z	N	Life time $\tau = \infty$	# of stable nuclides	+	# of long-lived stable nuclides	=	Total # stable nuclides
Even	Even	$_2^4$He	166	+	11	=	177
Even	Odd	$_2^3$He	57	+	3	=	60
Odd	Even	$_3^7$Li	53	+	3	=	56
Odd	Odd	$_1^1$H, $_3^6$Li	6	+	5	=	11
			282	+	22	=	304

13.3.5 Atomic number versus Number of stable isotopes

A plot of relative abundances of even-even nuclides versus (A), for A > 50, shows that the peaks are pronounced for nuclides with N = magic number (Table 13.2).

Table 13.2 Relative abundances of stable nuclides versus Z

Element	Ar	K	Ca	Sc	Ti	V	Cr	Ag	Cd	In	Sn	Sb	Te
Z	18	19	20	21	22	23	24	47	48	49	50	51	52
Stable Isotopes	3	2	6 ⇑	1	5	2	4	2	8	1	10 ⇑	2	8

Table 13.3 Sable isotopes of Z = even Elements

Nuclide	N	% acount	Element
$^{16}_{8}O$	8	99.75	O
$^{39}_{19}K$	20	93	K
$^{52}_{24}Kr$	28	83	Kr
$^{88}_{38}Sr$	50	82	Sr
$^{138}_{56}Ba$	82	72	Ba
$^{140}_{58}Ce$	82	89	Ce

13.3.6 Isotopic abundance of Even-Z Elements

At least two isotopes of equal abundances are present among Z = even elements in nature. Look at Table 13.3,

Even-Z element accounts for more than half the total amount of that element in nature.

13.3.7 Large level spacing in Excited states of Nuclei with one neutron more or less

For example, Cerium ($^{Ce}_{58}$) has 14 isotopes. It is known that it has

Even-Odd differences, and

Closed shell discontinuity.

13.3.8 Number of stable Isotones for Magic number nuclei

A perusal of Table 13.4 shows that around N = magic number there are more number of stable isotones.

Table 13.4 Number of stable Isotones versus N of the nuclide

N	18	19	20	21	22		48	49	50	51	52
Stable Isotones	5	0	5 ⇑	0	3		4	1	5 ⇑	1	4

(⇓ markers above N=20 and N=50)

13.3.9 Ground and first excited states of the even-A isotopes of $_{82}Pb$.

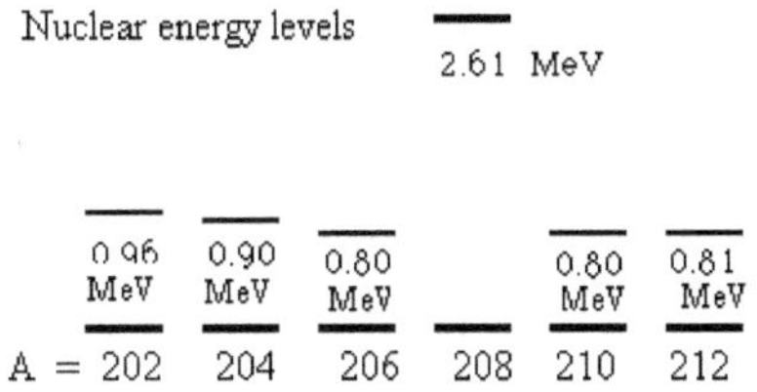

Nuclear energy levels ——— 2.61 MeV

0.96 MeV	0.90 MeV	0.80 MeV		0.80 MeV	0.81 MeV
A = 202	204	206	208	210	212

Fig 13.1 Energy levels of Even-A isotopes

For example, consider the $_{82}Pb$ element. It has six Even-A isotopes, viz., $_{82}^{202}Pb$, $_{82}^{204}Pb$, $_{82}^{206}Pb$, $_{82}^{208}Pb$, $_{82}^{210}Pb$, and $_{82}^{212}Pb$.

Excepting $_{82}^{208}Pb$, all the other 5 nuclides have all the same spin and parity 2^+ for the first excited states compared to 3^- for $_{82}^{208}Pb$. Further, $_{82}^{208}Pb$ is hard to excite, as it has a closed shell! Fig 13.1 shows these.

13.3.10 Energy of the first excited states of Even-Even nuclides

A plot of the energy of the first excited level of Even-Even nuclides against neutron number N reveals that all doubly-magic nuclides have relatively very large energies than other nuclides.

13.3.11 Electric Nuclear Quadrupole Moments

It is noteworthy that all magic-number- nuclides have Q = 0, or minimum, and are spherical. In shape (i.e., they have closed shell configuration); whereas nuclides above a closed shell structure have Q = - ve, and those below have positive Q.

13.3.12 Nuclear spin, I

From an examination of the Table of nuclear spins of all nuclides the Table 13.5 has been prepared.

Table 13.5 Nuclear spin (I) of Nuclides

Nuclide	Example	I
Even Z - Even N	$^{12}_{6}C$	0
Odd Z - Odd N	$^{14}_{7}N$	Integer
Odd A	$^{11}_{5}B$	Half-integer
Mirror nuclides	$^{16}_{8}O$	Equal

13.3.13 Nuclear Stability

To account for nuclear stability and all afore mentioned features one has to examine the treatment of nuclei on Wave Mechanics, to be described in the next subsection.

13.4 THE WAVE MECHANICS (Shell Model) OF NUCLEI

Any model of the structure of the nucleus has to provide a reasonable explanation of these characteristic numbers. We will examine how this is done in the Shell Model.

The Shell Model is based on the assumption that nucleons inside the nucleus are in definite states of energy and angular momentum. The notion of nucleons moving on orbits is somewhat at odds with a strongly interacting many particle system.

Nucleons are Fermi particles, having spin $I = \frac{1}{2}$.

13.4.1 Assumptions

Nucleons are independent particles, and are free to move inside nucleus of which they compose, along orbits around the centre-of-mass.

There is no massive central body to serve as an interaction centre to the nucleons because of short range nature of nuclear forces, either a square well potential (depth, $V_o \approx -50\ MeV$, because of the deuteron problem) for light nuclei and an infinite harmonic oscillator potential

$$V(r) = \tfrac{1}{2} k\ r^2 \qquad (13.4.1)$$

for heavy nuclei, so as to represent the nuclear density distribution.

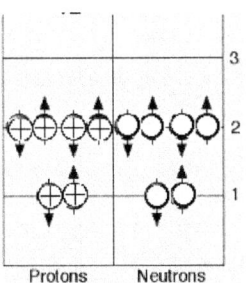

Fig 13.2

The Schrödinger's Equation (TISE) for a particle in a potential well holds good and solved. Stationary states of the system occur and are characterized by quantum numbers $n, \ell,$ and m_ℓ, whose significance is the same as in the case of orbital electrons in stationary states of an atom (See, for example, **Quantum Mechanics** by S. Devanarayanan, 2005) [10].

Nucleons of the two species (*i.e.* neutrons and protons) having spin $I = \frac{1}{2}$ occupy separate sets of states in a nucleus (Fig 13.2), because neutrons are neutral and protons interact electrically, as well as through specifically nuclear charge..

The tensor, or spin-orbit coupling, force is also to be taken into account, in addition to the central force (4).

13.4.2 Theory of Square well potential for Nucleus

The Nuclear Shell Model is based on the single particle (one nucleon) approximation. The effect of all other nucleons is smoothed out and represented by a potential which is then used in the Schrödinger Equation. The Coulomb repulsion between protons is neglected for the moment, so that the potential can be based on the nuclear matter distribution function

$$V(r) = -V_0 \frac{1}{[1 + e^{(r-b)/a}]} \qquad (13.4.2)$$

where $V_o \approx -60\ MeV$,

$b \square (1.25) A^{1/3}\ fm$,

$a \square 0.65\ fm$

which are obtained from scattering experiments. For a central potential (*i e*, spherical symmetry) the variables can be separated and the wave function written as

$$\psi(r) = \Re_{n\ell}(r) \, Y_\ell^m(\vartheta, \varphi). \tag{13.4.3}$$

The functions $Y_\ell^m(\vartheta,\varphi)$ are the same spherical harmonics that appear in the wave functions for the hydrogen atom, and the equation for the radial function $\Re_{n\ell}(r)$ an be written as

$$\frac{d^2}{dr^2}R_{n\,\ell}(r) + \frac{(2M)^{1/2}}{\hbar}\left[E - V(r) - \ell(\ell+1)\frac{\hbar^2}{2Mr^2}\right]R_{n\,\ell}(r) = 0 \tag{13.4.4}$$

where $E = (\text{kinetic} + \text{potential})$ energy.

$$\ell(\ell+1)\frac{\hbar^2}{2Mr^2} = \text{the centrifugal effect that the angular momentum has.}$$

Note how this term tends to raise the energy levels and push the wave function away from the origin.

For zero angular momentum a simple approximation to the solutions for the finite square well

$$V(r) \begin{cases} = -V_0, & \text{for } r < b \\ = 0, & \text{for } r > b \end{cases} \tag{13.4.5}$$

is of the form

$$r\Re_{n\ell}(r) \begin{cases} = A\sin(kr), & \text{for } r < b \\ = Be^{-Kr}, & \text{for } r > b \end{cases} \tag{13.4.6}$$

where $k \approx \frac{n\pi}{b}$ (exactly so for the infinite square well) and $K = -\frac{(2ME)^{1/2}}{\hbar}$. Then employing the usual technique of normalization and application of the condition of continuity to both the wave function and its derivative at $r = b$ the constants A, B and E can be found. As expected only certain discrete values of E are allowed. The ordering of the levels is shown in Table 13.6.

Table 13.6 The ordering of the levels

	1s	1p	1d	2s	1f	2p	1g	2d	3s
$2(2\ell+1)$	2	6	10	2	14	6	18	10	2
Total	2	8	18	20	34	40	58	68	70

As can be seen these totals are not in very good agreement with the 'magic' numbers that we have listed above.

13.4.3 Parabolic potential

When one solves the Radial wave equation (13.3.6) with

$$V(r) = \tfrac{1}{2} k r^2 \quad , \tag{13.4.7}$$

the solution is that for the harmonic oscillator, which is well known and gives the eigen values E_Λ given by

$$E_\Lambda = \hbar\omega(\Lambda + \tfrac{3}{2}), \tag{13.4.8}$$

for the 3-D case. $\Lambda = 0, 1, 2, 3, \square\square\square$, is the oscillator quantum number. It is known that for each Λ value there are ℓ-degenerate values.

$$\Lambda = (2n + \ell - 2), \quad \ell \leq \Lambda \tag{13.4.9}$$

In particular, ℓ = odd, when Λ = odd, and ℓ = even, when Λ = even.

Also total degenerate states N_T (# of nucleons),

$$N_T = \sum_0^\Lambda 2(2\ell + 1) \tag{13.4.10}$$

In the single particle potential shell model, the paired nucleons form an inert core for even-even A nucleus, whereas for the odd-odd nuclei have unpaired proton and neutron. The energy level diagram for the nucleon moving in the parabolic potential field is as shown in Table 13.7:

The energy levels appear in groups such as: 1s; 1p; 1d, 2s; 1f, 2p; *etc*. These grouped levels are degenerate, occupying the same energy state. It can be seen that the closed shells can be predicted as indicated in the last row of Table 13.7. Shell closures at 2, 8, and 20 are in good agreement with experimental evidence, and disagree above. This model is obviously not good enough. Since the first three closure number 2, 8, and 20 agree to the magic numbers, the harmonic potential for the nuclear forces holds good only for light nuclei!

Table 13.7   Single particle SM level scheme as per the infinite harmonic potential well

$\Lambda = (2n + \ell - 2)$	0	1	2	3	4	5	6
Degenerate states ℓ, odd, Λ=odd for each Λ, E/$\hbar\omega$ / even, Λ=even	3/2 0 1s	5/2 1 1p	3/2, 7/2 0, 2 2s, 1d	5/2, 9/2 1, 3 2p, 1f	3/2, 7/2, 11/2 0, 2, 4 3s, 2d, 1g	5/2, 9/2, 13/2 1, 3, 5 3p, 2f, 1h	3/2, 7/2, 11/2, 15/2 0, 2, 4, 6 4s, 3d, 2g, 1i
# particles in state $N_\Lambda = (\Lambda+1)(\Lambda+2)$	2	6	12	20	30	42	56
# particle accumulated CLOSURE NUMBER $N = (\Lambda+1)(\Lambda+2)(\Lambda+3)/3$	(2)	(8)	(20)	(40)	(70)	(112)	(168)

13.4.4 Spin-orbit Potential (Tonsorial Component of Nuclear Potential)

Other shapes for the potential does not improve the situation, and this was for a long time a puzzle in the development of the theory of the structure of nuclei. The missing ingredient in the potential is *spin-orbit* coupling which leads to a term of the form V(r) ℓ.s which is attractive for j = ℓ + 1/2 (ℓ and s parallel) and repulsive for j = ℓ - 1/2 (ℓ and s anti-parallel). The proposal of this extra term in the nuclear potential was made by Mrs. Maria Goeppert Mayer (1949) and by J.H.D. Jensen, Haxel & Suess (1949) independently includes a very special term called LS-potential. Because of this, Mayer-Jensen's shell model is sometimes called the *jj-coupling* shell model. Thereby the shell structure in nuclei was theoretically established and it became a firm foundation of the Nuclear Structure Theory.

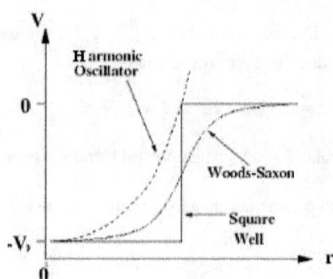

Fig 13.4   Different Nuclear Potentials

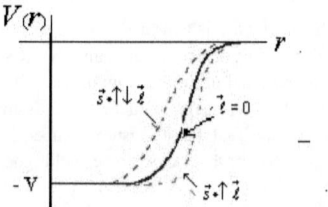

Effect of **S.L** coupling on Nuclear potential

Fig 13.5   Effect on the potential shape

The Spin-Orbit force is also required in order to explain polarization effects in *nucleon-nucleon* scattering. The effect on the potential shape is shown schematically in Fig 13.5.

Note that by using the cosine rule and recalling that the magnitude of the vector $\ell \to \sqrt{\ell(\ell+1)}$, etc it can be shown that:

$$(s,\ell) = [j(j+1) - \ell(\ell+1) - s(s+1)] / 2 \qquad (13.4.11)$$

For $\quad j = \ell + \tfrac{1}{2} \;\to\; (s,\ell) = \ell/2 \;$ and

$\qquad\quad j = \ell - \tfrac{1}{2} \;\to\; (s,\ell) = -(\ell+1)/2$

Thus the splitting in the energy levels, in the ℓth state, due to this difference in the potential is proportional to $(2l + 1)$, separated into two levels with relative spin, ↑ or ↓, by the S-L interaction. For the set of quantum numbers, n, ℓ, and m_ℓ, the total $j = \ell \pm \frac{1}{2}$, of the individual nucleons. The lower energy corresponds to $\vec{L}$ and $\vec{S}$ parallel.

Two cases exist:

13.4.4.1. For Light Nuclei:

L-S coupling holds good for only light nuclei till $^{16}_{8}O$.

$$\vec{L} = \sum \ell_i \qquad (13.4.12)$$

$$\vec{S} = \sum s_i \qquad (13.4.13)$$

$$\vec{J} \equiv I = \vec{L} \pm \vec{S} \qquad (13.4.14)$$

where the first p shell closes.

Magic number for square well nuclear potential

$$\Box = \frac{1}{3} n (n^2 - 1), \quad n = 2, 3, 4, (5). \qquad (13.4.15)$$

13.4.4.2 For most Nuclei

The heavier nuclei exhibit j-j Coupling. Here each nucleon characterized by j_i, ℓ_i, s_i are first coupled to form a total $j_i = \ell_i + s_i$ for that nucleon with angular momentum $j_i \to \sqrt{j_i(j_i + 1)}$. The various such components couple together to form the total angular momentum $\vec{J}$; so that

$$I \equiv \vec{J} = \sum_i j_i . \qquad (13.4.16)$$

A nucleon is thus designated by (n, ℓ, j), where $j = \ell \pm \frac{1}{2}..$

Magic number, for square well nuclear potential,

$$N = \frac{1}{3} n (n^2 + 5) , n = (4), 3, 6, 7. \qquad (13.4.17)$$

13.4.5 Additional Assumptions:

a) Nuclear Spin-Orbit interaction is not Electro-Magnetic in origin.

For a given ℓ, in the case of a nucleus, $\left| j = \ell + \frac{1}{2} \right\rangle$ state is more tightly bound than the $\left| j = \ell - \frac{1}{2} \right\rangle$ state. That is, as in Fig 13.6,

$$E_{|j=\ell+\frac{1}{2}>} < E_{|j=\ell-\frac{1}{2}>} \quad (13.4.18)$$

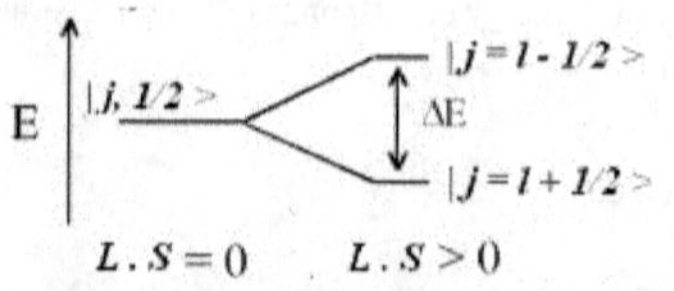

Fig 13.6 Interaction slitting nuclear energy levels.

This contrary to the electronic case of atoms!

b) The Energy separation between two states of the same ℓ, but different j

$$\bar{L} \cdot \bar{S} = \frac{1}{2}(J^2 - L^2 - S^2) = \begin{cases} \frac{1}{2}\ell \hbar^2 & \text{for } |j=\ell+\frac{1}{2}> \\ \frac{1}{2}(\ell+1)\hbar^2 & \text{for } |j=\ell-\frac{1}{2}> \end{cases} \quad (13.4.19)$$

The Energy separation between two states of the same ℓ, but different j is

$$\Delta E_{\ell j} \propto \frac{2\ell+1}{A^{2/3}} \quad (13.4.20)$$

The Harmonic oscillator and square well potential quantum numbers Λ and n are related so that

$$\Lambda = [2(n-1) + \ell] \quad (13.4.21)$$

Table 13.8 Sublevels of Nuclear energy state j for harmonic potential well

Nuclear state j	$s_{1/2}, p_{1/2}$	$p_{3/2}, d_{3/2}$	$d_{5/2}, f_{5/2}$	$f_{7/2}, g_{7/2}$	$g_{9/2}, h_{9/2}$	...
	1/2	3/2	5/2	7/2	9/2	
Maximum # protons or neutrons in state j [2j + 1]	2	4	6	8	10	12

As in the atomic case, a nucleonic state is designated by, for example, $\ell = 0$ by $s_{1/2}$ state, etc. where the subscript denotes the value of j.

The L-S coupling splits each nucleonic state of a given j-value into $(2j + 1)$ sublevels. According to the Pauli Exclusion Principle, the maximum number of nucleons in given (n, ℓ, j)- level or shell is $(2j+1)$, as illustrated in Table 13.8.

13.4.6 Shell Structure

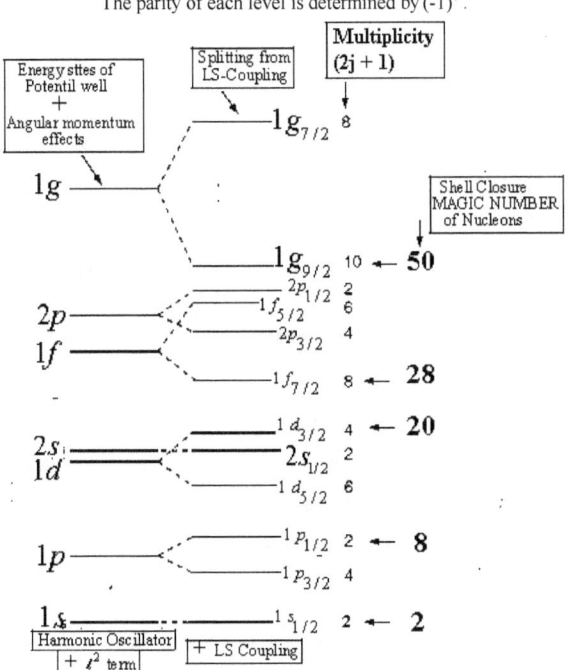

Fig 13.8 The general structure of the nucleon-energy levels in the Shell Model

Each shell, for a specified ℓ-value of a species, say neutron, can accommodate $[2(2\ell+1)]$ neutrons; and so for protons. Accordingly, the shell structure of nucleons is shown in Table 13.9.

The general structure of the nucleon-energy levels in the Shell Model is shown in an energy-level diagram (in Fig 13.8), where the successive states are in a very deep rectangular potential well as well as for the harmonic potential well for comparison..

Table 13.9 Shell structure of each Nuclear Species for state j, for harmonic potential well

SHELL	I	II	III	IIIa	IV	V
State j)	$1s_{1/2}$	$\begin{Bmatrix}1p_{3/2}\\1p_{1/2}\end{Bmatrix}$	$\begin{Bmatrix}1d_{5/2}, 2s_{1/2}\\1d_{3/2}\end{Bmatrix}$	$1f_{7/2}$	$\begin{Bmatrix}1p_{3/2}, 1f_{5/2}\\2p_{1/2}, 1g_{9/2}\end{Bmatrix}$	$\begin{Bmatrix}1g_{7/2}, 2d_{5/2}, 2d_{3/2}\\3s_{1/2}, 1h_{11/2}\end{Bmatrix}$ ..
# protons or neutrons in state j is [2j+1]	2	4, 2 6	6, 2, 4 12	8 8	4, 6, 2, 10 22	8, 6, 4, 2, 12 32
Total # particles CLOSURE NUMBER = $\sum (2j+1)$	(2)	(8)	(20)	(28)	(50)	(82)

Increasing Binding energy → E

13.4.7 Major Closed Shells

Thus j-j coupling plus the assumption that major shells close with $\left|j = \ell + \tfrac{1}{2}\right\rangle$, and the next shell begin with the corresponding $\left|j = \ell - \tfrac{1}{2}\right\rangle$ state. This scheme will predict the magic numbers up to and including 126, as the total number of protons / neutrons in a filled shell. The energy gaps at the magic numbers. The shell model is able to account for several nuclear phenomena in addition to predicting correctly magic numbers. Thus one has many simple and important deductions from the nuclear shell model.

They are:

The nuclear state is characterized by quantum numbers (n, ℓ, j), where $\left|j = \ell \pm \tfrac{1}{2}\right\rangle$. For nuclei with Z > 50, for a given l-value, the $\left|j = \ell + \tfrac{1}{2}\right\rangle$ is more tightly bound than $\left|j = \ell - \tfrac{1}{2}\right\rangle$.

$$\Delta E_{\ell j} \propto \frac{2\ell + 1}{A^{2/3}}, \qquad (13.4.22)$$

which increases with increasing l-value.

An even number of neutrons / protons having the same ℓ and j will always couple to give even parity, so j = 0, and magnetic moment = 0.

An odd number of identical nucleons with the same ℓ and j will always couple to give parity of the state odd, if l = odd, and even parity if l = even. $j \neq 0$, and µ = *magnetic moment* of single nucleon in $|j\rangle$ state.

There arises an additional binding energy, known as pairing energy, δ, associated with double occupancy of any state $|\ell, j\rangle$ by two identical nucleons. This δ = largest for $|\ell, j\rangle$ with the highest j-value. For an even nucleon,

$$\delta \propto \frac{2j+1}{A}. \qquad (13.4.23)$$

13.5 RELATIONSHIPS BETWEEN NUCLEAR SPIN (I) AND MAGNETIC MOMENT (μ)

Measurements of I and μ or the ground levels of more than 100 nuclides have been made. From these data it is known that, without exception, among stable nuclides for all Even-Even nuclides, $I = 0$, whereas for all other nuclides $I \neq 0$.

13.5.1.1 Even A – nuclides

Since each energy sublevel can accommodate two nucleons of the same species (one with spin ↑ and the other with spin ↓), only filled sublevels are therefore present when Z = even and N = even in a nucleus. This means all even-even nuclides have $I = 0$, in ground state.

Even-A nuclide obeys the Bose-Einstein Statistics.

13.5.1.2 Odd A- Nuclides

Odd Z-even N and Even Z – Odd N nuclides come under Odd-A nuclides.

Nuclides with odd-A obey the Fermi-Dirac Statistics.

Worked out Example 13.1

Examine if nucleus $^{11}_{5}B$ has spin $I = \frac{3}{2}$:

Solution: $\boxed{STEP\ *\ 1}$ For $^{11}_{5}B$, Z = 5 and N = 6. It will be seen from Fig 13.6 that the 6 neutrons fill all the 1s and $^{1}P_{3/2}$ states.

$\boxed{STEP\ *\ 2}$ On the other hand, its protons fill the same states except for one $^{1}P_{3/2}$ state. This means the nuclear spin I coincides with the j-value, viz., $I = \left| j = \ell + \frac{1}{2} \right|$ of the proton, which is the last odd nucleon. Since $^{1}P_{3/2}$ state has l = 1, $I = \left| j = \ell + \frac{1}{2} \right|$, for $^{11}_{5}B$.

$\boxed{STEP\ *\ 3}$ The parity of the p-state is $(-1)^1$ is negative, i.e., odd. Therefore the nuclear spin of $^{11}_{5}B$ is $\frac{3}{2}^{-}$.

13.5.2 Magnetic Moments (μ_j) of Odd-A Nuclides

Perhaps one of the most striking successes of the nuclear Shell Model is its ability to describe the general behaviour of nuclear magnetic moments of odd-A nuclides.

13.5.2.1 Calculation of μ_j:

Assumption 1: In odd-A nuclides, $\vec{\mu}_{\ell*}$ and $\vec{\ell}*$ are collinear; where according to the Kramer's Rule for quantum mechanical squares, $\vec{\ell}* \equiv \sqrt{\ell(\ell+1)}$.

Assumption 2: In odd-A nuclides, $\vec{\mu}_{s*}$ and $\vec{s}*$ are collinear; $\vec{s}* \equiv \sqrt{s(s+1)}$.

Assumption 3: $\vec{\mu}_{j*}$ and $\vec{j}*$ are collinear; where $\vec{j}* \equiv \sqrt{j(j+1)}$.

Both $\vec{\ell}*$ and $\vec{s}*$ precess about their resultant vector $\vec{j}*$ (Fig 13.10).

$$\vec{j}* = \vec{\ell}* \pm \vec{s}* \tag{13.5.1}$$

$\vec{j}*$ precesses about $\vec{j}\,^*_z$, the maximum value of $\vec{j}*$.

Though $\vec{\mu}_{\ell*}$ lies along $\vec{\ell}*$ and so also $\vec{\mu}_{s*}$ along $\vec{s}*$, but $\vec{\mu}_{j*}$ does not lie along $\vec{j}*$ so that

$\vec{j}* = \vec{\ell}* + \vec{s}* = |\vec{\ell}*| \pm |\vec{s}*|$. Therefore $\vec{\mu}_{\ell*}$ precesses around $\vec{j}$-vector so that only

$|\vec{\mu}_{\ell*}|$ (is the component of $\vec{\mu}_{\ell*}$ along $\vec{j}$ = a constant.

$$\vec{\mu}_{j*} - g_j \vec{j}* \; nm \tag{13.5.2}$$

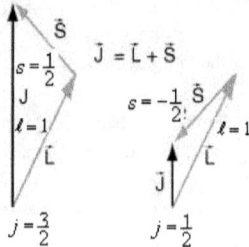

Fig 13.9 Vector couplings

All the other nucleons have their resultant angular momenta paired in opposite directions and therefore do not contribute to the resultant nuclear magnetic dipole moment.

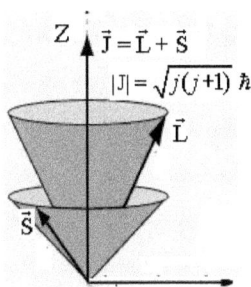

Fig 13.10 Vector S-L coupling

To evaluate $\vec{\mu}_{\ell^*}$

$$\cos(\vec{\ell}^*, \vec{j}^*) = \frac{\vec{\ell}^{*2} + \vec{j}^{*2} - \vec{s}^{*2}}{2\,\vec{\ell}^*\,\vec{j}^*} \qquad (13.5.3)$$

Similarly

$$\cos(\vec{s}^*, \vec{j}^*) = \frac{\vec{s}^{*2} + \vec{j}^{*2} - \vec{\ell}^{*2}}{2\,\vec{s}^*\,\vec{j}^*} \qquad (13..5.4)$$

But

$$\vec{\mu}_{j^*} = (g_\ell\, \vec{\ell}^* + g_s\, \vec{s}^*)\; nm \qquad (13.5.5)$$

The orbital gyro-magnetic ratios (i.e., orbital nuclear g-factor) as well as spin g-factors for protons and neutrons are listed in Table13.9.

Table 13.10	g-factors for protons and neutrons	
	Protons	Neutrons
1. Orbital g_ℓ	+1	0
2. Spin g_s	+5.5855	-3.8263

13.5.2.2 Case i): Parallel Spins

$$g_{j^*} = g_\ell\, \frac{\ell}{\ell + s} + g_s\, \frac{s}{\ell + s} \qquad (13.5.6)$$

whence in a magnetic field, B,

$$\Delta E = -\vec{\mu}_I \cdot \vec{B} \qquad (13.5.7)$$

where $\vec{\mu}_I = \vec{\mu}_\ell + \vec{\mu}_s$ (13.5.8)

Since
$$\langle \mu \rangle = |\vec{\mu}_{j^*}| = (\vec{\mu}_{j^*} \square \vec{j}^* / \vec{j}^2) \vec{j}_z$$
$$= [(g_\ell\, \vec{\ell}^* + g_s\, \vec{s}^*) \square \vec{j}^* / \vec{j}^2] \square (m = j)\, nm \qquad (13.5.9)$$

This means
$$(\vec{\mu}_I / nm) = (\langle \mu \rangle / nm) = (g_\ell\, \vec{\ell}^* \square \vec{j}^* + g_s\, \vec{s}^* \square \vec{j}^*) / (j + 1)$$
$$= (g_\ell\, \vec{\ell}^* + g_s\, \vec{s}^*) \qquad (13.5.10)$$

For a single particle,
$$j \equiv I,\quad g_j = g = \mu/I = \mu/j = \vec{\mu}_{j^*} / \vec{j}^* \qquad (13.5.11)$$

$$I = j = (\ell + \tfrac{1}{2}) \qquad (13.5.12)$$

Denoting by superscript "(+)" for parallel signs,
$$\vec{\mu}_I^{(+)} = g_\ell^{(+)} I = (g_\ell\, \ell + g_s\, s)$$
$$= (j - \tfrac{1}{2}) g_\ell + \tfrac{1}{2} g_s \qquad (13.5.13)$$

13.5.2.3 Case ii) Anti-parallel spins:

$$I = j = (\ell - \tfrac{1}{2}) \qquad (13.5.14)$$

Using the superscript "(-)" sign, to denote anti-parallel spins,
$$\vec{\mu}_I^{(-)} = g_\ell^{(-)} I = (g_\ell\, \ell + g_s\, s) \qquad (13.5.15)$$

$$\vec{\mu}_I^{(-)} = g_\ell\, I - (\mu_s - \tfrac{1}{2} g_\ell) \tfrac{I}{I+1}$$
$$= [(I + \tfrac{3}{2}) g_\ell - \tfrac{1}{2} g_s] \tfrac{I}{I+1} \qquad (13.5.16)$$

The results given by relations (13.5.13) and (13.5.16) can be used to derive $\vec{\mu}_I$ of nuclei for single particle.

13.5.3 Schmidt Formulae

According to single particle model of nuclei, with unpaired nucleon (Section 13.5.2),

$$\vec{\mu}_I = \vec{\mu}_{j=\ell \pm \frac{1}{2}}$$

$$\vec{\mu}_{j=\ell+\frac{1}{2}} = g\,I = g_\ell\,\ell + \mu_s \quad (13.5.17)$$

$$\vec{\mu}_{j=\ell-\frac{1}{2}} = g\,I = g_\ell\,I - (\mu_s - \frac{1}{2}g_\ell)\frac{I}{I+1}$$

$$= g_\ell \left[\frac{(\ell+1)(\ell-\frac{1}{2})}{(\ell+\frac{1}{2})} \right] - \mu_s \frac{(\ell-\frac{1}{2})}{(\ell+\frac{1}{2})} \quad (13.5.18)$$

From Dirac's concept of the electron spin, it is seen from Table 13.10, that $g_\ell = +1$, for protons ($g_s = -2.0$, for electron) so that

$$g_s^{(p)} = 2\,\mu_s^{(p)} = 2\,(2.79) = 5.588 \quad (13.5.19)$$

$g_\ell = 0$, for neutrons,

$$g_s^{(n)} = 2\,\mu_s^{(n)} = 2\,(-1.91) = -3.826 \quad (13.5.20)$$

Thus spin moments cause anomaly in $g_s^{(p)}$ and $g_s^{(n)}$.

One can, therefore, write

Table 13.11   Schmidt Formulae for Nuclear Magnetic Moment

I	Odd Z-even N nuclides, Odd Z	Even Z-odd N nuclides, Odd N
$(\ell+\frac{1}{2})$	$\vec{\mu}_I^{(o-e)} \equiv \vec{\mu}_j^{(o-e)} = \left(I + \frac{2.29}{I}\right)$ nm	$\vec{\mu}_I^{(e-o)} \equiv \vec{\mu}_j^{(e-o)} = \left(-\frac{1.91}{I}\right)$ nm
$(\ell-\frac{1}{2})$	$\vec{\mu}_I^{(o-e)} \equiv \vec{\mu}_j^{(o-e)} = \left(I - 2.29\frac{I}{I+1}\right)$ nm	$\vec{\mu}_I^{(e-o)} \equiv \vec{\mu}_j^{(e-o)} = \left(\frac{1.91}{I+1}\right)$ nm

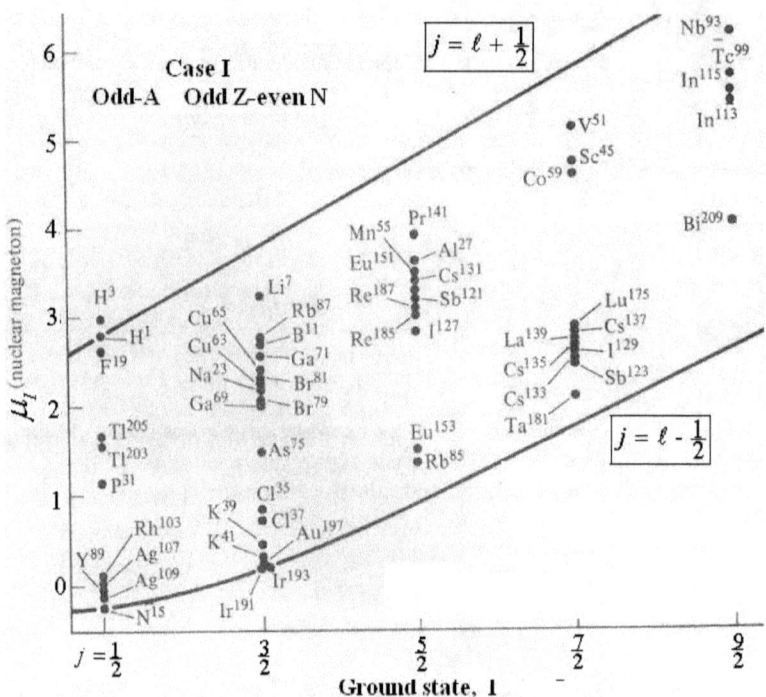

Fig 13.11 Magnetic moments versus angular momenta of Odd-Z nuclides, and Schmidt limits.

The set of two values of $\vec{\mu}_I$, for each value of I, depending on

$$I = j = \ell \pm \tfrac{1}{2}, \qquad (15.5.22)$$

and the set constitute the so-called Schmidt Limits. The measured values of $\vec{\mu}_I$ and I for the ground energy levels of nuclei, numbering more than hundred, are seen to fall generally the *Schmidt limits*. This gives two *Schmidt lines*, one the upper and the lower, for each set, in the plot of $\vec{\mu}_I$ versus I, as shown in Fig 13.11 and Fig 13.12. In these plots, "cross" symbols denote experimental values for $I = j = \ell \pm \tfrac{1}{2}$ and symbol "circle" correspond to $I = j = \ell \pm \tfrac{1}{2}$. Most of the magnetic moments are found to lie in between the upper and lower Schmidt line limits.

However, there is lack of precise agreement between the Sell Model and experiment!

13.5.3.1 Determination of Nuclear State

Worked out Example 13.2

Find the nuclear state of $^{59}_{27}Co$. Given that $I = \frac{7}{2}, \mu = 4.6.4nm$ and it falls in the upper Schmidt limit.

Solution. STEP * 1 Upper Schmidt limit indicates that $I = j = \ell + \frac{1}{2}$, giving $\ell = 3$.

STEP * 2 Z = 27, and N = 32, *i.e.*, the odd nucleon is the proton.

STEP * 3 $\ell = 3$ means the odd proton is in the f-state.

The complete state of $^{59}_{27}Co$ is $f_{7/2}$.

13.5.4 Total Angular Momentum of Even-A Nuclides

There are only six stable Even-A, odd Z-odd N nuclides. Experiments show that I > 0 for these, cases, *viz.*,

^2_1H, ^6_3Li, $^{10}_5B$, $^{14}_7N$, etc.

It is more difficult, than the odd-A nuclides, to determine the nuclear spin,

$$I = (j_p \pm j_n)$$

All integral values of I from

$$I = (j_p + j_n) \text{ to } |j_p - j_n| = \text{integer}$$

are allowed in principle. j_p and j_n correspond to the unpaired proton and neutron

Worked out Example 13.3

Find the state of the nucleons in ^2_1H which has an experimental value of $I = 1$.

Solution: STEP * 1 According to $I = (j_p + j_n)$ to $|j_p - j_n| =$ integer , STEP * 2 consider $I = (j_p + j_n) = 1$, this means $j_p = \frac{1}{2}$ and $j_n = \frac{1}{2}$.

STEP * 3 So the p and n are in parallel spins; and they are both in $s_{1/2}$ states.

13.5.5 Pairing-energy Effect

Another experimental evidence that can be explained by the Shell Model is the existence of almost no stable nuclides of the even-A (odd Z – odd N) type. Only 6 in number out of the total of 172 even-A nuclides are stable!

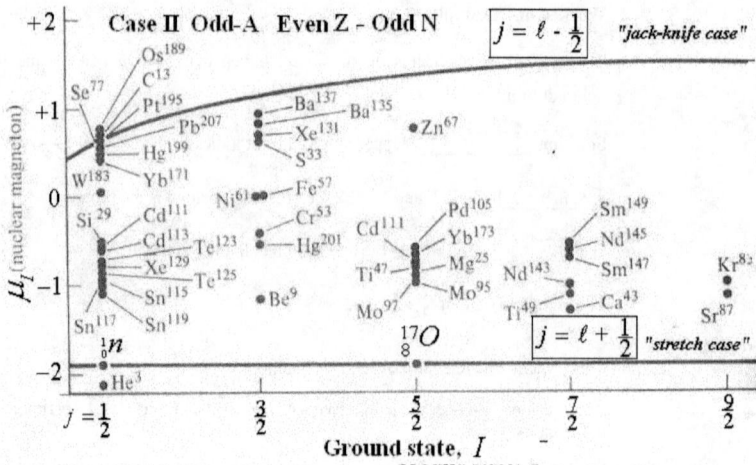

Fig 13.12   Magnetic moments versus angular momenta of Odd-N nuclides, and Schmidt limits

The difference between the actual nuclear interactions experienced by a nucleon from the average central-plus- spin-orbit force field (the Shell Model) is known as the residual interaction. This residual interaction is thought to contribute to nuclear stability, by what is termed a 'pairing-energy effect'.

Most nuclides in the odd Z-odd N class are radio-active, and they decay to even Z-even N nuclides (I = 0) due to 'pairing effect', in which

$$n \xrightarrow[\beta\text{-deacy}]{\tau = 12.8 \, m} p$$
(13.5.23)

13.5.5.1 Nordheim's Rules

If for the two odd nucleons,

$$j_1 + j_2 + \ell_1 + \ell_2 = \text{even}$$
(13.5.24)

the resultant spin $I = |j_1 - j_2|$.

It is possible to infer the state of the odd proton by comparison with a known odd-even (odd-A) nuclide having the same Z.

Worked out Example 13.4

The nuclear spin of $_3^7Li$ is $I = \frac{3}{2}$ ..I is seen to lie near the $I = j = \ell + \frac{1}{2}$ limit. Find the state of the odd nucleon.

Solution: $\boxed{STEP * 1}$ $I = j = \ell + \frac{1}{2} = \frac{3}{2}$; gives $\ell = 1$.

$\boxed{STEP * 2}$ The odd proton is therefore in the $p_{3/2}$ state.

Analogously, the state of the odd neutron can often be inferred.

L.W. Nordheim (1950) has pointed out that for at least 60 clear cases the two Empirical Rules can be applied. These two rules govern the coupling of the angular momenta of the proton and neutron with a tendency for parallel spin. For the ground state the two rules are:

a) The Strong Rule: If the odd proton and odd neutron belong to different Schmidt groups, *i.e.*,

$$j_p = \ell \pm \frac{1}{2} \text{ and } j_n = \ell \mp \frac{1}{2}, \text{ then}$$

$$j \equiv I = |j_p - j_n|.$$

b) The Weak Rule: If the odd proton and odd neutron belong to the same Schmidt groups, *i.e.*,

$$j_p = \ell \pm \frac{1}{2} = j_n, \text{ then}$$

$$j \equiv I \leq j_p + j_n.$$

Brennan and Bernstein have modified Nordheim's Rules by analyzing experimental results.

13.5.6 Double Scattering Experiment – Evidence for L-S Interaction in Nuclei

It is known that the nuclear Shell Model assumed conveniently the L-S interaction to explain certain features of the energy levels of nuclei. It is quite understood that the interaction between a magnetic field $\vec{B}$ and a magnetic dipole $\vec{\mu}$, i.e., $E_B = -\vec{\mu} \cdot \vec{B}$, implies a torque $\vec{\tau} = \vec{\mu} \wedge \vec{B}$, which tends to align $\vec{\mu} \cdot \vec{B}$. Analogously, a L-S interaction of the form, $E_{L-S} = -f(r)\vec{L} \cdot \vec{S}$, implies a torque that tends to align $\vec{S} \cdot \vec{L}$.

In the double scattering experiment, a beam of unpolarized protons (protons with spin ↑ and ↓) strike helium nuclei. The scattered protons, along path 1 and path 2, by the scatterer (helium nuclei) will act as the reference for orbital momentum $\vec{L}(\uparrow)$ and $L(\downarrow)$, due to the L-S interaction of the protons with helium nucleus. Next, beam1 is made to strike another helium nucleus causing it to split into two beams, denoted by (3) and (4), which have $\vec{L}(\uparrow)$ and $L(\downarrow)$, respectively; protons detected by detectors D3 and D4. Since beam1 contains more protons of ↑ than ↓, the interaction favours scattering through beam 3 over beam 4. So the detector D3 will register an intense current than D4. This is due to the presence of L-S interaction, as manifested by the polarization produced by scattering. In the absence of L-S interaction, detectors D3 and D4 would have recorded currents not of equal intensity.

13.5.7 Conclusions

The phenomena that could be explained satisfactorily by the Shell Model, include

1, Odd-A nuclides obey the Fermi-Dirac Statistics.

2. The discrepancy that the μ^- value of an odd-A nucleus does not fall on the Schmidt lines indicate that the moment of a nucleon inside a nucleus is not necessarily its free moment, and that there is distortion to the nucleonic structure

3. The Shell model of nucleus with j-j coupling gives only a satisfactory description of some experimental nuclear data.

4. The origin of discrete γ- spectra is explained.

5. Magnetic dipole moment

6. Systematic variations in nuclear quadrupole moment,

7. Nuclear energy levels,

8. Existence of nuclear isomers,

9. Discontinuities in nuclear binding energy,

10. Pairing energy of even-A nuclei,

11. Spins and parities of many unstable nuclei, which exhibits β- decay (especially odd- A nuclides).

12. But the Shell Model is mostly successful near magic number nuclei.

13.6 LIQUID DROP MODEL

The suggestion that nuclei have approximately constant density, as obtained from scattering experiments, enables calculation of nuclear radius by using that density as if the nucleus were a drop of a uniform liquid. In the liquid drop model, formulated by Niels Bohr, the nucleons are imagined to interact strongly with each other, like the molecules in a drop of liquid. The fact is that the forces on the nucleons on the surface are different from those on nucleons on the interior, where they are completely surrounded by other attracting nucleons. This is something similar to taking account of surface tension as a contributor to the energy of a tiny liquid drop.

Always each nucleon collides frequently with other nucleons within the nucleus. Its mean free path as it moves about will be substantially less than the nuclear radius. The liquid drop model allows one i) to correlate many facts about nuclear masses and binding energies E_B; ii) it is useful in explaining nuclear fission. and iii) It also provides a useful model for understanding a large class of nuclear reactions.

The mass of a nucleus defined by A and Z, denoted by $M(A,Z) \equiv {}^A_Z M$, is given by

$$M(A,Z) \equiv {}^A_Z M = [Z M_H + (A - Z) M_n - E_B / c^2] u \qquad (13.6.1)$$

where ${}^A_Z M$ = Mass of the nuclide, in mass units,

M_H = mass of ${}^1_1 H = 1.007825\ u$

M_n = mass of ${}^1_0 n = 1.008665\ u$

$E_B \equiv E_B(A,Z)$ = Binding energy of the nuclide.

The observed variation of binding energy per nucleon with mass number can be explained by the sum of the following three types of energies.

13.6.1 Volume Energy

The volume term

$$E_V = + a_V A. \qquad (13.6.2)$$

Since the exchange nuclear force is saturated, each nucleon contributes $E_B / A = \Box\ 8\ MeV$ is constant, for nuclei with $A \geq 16$, to the binding of the nucleus. $\qquad a_V = 15.753\ MeV$.

(13.6.3)

The interactions between the nucleons are essentially just between nearest neighbours. It is so,

And this energy $E \propto A$,

and nuclear volume $V \propto A$.

(If this were not so, $E \propto A^2$).

13.6.2 Surface Energy

Within the volume of a nucleus all the nucleons are equally attracted in all the directions. On the other hand, a nucleon at the surface of a nucleus interacts with fewer other nucleons and hence it is weakly bound from inside (equivalent to surface tension in a liquid drop). A correction to this effect, $\Delta E \propto 4\pi (r_o A^{1/3})^2$ / nucleonic cross section.

$$\therefore \quad \Delta E = a_S A^{2/3}, \text{ or } E_S = -a_S A^{2/3}, \quad (13.6.4)$$

where $a_S = 17.80\ MeV$. $\quad (13.6.5)$

13.6.3 Coulomb Effect

For $A = 60$, $E_B/A = \Box\ 8.7\ MeV$, which decrease to $7.3\ MeV$ for $A = 238$. This is due to mutual ES repulsion of the $Z(Z-1)/2$ pairs of nuclear protons, and contributes to the Coulomb term. For a spherical nucleus of radius $R = r_0 A^{1/3}$ with the charge spread evenly throughout the sphere, $\rho = Ze/[\frac{4}{3}\pi R^3]$, one can obtain the Coulomb energy as follows:

i) The additional work required to add a layer of thickness dr, to a uniformly charged spherical nucleus of radius r and density ρ, is calculated by assuming the nuclear charge located at the centre of the shell,

$$V_C = \int_0^R [\frac{4}{3}\pi r^3 \rho]\Box [4\pi r^2\ dr\ \rho]\frac{1}{r} = \frac{3}{5}\frac{1}{4\pi\varepsilon_0}\frac{Z^2 e^2}{R} \quad (13.6.6)$$

ii) This contains a spurious "self energy" term $\frac{3}{5}\frac{e^2}{R}$, for each proton, especially for light nuclei.. Subtracting this term for Z protons,

$$V_C = \left[\frac{3}{5}\frac{1}{4\pi\varepsilon_0}\frac{e^2}{r_0}\right]Z(Z-1)A^{-1/3}$$

$$V_C = +a_C\ Z(Z-1)A^{-1/3}, \text{ i.e.,}$$

$$E_C = -a_C\ Z(Z-1)A^{-1/3} \quad (13.6.7)$$

where $a_C = 0.617\ MeV$, with $r_0 = 1.4\ fm$ and

$a_C = 0.7103\ MeV$, with $r_0 = 1.2\ fm$. $\quad (13.6.8)$

Worked out Example 13.5

Obtain the condition for stability of a nuclide whose mass can be considered as the sum of the volume, surface and Coulomb effects. Estimate A for the nucleus for $Z = 20$.

Solution STEP * 1 Recall equation (13.6.9)

$$M(A,Z) = \left\{ZM_H + (A-Z)M_n - \left[-a_V A + a_S A^{2/3} + a_C Z(Z-1)A^{-1/3}\right]/c^2\right\}u$$

STEP * 2 Setting $\left.\dfrac{dM}{dZ}\right|_A = 0$,

STEP * 3 $Z_{Stable} = 0.66\, A^{1/3}$.   (13.6.10)

STEP * 4 $A = 28{,}000$.

The answer to Ex 13.5 conflicts experiment. This means at least one more correction id required for semi-empirical mass formula (13.6.8).

Fig 13.13 shows the three terms contributing to the Binding energy of a nucleus.

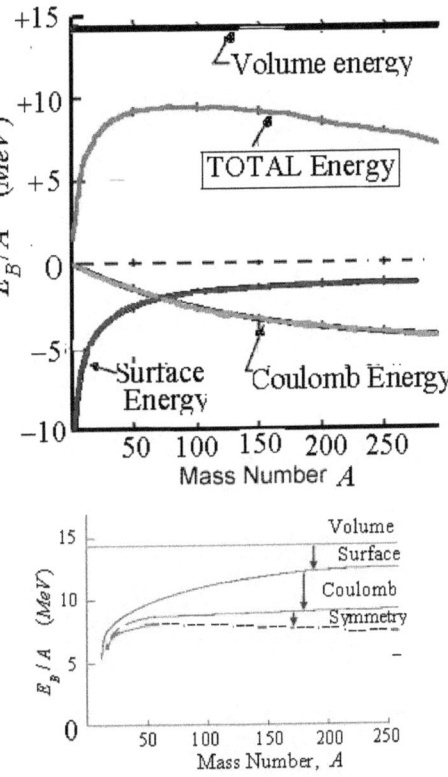

Fig 13.13 Semi-empirical mass formula: Volume, Surface and Coulomb terms

An empirical expression for the binding energy of nuclei was first proposed in 1935 by C. Von Weizsaecker modified in 1936 by Hans Albrecht Bethe and Robert Fox Bacher, which is referred to as the "semi-empirical mass formula" and the "Bethe-Weizsaecker formula".

13.6.4 Asymmetry Term (Isotope Effect)

The Pauli principle favours nuclei in which A=2Z, so the empirical model of binding energy contains a term of the form

$$\Delta E_b^{Pauli} = \frac{-(23.7 MeV)(A-2Z)^2}{A} \qquad (13.6.11)$$

The Pauli principle also favors nuclear configurations with even numbers of neutrons and protons. In the liquid drop model, this is included by using the even-odd nucleus as a reference and adding a correction term which is positive for even-even nuclei and negative for odd-odd nuclei.

Expressed in terms of the mass number A and the atomic number Z for an even-odd nucleus, the Weizsaecker formula is

$$\Delta E_b^{Even-Odd} = \underset{Volume\ term}{(15.75\ MeV)A} - \underset{Surface\ term}{(17.8\ MeV)A^{2/3}}$$
$$- \underset{Coulomb\ term}{\frac{(0.711 MeV)Z^2}{A^{1/3}}} - \underset{Pauli\ term}{\frac{(23.7 MeV)(A-2Z)^2}{A}} \qquad (13.6.12)$$

Worked out Example 13.6

Calculate the atomic number of the most stable nucleus of given A.

Solution

$$M(A,Z) = \left\{ \begin{array}{l} Z M_H + (A-Z) M_n \\ - \left[\begin{array}{l} -a_V A + a_s A^{2/3} + a_C Z(Z-1) A^{-1/3} \\ + a_A (Z - A/2)^2 / A \end{array} \right] / c^2 \end{array} \right\} u \qquad (13.6.13)$$

Setting $\left. \dfrac{dM(A,Z)}{dZ} \right|_A = 0$,

$$Z_{Stable} = \frac{A}{1.98 + 0.0155 A^{2/3}} \qquad (13.6.14)$$

For light nuclei, this results in $Z \cong \dfrac{A}{2}$

This result is in confirmatory with experiment!

13.6.5 Pairing Effect

Finally, proton and neutron levels are filled, by forming pairs of spin-up and spin-down, according to the Pauli's principle. Using the even Z-odd N as a reference, there are then correction terms to account for even-even and odd-odd nuclei, the even-even groupings of protons and neutrons being favored in stability. Enrico Fermi in 1950 introduced a small empirical correction term

$$E_{e-e} < E_{e-o} < E_{o-o}$$

$$E_{o-e} \begin{cases} = 0 & \text{for e-o/o-e, Odd-}A \\ -\delta - + a_p A^{-3/4} & \text{for e e, Even }A \\ -\delta - a_p A^{-3/4} & \text{for e o, Even }A \end{cases} \qquad (13.6.15)$$

where $a_p = 33.6\ MeV$.

An estimate of nuclear binding energy can then be obtained by first applying the even-odd formula.

For atomic mass number A = atomic number

$$Z = \frac{E_b^{even-odd}}{} = MeV$$

Then, if the nucleus is even-even or odd-odd, the appropriate correction can be made.

$$Z = \frac{E_b^{even-even}}{} = MeV$$

$$Z = \frac{E_b^{odd-odd}}{} = MeV$$

13.6.6 Total Binding Energy of the Nucleus

$$M(A,Z) = \left\{ \begin{array}{l} Z M_H + (A-Z) M_n \\ \left[-a_V A + a_s A^{2/3} + a_C Z(Z-1) A^{-1/3} \right] / c^2 \\ + a_A (Z - A/2)^2 / A \pm a_p A^{-3/4} \end{array} \right\} u \qquad (13.6.16)$$

Table 13.12 Constant of the Semi-empirical Mass Formula

	Constant	a_V	a_S	a_C	a_A	a_P
Set I	MeV	14.1	13	0.595	19.0	33.6
	$u\ (10^{-2})$	1.51	1.40	6.39	2.04	3.60
Set II	MeV	15.6	17.2	0.70	22.5	11
	$u\ (10^{-2})$	1.69	1.91	7.63	2.54	3.60

This expression is known as the Bethe-Weizsacker relation or the "*Semi-empirical Mass Formula*". The coefficients are determined by fitting to a suitably large data set of masses (hence semi-empirical) of a number of stable nuclides. Two sets of values are listed in Table 13.12

The equation of the E_B/A versus A curve is

$$\frac{E_B}{A} = \left[\begin{array}{c} 15.760\,A - 17.810\,A^{2/3} - 0.711\,Z(Z-1)A^{-1/3} \\ -23.702\,(Z-A/2)^2/A \pm 34\,A^{-3/4} \end{array} \right] MeV \qquad (13.6.17)$$

This equation matches within ±1% of the experiment. In the $^{12}_{6}C$ scale

$$M(A,Z) = \left\{ \begin{array}{c} 0.991175\,A - 0.000840\,Z \\ +0.0191114\,A^{2/3} + 0.0007626\,Z(Z-1)A^{-1/3} \\ +0.101750\,(Z-A/2)^2\,A^{-1} \pm 0.036\,A^{-3/4} \end{array} \right\} u \qquad (13.6.18)$$

One can note that of 342 (beta-) stable nuclei in the 1993 mass compilation, there are

a) 209 with even A, even Z;

b) 70 with odd A, even Z;

c) 59 with even A, odd Z and

d) Only 4 with odd N and Z ($^{2}_{1}H$, $^{6}_{3}Li$, $^{10}_{5}B$, $^{14}_{7}N$)..

Clearly pairing enhances stability (or binding energy). This can also be seen, for instance, in the neutron separation energies of neighbouring isotopes, etc.

Worked out Example 13.7

Find out the difference in mass between two nuclides $^{A}_{Z}X$ and $^{A}_{Z+1}Y$.

Solution $\quad \Delta M = \left\{ M_H - M_n + 2a_C\,Z\,A^{-1/3} \right\} u, \qquad (13.6.19)$

$\qquad\qquad = \left\{ -1.390 + 1.526\,Z\,A^{-1/3} \right\} 10^{-3}\,u$

13.6.6.1 Determination of the value of the coefficient of the Coulomb effect

From a determination of the mass difference between two mirror nuclides, such as $^{23}_{11}Na$ and $^{23}_{12}Mg$, (as in Ex 13.7) from experimental masses, using equation (13.6.19), the value of a_C can be obtained.

13.6.6.2 Determination of the radius constant of a nucleus

From the measured value of a_C (Section 13.6.6.1), and using the relation

$$a_C = \frac{3}{5} \frac{1}{4\pi \varepsilon_0} \frac{e^2}{r_0}; \qquad (13.6.20)$$

obtained from equation (13.5.6.6), one can evaluate r_0.

13.7.1 THE LINE OF STABILITY

Two kinds of stability of nuclei can be considered.

i) Dynamical Stability, which is the breaking up of nuclear system into two or more fragments. This is energetically impossible.
ii) β-instability: The transformation of $n \rightarrow p$ or $p \rightarrow n$ with the capture or emission of an electron and a ν_e is energetically impossible.

The stability conditions are based on two main factors.

a) The Symmetry Effect,
b) The Charge Effect, and
c) The Spin dependence of nuclear forces.

An examination of the Segre chart (N-Z diagram) for naturally occurring nuclides show that for stable nuclei, with low-A, have $N = Z = A/2$. All nuclides of the same N (isotones) appear in horizontal lines and all nuclides in vertical lines having the same Z are isotopes.

13.7.2 MASS PARABOLAS

It is implied that for greater binding energy per nucleon there requires greater stability. It is most convenient to explore this in the context of a set of isobars, i.e. a set of nuclides with the same A. These can transform into one another by various forms of beta decay.

The masses of the members of a set of isobars can be obtained by rearranging the semi-empirical mass formula (13.6.16):

$$M(A, Z)\, c^2 = \left\{ \begin{array}{l} Z M_H + (A-Z) M_n \\ \left[\begin{array}{l} -a_V A + a_s A^{2/3} + a_C Z(Z-1) A^{-1/3} \\ + a_A (Z - A/2)^2 / A \pm a_p A^{-3/4} \end{array} \right] \end{array} \right\} u$$

$$M(A,Z)\, c^2 = \alpha A + \beta Z + \gamma Z^2 \mp \delta \qquad (13.7.1)$$

where $\alpha = (M_n c^2 - a_V + a_A + a_S A^{-1/3})$, $\qquad (13.7.2)$

$\beta = [-4 a_A - (M_n - M_p) c^2]$ $\qquad (13.7.3)$

$\gamma = [4 a_A A^{-1} + a_C A^{-1/3}]$ $\qquad (13.7.4)$

This equation has the form of a parabola for fixed A (stable nuclide). This expression is a parabola in Z, opening upward. One can solve for the value of Z giving the greatest binding energy (smallest mass), i.e. the most stable isobar. Thus

$$\left.\frac{dM(A,Z)}{dZ}\right|_A = 0 \qquad (13.7.5)$$

yields $\quad Z_S = \dfrac{\beta}{2\gamma} = \dfrac{\left(\frac{A}{2} + (M_n - M_p)c^2 \frac{A}{8\,a_A} + a_C \frac{A^{2/3}}{8\,a_A}\right)}{\left(1 + \frac{1}{4}\frac{a_C}{a_A}A^{2/3}\right)} \qquad (13.7.6)$

Inserting the values for the coefficients and rearranging,

$$Z_S = \frac{A\,(1 + 0.0075\,A^{-1/3})}{1.983 + 0.01517\,A^{2/3}} \qquad (13.7.7)$$

This then gives the equation for the `*valley of stability*' on the N-Z chart of nuclides. Note that it is determined by an interplay between the Coulomb effect (makes Z a minimum) and the asymmetry term (makes N = Z).

13.7.3 The valley of β-stability

13.7.3.1 For odd A,

$\delta = 0$ both before and after a beta-decay (since either Z or N is odd both before and after decay). The equation gives a single parabola, as in Fig 13.13. In this case, we may find the most stable Z value for a given A by finding the minimum of the expression above:

The actual Z of lowest energy, Z_{Min}, is the integer value nearest the computed Z.

Since $A^{2/3}\ [= 0.714\,A^{2/3}\ MeV] > (M_n - M_p)c^2\ [= 0.78\ MeV]$,

$A \geq 2$ it is seen at once that

$$Z_{Min} \leq A/2 \text{ and } N \geq Z_{Min}. \qquad (13.7.8)$$

<u>Case 1</u>. Since the v_e is (nearly) massless, β^--decay is possible if

$$M(A, Z) > M(A, Z+1) \qquad (13.7.9)$$

<u>Case 2</u>. If $Z \geq Z_{Min}$, decay can occur by β^+-(positron) emission if
$$M(A, Z) > M(A, Z-1) + 2\,m_e \qquad (13.7.10)$$

<u>Case 3</u> EC process

if $\quad M(A, Z) > M(A, Z-1) + E_e$ (13.7.11)

where E_e = Ionization energy

For odd-A nuclides there can be only one isobar (stable) for which both theses conditions do not occur. But there are two exceptions, A =113 and A = 123 are no doubt have long-life ($\square$ 10^{12} y) are in the lower limit and have exceptionally small mass difference.

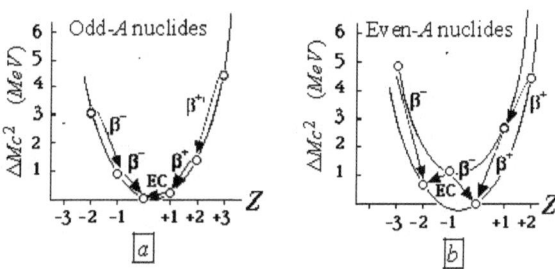

Fig 13.13 Stability of isobars and Mass parabolas

For comparison, the values of M/A are shown for unbound nucleons on the left at $Z/A = 0$ for $_0^1 n = 1.008665$ u and on the right at $Z/A = 1$ for $_1^1 H - E_C = 1.007071$ u.

13.7.3.2 For Even - A,

$\delta \neq 0$; therefore, there are two parabolas (Fig 5.13) generated by the equation for $M(A, Z)$ c^2.

13.7.3.3 Prediction of α – and β^- - emission properties using the Mass parabolas

Could for a given nuclide, say $^{258}_{92}U$, undergo α – or β^- - activity?

This can be examined as follows:

Case i) α – emission:

$$^{258}_{92}U \rightarrow {}^{254}_{90}Th + {}^4_2He$$

Inserting the values of A and Z, in the mass formula,

$$M(A, Z) \; c^2 = \alpha \; A + \beta \; Z + \gamma \; Z^2 \mp \delta \qquad (13.7.12)$$

$M(238, 92) - [\; M(238, 90) + M(4, 2)] \cong (0.004 \; u) \; c^2 = 3.8 \; MeV.$

This predicts an α-decay. Experimentally, this difference in mass gives 4.18 MeV.

Case ii) β^--emission:

$$^{258}_{92}U \rightarrow {}^{258}_{93}Np + \beta^-({}^{0}_{1}e)$$

Substituting the appropriate value of A and Z, in the mass formula,

$$[M(238, 92) - M(238, 93)] \, u \cong -(0.000025 \, u) < 0.548597 \, u$$

which is less than the electronic mass. So β^--emission is impossible.

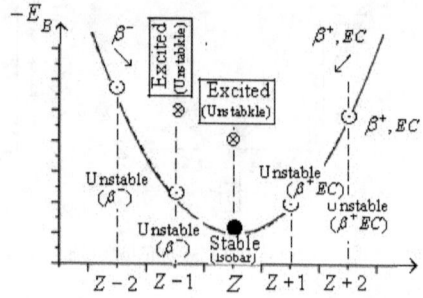

Fig. 13.14 Isobar cut across the valley of stability showing the different kinds of nuclei

Thus the usefulness of the mass formula in predicting radioactive properties of nuclides is illustrated (Fig 13.14). It can be obtained that the valley of the beta stability is

$$Z_{\beta \, stability} = \frac{2A}{4 + (a_c/a_a)A^{2/3}} \qquad (13.7.13)$$

13.8 ABUNDANCE OF NUCLIDES

Information relevant to nuclear stability rules and to nuclear structure are contained in the end products of the nucleo-synthesis. For such information is available in the abundance figures for nuclides with specific (A, Z), and received from the data on elements estimated from the isotope ratios seen in samples collected from terrestrial as well as meteorites. The important conclusions have been given below.:

1) Hydrogen and helium are the elements of highest abundance in the Universe,
2) Li, Be and B are deficient in the solar system, probably because of their nuclides find place in nucleo-synthesis.

3) A notable peak in the abundance of even-even nuclides occur around A = 56, which correlates to the maximum occurring at A = 56 in the E/A versus A curve. These are the most stable elements,
4) Distinct peaks are seen to occur at A value corresponding to N = 50, 82 and 126. Additional peaks are noticed at A = 80, 130 and 194. These have the same nuclear shell structure as those of the above nuclides.
   Additional evidence for shell structure in nuclei fro abundance data lie in the distribution of isotopes and isotones with reference to stability shown in the Segre Chart.
5) The number of stable and long-lived isotopes is greater for Z = 20, 28, 50, and 82 than for their neighbouring elements.
6) The number of stable and long-lived isotones is greater for N = 20, 28, 50, 82 and 126 than for their neighbouring N-values

13.9   FERMI GAS MODELS

Treating the nucleus in terms of Independent Particle Model (IPM) or by the Liquid Drop Model (SIM) has been examined in detail in Section 13.3 to Section 13.6. In this Section the most primitive IPM, *viz.*, the Fermi-Gas Model is briefly discussed. Consider a Fermi gas of nucleons confined to a nuclear volume V. Since neutrons and protons are distinguishable, we consider two independent gases; the particles do not interact and we neglect the Coulomb interaction in the case of protons; and we include a factor of two to account for the spin degeneracy (*i.e.*, two possible spin states in each Fermi gas level). The nucleus can be treated as a degenerate Fermi gas of nucleons. (Just as electrons in a metals). The N neutrons and Z protons ($N > Z$) of a nucleus $_Z^A X$ fill up the levels as shown schematically in Fig 13.15.

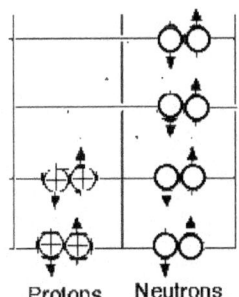

Fig 13.15 N neutrons and Z protons ($N > Z$) of a nucleus $_Z^A X$

The nucleons move throughout the volume V

$$V = \frac{4}{3}\pi r_0^3 A \qquad (13.9.1)$$

The neutron and proton wells have different depths and shapes because of Coulomb interaction between protons. The bottom of the proton well is slightly higher than that of the neutron well. The wells contain a finite number of energy levels. The number of levels filled with two particles (2 n) is (N - Z) / 2. and two protons (2 p) is Z / 2., since the nucleons fill as per the Pauli Exclusion Principle, one with spin-up and the other with spin-down.

13.9.1 Potential Energy

J. M. Blatt and Victor Frederick Weisskopf evaluated the potential energy between the nucleons as

$$\langle V \rangle = -a_V A + a_s A^{2/3} + C_1 T^2 / A \tag{13.9.2}$$

13.9.2 Kinetic Energy

The number of nucleons that can be contained in a certain volume of phase space is obtained by dividing that volume by the volume of one state in phase space, $(2\pi\hbar)^3$:

$$dn = 2 \left(\frac{V \, 4\pi p^2 \, dp}{(2\pi\hbar)^3} \right) p_F \tag{13.2.3}$$

These states will be filled with Z protons or N neutrons up to some maximum momentum p_F (for Fermi momentum).

The number of protons with momenta in the range p and $p + dp$ in a volume V is $2V \, 4\pi p^2 \, dp / (2\pi\hbar)^3$. Thus the number of protons in the lowest energy state,

$$Z = \int_0^{p_{F_p}} 2V \, 4\pi p^2 \, dp / (2\pi\hbar)^3 = \frac{V \, 8\pi p^2}{3 \, \hbar^3} p_{F_p}^3 \tag{13.9.4}$$

where $p_{F_p} = \hbar k_{F_p}$ = Fermi momentum for the protons.

$$p_{F_p} = \hbar k_{F_p} = \hbar \left(\frac{3 Z}{8\pi V} \right)^{1/3} = \hbar \left(\frac{3}{8\pi} \rho_p \right)^{1/3} \tag{13.9.5}$$

A similar expression holds for neutrons.

$$p_{F_n} = \hbar k_{F_n} - \hbar \left(\frac{3 N}{8\pi V} \right)^{1/3} - \hbar \left(\frac{3}{8\pi} \rho_n \right)^{1/3} \tag{13.9.6}$$

where $V = \frac{4}{3} \pi r_0^3 A$, $\rho_p = Z/V$, $\rho_n = N/V$ and $\rho_p = \rho_n$.

Thus the kinetic energy of a nucleon at the top of the Fermi distribution (at the Fermi surface) is about 35 *MeV*. Such a nucleon is still bound by about 8 *MeV*, so the shell model potential must be about 43 *MeV* deep. The kinetic energies of the protons and neutrons in ground state are, respectively,

$$T_p = \int_0^{P_{F_p}} \frac{2V \, 4\pi p^2}{(2\pi \hbar)^3} \frac{p^2}{2M_p} dp = \frac{4\pi}{5M_p} \left(\frac{3}{8\pi}\right)^{5/3} \left(\frac{\hbar^2}{V^{2/3}}\right) Z^{5/3} \quad (13.9.7)$$

$$T_n = \frac{4\pi}{5M_n} \left(\frac{3}{8\pi}\right)^{5/3} \left(\frac{\hbar^2}{V^{2/3}}\right) N^{5/3} \quad (13.9.8)$$

The average kinetic energy of a nucleon is

$$<T> = \left(\int_0^{P_F} E \, d^3p / \int_0^{P_F} d^3p\right) = \left(\frac{3}{5}\right) E_F \quad (13.2.9)$$

Total KE of all nucleons = $T_n + T_p$

Thus the energy of the nucleons in the nucleus is

$$E = Z<T_Z> + <T_N> = \left(\frac{3}{10M_n}\right) \frac{\hbar^2}{(r_o^2)} \left(\frac{9\pi}{4}\right)^{2/3} \frac{Z^{5/3}+N^{5/3}}{A} \quad (13.2.10)$$

Now $A = N + Z$, $Z = \frac{A}{2} - T$; for $T << A$, on can get

$$KE \propto 2 \frac{A}{2^{5/3}} \left[1 + \frac{20}{9} \frac{T^2}{A^2}\right] \quad (13.9.9)$$

$$\langle KE \rangle \simeq b A + C_2 \, T^2 / A \quad (13.9.10)$$

where $b = \frac{4\pi}{M} \left(\frac{3}{8\pi}\right)^{5/3} \left[\frac{\hbar^2}{\left(\frac{4}{3}\pi r_o^3\right)^{2/3}}\right] \frac{Z}{2^{5/3}} \quad (13.9.11)$

$$C_2 = \frac{4\pi}{M} \left(\frac{3}{8\pi}\right)^{5/3} \left[\frac{\hbar^2}{\left(\frac{4}{3}\pi r_o^3\right)^{2/3}}\right] \frac{Z}{2^{5/3}} \frac{20}{9} \quad (13.9.12)$$

The binding energy of the nucleus $^A_Z X$ is

$$B(A, Z) = -a_V A + b A + C \, T^2 / A + a_s \, A^{2/3} \quad (13.9.13)$$

Expand this expression around the symmetrical position $N = Z = A/2$ to obtain a contribution to the asymmetry energy. Let $t = N - Z$. Then, expanding in powers of t,

$$E \propto A + \frac{5\,t^2}{9A} + \square\square\square \tag{13.2.7}$$

This then justifies the form of the asymmetry term in the semi-empirical mass formula. From this one obtains $a_A \approx 11\ MeV$; about half the usual value. (The rest is accounted for in the dependence of the potential well on t).

13.10 THE COLLECTIVE MODEL (Models of Deformed Nuclei)

The Deformed Nuclear Model was developed by Aage Bohr (1952) and Aage Bohr & B.R. Mottelson (1953). It is a consistent scheme of combining the best features of both the liquid drop model and the shell model, to give one model that will predict all the observed nuclear phenomena. Partial success has been achieved in the endeavor.

13.10.1 Nucleus in which a small number of neutrons (or protons) orbit outside a core of closed shells that contain a magic number of neutrons (or protons).

The "extra" nucleons move in quantized orbits, in a potential well established by the central core, thus preserving the central feature of the independent particle model. These extra nucleons also interact with the core, deforming it and setting up "tidal wave" motions of rotation or vibration within it. These "liquid drop" motions of the core preserve the central feature of that model. This collective model of nuclear structure thus succeeds in combining the seemingly irreconcilable points of view of the liquid drop and independent particle models. It has been remarkably successful and is currently our best theory. Perhaps it represents the limits of what we can hope for in nuclear physics, given the absence of a theory

To explain the large quadrupole moments we return to the picture of the nucleus as a collective body. The basic idea of this Model is that interactions between the outer nucleons and the closed shell core lead to permanent deformation. This is expected to be a particularly strong effect midway between shell closures. In the case of a permanent deformation the single particle states have to be calculated in a non-spherical potential. The spacing of the energy levels then depends upon the magnitude of the distortion. This marriage of the single particle and the corporate models presents very difficult theoretical problems so we will just concentrate on the qualitative features.

As noted above doubly closed shell nuclei are very stable with a first excited state well removed from the ground state. One nucleon more or one less than this very stable configuration will exhibit single particle states. Nuclei further away from the closed shells should be easily deformed leading to excited states due to the vibrational motion of the core. In the region of half filled shells the nuclei are permanently deformed and consequently have large quadrupole moments and also rotational energy levels.

Nuclides	^{150}Sm	^{152}Gd	^{152}Sm	^{190}Os	^{200}Pt	^{200}Hg	^{206}Pb	^{207}Pb	^{208}Pb
Spectra	Vibrational		Rotational		Vibrational		2-particle	1-prticle	Magic
$Q(odd\ A)$	50		200-700-200		50		-	-	-

These points are illustrated in the table below which lists the type of energy levels observed for a wide range of even-even nuclei. The examples given cover those nuclei in the region just below ^{208}Pb ($Z = 82, N = 126$). As additional evidence the approximate sizes of the quadrupole moments of the odd A nuclei in this same region are also listed (in fm^2).

13.10.2 Rotational states of Deformed nuclei –

Just consider even-even nuclei and recall that the ground state is always 0+ in this case. Defining the z direction as the symmetry axis of the deformed nucleus and recalling that the angular momentum operator for the component along this axis is equal to $-i\hbar\ \partial/\partial\varphi$, then there are no rotational states about this axis because $\partial\psi/\partial\varphi$ is zero.

For rotational angular momentum L, the energy is just $L^2/2\Im$, where the effective moment of inertia is $\Im$. The energies are given quantum mechanically by the Schrödinger equation of the form $(L^2/2\Im)\psi = E\psi$ and since the operator $\hat{L}^2$, the normal angular momentum operator, the eigenvalues and eigenfunctions are given by

$$\hat{R}^2\ Y^M_J(\theta,\varphi) = J(J+1)\hbar^2\ Y^M_J(\theta,\varphi)$$

where J can normally be 0, 1, 2, ... but in this case since the ground state is even parity, only even J values are admissible (odd J spherical harmonics have odd or negative parity). Thus the energy levels are as given below.

$$E_J = J(J+1)\hbar^2/(2\Im),\ \text{with}\ J = 0, 2, 4, ...$$

For example $E_2 = 6\hbar^2/(2\Im)$ and other energy levels can be expressed in terms of this as

$$E_J = J(J+1)E_2/6$$

(see Table 13.15 of states given below in MeV).

Table 13.15 Table of states in MeV

Spin Parity	$^{164}_{68}Er$	$^{165}_{70}Yb$	$^{176}_{72}Hf$	$^{178}_{74}W$	$J(J+1)/6$
16+	-	-	3.15(31.5)	-	45.3
14+	-	-	2.56(25.6)	2.68(21.8)	35.0
12+	2.08(22.8)	2.17(21.3)	2.01(20.1)	2.13(17.3)	28.2
10+	1.52(16.5)	1.60(15.8)	1.50(15.0)	1.62(13.2)	18.3
8+	1.02(11.2)	1.10(10.8)	1.04(10.4)	1.15(9.3)	12.0
6+	0.61(6.7)	0.68(6.5)	0.64(6.4)	0.73(5.9)	7.0
4+	0.30(3.3)	0.33(3.2)	0.32(3.2)	0.38(3.1)	3.3
2+	0.09(1.0)	0.10(1.0)	0.10(1.0)	0.12(1.0)	1.0
0+	0(0)	0(0)	0(0)	0(0)	0

The predicted values are given in the last column of the Table, normalized to the lowest interval $0^+ \rightarrow 2^+$. They should be compared to the entries in brackets. The discrepancies at larger J values are due to the centrifugal stretching of the nucleus and the consequent increase in the moment of inertia $\Im$. Since

$$E_J = J(J+1)\hbar^2/(2\Im),$$

the higher J values have lower energies than expected. The extent to which these rotational states are intermingled with those due to other excitations is illustrated in the energy level diagram for $^{164}_{68}Er$ Fig 13.17.

$$E_J = J(J+1)E_2/6$$

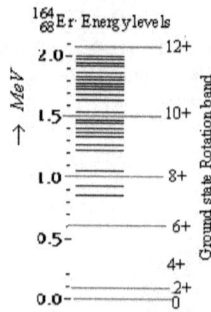

Fig 13.17 $^{164}_{68}Er$ rotational energy levels

13.10.2 Vibrational states

A picture of the nucleus is like a drop of incompressible fluid. Yet it is possible for the system to perform shape oscillations without change of density.

Fig 13.18 Nuclear Vibrational states

These changes from spherical symmetry take the form of surface standing waves which are proportional to the Legendre polynomials $Y^M_J(\theta,\varphi)$ with oscillating coefficients. Examples of quadrupole, sextuple and octupole oscillations are drawn below (Fig 13.18)..

The lowest order vibration look like a quadrupole and has $I = 2$. Thus there are five ($2I + 1$) possible values for m (ie five degrees of freedom). This is an eigenfunction of the total angular momentum and we say that this state has a single phonon of angular momentum 2. The second excited state has two phonons which can couple to form 0, 2, 4 units of angular momentum. Starting with the normal even-even nuclei ground state (0^+) the excited states are (2^+);   ($0^+, 2^+, 4^+$);   ($0^+, 2^+, 3^+, 4^+, 6^+$).

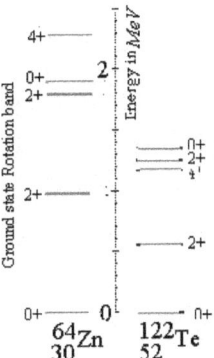

Fig 13.19 The lowest energy levels for $^{64}_{30}$Zn and $^{122}_{52}$Te

The energies are $E = (N + \frac{5}{2})\hbar\omega$ where N is the number of phonons. Note that the levels are equi-spaced just like the harmonic oscillator. This simple picture is by no means exact since energy levels due to different forms of excitation often overlap. In general vibrational levels are identifiable only up to excitation energies of 1 to 2 *MeV*. The lowest energy levels for $^{64}_{30}$Zn and $^{122}_{52}$Te are sketched to illustrate as in Fig 13.19.

As expected the excitation energy from the ground state to the two-phonon triplet ($0^+, 2^+, 4^+$) is about twice that to the one phonon singlet (2^+).

13.11 UNIFIED MODELS

13.11.1 Another approach to the structure of nucleus

Alternative approach to the structure of nucleus has been to unify the ideas of both nuclear shell model and the Collective model in the so-called '*Unified Model*', by David Rittenhouse Inglis (1953) and called the '<u>Cranking Model</u>'. Accordingly the nucleons in the ground state are moving in an ellipsoidal potential well, rather than the spherical one, and that the whole can undergo collective rotational motion.

13.11.2 NILSSON MODEL

S.G. Nilsson (1955) treated the motion of single particle in a deformed nuclear well, because some nuclei in ground state are deformed, which means a central potential which is no longer spherically symmetric well not be useful..

The *Unified Model* gives very complicated energy level schemes, but it has nevertheless been found that this description does help to account for a large number of experimental data.

13.11.3 OPTICAL MODEL

This model was developed by Feshbach, Porter & Victor F. Weisskopf in 1954. It is quite useful in providing an understanding in a broad sense of nuclear reactions. The principle used here is to consider the only the behaviour of the incident particle experiencing a potential well of the type employed in the nuclear shell model, but adding an imaginary term W, so that the potential is $(V + i\ W)$. The term W is included to account for any absorption of the incident particle by some form of reaction. (analogous to the imaginary term in the complex refractive index in optics to account for absorption). Thus the nucleus can be treated as a '*cloudy crystal ball*'.

REVIEW QUESTIONS

R.Q. 13.1 Illustrate the salient features of an atom (spin $s = \frac{1}{2}$) and a nucleus ($I = \frac{1}{2}$) in the energy level diagram. (Answer

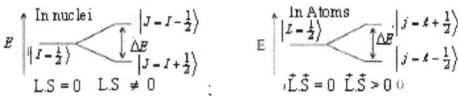

R.Q. 13.2 Determine the nuclear spin value for $^{13}_{6}C$ [Answer $I = \frac{1}{2}$]

RQ. 13.3. Find out the spin of the nucleus $^{11}_{5}B$. (Answer: $I = \frac{3}{2}$).

R.Q. 13.4 Find the spin momentum for $^{15}_{7}N$ and $^{17}_{8}O$. [Answer $I = \frac{1}{2}$ for $^{15}_{7}N$; $I = \frac{5}{2}$ for $^{17}_{8}O$].

R.Q.13.5 Calculate the magnetic dipole moments for nuclei: ^{2}H, ^{3}He, ^{4}He, ^{7}Li, ^{12}C, and ^{35}Cl [Answers: Using Table 5.9, 0.8574 nm, -2.1275 nm, 0 nm, 2.9789 nm, 3.2563 nm, 0 nm, 0.8218 nm].

R.Q. 13.6 Treat the nuclei in the single particle model and find out the nuclear magnetic moment for a nucleus with an unpaired nucleon. [Answer $\vec{\mu}_I = \vec{\mu}_{j = \ell + \frac{1}{2}}$].

R.Q. 13.7 What are the evidences for the existence of nuclear shells?

R.Q. 13.8 State briefly the theoretical basis of the nuclear shell model by Mayer and Jensen et al., and the coupling rules involved.

R.Q. 13.9 Give a brief account of the shell model of nucleus. Obtain a schematic energy level diagram, showing L-S splitting, up to magic numbers.

R.Q. 13.10 Find out the difference in mass between two nuclides $^{A}_{Z}X$ and $^{A}_{Z+1}Y$.

(Answer $\left\{ -1.390 + 1.526 Z A^{-1/3} \right\} 10^{-3}$ u).

R.Q. 13.11 What is the total binding energy of lide $^{250}_{100}Fm$?

[Answer, $E_B = A \left\{ 7.5 - \delta A^{-1} \right\}$ MeV].

R.Q. 13.12 Develop the Fermi gas model of the nucleus and obtain the expression for total energy of the nucleus. What is meant by nuclear temperature?

R.Q. 13.13 State the various contributions to the mass M(Z,A) of a nucleus in the semi-empirical mass formula.

R.Q. 13.14 Describe the Shell Model of the nucleus.

R.Q. 13.15 What is the origin of Pairing energy term in Mass formula?

RQ. 13.15 Derive the equations for the Schmidt lines in nuclear magnetic moments on the basis of shell model. Discuss how will you calculate the quadripole moment of $^{39}_{19}K$ on the basis of the shell model.

%%*%*%*%*%*

Chapter 14

PARTICLE ACCELERATORS

Chapter 14

PARTICLE ACCELERATORS

"When a man sits with a pretty girl for an hour, it seems like a minute.

But let him sit on a hot stove for a minute it's longer than an hour.

That's relativity!"   - Albert Einstein

14.1   Introduction

The first subatomic particle to be discovered was the electron in 1897. It is followed by the positively charged nucleus in 1911, and then the proton in 1919. The next particle discovered was the neutron in 1932. Particles from a radioactive source are emitted uniformly in all directions So to get a collimated beam of the particles is out of question an experimental physicist. Only about 0.25% of the particles emitted can be made available to strike a target.

Worked out Example 14.1

A source of 1000 MBq, is properly shielded and collimated. How many particles approx will be able to strike the target?

Solution: $\boxed{STEP * 1}$ Particle available from a source of radioactive material, when collimated and shielded = 0.25 %

$\boxed{STEP * 2}$ # particles in the problem $= \left(\frac{0.25}{100}\right)(1000 \times 10^6) = 2.5 \times 10^6$.

The highest energy available, say of alpha particles, is of the order of a few 10 MeV. And these have a limited range even in empty space. To overcome these limitations, particle accelerators have been designed that can produce a range of sub atomic particles such as electrons, protons and high speed nuclei of the desired atoms together with their anti-particles. Such *'atom smashers'* are now available with particle energies of a few keV to several TeV.

The majority of all the known radioactive substances does not occur naturally and must be artificially produced in nuclear reactions. The first transmutation of a nucleus by artificial means was accomplished by J.D. Cockcroft and E.T.S. Walton in 1932. Since that time a variety of machines capable of accelerating beams of charged particles (electrons, protons) with energies ranging from a few MeV to a few hundred BeV are produced in *accelerators*, having intensities as high as 10^{14} particles / s. Accelerators have been indispensable to perform i) the creation of new particles and new states, and ii) the study of detailed structure of subatomic systems.

For physics, one consequence was a spectacular change in the scale of instrumentation. Nobody looking at Anderson's little cosmic-ray cloud chamber in 1932 could possibly have imagined today's gargantuan neutrino detectors. Similarly, nobody looking at Lawrence's 11-inch cyclotron in 1931 could have imagined the 85 *GeV* Large Hadron Collider (LHC) with its 27-*km* circumference at CERN in Geneva, Switzerland, 50 *GeV* collider SLC at Stanford, and Fermi lab's (I *TeV*, $1\frac{1}{2}$ mile diameter) Tevatron at Batavia.

Worked out Example 14.2

Estimate the KE of a proton to observe structural details of dimension $d=1\,fm$.

Solution: $\boxed{STEP\ *1}$ From $\lambda_{De\,Broglie} = \frac{h}{p}$, (14.1.1)

and $\frac{\lambda_{De\,Broglie}}{2\pi} \leq d$, (14.1.2)

$\boxed{STEP\ *2}$ one gets $p \geq \frac{h}{d}$ (14.1.3)

$\boxed{STEP\ *3}$ $\frac{KE}{M_p c^2} \geq \frac{1}{2}\left(\frac{\lambda_{Compton}}{d}\right)^2 \geq \frac{1}{2}\left(\frac{0.210\,fm}{1\,fm}\right)^2 = 0.02$, whence

$KE = 0.02 \times 938.2723\,MeV \approx 20\,MeV$!

$\boxed{STEP\ *4}$ Particle beams with such an energy of 20 *MeV* are neither available in nature nor can be produced by a single ES generator.

14.2. ELECTROSTATIC GENERATORS

14.2.1 Cockcroft-Walton Type

The first particle accelerator (Fig 14.1),, a Direct Current Generator, based on disintegration of nuclei by artificially accelerated particles, was performed by John D. Cockcroft & E.T.S. Walton (1932). Along with the discovery of the neutron by Sir James Chadwick, this opened up a new era of Nuclear Physics. The principle is as follows. Ions from a source enter an evacuated tube in which they are accelerated.. The capacitors C_1, C_2, and C_3 are connected in series, with a supply of voltage V across C_1. Two more capacitors, C_4, and C_5, can be switched into the circuit in either of the two ways. It is assumed that all the capacitors have the same value. By repeating switching up and down positions of the switch position the capacitors are charged more and more and the output across C_1, C_2, and C_3 becomes 3V. In practice the number of capacitors in the chain can be greatly increased, and the switches can be replaced by rectifiers if an input a.c. voltage source is used. Particles of charge q, velocity v and mass M in p.d of V volts, $\frac{1}{2}Mv^2 = qV$, give outputs of □ 1 *MV* or less can be obtained.

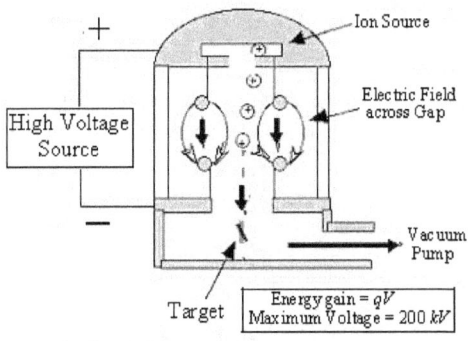

Fig 14.1 Cockcroft-Walton type Accelerator

This device was the first to use a beam of deuterons and production of ^3_1H and ^3_2He accordingly.

$$^2_1H\,(^2_1H,\,^1_1H)\,^3_1H + 4.04\,MeV \tag{14.2.1}$$

One such application is the production of neutrons by

$$^2_1H\,(d,\,^1_0n)\,^3_2He + 3.26\,MeV \text{ reaction, or} \tag{14.2.2}$$

$$^3_1H\,(d,\,^1_0n)\,^4_2He \text{ reaction.} \tag{14.2.3}$$

Both these reactions give reasonable yields at voltages as low as 150 keV.

The first element disintegrated was lithium by ☐ 150 keV protons by the reaction

$$^1_1H + 0.15\,MeV + ^7_3Li \rightarrow ^4_2He + ^4_2He + 17.5\,MeV \tag{14.2.4}$$

followed by $\quad ^1_1H + ^7_3Li \rightarrow ^8_4Be + \gamma\,(17\,MeV) \tag{14.2.5}$

$$^1_1H + ^{11}_5B \rightarrow 3\,^4_2He \tag{14.2.6}$$

Worked out Example 14.3

Electrons from an electron gun in a TV picture tube (CRT) accelerated by a potential difference of 3 kV. Find out the kinetic energy of electrons and the speed at which they strike the screen.

Solution: $\boxed{STEP * 1}$ $KE = (3 \times 10^3 V)(1.6 \times 10^{-19} C) = 4.8 \times 10^{-16} J$;

$\boxed{STEP * 3}$ $v = \sqrt{\dfrac{2 \times KE}{M}} = \sqrt{\dfrac{2(4.8 \times 10^{-16} J)}{9.1 \times 10^{-31} kg}} = 3.2 \times 10^7 m\, s^{-1}$

14.2.2 The Van de Graaff Electrostatic Generator

A second ingenious high voltage ES generator was developed by Robert Jemison Van de Graaff in 1932. The principle is based on the fact that there is no electric field inside a charged conducting sphere, so that as and when charge is introduced inside the sphere, the potential (V) to which it was charged increases, to a maximum of 12 *MV* (The BARC machine developed 5.5 *MV*). The charging is accomplished by a moving belt onto which positive charge has been sprayed at the position of spray points. The charge is removed at the high voltage terminal by a set of charge-remover points. Inside this terminal is a source of positive ions such as protons or deuterons. These ions are formed in the high-voltage terminal; they are allowed to leave the sphere and fall through the accelerating tube toward a target at ground potential. The particle receives energy equal to

$$E = n e V \qquad (14.2.7)$$

where n and V denote the number of electrons removed from the ion and the voltage at the high-voltage terminal.

The generator, shown in Fig 14.2, can be used to accelerate electrons, protons and neutrons. The reaction

$$^{7}_{3}Li \left(^{1}_{1}H + 1.85\, MeV,\, ^{1}_{0}n + 1.62\, MeV \right) ^{7}_{4}Be \qquad (14.2.8)$$

produces mono-energetic neutron beams.

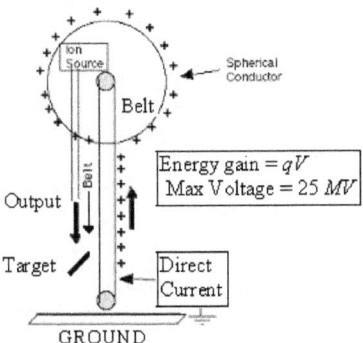

Fig 14.2 Van de Graff Generator

14.2.3 TANDEM Accelerator

The type of generator called Tandem Van de Graffs allows the output energy of ions can be doubled or tripled. Two Van de Graff Generators are joined back-to-back position in a long high pressure tank, with the ion source at one end and producing a beam of negative ions, example, H^-, which are accelerated toward the middle of the tank, where they are stripped off their two electrons by passage through a gas containing canal. The positive ions are formed from the negative ions, and are accelerated away from the HV potential. Protons beam of 1.5 μA and of 20 MeV energy can be obtained.

14.3 CYCLOTRON (Fixed frequency)

An alternative to higher potentials is the ingenious method of using a comparatively low voltage over and over was employed by E.O. Lawrence & M.S. Livingston (1932) developing the popular device called the *cyclotron*. In this device the charged particle is made to move in a circular path by applying a magnetic field and is given a small increase in energy during each cycle.

It consists essentially of a short hollow cylinder divided into two "dees" (D-shaped chambers) or sections, D_1 and D_2. These dees are placed between the pole pieces of a very large electro-magnet (Fig 14.3).

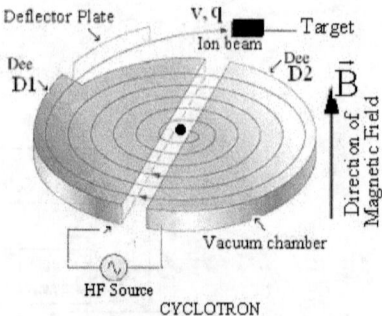

Fig 14.3 The Cyclotron

Principle of operation is that the frequency (ω) of circulation of ions in the magnetic field (B), in the non-relativistic range, is independent of the energy of the ions of charge (q) and mass (m) traveling with velocity v.

According to Newton's II Law,

$$q\,\bar{v}\wedge\bar{B} = \frac{m v^2}{\rho} \qquad (14.3.1)$$

where ρ denotes the orbit radius.

In order to meet the proper frequency of the alternating current, the time t required to traverse one half circle must be computed.

$$t = \frac{\text{distance}}{\text{velocity}} = \frac{\pi \rho}{v} \qquad (14.3.2)$$

Then one can write the angular frequency of the particle beam to be $\omega = \frac{2\pi}{2t}$

$$\omega = \frac{q\,B}{m} = \text{constant}. \qquad (14.3.3)$$

The terminals of the Dees D_1 and D_2 are connected to a HF ac source of $\Box\, 10^6$ Hz. The positive ions will receive an additional momentum, *i.e.*, get accelerated toward one of the Dees, go across the gap between them each time the beam traverses a half circle, and increase its radii, each time it goes from one dee to the other.

In a typical cyclotron the magnet weighed 100 tons, 36-inch diameter pole pieces, deuterons injected at 8 *MeV*, 50 μA, with energy consumption of 10 *MHz* and 100 *kW*. The Cyclotron at Berkeley (1957) produces 10 *GeV*.

Spring-8 synchrotron near Kobe is the world's largest third-generation synchrotron radiation facility. The CIME cyclotron at GANIL in France, operated at 9.6 – 14.5 *MHz* and produces beams of unstable nuclei with energies 1.7 – 25 *MeV*, delivering maximum beam energy of about 1 *CeV*.

Worked out Example 14.4

Find both the momentum and velocity of a proton moving in a perpendicular magnetic field 0.5 T and radius of curvature of 5 cm. Given mass of proton, 1.6×10^{-27} kg.

Solution: STEP * 1 Use $p = B q \rho$,

STEP * 2 $p = (0.5\ T)(1.6 \times 10^{-19}\ C)(5.0 \times 10^{-3}\ m) = 4.0 \times 10^{-21}\ kg\ m\ s^{-1}$;

STEP * 3 $v = \dfrac{p}{M} = \dfrac{4.0 \times 10^{-21}\ kg\ m\ s^{-1}}{1.6 \times 10^{-27}\ kg} = 2.4 \times 10^{6}\ m\ s^{-1}$.

14.4 **LINAC (Linear Accelerator)**

It is a type of resonance accelerator. The idea of accelerating positive ions by the aid of RF field (using the resonance principle) was first put forward by Ising in 1925, and the first device was developed in 1928 by Wideroe. The ions are injected

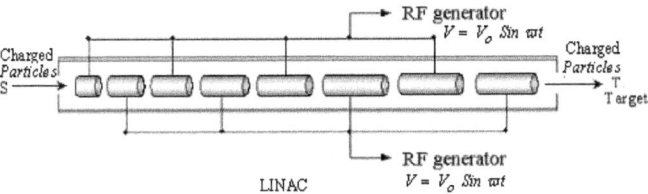

Fig. 14.4   LINAC

along the axis of a series of co-axial metal drift tubes (electrodes) and connected alternately to one or the other of the two terminals of a RF generator, as in Fig 14.4.

Each successive drift tube has proper length calculated from the particle's velocity and the frequency of the AC. The length of adjacent drift tube is such that the ion moves across the gap after exactly on half-period of the RF wave (Fig 14.5). So as it get accelerated. The condition for the synchronism is, in general,

Half period $\dfrac{1}{2f} = \dfrac{L_S}{v_S}$,   (14.4.1)

$$L_s = \frac{v_s}{2f}, \text{ and} \tag{14.4.2}$$

$$\frac{1}{2} m\, v_s^2 = n\, e\, V_0 + C \text{ (a constant)}, \tag{14.4.3}$$

where L_s = Length of the sth drift tube including the gap,

v_s = velocity of the ions inside the sth drift tube,

f = Frequency of the RF generator,

V_0 = Average potential difference that the ion experiencing in passing a gap,

n = Degree of ionization of the ion,

$$L_s = \frac{1}{2f} \sqrt{2\,(n\, e\, V_0 + C \text{ (a constant)})\,/\,m} \tag{14.4.4}$$

This is a standing wave device, because

$$L_s = \lambda \sqrt{(n\, e\, V_0 + C \text{ (a constant)})\,/\,2\,m\,c^2} \tag{14.4.5}$$

Linear accelerators are of two types, electron and proton (or other heavy particles).

Examples: Several linacs use traveling-wave devices giving □ 3 GHz and operating at energies 10 MeV to 20 GeV.

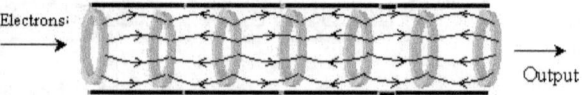

Fig. 14.5 Ion movement between gaps of adjacent drift tubes

The Stanford linac (1954) had length 91.5 m 700 MeV. The electron linac (SLAC) at Stanford (1961) produces 20 GeV electrons with 12 μA current; it has a length of □ 3 km; and uses 10 MHz micro-waves.

The Stanford Mark III machines generate electrons of 1 BeV; Mark IV gives 45 BeV, but it has a length of 10,000 ft. The proton linac at Berkeley (1947) had a set of 46 drift tubes of steel chamber containing a 12-sided copper tube of 40 ft long and 1 m diameter, working at a resonant frequency of 202.5 MHz. The proton beam of 4 MeV from a Van de Graaff generator is injected to obtain a proton beam of 35 MeV, and power 2.15 MW. A proton linac, known as meson factory, at Los Alamos generated particle of 800 MeV energy.

14.5 SYNCHRO - CYCLOTRON

In cyclotrons the resonance frequency f = constant. But when developing further large cyclotrons, the particles reach $v \approx c$, $f \neq$ constant, and the cyclotron fails to work. In 1946 E.O. Lawrence overcame this difficulty, and used *frequency modulation* (FM).

In
$$q \, \bar{v} \wedge \bar{B} = \frac{m v^2}{\rho}, \qquad (14.5.1)$$

$$\rho = \frac{m_o \, v}{q \, B \sqrt{1 - v^2/c^2}} \qquad (14.5.2)$$

and
$$\omega = \frac{q \, B \sqrt{1 - v^2/c^2}}{m_o} \qquad (14.5.3)$$

Therefore, $\omega \propto 1/$ (Particle energy) f must be decreased in synchronism with ω, i.e., motion of the particles, which means the value of B should be increased. This is achieved by using the principle of *phase stability*, by Edwin M. McMillan and by V. Veksler (1945).in their Synchro-cyclotron, a machine in which f is decreased while B = constant. The ions are accelerated far into the relativistic velocity range during a fraction of each period of modulation (phase stability is obtained). Thus the ions are produced in bursts rather than continuously.

The largest Synchro-cyclotron at California (1946) accelerates deuterons to 190 *MeV* or α- rays to 380 *MeV* or protons to ▯ 720 *MeV*. It uses $B = 1.5 \, Wb \, m^{-2}$, $\rho = 92$ *inches*, $\omega = 10 \, MHz$, the frequency of modulation produces ▯ 100 surges per *s*. Each ion performs about 10,000 revolutions in a period of about 1 *ms*.

Worked out Example 14.5

A proton cyclotron of maximum radius 0.5 *m* employs a field of 1.5 T. Estimate the cyclotron frequency and the kinetic energy of the emerging proton. Given proton mass 1.67×10^{-27} *kg*

Solution: STEP * 1 $f_{cycl} = \frac{B \, q}{2 \pi \, M} = \frac{(1.5T)(1.6 \times 10^{-19} C)}{2\pi (1.67 \times 10^{-27} kg)} = 23 \, MHz$;

STEP * 2 $KE = \frac{1}{2} \left(\frac{q^2 \, B^2 \, \rho^2}{M} \right) = \frac{1}{2} \left(\frac{(1.6 \times 10^{-19} C)^2 \, (1.5T)^2 \, (0.5m)^2}{1.67 \times 10^{-27} kg} \right) = 4.3 \times 10^{-12} \, J$

$= \frac{(4.3 \times 10^{-12} J)(1 \, eV)}{(1.6 \times 10^{-19} J)} = 26.9 \, MeV$

14.6 VEC

A variable energy cyclotron (VEC) was installed at Kolkatta by BARC in 1977.

14.7 BETATRON

The betatron is an electric transformer, in which the secondary winding is replaced with a ring or dough-nut shaped vacuum chamber, and the electrons are. The evacuated doughnut glass chamber is between the pole pieces of a powerful electromagnet, designed for AC current (in contrast to a cyclotron), as shown in Fig 14.6.

To produce high energy electrons □ $300 MeV$, in 1940 Donald Kerst constructed the *betatron*. It is best thought of as a transformer, with a ring of electrons (dough-nut shaped vacuum chamber, and the electrons) as the secondary coil. They are subjected to the force exerted by the induced electric field. The evacuated doughnut glass chamber is between the pole pieces of a powerful electromagnet, designed for AC current (in contrast to a cyclotron). The magnetic field used to make the electrons move in a circle is also the one used to accelerate them, although the magnet must be carefully designed so that the field strength B_{orb} at the orbit radius is equal to half the average field strength $\ddot{B}$ linking the orbit:

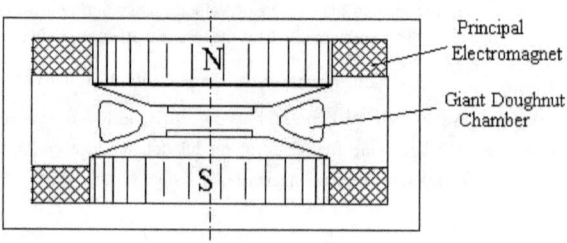

Fig. 14.6 Betatron

$$B_{orb} = \frac{\ddot{B}}{2} \qquad (14.7.1)$$

If the magnetic field increases, there is a changing flux linking the loop of electrons and so an induced e.m.f. which accelerates the electrons. As the electrons get faster they need a larger magnetic field to keep moving at a constant radius, which is provided by the increasing field – the effects are proportional, so the field is always strong enough to keep the electrons in orbit!

The field is changed by passing an alternating current through the primary coils and particle acceleration occurs on the first quarter of the voltage sine wave's cycle. Although the last quarter of the cycle also has a changing field that would accelerate the electrons, it is in the wrong direction for them to move in the correct circle! The target is bombarded with pulses of particles at the frequency of the ac supply

When an alternating magnetic field is applied parallel to the axis of the tube, two effects are produced:

1) Electrons gain additional energy, due to acceleration by the sinusoidal induced emf, (V) production of an emf in its orbit by the changing magnetic flux Φ, and

2) Electrons are maintained in circular motion due to the radial force effected by the magnetic field.

1) The magnetic flux Φ through the electron orbit has to be chosen such that the motion of electrons will be in stable orbit of radius ρ.

Applying Faraday's Law induced emf V

$$V = \frac{d\Phi}{dt} \qquad (14.7.2)$$

and considering the work done dW on an electron by the tangential force over an infinitesimal distance ds, and integrating for a revolution,

$$\bar{F} = \int_0^{2\pi\rho} \frac{dW}{ds} = \left(\frac{e}{2\pi\rho}\right) \frac{d\Phi}{dt} \qquad (14.7.3)$$

Applying Newton's II Law, and equating one gets

$$d(m\ v) = \left(\frac{e}{2\pi\rho}\right) d\Phi \qquad (14.7.4)$$

The electrons will traverse out of their circular path if hey are not constrained.

2) The centrifugal force due to the magnetic flux normal to the electron orbit is

$$q\ \bar{v} \wedge \bar{B} = \frac{m\ v^2}{\rho} \qquad (14.7.5)$$

$$m\ v = B\ e\ \rho \qquad (14.7.6)$$

For stable orbits r = constant.

Equating (14.7.6) and (14.7.6),

$$e\ \rho\ dB = \left(\frac{e}{2\pi\rho}\right) d\Phi \qquad (14.7.7)$$

Integrating

$$\int_0^\Phi d\Phi = \Phi_{average} = 2\pi\rho^2 \int_0^B dB = B_{Orbit} \qquad (14.7.8)$$

The instantaneous relationship between Φ and B is found,

$$\Phi_{average} = 2(\pi\rho^2 B_{Orbit}) = \Phi_{Orbit} \tag{14.7.9}$$

This is the limiting or tuning condition for the Betatron.

When the electrons have acquired the desired amount of energy, a capacitor is discharge through two coils of wire one directly above and the other directly below the stable orbit region produce a pulse. This disturbs the stability condition and gives the output beam to strike a properly placed target from which the X-ray beam will emerge.

Example The betatron at Illinois had $\rho = 1.22\ m$, $B_{Orbit} = 9.2\ kG$, output = 315 MeV, Pulse repetition rate = 6 pulses / s.

14.8 SYNCHROTRON

The *synchrotron* has many features similar to those of the cyclotron; but it does not have a giant magnet. In 1945 Edwin M. McMillan and independently V. Veksler proposed it. The *phase stability* is achieved by varying the magnetic field B while the frequency may or may not be varied. Particles from a linac are injected to the synchrotron, and travel along a big circle in a doughnut vacuum chamber between a ring-shaped magnetic field and get accelerated in many successive steps, applying phase stability.

The total energy (E) of the particle (velocity, v_i)

$$E^2 = p^2 c^2 + m_0^2 c^4 \tag{14.8.1}$$

have frequency of rotation $\Omega = v_i / \rho$, in the relativistic range becomes

$$\Omega = \frac{p_i c^2}{E_i \rho} \tag{14.8.2}$$

As the particles are constrained at the radius ρ, in a magnetic field B,

$$\frac{\bar{B} \wedge \bar{v}_i q}{c} = \frac{m v_i^2}{\rho} \tag{14.8.3}$$

$$\therefore \quad B = \frac{p_i c}{|q|\rho} \tag{14.8.4}$$

The magnet (is a dipole magnet) thus works as a momentum selection device, since $p_i = B|q|\rho/c$.

Once the RF power (ω) is tuned on, the situation changes, and synchronization is obtained hen $\omega = k\,\Omega$, where k = an integer. Hence the *cyclotron resonance condition* is

$$\omega = \frac{k\,c}{\rho} \Box \frac{p_i\,c}{E_i\,\rho} \tag{14.8.5}$$

It means as E increases, the applied RF (ω) must increase up to the point where the particles are fully relativistic, and any further increase in E does not produce further increase in m, and $v \approx c$, and ω becomes a constant. When

$$\omega = \Omega = \frac{q\,B}{m_o\,c} = \text{constant} \tag{14.8.5}$$

one has $\omega \propto B$, i.e., to increase the total energy E or mass m of the ion, B must be increased in the same ration. Only when this condition is fulfilled, the particles (pre-accelerated from a linac) get accelerated.

The Brookhaven machine, called "cosmotron", protons are given energy of 3 BeV, while they perform about 3 million revolutions and, within a second, travel 6 times the circumference of the earth. Similar machines also operate at Berkeley and near Moscow. And other places. The Super Proton Synchrotron (SPS) at CERN (Geneva) has a circumference of ar5ound 6 km and can produce proton beam of energy 450 GeV. The Large Electron-Positron Collider (LEPC) has a circumference of 27 *km* can accelerate electron (and positrons) up to 60 *GeV*. The Superconducting Super Collider (SSC) in the USA has a circumference of 87 *km* and can produce proton and antiproton beams with around 20,000 *GeV*.

Worked out Example 14.5

In the Fermi Lab Tevatron the protons have momentum 1 TeV/c. What is the strength of the field to hold these protons in the orbit of radius 1 *km* ? Find the frequency of the proton in the orbit/. Given the mass of proton $M_p = 1.67 \times 10^{-27} kg$.

Solution: $\boxed{STEP\ *\ 1}$ $\rho = \left(\frac{Mv}{Bq}\right)$, $\rho = \left(\frac{Mv}{Bq}\right)$;

$\boxed{STEP\ *\ 2}$ $1\,TeV/c = (10^{12})(1.6 \times 10^{-19}C)/(3 \times 10^8 m\,s^{-1}) = 5.3 \times 10^{-16} kg\,m\,s^{-1}$;

$\boxed{STEP\ *\ 3}$ $B = \frac{(5.3 \times 10^{-16} kg\,m\,s^{-1})}{(1.6 \times 10^{-19}C)(1000\,m)} = 3.3\,T$; period $= 2000\,\pi/c = 21\mu\,s$..

14.8.1 Betatron Synchrotron

A synchrotron when accelerates electrons it is called a betatron-synchrotron. According to James Clerk Maxwell's electrodynamics, an accelerated charge must radiate. In a synchrotron a particle is forced to remain a circular path undergoes continuously centripetal acceleration, and so emits EM radiation. Schwinger (1946) has calculated that the power (S) radiated by an accelerated particle in circular trajectory is

$$S = \frac{c}{6\pi\varepsilon_0} \frac{e^2}{\rho^2} \left(\frac{(v/c)^4}{[1-(v/c)^2]^2} \right) \tag{14.8.6}$$

Since $v \approx c$ and $\Omega = \omega = \dfrac{p\,c^2}{E\,\rho}$, one gets energy E radiated during one revolution

$$E_{1\ reovolution} = S\,\frac{2\pi}{\omega} = \frac{1}{3\varepsilon_0}\frac{e^2}{\rho}\,\omega\,(v/c)^3 \left(\frac{E}{m_0 c^2} \right)^4 \tag{14.8.7}$$

This *radiation loss* is appreciable, and it sets a practical limit to the size of betatron-synchrotron. It has been seen that the total loss of 4.4 *MeV* / revolution for a 6 *BeV* betatron-synchrotron of $\rho = 26\ m$ has been seen.

14.8.2 Proton Synchrotron (Bevatron, or Cosmotron)

Energetic protons, say 50 *MeV*, from other accelerators, say linac, are injected into the vacuum chamber of a proton-synchrotron. The protons require energies much higher than electrons to reach relativistic range. The Brookhaven device has $\rho = 128\ m$ and could accelerate the initial 50 *MeV* protons to 33 *BeV*. The Fermi National Lab, Batavia uses a 500 *BeV* proton-synchrotron (diameter 2 *km*), whereas the Dubna device gives 10 *BeV*. The Serpukhov near Moscow possesses a 76 *BeV* device.

For equal radii and equal total energies, the betatron- and proton- synchrotrons have their radiation losses given by

$$\frac{S_{Beta-Sync}}{S_{Prot-Sync}} = \left(\frac{M_p}{m_e} \right)^4 \approx 10^{13} \tag{14.8.8}$$

TABLE 14. Proton Synchrotrons

Machine	Beam energy (GeV)
1. KEK, Tokyo	12
2. PS, CERN, Geneva	28
3. AGS, Brookhaven	32
4. Serpukhov	76
5. SPS, CERN	450
6. Tevatron-II, Fermilab	1000

This shows that the radiation loss is negligible in the case of proton-synchrotrons.

In Table 14.1 are listed the various proton synchrotrons and beam energies of some typical machines.

14.8.3 Alternating-Gradient Synchrotrons

In the early 1950s E.D. Courant, M.S. Livingston & H.S. Snyder and independently N.C. Christofilos evolved the principle of '*strong focusing*' which enabled significant enhancement of the energy limit of conventional devices.

The AG synchrotron differs from the ordinary synchrotron in that the pole faces of the C-shaped magnets are shaped so that half of them produce inhomogeneous magnetic fields with an inward gradient while the other half produce magnetic fields with an outward gradient. These two types of magnets are placed alternately around the accelerating chamber. These advantages provide better focused beam of accelerated particles. This is analogous to a lens system consisting of alternating converging and diverging lenses. The focal length F of the lens system is

$$\frac{1}{F} = \frac{1}{f_1} + \frac{1}{f_2} - \frac{S}{f_1 f_2}$$

where S is the separation between the two lenses f_1 & f_2.

A 6-*BeV* electron machine was installed in the MIT and at Harvard in 1962.

14.9 COLLIDING BEAM ACCELERATORS

In a fixed target collision, there is a large initial momentum, and so there must be a large final momentum and thus kinetic energy. But not all the KE is available to become mass. In a colliding beam set-up, the initial total momentum is zero.

TABLE 14. COLLIDERS

Machine	Accelerated particles
1. CESR, Cornell, NY	e^+ (6 GeV) + e^- (6 GeV)
2. PEP, Stanford	e^+ (15 GeV) + e^- (15 GeV)
3. TRISTAN, Tokyo	e^+ (32 GeV) + e^- (32 GeV)
4. SLC, Stanford	e^+ (50 GeV) + e^- (50 GeV)
5. LEP, CERN, Geneva	e^+ (60 GeV) + e^- (60 GeV)
6. SppS, CERN, Geneva	p (450 GeV) + $\bar{p}$ (450 GeV)
7. Tevatron 1, Fermilab, Batavia	p (1000 GeV) + $\bar{p}$ (1000 GeV)
8. HERA, Hamburg	e^- (26 GeV) + p (820 GeV)
9. UNK, Serpukhov	p (3000 GeV) + $\bar{p}$ (3000 GeV)
10. LHC, CERN	e^- (50 GeV) + p (8000 GeV)
	p (8000 GeV) + $\bar{p}$ (8000 GeV)
11. LEP-II, CERN	e^+ (100 GeV) + e^- (100 GeV)
12. SSC, Texas	p (20000 GeV) + $\bar{p}$ (20000 GeV)

For a single particle the result (unrealistic though it might be), the particle would be stationary and all of the initial KE would have been converted to mass. A collider wasted no energy. Geneva, Switzerland. The aim is to prove / disprove theories of nuclear physics, and especially to detect the theoretical Higgs boson, and to advancing our understanding of the laws of physics as we know to date. The principle used in the device is to prepare the collision of two beams of protons of kinetic energy at least 7 *TeV*.

REVIEW QUESTIONS

R.Q. 14.1   Describe the main features of a cyclotron. What limits the energy of of particles in a cyclotron?

R.Q. 14.2   A cyclotron is designed for protons and has dees 0.6 m and applied perpendicular flux density of 1.25 T. Calculate the f_{cycl}, the period, the maximum velocity of the proton beam and their energy in MeV. [Answers: $19\ MHz, 5.2 \times 10^{-8}\ s,\ 3.6 \times 10^7\ m\ s^{-1},\ 7\ MeV$].

R.Q. 14.3   Describe the main features of a Synchrotron. What is the mechanism that limits the performance of the instrument?. In what way a synchrotron differs from a Linac?

R.Q. 14.4   Find the total energy and the frequency of the beam of protons accelerated in a cyclotron with B = 16 kG and radius 50 cm. [Answer $KE \approx 31\ MeV,\ v = 24.3\ MHz$].

R.Q. 14.5   A Cyclotron dees of radius of curvature ρ is used to accelerate ion of mass M and charge q., undergoing acceleration through an equivalent voltage, V, in magnetic field, B. Show that a) $V = \frac{1}{2} B^2 \rho^2 \frac{q}{M}$, b) energy of the particle is $E = 3.12 \times 10^5\ B^2 \rho^2 \frac{q^2}{M}$ MeV ,and energy of the proton, if $\rho = 60$ cm, B = 10 kG. [Answer: $KE \approx 17.2\ MeV$].

R.Q. 14.6   A deuteron is moving in a field of 17 kG in resonance with the applied dee frequency of 12 MHz. Calculate a) the KE in MeV, b) The deuteron momentum, c) The velocity of the deuteron, d) the orbital radius. [Answers   a) 185 MeV, b) 855 MeV / c, c) 0.42 c, d) 1.63 m.].

R.Q. 14.7   Calculate the following quantities of a proton in a Cyclotron. Given, B = 17 kG, f = 18 MHz. a) Energy, b) Momentum, c) Linear velocity and d) Radius of orbit. [Answers a) 4215 MeV, b) 974 MeV / c, c) 0.72 c, d) 1.92 m.].

R.Q. 14.8   Describe the construction of a Geiger Muller tube. A gamma ray source, when placed before a Geiger tube, shows a count rate of $640\ s^{-1}$. After two hours the source shows a count rate of $40\ s^{-1}$. Calculate the half-life period of the gamma ray source.

R.Q. 14.9   Explain the principle of operation of a cloud chamber. Describe and explain the difference between the tracks formed in a cloud chamber by alpha and beta particles

R.Q. 14.10   What is the function of a magnetic field in a cyclotron? Does the magnetic field increase the kinetic energy of the charged particles?

R.Q. 14.12   Describe the construction by the help of labelled sketches of a cyclotron and explain its working.

RQ. 14.13   Explain the principle of the construction of a cyclotron and the modifications necessary in it for obtaining very high energy particles.

RQ. 14.16 In a cyclotron, protons spiral out at 15 cm radius in a Magnetic field of 1.25 T. a) T what frequency the AC voltage is used to accelerate the beam in the gap ?, and b) At what energy protons are accelerated.? Given: $M_p = 1.6726 \times 10^{-27} kg$, $e = 1.6021 \times 10^{-19} C$, $1 eV = 1.6021 \times 10^{-19} J$ (Answers: a) $3.91 \times 10^7 Hz$, b) 1.6 MeV).

&&*&*&*&*

Chapter 15

NUCLEAR REACTIONS

Chapter 15

NUCLEAR REACTIONS

"To be confused about what is different and what is not, is to be confused about everything"

- David Bohm

"For the wise all 'things are wiped away" - Gautama Buddha

15.1 INTRODUCTION

15.1.1 Nuclear Reaction – What does it mean?

The process called *nuclear reaction* is any encounter between nuclei, or between a nucleus and a nuclear particle, that results in a rearrangement of their constituent parts. This results in rearrangement of nucleons within the nucleus. This is analogous to a chemical reaction in which rearrangement of atoms resulting molecules. When energetic particles from an accelerator impinge upon matter there is the possibility of a nuclear reaction. This is a quite generally used way of representing a nuclear reaction. The nuclear *projectile* (m_i) is written first, then the *target* (M_i) nucleus which passes into the final *daughter* or *residual* (M_f) nucleus plus *ejected* (m_f) particle. A more *compact* (abbreviated) *form* which is frequently used is

$$M_i \left(m_i , m_f \right) M_f \qquad (15.1.1)$$

where the light weighted (projectile and ejected) particles are enclosed in parentheses between the target and daughter nuclei. This latter is convenient as the bracketed part can be used by itself to refer to a particular class of reactions like (α, p) or (n, γ).

Depending on the type of target, energy and nature of the incident (projectile) particle, many different nuclear processes can occur, when the projectiles come close enough to interact with the target nuclei.

a) In this process the incident particle may

i) Simply change direction,

ii) Lose some of its energy,

iii) Be completely absorbed by the target,

i) Perhaps particles which are altogether different may be knocked out of the target nucleus,

v) Be captured and a γ-ray emitted.

b) The target nuclei, after bombardment, are usually different from what they were before bombardment. They may change mass number (A) or atomic number (Z), or both. Such a change in the target nucleus resulting from its interaction with a bombarding (striking / incident, projectile) particle, is called a **transmutation reaction** or simply *nuclear reaction*.

15.1.2 The First Nuclear Reaction

The first such reactions were observed by Ernest Rutherford (1919) using α-particles from RaC' ($^{214}_{84}Po$) source within a gas chamber, filled with a gas, at certain pressure. A product of the α-particle scattering experiments which led to the nuclear model of the atom was the transmutation of elements in nuclear reactions such as

$$^{4}_{2}He + ^{14}_{7}N \rightarrow ^{17}_{8}O + ^{1}_{1}H \qquad (15.3.2)$$

i.e., $^{14}_{7}N \left(^{4}_{2}He, ^{1}_{1}H \right) ^{17}_{8}O$ \qquad (15.3.3)

When the gas pressure was reduced some of the α-particles pass through the gas and then through the thin silver foil and to the ZnS Screen detector. If the gas were either oxygen or CO_2, the scintillations disappeared when the gas pressure was increased to a critical value consistent with the known ranges of α-particles in these gases and the Ag-foil. However, when the chamber was filled with nitrogen and the pressures were raised to a value when all α-particles should be absorbed in gas plus foil, occasional scintillations were still observed. These could not possibly due to α-particles, but some more energetic particle originating in the nitrogen. Rutherford measurement showed the range of these secondary particles was at least 28 cm. Magnetic deflections of these particles showed that they were high-speed protons. So the transformation producing protons had to be

$$^{4}_{2}He + E_1(=7.683\ MeV) + ^{14}_{7}N \rightarrow ^{17}_{8}O + ^{1}_{1}H + E_2\ (=6.5\ MeV) (15.3.4)$$

since

$$^{4}_{2}He = 4.00387\ u,\ ^{14}_{7}N = 14.00752\ u,$$
$$^{17}_{8}O = 17.00453\ u,\ E_1 = 7.683\ MeV \equiv 8.25 \times 10^{-3}\ u.$$

giving $E_2 = 6.97 \times 10^{-3}\ u \equiv 6.5\ MeV$.

To conserve momentum, the lighter proton receive most of the 6.5 MeV of KE.. Protons of this energy are known to have ranges 0f around 40 cm in standard air, in contrast to 7 cm for α-particles.

15.2. NUCLEAR REACTIONS BASED ON THE CN MODEL

Provided the energies involved are not high ($\Box$ 15 MeV), many nuclear reactions actually involve two separate stages according to Niels Bohr (1936).

15.2.1.1 First step:

An incident particle bombards a target nucleus and the two combine to form a *compound nucleus* (CN), or intermediate nucleus. This compound nucleus will be in a highly excited state. There are several conservation laws which are believed to be obeyed universally in all nuclear reactions. No reactions are observed in which a proton disappears unless a neutron appears, and the vice-versa.

However, the compound nucleus may be formed in a variety of ways, as shown below:

Target nucleus + Incident particle + Compound nucleus   (15.2.1)

$$\begin{aligned}
{}^{14}_{7}N + {}^{1}_{0}n &\to {}^{14}_{7}N^{*} \quad (10.5\ MeV) \\
{}^{13}_{6}C + {}^{1}_{1}H &\to {}^{14}_{7}N^{*} \quad (7.5\ MeV) \\
{}^{12}_{6}C + {}^{2}_{1}H &\to {}^{14}_{7}N^{*} \quad (10.3\ MeV) \\
{}^{11}_{6}C + {}^{3}_{1}H &\to {}^{14}_{7}N^{*} \quad (22.7\ MeV) \\
{}^{11}_{5}B + {}^{3}_{2}He &\to {}^{14}_{7}N^{*} \quad (20.7\ MeV) \\
{}^{10}_{5}B + {}^{4}_{2}He &\to {}^{14}_{7}N^{*} \quad (11.6\ MeV)
\end{aligned} \quad (15.2.2)$$

Compound nuclei have life times

$$\tau_{1/2} \cong 10^{-16} s \qquad (15.2.3)$$

which, while so short as to prevent actually observing such nuclei directly, are never-the-less long compared to the

nuclear time $\tau_{nt} \cong 10^{-21}\ s$ (15.2.4)

which is the time taken by a high energetic (several *MeV*) to pass through a nucleus.

15.2.1.2 Second step:

After a relatively long period of time (typically from $10^{-19}\ s - 10^{-15}\ s$) and independent of the properties of the reactants, the compound nucleus disintegrates, usually into an ejected small particle and a product nucleus.

The compound nucleus may be *de-excited*. The decay may take place in one or more different ways, depending on its energy of excitation.

Compound nucleus (CN) → Recoil nucleus + Ejected particle   (15.2.5)

To illustrate this let the CN ${}^{14}_{7}N^{*}$ with excitation energy of, say 12 *MeV*, can decay via the different *decay modes*:; each mode is known as a **channel**.

$$\begin{aligned}
{}^{14}_{7}N^{*}(12\ MeV) &\to {}^{13}_{7}N + {}^{1}_{0}n \\
{}^{14}_{7}N^{*}(12\ MeV) &\to {}^{13}_{6}C + {}^{1}_{1}H \\
{}^{14}_{7}N^{*}(12\ MeV) &\to {}^{12}_{6}C + {}^{2}_{1}H \\
{}^{14}_{7}N^{*}(12\ MeV) &\to {}^{10}_{5}B + {}^{4}_{2}He, \\
\text{or,}\ {}^{14}_{7}N^{*}(12\ MeV) &\to {}^{14}_{7}N + 1\ \text{or}\ 2\ \gamma\text{-photons}
\end{aligned} \quad (15.2.6)$$

But it cannot disintegrate by the emission of a triton (${}^{3}_{1}H$) or a ${}^{3}_{2}He$-particle, since it does not have enough energy to liberate them. Usually a CN in a specific excited state only can decay through a particular decay mode.

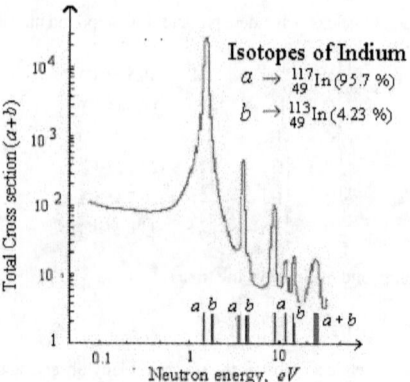

Fig 15.1 Indium total cross section for neutrons

The CN model has been very successful in explaining nuclear reactions for incident particle energies up to $15\ MeV$ and target of $A > 10$.

The most obvious evidence for long lived CN (intermediate) states in nuclear reactions is the strongly resonant nature of nuclear cross-sections. This is illustrated in Fig 15.1 which shows the indium total cross section for neutrons.

The energy of these long lived states is defined to a few eV and if we apply the uncertainty relation $\Delta E \Delta t \sim \hbar$ it can be seen that this implies a lifetime of $\sim 10^{-16}\ s$ which is very long compared to the time it takes a nucleon to traverse a nucleus $\sim 10^{-22}\ s$. The second type of reaction does not involve the formation of an intermediate state and so in this case the characteristic time of interaction is more like $10^{-22}\ s$. Also variations in the cross-section as a function of energy are more likely to be spread over a few MeV.

15.2.2 Compound Nucleus (CN)

For a short range (or abrupt sided) potential (Fig 15.2), there exist quasi-bound or virtual single particle states which have positive energy. A long range potential like the coulomb potential has no such states.

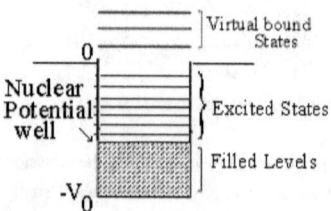

Fig 15.2 Virtual bound state for a short range nuclear potential well

In the simplified picture of the formation of the compound nucleus the positive energy projectile particle is momentarily trapped in one of the single particle virtual states. The collision with nucleons inside the nucleus it shares its energy with them, causing some of them into excited states and itself dropping into one of these states due to the energy it loses. Thus the Compound Nucleus is formed from many particles excited. At this stage all memory of the original mode of excitation is lost. At a later time a possibility for decay occurs when the energy of excitation is once more concentrated in a single or few particle virtual states. In the evolution of the many particles excited states several such fleeting decay configurations may occur before one of them is realized and the products separate out.

The problem of nuclear reactions by Quantum Mechanics using the Shell Model potential to represent the nucleus then there is no mechanism for changing the energy of the incident particle described above. There is no way of absorbing the bombarding particle into the compound nucleus.

The use of a complex potential in what is called the *Optical Model* Is used. This is a simple model which is used to deal in a general way with scattering processes when absorption is involved. Consider a traveling wave moving in a potential V then this plane wave function is written

$$\Psi = e^{i\,kx} \tag{15.2.7}$$

where $k = \dfrac{8\pi^2 m\,(E-V)^{1/2}}{h^2}$ (14.2.8)

If the potential V is replaced by V + iW then k also becomes complex and the wave function can be written

$$\Psi = e^{i\,k_1 x}\, e^{i\,k_2 x}, \tag{15.2.9}$$

where $k = k_1 + i\,k_2$

This is a traveling wave the amplitude of which is decreasing as it progresses - it is being absorbed. In most cases V » W and under this assumption we can make an estimate of the mean free path ($1/2k_2$) - that is the distance over which the intensity ($\Psi^*\Psi$) is attenuated by e^{-1} ..

$$k \simeq \dfrac{8\pi^2 m\,(E-V)^{1/2}}{h^2}\{1 - i\,W/mv^2\} \tag{15.2.10}$$

replacing the kinetic energy (E - V) by the expression $\tfrac{1}{2}mv^2$ gives

$$k \simeq \dfrac{2\pi\,mv}{h} - \dfrac{2\pi\,W}{hv} = k_1 + i\,k_2 \tag{15.2.11}$$

The real part of this expression is just $2\pi\,\lambda$ where λ is the de Broglie wavelength of the incident particle.

The mean free path ($1/2k_2$) can be obtained from the imaginary part as $\dfrac{h\,v}{4\pi W}$ or the mean free time $dt \simeq \dfrac{h}{4\pi W}$. To describe diffraction and scattering phenomena with the optical model requires an imaginary potential of a few MeV and this is entirely consistent with the lifetime of the virtual single particle state before absorption of about $10^{-22}\,s$.

Writing the compound nucleus reaction as

$$a + X \rightarrow C^* \rightarrow Y + b \qquad (15.2.12)$$

$$\sigma_{ab} = \sigma_a \Gamma_b / h \qquad (15.2.13)$$

the cross-section can be expressed as where if the lifetime of the b single particle state is τb. The decay probability is then given by

$$\lambda_b = 1/\tau_b = \Gamma_b/\hbar \qquad (15.2.14)$$

Γ_b is the so called partial width and Γ is the total width or $\hbar$ times the total decay probability. The cross-section for the formation of the compound nucleus has just been written as σ_a

The behaviour of the cross-section with energy depends on the relative sizes of Γ and the spacing between the energy levels. For low excitation of a nucleus the energy levels are relatively well spaced and the cross-section exhibits resonance behaviour while at higher energies of excitation the width Γ will overlap several energy levels and the cross-section varies much more slowly with energy. This is the so called continuum region. The energy at which the transition from resonance to continuum behaviour occurs depends upon A. For $A \sim 20$ it occurs at about 10 MeV while for $A \sim 200$ the onset of the continuum is much lower in energy at about 100 keV.

For the case of well separated levels an individual state will decay as $e^{-\Gamma_b/\hbar}$ - remember that Γ is called the total width and is equal to $\hbar$ times the total decay probability. Thus the wave function can be written

$$\Psi = \Psi(r)\, e^{-i E_o t/\hbar}\, e^{-\Gamma t/2\hbar} \qquad (15.2.15)$$

where the first exponential gives the normal oscillatory time dependence of the wavefunction with E_o as the energy above the ground state. The second exponential gives the decay of the state and the factor 1/2 in the exponent is to ensure the correct decay of the state which is proportional to $|\Psi|^2$. As the state is decaying it is not a solution of the Schrödinger equation with a static potential - it is not a so called stationary state but can be considered as a superposition of such states

$$\Psi = \int A(E)\, e^{-i E_o t/\hbar}\, dE \qquad (15.2.16)$$

The function $A(E)$ can be obtained by Fourier transform which yields

$$|A(E)|^2 = \dfrac{|\Psi|^2}{\{4\pi^2\,[(E - E_o)^2 + \Gamma^2/4]\}} \qquad (15.2.17)$$

This means that the cross-section is expected to have the form

$$\sigma_a = \frac{C}{[(E-E_0)^2 + \Gamma^2/4]} \tag{15.2.18}$$

The similarity between this expression and resonance curves met in other branches of physics (eg., ac electrical circuits) justifies the use of the same terminology here and in particular indicates why Γ is called the width.

The constant C depends upon the phase space available to the incident particle and the statistical weight $g = (2\ell + 1)$, where ℓ is the orbital angular momentum of the initial state. We will just quote the result

$$C = \hbar^2 \Gamma \Gamma_a \, g / (4\pi p^2) \tag{15.2.19}$$

where p is the momentum of the incident particle and Γ_a is the partial width for the decay back into the initial state.

Finally bringing the various factors together the cross-section can be written

$$\sigma_{ab} = \frac{\lambda^2 \, g \Gamma_a \Gamma_b}{4\pi \, [(E-E_0)^2 + \Gamma^2/4]} \tag{15.2.20}$$

where λ = h/p the de Broglie wavelength of the incoming particle.
This is the Breit-Wigner formula for the single level reaction cross-section. For example - in the case of elastic scattering ($\Gamma_a = \Gamma_b$) at the maximum of the resonance ($E - E_0$) the cross-section is

$$\sigma_{el} = \frac{\lambda^2 \, \Gamma_a^2 \, (2\ell+1)}{\pi \, \Gamma^2} \quad \Gamma = \Gamma_a \tag{15.2.21}$$

and if no other processes are possible ($\Gamma = \Gamma_a$) this further reduces to

$$\sigma_{el} = \frac{\lambda^2 \, (2\ell+1)}{\pi} \tag{15.2.22}$$

the maximum possible elastic cross-section.

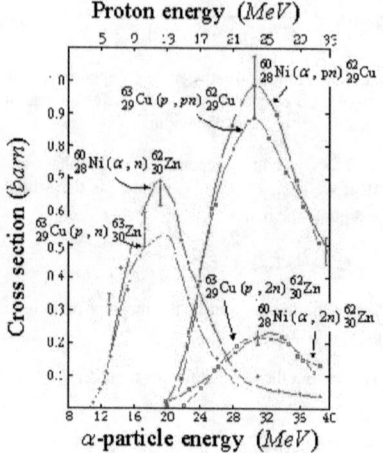

Fig 15.3 The compound nucleus $^{64}_{30}Zn$ and yields of its decay products

For the inelastic cross-section ($\Gamma_b = T - \Gamma_a$) at the maximum of the resonance ($E - E_o$) its value is

$$\sigma_{inel} = \frac{T^2 \Gamma_a (T - \Gamma_a)(2\ell+1)}{\pi T^2} \quad (15.2.23)$$

This latter has a maximum value of

$$\sigma_{inel} = \frac{\lambda^2 (2\ell+1)}{4\pi} \quad (15.2.24)$$

i.e., when

$$\Gamma_a = \Gamma/2$$

Consider that the separation of the energy levels is much smaller than the total width. This results many levels contributing to a given process. This mixture of levels gives a good separation of the decay of the compound nucleus from the mode of excitation. In this case there is good supporting evidence of the existence of a compound nucleus independent of the initial state.

This is illustrated in the Fig 15.3 of the yields of decay products from the compound nucleus ^{64}Zn formed by two different routes.

15.3. DIRECT REACTIONS

When there is direct interaction with a single nucleon rather than with the nucleus as a whole, the process is known as a *Direct reaction*

a) Inelastic Scattering:

Under conditions when single collisions between the incoming particle a and a nucleon in the target nucleus A in which the incident particle emerges with reduced energy.

$$A + a \rightarrow A^* + a \qquad (15.3.1)$$

The incoming particle does not really enter the nucleus but excites it in passing.

b) The incident particle a enters the nucleus and exchanges charge with one of the neutrons, form a nucleus B'

$$A + a \rightarrow B^* + b \qquad (15.3.2)$$

c) *"Knock-out"* reaction:

With the incident particle a carrying off most of the energy, one can write

$$A + a \rightarrow B + a + b \qquad (15.3.3)$$

B is the residual nucleus remaining after the nucleon b has been knocked out.

c) *"Stripping"* reaction:

A composite incident particle, say a deuteron, is stripped of one of its component nucleons which remain in the target nucleus and the remaining nucleon(s) escape.

d) *"Pick-up"* reaction:

As in c) but conversely, the incident particle, usually a nucleon, picks up another nucleon from the target nucleus and carries it away, emerging as a deuteron. e.g., (n, d).

In direct reactions, the incident particle has several *MeV* of energy, takes place in a time of the order of that taken by a nucleon to cross a nucleus ($R/c = 10^{-22} s$).

e) CN reaction:

In a compound nucleus (CN) reaction i) it is expected that all sense of direction of the bombarding particle would be lost and ii) that the produced particles would boil off with an essentially isotropic distribution in the CM frame. The CN exists for times in the range $= 10^{-14} s$ to $10^{-20} s$.

$$A + a \rightarrow C^* \rightarrow B + b \qquad (15.3.4)$$

where C* represents the excited CN.

15.4 WHY SHOULD ONE INVESTIGATE ON NUCLEAR REACTIONS?

A specific reaction is studied by determining the angles and kinetic energies of the reaction products (the kinematic variables). Particle and radiation detectors designed for the expected charge and energy of each product are arranged around the target.

The reaction cross section is the primary quantity of interest for a specific set of kinematic variables. The cross section is a measure of the probability for a particular reaction to occur. This quantity, σ, which has the dimension of area, is measured in unit *barn* by the experimental ratio

$$\sigma = \frac{\text{Number of reaction particles emitted}}{(\text{Number of beam particles per unit area})(\text{Number of target nuclei within the beam})} \quad (15.4.1)$$

The cross section can also be calculated from a mathematical model of the nucleus by applying the rules of quantum mechanics. Comparing the measured and calculated values of the cross sections for many reactions validates the assumptions of the nuclear model.

Table 7-1 shows some of the many types of nuclear reactions and what they teach us about nuclei and nuclear energy.

The properties of nuclei are measured by using nuclear reactions and nuclear scattering. Reactions that exchange energy of nucleons can be used to measure the energies of binding and excitation, quantum numbers of energy levels, and transition rates between levels. A particle accelerator (see Chapter 14) produces a beam of high-velocity charged particles (electrons, protons, alphas, or "heavy ions"), which then strikes a target nucleus. Nuclear reactions can also be produced in nature by high-velocity particles from cosmic rays, for instance in the upper atmosphere or in space. Beams of neutrons can be obtained from nuclear reactors or as secondary products when a charged-particle beam knocks out weakly-bound neutrons from a target nucleus. Nuclear reactions can also be produced by beams of photons, mesons, muons, and neutrinos.

In order for a nuclear reaction to occur, the nucleons in the incident particle, or projectile, must interact with the nucleons in the target. Thus the energy must be high enough to overcome the natural electromagnetic repulsion between the protons. This energy "barrier" is called the Coulomb barrier (Fig 15.4). If the energy is below the barrier, the nuclei will bounce off each other. Early experiments by Rutherford used low-energy alpha particles from naturally radioactive material to bounce off target atoms and measure the size of the target nuclei.

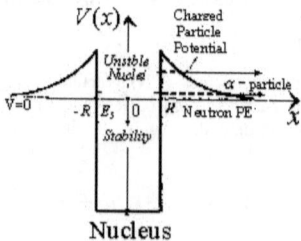

Fig 15.4 Nuclear Potential

When there is collision between the incident particle and a target nucleus, either

i) the particle beam scatters elastically leaving the target nucleus in its ground state or
ii) the target nucleus is internally excited and subsequently decays by emitting radiation or nucleons.

A nuclear reaction is described by identifying the incident particle, target nucleus, and reaction products.

Example: When a neutron strikes a nitrogen nucleus, ^{14}N, to produce a proton, ^{1}H, and an isotope of carbon, ^{14}C, the reaction is written as

$$^{1}_{0}n + ^{14}_{7}N \rightarrow ^{14}_{6}C + ^{1}_{1}H \qquad (15.4.2)$$

Sometimes the reaction is abbreviated as

$$^{14}_{7}N \, (n,p) \, ^{14}_{6}C \qquad (15.4.3)$$

15.4.2 Reaction Cross-section

Nuclear reactions can be tackled from three points of view:

1. Nuclear reaction dynamics,

2. Nuclear cross-section, and

3. Mechanism of nuclear reactions.

15.4.3 Nuclear Reaction Dynamics

The CN model is helpless in explaining some nuclear reactions. One example is the so called **stripping** reaction.

$$^{24}_{12}Mg + ^{2}_{1}H \rightarrow ^{1}_{1}H + ^{25}_{12}Mg$$
$$\text{or,} \quad ^{24}_{12}Mg \, (d,p) \, ^{25}_{12}Mg \qquad (15.4.3)$$

Experimental evidence indicates that in this process, when the deuteron comes very close to the target nucleus, the neutron is captured and the proton is repelled without the formation of the CN.

Nuclear reactions are also essentially treated as **collision processes** in which several conservation laws are believed to be obeyed. The Rutherford –Soddy Law for balancing nuclear reactions are:

i) Conservation of electric charge, and

ii) Conservation of nucleon (i.e., mass) number.

15.4.4. Conservation Conditions and Nuclear reactions

Any reaction equation has to obey a number of conservation conditions: They correspond to Classical Mechanics and Electro-dynamics (*viz.*, energies, mass, momentum, angular momentum, and electric charge) and conservation laws unique to Nuclear Physics (to be discussed in Chap 8).

1. The mass number A and the charge Z must balance on each side of the reaction arrow. Thus in the example

$$^{1}_{0}n + ^{14}_{7}N \rightarrow ^{14}_{6}C + ^{1}_{1}H \qquad (15.4.4)$$

the sum of superscripts (1 + 14) on the left equals the (14 + 1) on the right to balance the A's. The (0 + 7) subscripts on the left equal the (6 + 1) on the right to balance the Z's.

2. The total energy before the reaction must equal the total energy after the reaction. The total energy includes the particle kinetic energies plus the energy equivalent of the particle rest masses, $E = mc^2$.

3. Linear momenta before and after the reaction must be equal. For two-particle final states this means that a measurement of one particle's momentum determines the other particle's momentum.

4. Quantum rules govern the balancing of the angular momentum, parity, and isospin of the nuclear levels and

5. Rules like conservation of nucleon or baryon number, lepton number, strangeness, hypercharge

15.5 SCATTERING

If the incoming and outgoing particles are the same, the nuclear process is called **scattering**.

Elastic scattering: In elastic scattering both the kinetic energy and the total energy are conserved.

Inelastic scattering: If the nucleus is left in a different state then it is called inelastic scattering.

15.5.1 Nuclear reactions are classified as:

Exothermic

$$(\text{Total Kinetic energy})_{Initial} < (\text{Total Kinetic energy})_{Final} \qquad (15.5.1)$$

Endothermic

$$(\text{Total Kinetic energy})_{Initial} > (\text{Total Kinetic energy})_{Final} \qquad (15.5.2)$$

15.5.2 Q of a Nuclear Reaction

Q of a nuclear reaction $M_i\left(m_i, m_f\right) M_f$ is given by the expression

$$Q = \left[(M_i + m_i) - (M_f + m_f)\right] gm \cdot c^2 \qquad (15.5.3)$$

or $\qquad Q = \left[(M_i + m_i) - (M_f + m_f)\right] u \cdot 931.49432 \, MeV \qquad (15.5.4)$

15.5.3 Types of Nuclear Reactions.

Depending on the circumstances, it is convenient to classify nuclear reactions by the type of bombarding particle, bombarding energy, target, or reaction product.

1. <u>Type of projectile</u>:

They are:

a) Charged particle reactions, produced by
$p\ (^1_1H),\ d\ (^2_1H),\ \alpha\ (^4_2He),\ ^{12}_{6}C,\ ^{16}_{8}O,\ldots,\ldots,$

b) Neutron reactions: produced by ^1_0n. Since neutrons are uncharged there is no Coulomb barrier to overcome at the nuclear boundary of target and so neutrons of very low energy easily penetrate and interact with the nuclei that it bombards.

c) Photo-nuclear reactions: produced by γ-rays, $^{14}_{7}N\ (\gamma,p)\ ^{13}_{6}C$, or $^{13}_{6}C^*$

Electron-induced reactions

2. <u>Bombarding energy</u>:

They are

a) Thermal energies $\frac{1}{40}\ eV$

b) Epithermal energies $\approx 1\ eV$

c) Slow neutron $\approx 1\ keV$

d) Fast neutron ≈ 0.1 to $10\ MeV$

e) Low energy charged particles ≈ 0.1 to $10\ MeV$

f) High energies $\approx 10\ MeV$ and higher.

3. <u>Based on targets</u>

They are

a) Light nuclei: $A \leq 40$

b) Medium weight nuclei: $40 \leq A \leq 150$

c) Heavy nuclei: $A \geq 150$.

4. a) <u>Elastic Scattering</u>

If the light reaction product is identical to the projectile, and has identical energy (C-System of coordinates)

Example: $^{14}_{7}N\ (p,p)\ ^{14}_{7}N$

b) <u>Inelastic scattering</u>:

If the light reaction product is identical to the projectile, and only C-System energy is different, Example, $^{14}_{7}N\,(p,p')\,^{14}_{7}N^*$.

5) Capture reaction

If only γ-rays are emitted: Example of proton capture reaction,

$$^{14}_{7}N\,(p,\gamma)\,^{15}_{8}O,\text{ or }^{15}_{8}O^*,$$

and photonuclear reaction,

$$^{14}_{7}N\,(\gamma,p)\,^{13}_{6}C,\text{ or }^{13}_{6}C^*$$

6) *Spallation* (or Fission)

If the product nuclei which have masses comparable, it is *Spallation* (or Fission)

Example, $\quad ^{14}_{7}N\,\left(n,\,^{6}_{3}Li\right)\,^{9}_{4}Be,\text{ or }^{9}_{4}Be^*$

7) Coulomb Excitation

A charged particle with energy well below the Coulomb barrier cannot interact directly with the target nucleus through the nuclear force. However, sometimes it is possible for the charged particles to excite a target by without penetrating the nuclear radius, through the electromagnetic force. Heavy ions, because of their high charge, are particularly effective in this respect.

Example: $\quad ^{9}_{4}Be\,\left(^{6}_{3}Li,n\right)\,^{14}_{7}N,\text{ or }^{14}_{7}N^*$

Worked out Example 15.1

Determine the energy change in the conversion of Na-23 to magnesium by proton bombardment $^{23}_{11}Na\,(p,n)\,^{23}_{12}Mg$ Given: $M(23,11) = 22.99714\,u$, , $M_n = 1.00898\,u$, $M_p = 1.00812\,u$.

Solution: $\boxed{STEP\,*\,1}$ $Q = 24.00526\,u - 24.01043\,u = -0.00517\,u = -4.79\,MeV$;

$\boxed{STEP\,*\,2}$ This value for Q means for the reaction to take place the protons must have a minimum energy of 4.79 *MeV*.

15.6 CONTINUUM THEORY OF NUCLEAR REACTIONS

(Resonance Reactions, Breit-Wigner Dispersion Formula)

In the CN Model of nuclear reactions the unbound states of nuclei is treated as if they formed a structureless continuum. This means the several numbers of discrete nuclear states exist and they form a continuous spectrum. Each of these supposed to be discrete states is unstable and is characterized by certain *level width*. When the individual level widths of the numerous discrete states are much larger than the level separation (spacing) between adjacent states, the result is the compound nucleus continuum.

In the direct reactions, the bound states are stable against particle emission, the individual level has mean life, say, $\tau_{1/2} = 1\ ps$ has $\Delta E = \Gamma \cong 10^{-3}\ eV$ ☐ typical spacing between levels of bound states. So theses discrete states are specified by definite wave functions.

Resonance is the phenomenon observed in nuclear reactions caused by bombardment of nuclei by particles both charged and uncharged. The resonance region is between the two cases discussed above, viz., the discrete region and the CN region. They are characterized by large cross sections (high probability for formation) and very small level widths, because there are only two modes of decay available to it. These two modes are:

The re-ejection of the incident particle, takes place as in

Elastic $\quad\quad\quad\quad {}^{208}_{82}Pb\ (n,\ n)\ {}^{208}_{82}Pb$, or

Inelastic scattering, $\quad {}^{208}_{82}Pb\ (n,\ n')\ {}^{208}_{82}Pb^*$, or

γ-emission, $\quad\quad\quad {}^{208}_{82}Pb\ (n,\ \gamma)\ {}^{208}_{82}Pb$.

Using square-well nuclear potential for the captured particle, qualitative analysis was done. Accordingly, in a single isolated resonance of energy E_R, and level width, $\Delta E = \Gamma = \hbar / \tau$.

The probability $P(E)\ dE$ to occur the system in the energy interval E and $(E + dE)$ in the vicinity of the reaction level E_r is

$$P(E)\ dE = \frac{dE}{\left[(E - E_r)^2 + (\Gamma/2)^2\right]} \quad\quad (15.6.1)$$

The reaction cross section σ_r is known, from partial wave analysis in Quantum Mechanics,

$$\sigma_{SC} = \sum_{\ell=0}^{\infty} 2\pi \lambda^2\ (2\ell + 1)\ (e^{-2ik_\ell t} - 1) \quad\quad (15.6.2)$$

$$\sigma_r = \sum_{\ell=0}^{\infty} \pi \lambda^2\ (2\ell + 1)(1 - |\eta_\ell|^2) \quad\quad (15.6.3)$$

where the complex coefficient $\eta_\ell = e^{2i\delta_\ell}$, $(1 - |\eta_\ell|^2) \to 4\sin^2\delta_\ell$

$\ell = \ell^{th}$ partial wave, δ_ℓ = phase shift of ℓth partial wave, and total cross section

$$\sigma_T = (\sigma_{SC} + \sigma_r) \quad\quad (15.6.4)$$

For a scattering resonance to occur, $\eta_\ell = -1$, $\delta_\ell = \pi/2$, Applying Taylor's series expansion to $\cot \delta_\ell$, at E_R the resonant level, and defining the width

$$\Gamma = 2 \left(\frac{\partial \delta_\ell}{\partial E} \right)^{-1}_{E=E_R} \tag{15.6.5}$$

and considering only up to the second order terms in the expansion, one gets

$$\cot \delta_\ell = - \frac{(E - E_r)}{(\Gamma/2)} \tag{15.6.6}$$

At resonance Γ = FWHM which causes the cross section to fall slowly to half the central value at $(E - E_r) = \pm (\Gamma/2)$, and this occurs when $\cot \delta_\ell = \pm 1$. In other words,

$$\delta_\ell = \pi/4, 3\pi/4, \ldots,$$

$$\sin \delta_\ell = \frac{(\Gamma/2)}{\left[(E - E_R)^2 + (\Gamma/2)^2 \right]^{1/2}} \tag{15.6.7}$$

For <u>elastic scattering</u>, substituting this value in the expression (7)

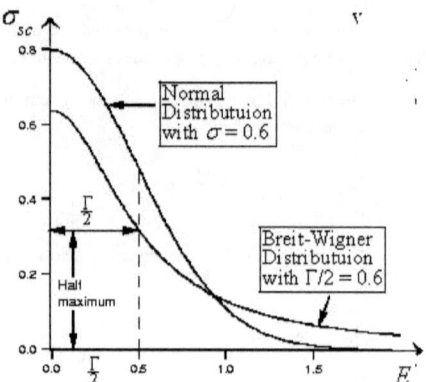

Fig 15.5 Breit-Wigner distribution compared with Normal distribution

$$\sigma_{SC} = \frac{\pi}{k^2} (2\ell + 1) \frac{(\Gamma/2)}{\left[(E - E_R)^2 + (\Gamma/2)^2 \right]} \tag{15.6.8}$$

This is the ***Breit-Wigner formula*** for resonance (Fig 15.5).

Factor Γ in the denominator is related to the decay width of 'resonant state', $\Gamma = \hbar / \tau$.

There are entrance and exit channels / modes for the resonance reaction, and allowing for 'partial entrance' and 'partial exit' widths, such that $\Gamma = \sum_{i=0}^{\infty} \Gamma_i$.

Γ^2-factor in the denominator is directly related to the formation of the resonance and to its probability to decay into a particular exit channel. For elastic scattering, the entrance and exit channels are identical, so that for a reaction

$$a + X \rightarrow (CN) \rightarrow a + X \qquad (15.6.9)$$

$$\Gamma = \sum (\Gamma_{entr} + \Gamma_r) = \sum (\Gamma_{entr} + \Gamma_{exit}) \qquad (15.6.10)$$

15.7 Method of Mass Determination from the Energetics of Nuclear Reactions

Consider that the reaction is expressed as

$$a_i + A_i \rightarrow a_f + A_f \qquad (15.6.11)$$

a_i denotes bombarding particle of mass m_i

a_f denotes residual particle of mass m_f

A_i denotes target nucleus of mass M_i and

A_f denotes product nucleus of mass M_f.

There is a conversion of mass into KE (T) in such reactions.

$$Q = (m_i + M_i - M_f - M_f)c^2 \qquad (15.6.12)$$

$$= (T_{Af} + T_{af} - T_{ai}) \qquad (15.6.13)$$

If m_i, m_f and M_i are known and one would like to determine M_f, then Q must be measured, In general, it is neither easy nor convenient to measure the KE of the product nucleus. One, therefore, can express an equation for momentum conservation and use the equation to eliminate T_{Af}. To further simplify, limiting to the non-relativistic range,

$$Q = \left\{ \begin{array}{l} T_{Af}(1 + \dfrac{m_f}{M_f}) - T_{ai}(1 - \dfrac{m_i}{M_f}) \\ - \dfrac{2}{M_f}\sqrt{T_{ai} \, T_{af} \, m_i \, m_f} \, \cos\theta \end{array} \right\} \qquad (15.6.14)$$

It can be noted that this expression does not depend on the exact value of mass m_i and KE of M_f; but depend on $(m_f / M_f), (m_i / M_f)$ and $(m_i \, m_f / M_f)$, which are all of the order of 10^{-4}. One can now calculate the value for M_f. Using this value Q can be iterated to the limits of experimental accuracy I one step.

Q = +ve, for **exo-ergic** and (16.6.15)

Q = -ve for **endo-ergic** reactions. (15.6.16)

15.7 DIFFERENTIAL SCATTERING CROSS-SECTION, $\sigma(k,\theta) \, d\Omega$:

Experimental data in a scattering experiment are commonly expressed in terms of the *Differential Scattering Cross-Section*, defined as $\sigma(k,\theta) \, d\Omega$. Fig. 15.6 illustrates the various quantities used in getting an expression for $\sigma(k,\theta) \, d\Omega$.

$$\sigma(k,\theta) \, d\Omega = \frac{[\text{\# of particles scattered into at angle}]}{[\text{\# of particles incident / unit area}]} \quad (15.7.1)$$

$$\sigma(k,\theta) = d\sigma(k,\theta)/d\omega = \{dN_{sc}/d\Omega\} / \{dN_{inc}/d\Omega\} \quad (15.7.2)$$

where dN_{sc} = # of particles that traverse elemental area dA normal to $\hat{k}$ in the incident packet,

dN_{sc} = # of particles scattered into the cone subtended by the solid angle, $d\Omega$,

$$d\Omega = Sin\theta \, d\theta \, d\phi . \quad (15.7.3)$$

15.7.2 Total Scattering Cross section:

If the potential has spin dependence, i.e. azimuthal dependence, the *Total Cross-Section*, σ, is by definition,

$$\sigma_T = \sigma(k) = \int_{-1}^{+1} d(Cos\,\theta) \int dr \int_0^{2\pi} d\phi \, (d\sigma/d\Omega) \quad (15.7.4)$$

$= N(\theta) / (\hbar \overline{k}/m) \; A = |f(\overline{k}', \overline{k}, \theta)|^2 \; [\hbar \overline{k}/m r^2]$

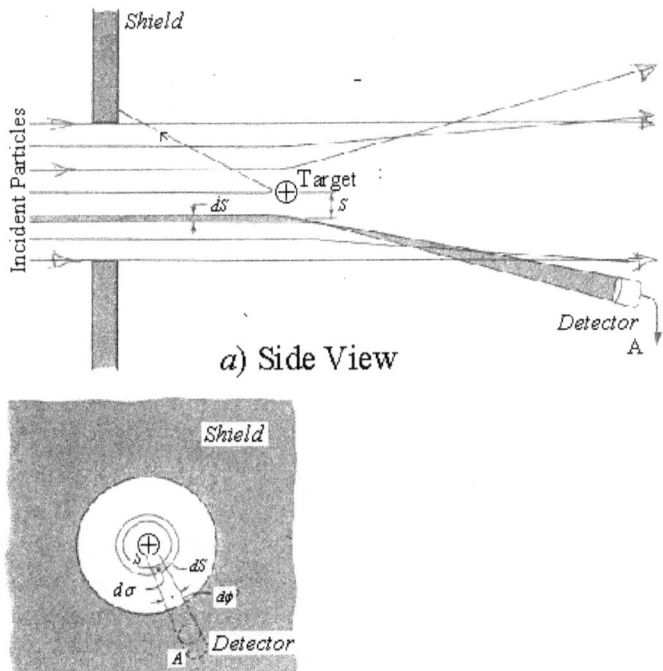

Fig.15.6 The small part of the beam incident falls on the detector defines the element of cross section, $d\sigma$.

It can be shown that

$$dN(\theta) = N(\theta)\,|f(\bar{k}', \bar{k})|^2\,[\hbar\bar{k}/m]\,(dA/r^2) = |f(\bar{k}', \bar{k})|^2\,[\hbar\bar{k}/m]\,d\Omega$$

Equation (15.7.2) yields

$$(d\sigma/d\Omega) = |f(\bar{k}', \bar{k}, \theta)|^2 \tag{15.7.5}$$

$$\sigma_T \equiv \sigma(k) = \int (d\sigma/d\Omega)\,d\Omega = \int |f(\bar{k}', \bar{k}, \theta)|^2\,d\Omega. \tag{15.7.6}$$

i.e., $|f(\bar{k}', \bar{k}, \theta)|$ is directly proportional to the experimentally observed Cross-Section.

15.7.3 Scattering cross sections in the Centre-of-Mass (CM) and Laboratory (L) Frames: What is the significance of Centre-of-Mass?

The principles of conservation of linear and angular momenta are known as applied to *particles*. But what is the significance of these two laws for a macroscopic body or for systems of particles? To apply these laws with complete generality, one must find the point (in the system) which moves in accordance with these laws. This is the Centre-of-Mass (CM) of the system. When considering the implications of the conservation laws in the motion of a system of bodies, one must always refer to the motion of the CM of that system.

15.7.3.1 CM and the Laws of Dynamics

These are:

Newton's 1st Law: The CM of a system continues in rest or move in constant velocity.

If an external force is applied to the system, the CM undergoes acceleration.

In the absence of external forces, the total angular momentum of the system around the CM remains constant.

15.7.3.2 CM and Nuclear Reactions

Most of the knowledge about atomic nuclei and their interactions have come from experiments in which energetic particles collide with target nuclei. It is instructive to calculate in detail the relationship between two quantities, viz., *threshold KE* (i.e., minimum KE) and the *Q-value* (obtained from mass difference). This calculation much simplified by the use of the Centre-of-Mass (CM) coordinate system (known as the C-system) [which moves with respect to the laboratory (i.e., the CM of the colliding particles)], so that the total momentum in this system is zero, rather than the usual L-system of coordinates (which is held fixed in the laboratory).

15.7.3.3 Relationship between the L-system and the CM-system:

The L-system (Laboratory system) is fixed in space, with the target particles (before recoil) at rest. This is the system in which all the angles actually measured in laboratory in performing an experiment are referred. In the C-system; the scattering can be described as taking place from a fixed scattering centre (CM), two bodies remaining collinear with this centre with both moving either toward or away from this centre with equal momentum.

Note that

a) The scattering experiments are performed in the L-frame, and

b) The theory is conveniently worked out in the C-system,

It is, therefore, necessary to be able to transform scattering results from the L- frame to the C--frame and the vice-versa.

15.7.4 Relation between the L- and C-systems:

Consider non-relativistic particles, *i.e.*, in the L-frame

$$\text{particle velocity, } v \ll c, \text{ or} \tag{15.7.7}$$

the kinetic energy (KE) of each body is such that

$$KE \ll \text{rest mass energy} \tag{15.7.8}$$

m_1 = mass of the projectile particle positioned at x_1,

v_i = velocity of the projectile moving towards the target,

M_2 = mass of the target at rest, at x_2,

$v_2 = 0$, velocity of the target,

15.7.4.1 Before Collision:

(i) For motion in the L-frame (Fig 15.7), the conservation of momentum requires
If the CM is located at position X,

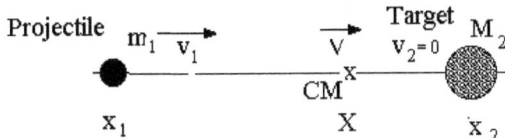

Fig 15.7   Motion in L-system

$$(m_1 + M_2) X = (m_1 x_1 + M_2 x_2), \qquad (15.7.9)$$

$$(m_1 + M_2) V = (m_1 v_1 + M_2 v_2) = 0 \qquad (15.7.10)$$

where $v_1 = \dfrac{dx_1}{dt}$, $v_2 = \dfrac{dx_2}{dt}$, and $V = \dfrac{dX}{dt}$

whence the CM velocity,

$$V = \dfrac{m_1}{m_1 + M_2} v_1 \qquad (15.7.11)$$

Total kinetic energy, in the L-system,

$$T_L = \tfrac{1}{2} m_1 v_1^2, \qquad (15.7.12)$$

(ii) For motion in the C-frame;

An observer located at X, the CM, (Fig 15.8) and moving with it, experiences momentum,

$$p = m_1 (v_1 - V) - M_2 V \qquad (15.7.13)$$

$$p = \left[m_1 v_1 + m_1 \left(\frac{m_1}{m_1 + M_2} \right) v_1 \right] - \left[M_2 \left(\frac{m_1}{m_1 + M_2} \right) v_1 \right] \qquad (15.7.14)$$

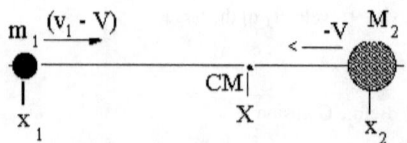

Fig 15.8   Motion in C-system

$$p = m_1 \left(\frac{M_2}{m_1 + M_2} \right) v_1 - M_2 \left(\frac{m_1}{m_1 + M_2} \right) v_1 = 0 \qquad (15.7.15)$$

Total kinetic energy, in the CM-system,

$$T_C = \tfrac{1}{2} m_1 (v_1 - V)^2 + \tfrac{1}{2} M_2 V^2 . \qquad (15.7.16)$$

On manipulation,

$$T_C = T_L - \tfrac{1}{2} (m_1 + M_2) V^2 \qquad (15.7.17)$$

$$\frac{T_C}{T_L} = \frac{M_2}{m_1 + M_2} . \qquad (15.7.18)$$

15.7.4.2   After Collision: (Elastic)

In the C-frame, each particle is scattered at the same angle, θ; thus conservation of KE requires conservation of velocities and each particle has the same KE after the collision that it had before the collision. Using single quote symbol (i.e., primed) to the quantities used before the collision,

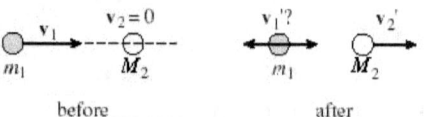

before   after

Elastic collision

Fig 15.9

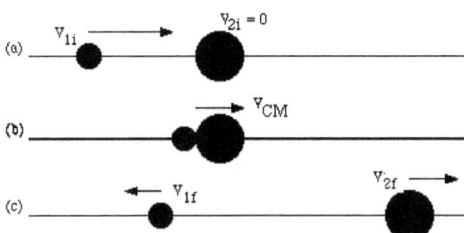

Fig 15.10. Head-on-Collision, in C-system

Total initial momentum = total final momentum.

i.e., $\quad m_1 v_1' = M_2 v_2'$. $\hspace{4cm}$ (15.7.19)

Total initial k. E. = Total final k. E.

$$\tfrac{1}{2} m_1 (v_1 - V)^2 + \tfrac{1}{2} M_2 V^2 = \tfrac{1}{2} m_1 v_1'^{\,2} + \tfrac{1}{2} M_2 v_f'^{\,2} \quad (15.7.20)$$

The only solution, which satisfies both these equations, is

$$v_1' = v_1 - V \hspace{4cm} (15.7.21)$$

and $\quad v_2' = V$. $\hspace{5cm}$ (15.7.22)

Thus the velocity of each particle in the L-system is obtained by adding V vectorially to the velocity of each in the CM-frame. On the other hand, the speeds of the particles in the CM-frame will be unchanged by elastic collisions, but the observer sees the direction of each changes by the same amount φ.

15.7.4.3 Inelastic Collision:

In the macroscopic case the part of the energy dissipated in an inelastic collision (Fig 15.11) usually manifests as heat and sound.

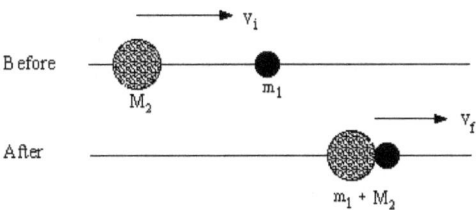

Fig 15.11 Inelastic Collision

On the other hand, in the case of subatomic particles, the lost KE is used to excite the target nuclei which in turn emit γ-rays. If enough energy is involved mesons may be ejected.

$$(v_1 - V) \sin\theta_C = v_1' \sin\theta_L, \quad (15.7.23)$$

$$V + (v_1 - V)\cos\theta_C = v_1' \cos\theta_L, \quad (15.7.24)$$

and $\quad \tan\theta_L = (\sin\theta_C) / [V/(v_1 - V) + \cos\theta_C] \quad (15.7.25)$

This relation holds good for both Classical and Quantum Mechanics, since the conservation of energy and momentum are both fundamental to Classical as well as Quantum Mechanics.

$$\tan\theta_L = (\sin\theta_C) / [\eta + \cos\theta_C] \quad (15.7.26)$$

$$\theta = \frac{m_1}{M_2}. \quad (15.7.27)$$

If $\quad m_1 \ll M_2, \; \theta_L \approx \theta_C \quad (15.7.28)$

Table 15.1. Physical quantities in L-Frame and C-Frame

Quantity	L-Frame	C-Frame
1. Momentum of m_1	$m_1 v_1$	$\frac{m_1}{m_1+M_2} v_1$
2. Momentum of M_2	0	$-\frac{m_1}{m_1+M_2} v_1$
3. Total energy	$T_L = \frac{1}{2} m_1 v_1^2$	$T_C = T_L - \frac{1}{2}(m_1 + M_2) V^2$

15.7.5 Relationship of cross sections in the L- and C-systems

The relation between cross sections is just given by the transformation of angles since the integrated fluxes must be conserved. From the relation between scattering angles in the L- and C-systems, one obtains

$$d\Omega_{CM} = \sin\Theta \, d\Theta \, d\Phi \quad (15.7.29)$$

$$d\Omega_{CM} = \frac{\sqrt{(m/M)^2 + 2(m/M)\cos\Theta + 1}}{1 + (m/M)\cos\Theta} \, d\Omega_L \quad (15.7.30)$$

Threshold Energy, T_{th}

Any endoergic reaction ($Q < 0$, i.e., $T_{final} < T_{initial}$) always requires in the L-system, a certain minimum initial KE, called the **threshold kinetic energy**, T_{th} for the reaction to take place.

In the C-system, this occurs for head-on-collisions for

$$T_C = Q \equiv T_{th} \quad (15.7.31)$$

$$Q = \frac{M_2}{m_1 + M_2} T_L \quad (15.7.32)$$

$$T_{th} = (-Q)\left(1 + \frac{m_1}{M_2}\right) \tag{15.7.32}$$

This expression also indicates that only a fraction of the bombarding energy T_1 is available for the reaction.

Worked out Example 15.2

What is threshold kinetic energy for the $^6_3Li\,(\alpha,p)\,^9_4Be$ reaction? Given: $M_\alpha = 4.003879\ u$, $M_p = 1.008145\ u$, $M(9,4) = 9.015030\ u$, and $M(6,3) = 6.016970\ u$.

Solution ⎡STEP * 1⎤ $Q = -0.002326\ u = -2.12\ MeV$,

⎡STEP * 2⎤ $T_{th} = 3.53\ MeV$ ..

Worked out Example 15.3

A ball of mass $m_1 = 0.1$ kg traveling with a velocity $v_1 = 0.5$ m/sec collides head on with a ball of mass $m_2 = 0.2$ kg which is initially at rest. Calculate the final velocities, v_1' and v_2', in the event that the collision is elastic, in the laboratory frame of reference.

Solution: ⎡STEP * 1⎤ $v_1' = -0.17\ m\ s^{-1}$, $v_2' = +0.33\ m\ s^{-1}$.

⎡STEP * 2⎤ Using, $v_1' = \dfrac{m_1 - m_2}{m_1 + m_2} v_1$, $v_2' = \dfrac{m_1}{m_2}(v_1 - v_1')$

Worked out Example 15.4

A ball of mass $m_1 = 0.1$ kg traveling with a velocity $v_1 = 0.5$ m/sec collides head on with a ball of mass $m_2 = 0.2$ kg which is initially at rest. Calculate the final velocities, v_1' and v_2', in the event that the collision is elastic, in the C-frame of reference.

Solution: ⎡STEP * 1⎤ $V_{CM} = \dfrac{m_1}{m_1 + m_2} v_1$, $V_{CM} = \hat{i}\ 0.17\ m\ s^{-1}$;

⎡STEP * 2⎤ $u_1 = v_1 - V_{CM}$, $u_1 = \hat{i}\ 0.33\ m\ s^{-1}$;

⎡STEP * 3⎤ $u_2 = v_2 - V_{CM}$, $u_2 = -\hat{i}\ 0.17\ m\ s^{-1}$;

⎡STEP * 4⎤ $v_1' = u_1' + V_{CM}$, $u_1' = -\hat{i}\ 0.17\ m\ s^{-1}$;

STEP * 5 $v_2' = u_2' + V_{CM}$, $v_2' = \hat{i}\, 0.33\ m\ s^{-1}$.

15.7.6 Another View of Reaction Cross sections, σ

It is convenient to describe a nuclear reaction in terms of a cross section, σ. It will express the probability that a projectile will interact in a certain way with a target particle in a certain reaction. It is simply the effective target area given by each target nucleus to the bombarding particle.

Integrated cross section σ is obtained as follows:

$$\text{Fractional sensitive area } f = \frac{\#\text{ of interacting particles }(-dN)}{\#\text{ of incident particles }(N)} = \frac{\text{Total sensitive area}}{\text{Total area in target}}$$

(15.7.33)

Case i) For infinitesimally small thickness of the slab

$$f = \frac{nA\,\sigma\,dx}{A} = n\,\sigma\,dx \qquad (15.7.34)$$

Case ii) For a finite thick slab,

$$f = \int_{N_0}^{N}\left[n\,\sigma\int_0^x dx\right] \qquad (15.7.35).$$

One gets $N = N_0\, e^{-n\sigma x}$ (15.7.36)

Number of particles that has undergone interaction,

$$(N_0 - N) = N_0\left(1 - e^{-n\sigma x}\right) \qquad (15.7.37)$$

This becomes in the limit of $x \to$ small,

$$(N_0 - N) \Rightarrow N_0\, n\, \sigma\, x \qquad (15.7.38)$$

and

$$\frac{dN}{N} \approx \frac{(N_0 - N)}{N_0} \approx n\,\sigma\,x \qquad (15.7.39)$$

One can define absorption coefficient,

$$\alpha = n\,\sigma \quad (\text{in } b) \qquad (15.7.40)$$

Relationship of cross sections in the LAB and CM systems, in terms of Solid angles

The relation between cross sections is just given by the transformation of angles since the integrated fluxes must be conserved. From the relation between scattering angles in the LAB and CM systems, we obtain

$$d\Omega_{CM} = \sin\Theta \, d\Theta \, d\Phi$$

$$d\Omega_{CM} = \frac{\sqrt{(m/M)^2 + 2(m/M)\cos\Theta + 1}}{1 + (m/M)\cos\Theta} \, d\Omega_L \qquad (15.7.41)$$

15.7.7 Mean Free Path

For a projectile beam of intensity I_o going through a dilute gas of density N of thickness dx, the current of projectiles scattered out of the beam is $dI = -I\, n\, \sigma\, dx$. If the target is "thin" so that the particle is not scattered back into the incident beam direction,

$$I = I_o \, e^{-n\sigma x} \qquad (15.7.42)$$

The number of times a particle traverses in the slab (Δx) without interacting,

$$H = \frac{1}{n\,\sigma\,\Delta x} \qquad (15.7.43)$$

Average distance the particle traverses the slab without having interaction with the target

$$\ell = \frac{\Delta x}{n\,\sigma\,\Delta x} = \frac{1}{n\,\sigma} \qquad (15.7.44)$$

The mean free path (MFP) is the mean distance between collisions:

$$\ell = \frac{1}{n\,\sigma} \qquad (15.7.45)$$

15.7.8 Differential Cross section

The product (light mass) particles in many nuclear reactions are not produced in an isotropic manner, with respect to the direction of incident beam. Therefore defining a *differential cross section*, (say, elastic) $d\sigma_{el}/d\Omega$, (Fig 15.12), in terms of

dN = Number of light reaction product particles,

$d\Omega$ = A small solid angle at some angle θ.

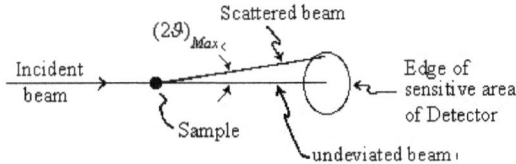

Fig 15.12 Differential Cross section

$$\frac{1}{I}\frac{dN}{d\Omega} \Big/ \text{target nucleus} = \frac{dN/d\Omega}{(I/A)(n\,A\,\Delta x)} \qquad (15.7.46)$$

and $\quad \sigma_{el} = \int\limits_{\text{All space}} \left[\dfrac{d\sigma_{el}}{d\Omega} \right] d\Omega \quad$ (15.7.47)

15.7.9 Reaction Rate (RR)

The number of nuclear reactions taking place in unit time is defined as the reaction rate (RR). If $\Phi = $ # crossing $\text{cm}^{-2}\ \text{s}^{-1}$, A = area of the target sheet,

$$RR = \Phi\, n\, \sigma\, A\, x \qquad (15.7.48)$$

Activity (*i.e.*, disintegration rate), R, of the CN formed is

$$R = \left|\dfrac{dN}{dt}\right| = \lambda N = \dfrac{0.693}{\tau_{1/2}} N = \dfrac{0.693}{\tau_{1/2}} [(RR)(\text{time})] \qquad (15.7.49)$$

Table 15.2 Nuclear Reaction Types

Reaction	What is learned
Nucleon-nucleon scattering	Fundamental nuclear force
Elastic scattering of nuclei	Nuclear size and interaction potential
Inelastic scattering to excited states	Energy level location and quantum numbers
Inelastic scattering to the continuum	Giant resonances (vibrational modes)
Transfer and knockout reactions	Details of the Shell Model
Fusion reactions	Astrophysical processes
Fission reactions	Properties of Liquid-drop Model
Compound nucleus formation	Statistical properties of the nucleus
Multi-fragmentation	Phases of nuclear matter, Collective Model
Pion reactions	Investigation of the nuclear "glue"
Electron scattering	Quark structure of nuclei

Worked out Example 15.5

A sheet of ^{59}Co is bombarded with a beam of thermal neutrons of flux 10^{14} particles $\text{cm}^{-2}\ \text{s}^{-1}$ for a period of 1 minute. The reaction taking place is $^{59}\text{Co} + {}^1_0 n \rightarrow {}^{60}\text{Co}^*\ (\tau_{1/2} = 5.26\,y) \rightarrow {}^{60}\text{Ni} + \beta^- + \bar{\nu}_e$. Find the number of $^{60}\text{Co}^*$ nuclides formed after irradiation, assuming that none of $^{60}\text{Co}^*$ has decayed during the irradiation process, and its activity.. Given, $\rho_{Co} = 8.71\ \text{gm cm}^{-3}$, $M(59, 27) = 58.933\ \text{gm mole}^{-1}$, and $\sigma = 19\ b$.

Solution:

STEP * 1   RR = [Φ = 10^{14} cm^{-2} s^{-1}] (n = = 0.915 x10^{23} cm^{-3}) (σ = 19b) ($A x$ = 0.0115 cm^3)

= 19.95 x10^{11} s^{-1};

STEP * 2   N = 12 x10^{13} ^{60}Co*.

STEP * 5   R = 5 x 10^5 dis s^{-1}

Table 15-2 shows some of the many types of nuclear reactions and what they teach us about nuclei and nuclear energy

The elastic scattering cross sections of protons and neutrons on a proton target give the essential data to reconstruct the nucleon-nucleon. A complete theory of nuclear structure and dynamics must start with this elemental interaction.

The systematics of nuclear sizes, shapes, binding energies, and other nuclear properties are the data that nuclear models are challenged to explain. The Shell Model has combined the large body of nuclear data into a coherent theory of nuclear structure. Most of this data was the result of elastic and inelastic scattering of electrons, protons, and neutrons from nuclei found in the chart of the nuclides.

At high enough excitation energies, a nucleus can undergo a series of normal modes of collective oscillations called giant resonances. The nucleus rings like a bell at distinct frequencies with all the nucleons participating and sharing the excitation energy.

Measurements of cross sections for nuclear reactions as well as for various elastic scattering processes, have played a crucial role in research in nuclear physics.

Worked out Example 15.6

Obtain an expression for the Coulomb Cross section (Charged particle Cross section)

Solution: STEP * 1 The elastic scattering of low energy charged projectiles is determined purely by the Coulomb forces, and from Rutherford's theory it is known that the deflection θ of the proton approaching a nucleus (Ze) is

$$\theta = \frac{Z e^2}{4\pi \varepsilon_0} \frac{2}{b p v}$$

p = momentum, v = velocity of the proton, b = impact parameter, the distance that the proton would pass the nucleus in the absence of interaction.

$$d\theta = \frac{2Z e^2}{4\pi \varepsilon_0 p v} \frac{db}{b^2}$$

$$d\sigma_e = 2\pi \left(\frac{2Z e^2}{4\pi \varepsilon_0 p v}\right)^2 \frac{d\theta}{\theta^3}$$

The differential cross section,

$$\frac{dN}{d\Omega} = \frac{D^2}{16 \sin^4 \frac{\theta}{2}}$$

STEP * 2 | $D \approx 7.6 fm$, for $^{59}_{27}Co$ (5.2.MeV p, p) $^{59}_{27}Co$, $\frac{D^2}{16} = 0.036\ b$; this breaks down as $D \square R$.

for small angles, $\frac{d\sigma_{el}}{d\Omega} = \frac{1}{2\pi \sin\theta} \frac{d\sigma_{el}}{d\theta} = \left(\frac{2Z e^2}{4\pi \varepsilon_0 p v}\right)^2 \frac{1}{\theta^4}$

STEP * 3 | $\sigma_t = \int_{All\ space} \left[\frac{d\sigma_{el}}{d\Omega} d\Omega\right] = \infty$, and

STEP * 4 | hence σ has no meaning for charged particles.

Measurements of cross sections for nuclear reactions as well as for various elastic scattering processes, have played a crucial role in research in nuclear physics.

15.7.9 The Role of the Coulomb Barrier in for Neutral and Charged Particle

Every nucleus is surrounded by an ES potential barrier that opposes both the entry and escape of positively charged particles in nuclei. Neutrons are not facing this problem, and are therefore more readily absorbed and ejected by nuclei than protons, deuterons, α-, and other charged particles. (Fig 15.13).

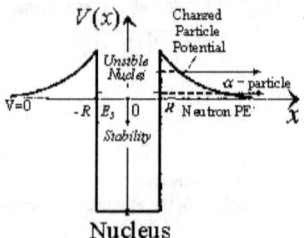

Fig 15.13. Absence of Coulomb barrier for Neutrons

Hence neutrons of very low energy easily penetrate matter and interact with nuclei. In the limit $E \to 0$, only elastic scattering and exothermic nuclear reactions can take place.

This is the reason for the entirely different cross sections versus energy curves for protons and neutrons. (Fig 15.14).

When a nuclear reaction is possible at very low energies, the reaction rate (RR) will be independent of E. $E \propto \rho_n$, beam density, near the target. The cross section σ_{exoth} is given by

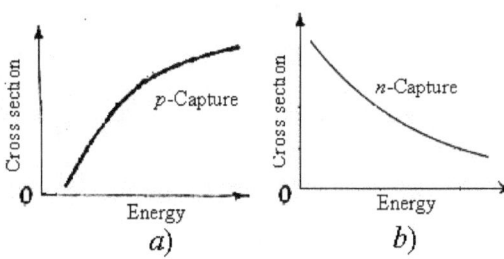

Fig 15.14. Diagrams showing n- and p- capture cross sections are different

$$({}_0^1 n \text{ flux}) \cdot \sigma_{exoth} = \text{RR per nucleus} \qquad (15.7.50)$$

$$ {}_0^1 n \text{ flux} = \rho_n \text{ beam density } v_1 \qquad (15.7.51)$$

Thus $\therefore \quad \sigma_{exoth} = \dfrac{(\text{constant})}{v_1, \text{ velocity of } {}_0^1 n} \qquad (15.7.52)$

At low energies.

If the low energies fall in a resonance region, and the Breit-Wigner formula gives

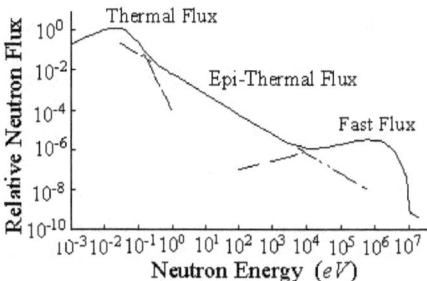

Fig 15.15 Neutron Flux versus Energy

$$\sigma_{SC} \approx \frac{1}{k} \frac{\text{(contant)}}{\left[E^2 + (\Gamma/2)^2 \right]} \qquad (15.7.53)$$

$$\hbar k = \frac{m_1}{m_1 + M_2} v_1 \qquad (15.7.54)$$

15.8 HIGH ENERGY REACTIONS

The incident projectiles in high energy reactions have energies in the range $> 500\ MeV$ and the upper limit is whatever highest energy available from the sources.

15.8.1 Photon-induced nuclear reactions

Photons of high energy (γ quanta) can excite nuclear states. Since the photons transfer mainly spin 1 and, to a lesser extent, spin 2, low-spin states are excited. These states deexcite by emitting γ transitions of discrete energies - according to the spacings between the excited states and lower lying states (see schematic figure). As this emission is induced by a photon of the appropriate (resonant) energy, the process is called '*Nuclear Resonance Fluorescence*". The properties of the γ transitions allow the quantum numbers of the excited states to be deduced:

a) Energy of the emitted photon $\longrightarrow$ excitation energy, E.

b) Intensity of the emitted photons $\longrightarrow$ Width, Γ.

c) Angular distribution of the emitted photons $\longrightarrow$ Spin, J.

d) Polarization of the emitted photons $\longrightarrow$ Parity, π.

The excited states with spin $J = 1$ are referred to as "dipole excitations". These excitations may involve collective excitations of the nucleus such as vibrations. The study of these excitations and the discovery of so far unknown types of nuclear motion is important for the understanding of many-body quantum systems, which play a role also in other fields of modern physics, e.g. in solid-state physics.

If an impinging photon has energy which exceeds the separation energy of $^1_0 n$ or $^1_1 H$ of a particular nucleus, then the photon can induce reactions with emission of particles. If the energy is high enough, even composite particles like α- particles can be emitted. Following the particle emission, various final nuclei are produced which may be unstable (radioactive) and decay *via* β decay.

Nuclear fission can be induced by photons when irradiating appropriate isotopes as *e.g.* $^{238}_{92}U$. This process enables the investigation of fission products. Among the fission products there are very neutron-rich isotopes far from stability, which can hardly be produced by means of other nuclear reactions. Information about the properties of these nuclei is very important *e.g.* for the understanding and modeling the processes of creating heavy elements in the Universe

15.8.2 HIgh Energy and Nuclear Physics

There evolved in 1950s two directions:

i) Studied nuclear structure and reactions and
ii) Investigated several new particles being discovered Cosmic Rays in air showers or by using high energy proton beams from new accelerators. This new field, called elementary particle physics (or high energy physics), focused on new phenomena such as pions and "strange" particles, and soon discovered a variety of other new particles.

The object has been to:

1) understand the structure of protons and neutrons, as well as mesons, in terms of the underlying quark and gluon structure;
2) understand nuclear structure and reactions, especially nuclei at the limits of mass (*super heavy* elements), excitation energy, angular momentum, and neutron-to-proton ratio;
3) study the properties and phases of nuclear matter;
4) study fundamental symmetries of nature as they are apparent in nuclei and nuclear reactions; and study the nuclear dynamics of the sun, supernova, other astrophysical objects, and nucleo-synthesis, which includes understanding the properties of neutrinos.

Study on nuclear physics till 1970s was performed mainly using Van de Graaff accelerators or cyclotrons for the study of nuclear structure and reactions. Now, most of the programme is focused at a few large accelerator, say LHC at CERN, with large collaborations of both national laboratory and university scientists. The nuclear physics programme continues to provide much of the expertise, manpower, and basic nuclear science for leading edge advancements in medical imaging, diagnostic tools, and radiation therapies.

Investigations on the basic constituents of matter and the fundamental interactions have been very important. These constituents include six types of quarks that are the building blocks of the hadrons, such as the proton or neutron, and six leptons, the most familiar of which is the electron. At this energy frontier the search for the *Higgs boson*, whose discovery would lead to a better understanding of the origin of mass, is a major campaign of the programme hereafterin the next few years. Another fundamental line of investigation is CP violation, the matter antimatter asymmetry in the weak nuclear interaction. This tiny asymmetry is believed to explain the predominance of matter over antimatter in the universe.

REVIEW QUESTIONS

R.Q. 15.1 Determine the Q-value for the reaction ^{23}Na (α, p) ^{26}Mg, taking the isotopic value from Standard Tables. Given: M(23,11) = 22.99714 u, M(26,12) = 25.99080 u, M_p = 1.00812 u , M_α = 4.00387 u (Answer Q = 27.00101 u - 26.99892 u = + 0.00209 u = + 1.95 MeV; Energy is released; Exothermic reaction).

R.Q. 15.2 For the reaction given by $^9_4$Be (α, n) $^{12}_6$C, determine its Q value.. (Answer Q = +5.7 eV).

R.Q. 15.3 For the reaction given by $^9_1$D $(^9_1$D, $n)$ $^3_2$He, determine its Q value. (Answer. Q = +3.3 eV).

R.Q. 15.4 For the reaction given by ^{12}B (p, n) ^{12}C, determine its Q value. (Answer Q = +12.6 MeV.).

R.Q. 15.5 For the reaction given by ^{12}B $(p, ^3H)$ ^{10}B, determine its Q value. (Answer Q = - 6.3 eV.).

R.Q. 15.6 For the reaction given by ^{27}Al (γ, p) ^{26}Mg, determine its Q value. (Answer Q = - 8.3 MeV.).

R.Q. 15.6 A neutron of energy 10 MeV was used in the reaction ^{19}F (n, p) ^{19}O, it was found that Q = - 3.9 MeV. If the ejected proton was detected at an angle 90° to the incident direction, what will be its energy? Given: M_p = 1.0078 u, M_n = 1.0087 u, and M(19,8) = 18.99 u .(Answer T_{af} = 5.3 MeV .).

R.Q. 15.7 Determine the position of Centre-of-Mass of the system shown in Fig 15.5.

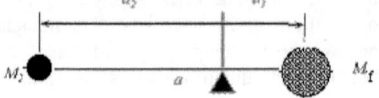

R.Q. 15.8 If the ^{19}F (n, p) ^{19}O reaction is known to have Q = - 3.9 MeV, what is the threshold energy? Given: M(19,9) = 18.9984 u , M_p = 1.0078 u, M_n = 1.0087 u, (Answer T_{th} = 4.1 MeV .).

R.Q. 15.9 For the reaction $^2_1$H $(^2_1$H, $^1_0n)$ $^3_2$He, find if threshold energy is required or not. Given: M_d = 2.014102 u, M_n = 1.0086655 u, and M(3,2) = 3.016029 u . (Answer: Δm = + 0.003510 u. Thus no threshold energy is required, the reaction can take place.)

R.Q. 15.10 For the reaction ^{7}Li (p, α) ^{4}He, determine the Q-value = released energy.

Given: $M(7,3) = 7.016003\ u$, $M_\alpha = 4.002602\ u$, $M_p = 1.007825\ u$.(Answer: Q = [(7.016003) + (1.007825) – 2(4.002602)] c^2 (931.5 MeV/uc^2) = + 17.35 MeV = energy released)

R.Q. 15.11 For the reaction, $^{18}O\ (p, n)\ ^{18}F$, a minimum input energy Q-value =-2.453 MeV is needed; what is the mass of the product nucleus ^{18}F ? Given $M(18,8) = 17.999160\ u$, $M_p = 1.007825\ u$, $M_n = 1.008665\ u$, (Answer $M(18,9) = 18.000953\ u$.).

R.Q.15. 12 A beam of α-particles consisting of 10^{10} particles/cm^2 bombards a target area, 1 cm^2, in the reaction A(α, p)B. If there are 10^{16} target nuclei and 10^2 reactions occur, what will be the reaction cross section? (Answer $\sigma = (\frac{dN}{N})\frac{1}{10^{16}} = 10^{-24}\ cm^2$)

R.Q. 15. 13 Consider the reaction $^{197}Au(n, \gamma)\ ^{198}Au$. Find the number of ^{198}Au /s/cm^2 formed, when the gold sheet ($\rho(Au) = 19.3\ gm\ cm^{-3}$, $M(197, 79) = 197.2\ u$) of thickness $3 \times 10^{-4} m$ is bombarded with a slow neutron beam of current density 10^7 neutrons $s^{-1}\ cm^{-2}$. Given the thermal neutron capture cross section for ^{197}Au is $\sigma(n,\gamma) = 94\ b$. (Answer $n_{Au} = 5.89 \times 10^{22}\ cm^{-3}$, $n\ \sigma(n,\gamma)\ \Delta x = 0.1686$, $(N_0 - N) = 1.55 \times 10^6$ neutrons $cm^{-2}\ s^{-1}$)

R.Q. 15.14 Using the two nuclear reactions $^{27}Al(p, \alpha)\ ^{24}Mg$ and its inverse, viz., $^{24}Mg(\alpha, p)\ ^{27}Al$, determine the energy levels of ^{28}Si. (Answer Q-values are, in the C-system, 11.588 MeV and 9.990 MeV, respectively).

R.Q. 15.15 Obtain a relation between scattering cross sections in L- and C-frames.

R.Q. 15.16 When formed, the compound nucleus will be in a state of extreme excitation Write down typical values of excitation energy and maximum angular momentum of the Compound Nucleus. (Answer: Around 40 MeV, with a maximum angular momentum of around 25–35 $\hbar$.).

R.Q. 15.17 Outline the different types of neutron induced nuclear reactions.

R.Q. 15.18 Develop Breit-Wigner dispersion formula for nuclear resonance scattering and obtain the expression for the cross-section. Bring out the implications of the same.

R.Q. 15.19 Calculate the threshold energy for $^{14}N\ (n, \alpha)\ ^{11}B$ reaction. $M_n = 1.0086655\ u$, $M_\alpha = 4.00387\ u$, $M(11,5) = 11.00931\ u$, $M(14,7) = 14.00307\ u$.

R.Q. 15.20 Discuss about Partial Wave Analysis of nuclear reactions.

RQ. 15.21 Write an essay on nuclear reactions.

RQ. 15.22 Determine the q-value for the reaction $^{23}Na\ (\alpha, p)\ ^{26}Mg$. Given the isotopic masses of Na and Mg cited are: 22.99714u and 25.99080 u., respectively.

RQ. 15.23 Complete the nuclear reaction equations listed below:

a) $^{14}_{7}N + ^{4}_{2}He \rightarrow ^{17}_{8}O + ?$, b) $^{9}_{4}Be + ^{4}_{2}He \rightarrow ^{16}_{6}C + ?$, c) $^{9}_{4}Be\ (p,\alpha)\ ?$,

d) $^{30}_{15}P \rightarrow ^{30}_{14}Si + ?$, e) $^{3}_{1}H \rightarrow ^{3}_{2}He + ?$, f) $^{43}_{20}Ca\ (\alpha, ?)\ ^{46}_{21}Sc$

(Answers: a)) $^{1}_{1}H$, b) $^{1}_{0}n$, c) $^{6}_{3}Li$, d) $^{0}_{+1}e$, e) $^{0}_{-1}e$, f) p.)

RQ. 15.24. The nuclide $^{63}_{30}Zn$ undergoes positron decay, after the following nuclear reaction, $^{63}_{29}Cu\ (p,n)\ ^{63}_{30}Zn$. Given the energy of the positron to be $2.36\ MeV$ a) Estimate the Q of the reaction. b) For the reaction $^{63}_{29}Cu\ (d,2n)\ ^{63}_{30}Zn$, $Q = -6.38\ MeV$, find out the binding energy of the deuteron. (Answers: a) $Q = -4.165\ MeV$; b) $Q = (6.38 - 4.165)\ MeV$).

&&*&*&*&*

Chapter 16
FISSION OF A NUCLEUS

Chapter 16

FISSION OF A NUCLEUS

"We cannot solve the problem that we have created with same thinking that created them" -   A. Einstein (The American Museum of Natural History,

NYC- seen by the author).

"Tell me what has politics to do with truth, goodness and beauty"

-Robert Oppenheimer

"For the present I believe that the war will be over long before the first atom bomb is built."   – W.Heisenberg,1939

"Atomic power can cure as well as kill. It can fertilize and enrich a region as well as devastate it. It can widen man's horizons as well as force him back into the cave."-
Alvin M. Weinberg, 1944

16.1   INTRODUCTION

16.1.1 Introduction

One of the major requirements for sustaining human progress is an adequate source of energy. The current largest sources of energy are the combustion of coal, oil and natural gas. They will last quite a while but will probably run out or become harmful in tens to hundreds of years. Solar energy will also work but is not much developed yet except for special applications because of its high cost. This high cost as a main source, *e.g.* for central station electricity, is likely to continue, and nuclear energy is likely to remain cheaper. A major advantage of nuclear energy (and also of solar energy) is that it doesn't put carbon dioxide (CO_2) into atmosphere. How much of an advantage depends on how bad the CO_2 problem turns out to be ?

Fission of atomic nuclei is one of the most interesting phenomena in Nuclear Physics with far reaching consequences Discovered in 1934, it paved the way for the modern nuclear reactor, a bountiful source of energy. Fundamentally, it is a unique illustration of Albert Einstein's $E = mc^2$ formula. In this Chapter I will give an account of the theory of fission and its peaceful application. It has been already put to use in the form of explosives and produced non-conventional electric power. From an engineering viewpoint, it is one of the most important nuclear processes.

16.1.2 Transuranic Elements

Enrico Fermi *et al.* in 1934 tried to produce elements of Z > 92, by bombarding with neutrons the element Uranium. According to previous beliefs when Uranium is bombarded with $_0^1n$, *transuranic* elements must be produced.

$$_{92}U + {_0^1}n \rightarrow \text{Transuranic elements} \qquad (16.1.1)$$

16.1.3 Discovery of Fission of Nucleus

Paul Harteck Otto Hahn and Fritz Strassmann (1938) discovered that *alkaline earth* metals ($^{139}_{56}Ba$ and $^{140}_{57}La$) are produced when Uranium is irradiated with slow ^1_0n, in radio-chemical experiments..

$$_{92}U + {}^1_0n \rightarrow \text{Alkaline rare earth metals}. \qquad (16.1.2)$$

An important and special type of nuclear reaction phenomenon is *nuclear fission*.

Lise Meitner and Otto R. Frisch (1939) suggested that by absorption of a neutron ^1_0n, the uranium nucleus, X (Tail of the arrow in Fig 16.1) becomes sufficiently excited and then splits into a pair of medium mass fragments, b & Y (Head of the arrow in Fig 16.1) of approximately equal mass and several ^1_0n. Such a reactions process

$$X(a, b)Y \qquad (16.1.3)$$

is called *nuclear fission*, if b and Y are comparable in masses.

It is known from Chapter3 that in the neighborhood of ^{56}Fe nuclei have the greatest binding energy per nucleon, E_B/A (Fig 16.1.).

In principle, therefore, nuclear potential energy can be released into KE and made available as thermal energy by forming nuclei closer in mass to iron, by heavy nuclei by fission. This Chapter is devoted to the physics of nuclear fission and its application to power reactors. Reports reveal that in 1997 some 430 nuclear power stations have been operating world-wide.

16.1.4 Reactants, Fissile Nucleus and Fission Products

An example of the fission of $^{235}_{92}U$ is:

$$^1_0n + {}^{235}_{92}U \rightarrow {}^{141}_{56}Ba + {}^{92}_{36}Kr + 3({}^1_0n) \qquad (16.1.4)$$

Here the $^{235}_{92}U$ nucleus is known as the *fissile nucleus* whereas the $^{141}_{56}Ba$ and $^{92}_{36}Kr$ are called the *fission products*. The fission products resulting from the fission of a particular fissile nucleus are not unique. For example,

$$^1_0n + {}^{235}_{92}U \rightarrow {}^{140}_{54}Xe + {}^{94}_{38}Sr + 2({}^1_0n) \qquad (16.1.5)$$

The fission products are radioactive. The components on the left hand side of eq (16.1.4) are called the *reactants*, whereas those on the right hand side are the *products*

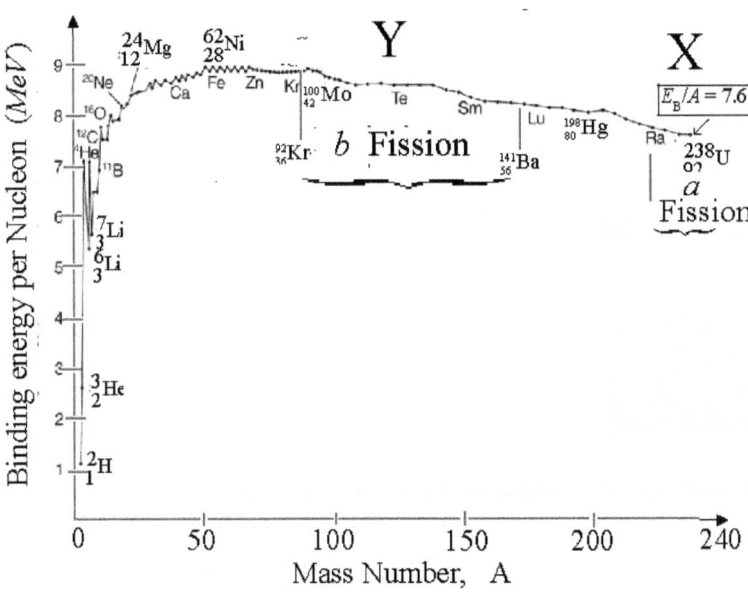

Fig 16.1 E_B/A versus A curve

16.2   ENERGY AVAILABLE FOR FISSION (Theory of Fission)

16.2.1 Calculation of Fission Energy for $^{235}_{92}U$

Nuclear fission results in the release of enormous quantities of energy. It is necessary to be able to calculate the amount of energy that will be produced. The logical manner in which to pursue this is to first investigate a typical fission reaction such as the one given below.

$$^{1}_{0}n + ^{235}_{92}U \rightarrow ^{140}_{55}Cs + ^{93}_{37}Rb + 3(^{1}_{0}n) \qquad (16.2.1)$$

It can be seen that there are two fission fragments, $^{93}_{37}Rb$ and $^{140}_{55}Cs$, and some neutrons $^{1}_{0}n$. Both fission products then decay by multiple β^- emissions as a result of the high neutron-to-proton ratio possessed by these nuclides.

Method 1:   In most cases, the resultant fission fragments have masses that vary widely. The most probable pair of fission fragments, for the thermal fission of the fuel $^{235}_{92}U$, have masses of about 95 and 140. Referring now to the binding energy per nucleon E_B / A curve (Fig 16.2), one can estimate the

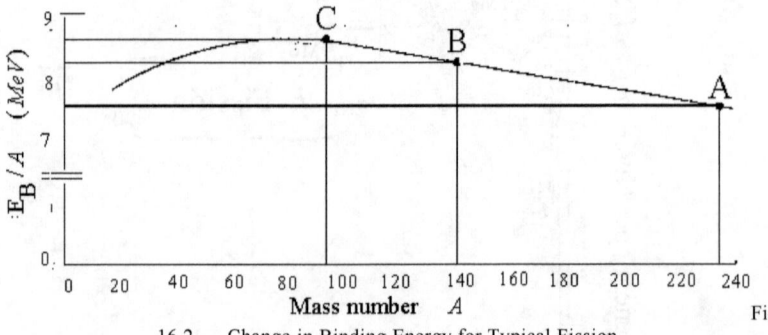

Fig 16.2 Change in Binding Energy for Typical Fission

amount of energy released in this "typical" fission by plotting this reaction on the curve and calculating the change in binding energy (ΔE_B) between the reactants on the left-hand side of the fission equation and the products on the right-hand side. Plotting the reactant and product nuclides on the curve shows that the total binding energy of the system after fission is greater than the total binding energy of the system before fission. When there is an increase in the total binding energy of a system, the system has become more stable by releasing an amount of energy equal to the increase in total binding energy of the system. Therefore, in the fission process, the energy liberated is equal to the increase in the total binding energy of the system.

TABLE 16.1 Binding energies calculated from E_B/A Curve

Nuclide	B.E per nucleon E_B/A	Mass number A	Binding energy $(E_B/A)A$
$^{93}_{37}Rb$	8.7 MeV	93	809 MeV
$^{140}_{55}Cs$	8.4 MeV	140	1176 MeV
$^{235}_{92}U$	7.6 MeV	235	1786 MeV

Fig 16.2 graphically depicts that the binding energy per nucleon E_B/A for the products C ($^{93}_{37}Rb$) and B ($^{140}_{55}Cs$) is greater than that for the reactant A ($^{235}_{92}U$). The total binding energy for a nucleus can be found by multiplying the binding energy per nucleon by the number of nucleons. The results are as listed in Table 16.1.

The energy released will be equivalent to the difference in binding energy (ΔE_B) between the reactants and the products.

$$\Delta E_B = E_B(\text{Products}) - E_B(\text{Reactants})$$
$$= E_B(Rb-93) + E_B(Cs-140) - E_B(U-235)$$
$$= (809\ MeV + 1176\ MeV) - 1786\ MeV \qquad (16.2.2)$$
$$= 199\ MeV$$

Method 2: The energy liberation during the fission process can also be explained from the standpoint of the conservation of mass-energy. During the fission process, there is a decrease in the mass of the system. There must, therefore, be energy liberated equal to the energy equivalent of the mass lost in the process. This method is more accurate than the previously illustrated method (*Method* 1) and is used when actually calculating the energy liberated during the fission process.

Again, referring to the "typical" fission reaction,

$$^1_0n + ^{235}_{92}U \rightarrow ^{140}_{55}Cs + ^{93}_{37}Rb + 3\,(^1_0n) \qquad (16.2.1)$$

The instantaneous energy, E_{inst}, is the energy released immediately after the fission process. It is equal to the energy equivalent of the mass lost in the fission process. It can be calculated as shown below:

Mass of the Reactants		Mass of the Products	
$^{235}_{92}U$	235.043924 u	$^{93}_{37}Rb$	92.91699 u
^1_0n	1.008665 u	$^{140}_{55}Cs$	139.90910 u
		$3\,(^1_0n)$	3.02599 u
	236.052589 u		235.85208 u

Mass difference = 236.052589 u - 235.85208 u = 0.200509 u $\qquad$ (16.2.3)

This mass difference can be converted to an energy equivalent

$$E_{inst} = (0.200509\text{ u})\left(\tfrac{931.5\ MeV}{u}\right) = 186.8\ MeV \qquad (16.2.4)$$

The total energy released per fission will vary from the fission to the next depending on what fission products are formed, but the average total energy released per fission of $^{235}_{92}U$ with a thermal neutron is about 200 *MeV*.

Worked out Example 16.1

A nucleus $^{238}_{92}U$ slits into two equal-size nuclei. Give the Binging energy per nucleon for $^{238}_{92}U$ is 7.6 *MeV*, and that for nuclei of half the mass of uranium, obtain the energy released in the process.

Solution:: Step # 1 $\qquad$ Given, E_B/A =7.6 *MeV*, for Uranium-238, E_B/A =8.6 *MeV* for 120 mass nuclei. A =238 nucleons.

Step # 2 $\qquad$ Energy of release by each nucleon in Uranium = (8.6 – 7.6) *MeV* = 1 *MeV*

Energy liberated in the spitting = (1 *MeV*)(238) = 238 *MeV*.

16.2.2 Energetic of Fission Fragments

Most of the energy that is released during fission appears as the KE of the fission fragments as listed in Table 16.2. Fowler and Rosen (1947) measured the energy distribution using ionization chamber. Leachman (1952) measured the velocities of fission fragments from slow neutron fission of the three nuclides by a 'time of flight' method.

16.2.2.1 The Kinetic energy (KE) of the Fission Products

As illustrated in the preceding example, the majority of the energy liberated in the fission process is released immediately after the fission occurs and appears as the KE of the fission fragments(Fig 16.3), KE of the fission $_0^1n$, and energy of instantaneous γ-rays. The remaining energy is released over a period of time after the fission occurs and appears as KE of the $\beta-$, neutrino (ν) and decay γ – rays. Table 16.2 shows the total energy contributes to all forms of KE of fission fragments, $_0^1n$, γ-rays, β-rays and the ν_e neutrinos, for induced fission of three fissile materials.

Table 16.2 Average Total energy released in Slow $_0^1n$ – Induced Fission.in

	$^{233}_{92}U$	$^{235}_{92}U$	$^{239}_{94}Pu$
1. E_{bust} KE of fission fragments	168.2 MeV	169.1 MeV	175.8 MeV
2. E_{bust} KE of prompt $_0^1n$	4.9 MeV	4.8 MeV	5.9 MeV
3. E_{bust} KE of prompt γ-rays	7.7 MeV	7.0 MeV	7.8 MeV
4. Decay Energy of β-particles	5.2 MeV	6.5 MeV	5.3 MeV
5. Decay Energy of $\bar{\nu}$-particles	6.9 MeV	8.8 MeV	7.1 MeV
6. Decay E of delayed γ-rays	5.0 MeV	6.3 MeV	5.2 MeV
TOTAL Energy of a fission event	197.9 MeV	202.5 MeV	207.1 MeV

Niels Bohr and John Archibald Wheeler (1939) worked out a theory of the fission with the help of the liquid drop model of nuclei. The fissionable nucleus is treated as a perfect spherical liquid drop. The pertinent terms in the C.Von Weizsacker mass formula, responsible for the shape of the nucleus, of radius R, are the surface energy and the Coulomb energy [Equation (16.2.5)]

$$E = \left[a_S \, A^{2/3} + a_C \, Z(Z-1) \, A^{-1/3} \right] \quad (16.2.5)$$

16.2.2.2 Reaction Energy or Q value in *Binary Fission* (Symmetric Fission)

Although the two fission fragments are strictly not of equal mass, for simplicity, let the fission process is symmetric (i.e., two fragments of equal mass). To make it clear,

$$M(A,Z) \xrightarrow{\text{Splits into}} M(\tfrac{A}{2}, \tfrac{Z}{2}) + M(\tfrac{A}{2}, \tfrac{Z}{2}) \quad (16.2.6)$$

Energy release Q_f in a single fission is

$$Q_f = M(A,Z) - 2\, M(\tfrac{A}{2}, \tfrac{Z}{2}) \quad (16.2.7)$$

Applying the semi-empirical mass formula (16.2.6)

$$Q_f = -(0.26)\, a_S\, A^{2/3} + (0.37)\, a_C\, Z(Z-1)\, A^{-1/3} \qquad (16.2.8)$$

In the case of $^{238}_{92}U$, this yields

$$Q_f = (-130 + 300) = 170\ MeV \qquad (16.2.9)$$

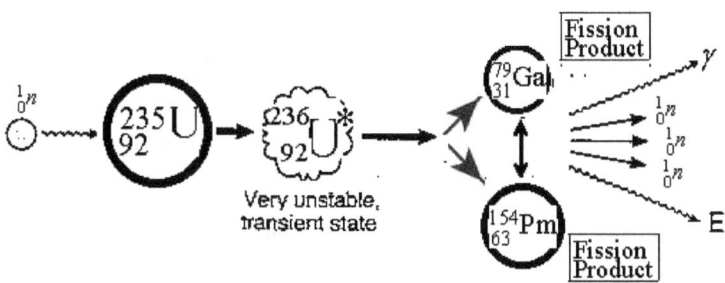

Fig 16.3 Fission Process

16.2.3.2 E_B/A stability plot

$$Q_f = 238\,(\text{Value of } E_B/A \text{ for } \tfrac{A}{2} \approx 120) - \text{Value of } E_B/A \text{ for } ^{238}_{92}U \text{ rgion} \qquad (16.2.10)$$

This yields $\quad Q_f = 238\,(8.5\ MeV - 7.6\ MeV) \approx 200\ MeV$

16.2.6 Difference in Fissionability of $^{235}_{92}U$ and $^{238}_{92}U$

(i) In the case of $^{238}_{92}U$,

$$E_{exc}\,(=4.8\ MeV) < E_A\,(=6.6\ MeV),\ \text{and}$$

(ii) for $^{235}_{92}U$

$$E_{exc}\,(=6.5\ MeV) > E_A\,(=6.2\ MeV)$$

The difference is illustrated in Fig 16.4

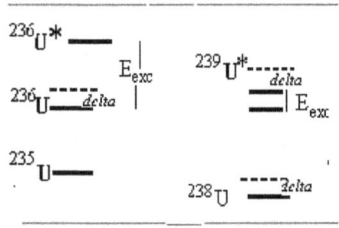

Fig 16.4. Difference in fissionability of $^{235}_{92}U$ and $^{238}_{92}U$

This is the primary difference between the two fissile nuclei. The dashed levels in Fig 16.5 denote the nuclear energies in the absence of pairing energy (*delta*).

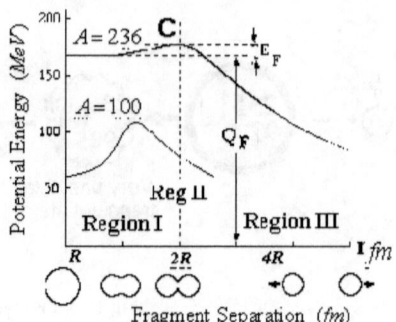

Fig 16.5 Fission mechanism

16.3   EXISTENCE OF FISSION BARRIER

This is best understood by reversing the fission process. The mutual PE of the two binary fission fragments as they are brought close together and coalesce to form the parent is shown by R.D. Evans as shown in Fig 16.6.

16.3.1 Excitation Energy E_{exc}

When a target nucleus, say $^{238}_{92}U$, captures a neutron to form the compound nucleus $^{239}_{92}U^*$, the excitation energy

$$E_{exc} = [M(^{239}_{92}U^*) - M(^{238}_{92}U)] c^2 \text{ (in } MeV) \tag{16.3.1}$$

16.3.2 Activation Energy E_A

The activation energy is the energy needed to overcome the fission barrier, in the PE curve. In the case of $^{236}_{92}U$, $E_A = 6.2\ MeV$. This means the energy required to excite the $^{236}_{92}U$ into fissionable state (activation energy) is exceeded (Exercise 16.1) by the energy one gets by adding a neutron to $^{235}_{92}U$. On the other hand, in the case of $^{238}_{92}U$.

$$E_{exc}(= 4.8\ MeV) < E_A\ (= 6.6\ MeV).$$

16.3.3 Bohr-Wheeler Criterion for Fission

By substituting the value of the constants in the equation (16.3.1), Niels Bohr and J.A. Wheeler found that

$$\left(Z^2/A\right) > 47.8, \quad Q_F \text{ decreases by distorting the nucleus.} \qquad (16.3.2)$$

The way one roughly determines the fissionability of an isotope is $\left(Z^2/A\right)$. When $\left(Z^2/A\right) \approx 45$, the electrostatic repulsion of the protons is equal to the binding energy, and the atom no longer has anything holding it together.

The *fissionability parameter* of a given isotope, find $\dfrac{\left(Z^2/A\right)}{45} \square\ ?$.

$$\dfrac{\left(Z^2/A\right)}{45} = 1 \Rightarrow \text{Total instability} \qquad (16.3.3)$$

$$\dfrac{\left(Z^2/A\right)}{45} < 1 \Rightarrow \text{Relative instability of isotope}$$

$$\dfrac{\left(Z^2/A\right)}{45} \square\ 0.8 \Rightarrow \text{for } {}^{238}_{92}U$$

16.3.4 Fission Mechanism (Theory of Fission)

Fig 16.7 depicts the supposed fission mechanism.

i) Before collision: Coulomb repulsion term leads to deformation of the nuclear shape, whereas the surface term overcomes Coulomb repulsion and keeps the nucleus in spherical shape.

ii) After collision: When the nucleus ${}^{235}_{92}U$ (liquid drop) absorbs a particle ${}^1_0 n$ or a γ-ray, it is suitably excited by forming ${}^{236}_{92}U^*$, and it may oscillate in a variety of ways: say spherical nucleus successfully becomes an oblate spheroid, a sphere, a prolate spheroid, a sphere, an oblate spheroid again, and so on (Fig. 16.6) as time advances. This is due to the Bohr-Wheeler criterion, viz., $\left(Z^2/A\right) > 47.8$, when the Q_f decreases by means of distortion.

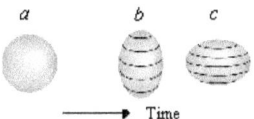

Fig 16.6 Fission Mechanism

iii) The restoring force of its surface tension always returns ${}^{236}_{92}U^*$ to spherical shape,

v) But the inertia of the moving nucleons (liquid molecules) causes the drop to overshoot sphericity and go to the opposite extreme distortion.

v) While the surface terms maintains the oscillation of the spherical shape, the nucleus is also subject to disruptive forces due to ES repulsion (Coulomb term) due to the mutual repulsion of its protons. If the degree of distortion is small, the excited nucleus $^{236}_{92}U^*$ emits γ-ray, to get spherical shape.

vi) (n, γ) reaction: When the excitation energy E_{exc} of the nucleus is larger the degree of deformation is sufficiently great (Fig 16.7). Even in this case there is some probability that it may return to the original shape, after de-excitation emitting γ-rays. This is (n, γ) reaction.

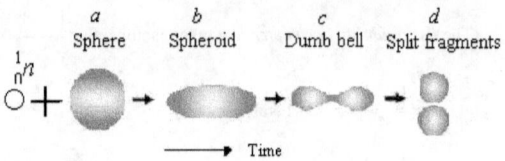

Fig 16.7 Binary fission, the mechanism

vii) (n, f) reaction: But if the E_{exc} is large enough for very high deformation and the nucleus splits into two fragments resulting in (n, f) reaction. These fission fragments are usually of nearly equal masses, $^{139}_{56}Ba$ and $^{140}_{57}La$, and this event takes place within $t \approx 10^{-22}$ s. $^{233}_{92}U$, $^{235}_{92}U$, and $^{239}_{94}Pu$ are the fissile nuclei falling under this category, viz., the (n, f) reaction.

viii) Sooner $t \approx 10^{-14}$ s, the two fragments having larger (N/Z) ratio eject 2 or 3 neutrons ('*prompt*' $^{1}_{0}n$) and subsequently β- rays, and attains stability. The fragments will also emit neutrons, β- and γ-rays in a period $t \approx 10^{-10}$ s. These are known as '*delayed*' $^{1}_{0}n$, β-, γ- rays.

$$Z_{Stab} = 0.65 \, A^{1/3}. \tag{16.3.6}$$

16.3.5 Fission Life-time

$$t \approx 10^{-22} \, s \tag{16.3.7}$$

13.4.1 SPONTANEOUS FISSION (SF)

When the fissile nucleus has $(Z^2/A) > 47.8$, $Q_F = (E_S + E_C)$ decreases by distortion, and **spontaneous fission** (SF) occurs, within the characteristic nuclear time $t \approx 10^{-22}$ s.

The occurrence of spontaneous fission of a heavy nucleus (nat U) was first observed by Petrzhak and Flerov (1940) and Emilio Gino Segre (1946), but reported in 1952. The substance under investigation is deposited in a thin layer on a platinum disc and placed inside an ionization chamber, connected to a linear amplifier. The ionization pulses produced by the fission fragments were counted. Thus $^{230}_{90}$Th and $^{241}_{96}$Am were studied. Hanna et al. (1951) first observed it in $^{242}_{98}$Cm. The even-numbered isotopes of plutonium (^{238}Pu, ^{240}Pu, and ^{242}Pu) undergo SF at a rate of 1100, 471 and 800 SF/g-s, respectively. Like (, n) neutrons, SF neutrons have a broad energy spectrum. SF neutrons are time-correlated (several neutrons are produced at the same time), with the average number of neutrons per fission being between 2.16 and 2.26. Uranium isotopes and odd-numbered plutonium isotopes undergo spontaneously fission at a much lower rate (0.0003 to 0.006 SF/g-s).

16.4.2 INDUCED FISSION

When a *fissile* nucleus has $\frac{Z^2}{A} < 47.8$, it resists fission as $E_S > E_C$. This resistance, however, has the character of a barrier only. This barrier is the nature of a 'hump' at C, indicated in the PE versus distortion plot of Fig 16.7.

Most *odd-A* nuclei can fission with thermal neutrons, whereas most fissile materials of even-even nuclei can fission only with fast neutrons. This is because of the *pairing term*, δ.

Fissions can be induced in ^{239}Pu, ^{235}U, and ^{238}U by neutron interrogation of the sample with an external neutron source. Like SF neutrons, they have a broad energy spectrum and are time-correlated.

16.4.3 Critical value for Spontaneous Fission

$$\left(Z^2/A\right) > 47.8 \tag{16.3.8}$$

for a *fissile* nucleus is termed the '*critical value*' for spontaneous fission.

16.4.3 n-capture and the Compe ting Processes

A n-capture leads to two *competing processes*:

a) Radiative capture (n, γ) occurs for about 15% only.

b) Fission, (n, f), viz., U $(n, 2n)$ XY and U $(n, 3n)$ XY This occurs for 85% events.

Thus adding (input) a few *MeV* energy ($E_A = \square$ *MeV*), can cause the energy release of $Q_F \approx 200\ MeV$!

Example, $^{235}_{92}$U + $^1_0 n$ (Slow) $\rightarrow [^{236}_{92}U^*] \rightarrow X + Y + \nu\ ^1_0 n$ (fast)

$$\tag{16.4.1}$$

Table :16.3   # ν of $^1_0 n$ released in $^{235}_{92}$U $(n, \nu n)$ XY

1. $^{235}_{92}$U + $^1_0 n$ (Slow) $\rightarrow [^{236}_{92}U^*] \rightarrow X + Y + \nu\ ^1_0 n$ (fast)	$\nu = 2.47$
2. $^{233}_{92}$U + $^1_0 n$ (Slow) $\rightarrow [^{236}_{92}U^*] \rightarrow X + Y + \nu\ ^1_0 n$ (fast)	$\nu = 2.51$
3. $^{239}_{94}$Pu + $^1_0 n$ (Slow) $\rightarrow [^{240}_{94}Pu^*] \rightarrow X + Y + \nu\ ^1_0 n$ (fast)	$\nu = 2.89$

The fissionable (*fissile*) materials as possible target nuclides are:

Uranium ore is
U_2O_3
Natural Uranium: "yellowcake"

Nat U composed of [$^{234}_{92}U$ (0.0057) + $^{235}_{92}U$ (0.7204) + $^{238}_{92}U$ (99.2739) (16.4.2)

In brackets are shown the % abundance of the isotope.

16.4.4 Thermal Neutron Cross section σ for Natural Uranium

Table 16.4 lists the values of the thermal neutron cross section σ for nat U fission and neutron capture and scattering

Table 16.4 Thermal ^1_0n cross section σ for Uranium

	$^{235}_{92}U$	$^{238}_{92}U$	Nat U
1. Fission	549 b	0 b	3.92 b
2. Capture	101 b	2.80 b	3.5 b
3. Scattering	8.2 b	8.2 b	8.2 b

16.4.5 Other Products of Fission (fall out or β^- - and γ- radio-activity)

There are more than 30 different modes of fission, in each of which a different pair of fragment nuclei is formed. Ba, La, Br, Mo, Rb, Sb, Te, Kr, I, Xe, Cs are some of them.

Fragment nuclei have their Z range from $30 \leq Z \leq 63$; $70 \leq A \leq 160$.

A fission chain is a series of product nuclei with the same mass number A. An example is

$$^{90}_{35}Br \xrightarrow[\tau_{1/2}=1.6\,s]{\beta^-} {}^{90}_{36}Kr \xrightarrow[23\,s]{\beta^-} {}^{90}_{37}Rb \xrightarrow[2.9\,m]{\beta^-} {}^{90}_{38}Sr \xrightarrow[28\,y]{\beta^-} {}^{90}_{39}Y \xrightarrow[64\,hr]{\beta^-} {}^{90}_{40}Zr \text{ (Stable)}$$

(16.4.3)

$$^{90}Kr \xrightarrow[33s]{\beta^-} {}^{90}Rb \xrightarrow[2,7min]{\beta^-} {}^{90}Sr \xrightarrow[28year]{\beta^-} {}^{90}Y \xrightarrow[64h]{\beta^-} {}^{90}Zr(stable)$$

(16.4.4)

$$^{143}Ba \xrightarrow[0,5min]{\beta^-} {}^{143}La \xrightarrow[12min]{\beta^-} {}^{143}Ce \xrightarrow[33h]{\beta^-} {}^{143}Pr \xrightarrow[13,7d]{\beta^-} {}^{143}Nd(stable)$$

(16.4.5)

The times under the arrows indicate the half-life of the corresponding β^- decay

60 such chains are known to occur. That is why there is a production of considerable amount of β^-- and γ-radio-activity in the phenomenon of fission. This is the source of most of the "fall out" from nuclear weapons.

16.4.6 Fission Yield, Y(A)

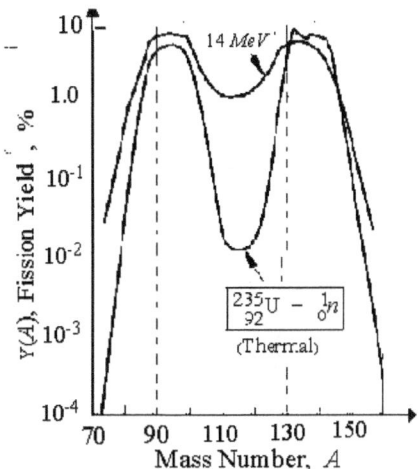

Fig 16.8 Mass number, A, versus % yield(A) plotted in the logarithmic scale

In addition to the neutrons, fission process usually results in two heavy fragmented nuclei, which carry most of the energy released in the form of KE. For reasons that are not clearly understood, one of the fragments from thermal neutron fission is actually heavier than the other.

The fission yield, Y(A) is defined as

$$Y(A) = \frac{\text{\# of nuclei with mass number } A \text{ formed per fission } N_A}{\text{Total \# of Fissions } N_F} \; 100\% \quad (16.4.6)$$

$^{235}_{92}$U ($\tau_{1/2}$ = 703.8 Myrs) Fission Yield vs. Mass Number is plotted in Fig 16.9

(Rev. Mod. Phys. 18, (1946) 539).

16.4.7 Spectrum of Fission Neutrons

The thermal neutron induced fission of $^{235}_{92}U$ is depicted schematically in Fig 16.9. The energy E distribution of ^1_0n is well represented mathematically by

$$N_n(E) = C\, Sinh(2E)^{1/2}\, e^{-E} \qquad (16.4.7)$$

$N_n(E)$ = The number of ^1_0n with a given energy E,

C = Empirical Constant.

The curve shows a maximum of $N_n(E)$ at 0.75 MeV.

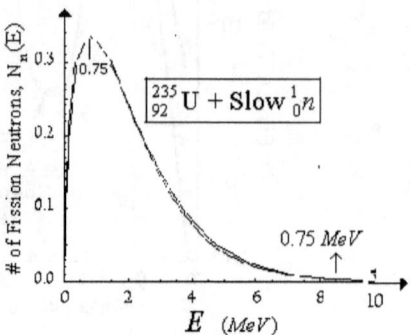

Fig 16.9   Spectrum of Fission neutrons

16.4.8 Prompt and Delayed Neutrons by Fission Fragments

A nucleus which has an excess of ^1_0n may decay by β^- - or by neutron emission. Neutron emission would most likely occur if, in the process of β^- - decay, the product nucleus is left in an excited state with energy in excess of the binding energy of a ^1_0n, in that nucleus. The emission is a '*prompt*' ^1_0n, if the emission is within $t \approx 10^{-14}$ s The neutron emitter may have the same $\tau_{1/2}$ for ^1_0n as the β^- - decay of the parent nuclide, i.e., several seconds. Such neutrons are known as the '*delayed*' ^1_0n. This is schematically shown in Fig 16.10.

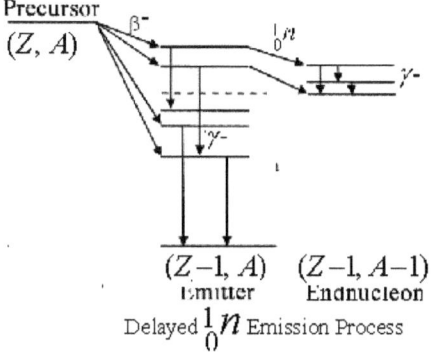

Fig 16.11: Schematic illustration of the delayed $_0^1n$ emission process.

On the right of Fig 16.10 the precursor nucleus (Z,A), in its ground state, β^- - decays to excited states of the possible $_0^1n$ emitting nucleus$(Z+1,A)$. The most excited levels of this nucleus may be above the $_0^1n$ binding energy, and thus, emit $_0^1n$, leaving a residual nucleus $(Z+1, A-1)$. But these delayed neutrons will be < 1% of the number of prompt neutrons. The energy of most of the prompt neutrons emitted will be $\square$ 0.8 - 2 MeV.

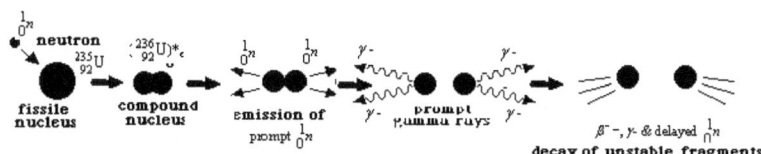

Fig 16.11 Fission of a $^{235}_{92}U$ nucleus

16.4.9 Transuranic Elements

Edwin M. McMillan and Philip H. Abelson discovered the first transuranic element neptunium,

$$^{238}_{92}U + {}_0^1n \rightarrow [{}^{239}_{92}U^*] \rightarrow {}^{239}_{92}U + \gamma$$
$$^{239}_{92}U \xrightarrow[23.5\,m]{\beta^-} {}^{239}_{93}Np + \beta^- \qquad (16.3.13)$$

$$^{239}_{92}U \xrightarrow[23.5\,m]{\beta^-} {}^{239}_{93}Np \xrightarrow[2.3\,d]{\beta^-} {}^{239}_{94}Pu + \beta^-$$
$$^{239}_{94}Pu \xrightarrow[24000\,y]{\alpha} {}^{235}_{92}U + {}^{4}_{2}He \qquad (16.3.14)$$

16.4.10 TERNARY FISSION

R. Present (1941) put the possibility for tri-partition fission. This is thought to take place only 4.3 events for every 10^6 binary fissions.

16.5 NUCLEAR REACTORS

Nuclear fission differs from other nuclear reactions in two ways:-

i) For each neutron $_0^1n$ absorbed by the *fissile* nucleus (fuel) $v > 1$ $_0^1n$ is emitted (Table 16.2).

ii) A large amount of energy is released ($\Box$ 200 *MeV*) in each fission and is available in the form of heat for power production.

16.5.1 Fission Chain Reaction

There are two aspects of fission that make it important from application point of view:

1. The fission reaction is **exothermic**; about 200 *MeV* of energy is released per *fission event*. This reaction energy is larger than by a factor of several 10^6 times typical chemical reactions. A typical fuel rod in a nuclear power plant contains about 3% fissile material; this means an energy density of about 10^6 times greater than fossil fuels like petrol (gasoline).
2. The incident particle $_0^1n$ in a (n, *f*)) reaction is contained among the products of each fission event, which is available for a subsequent fission event, and this self-sustaining process is continued time after time, then a *chain reaction* occurs.
In an *induced fission*, for each neutron $_0^1n$ absorbed by the fissile nucleus the # v of $_0^1n$ emitted $v > 1$ (Table 16.2).

This suggests a '*self-sustaining*' sequence of fissions, called *chain fission reaction* (Fig 16.12) is in principle possible. Whether the chain reaction remains steady, builds up, or dies down, depends on the competing processes, *viz.*,

a) Non-fission capture of the $_0^1n$ by the fissionable material (fuel),

b) $_0^1n$-capture by other materials in the system; (n, γ) reactions.

c) There is leakage of neutrons through the surface of the system.

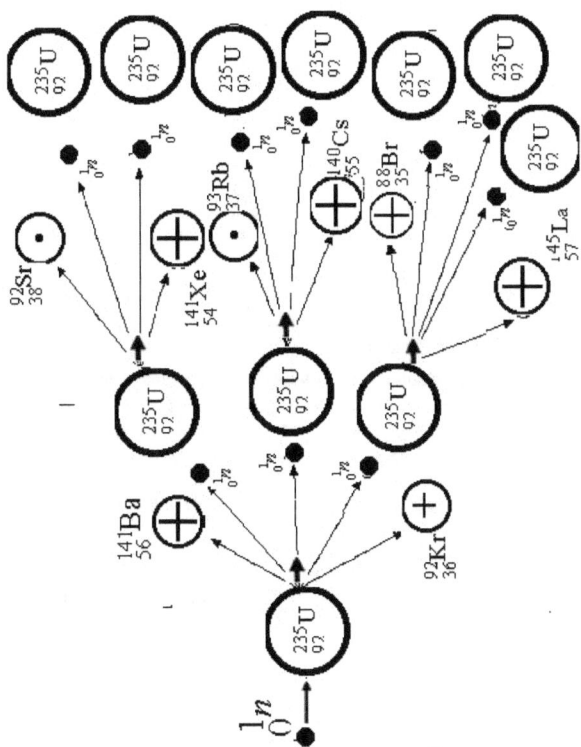

Fig 16.12. Neutron induced fission chain reaction

Suppose a small amount of the fuel is bombarded with $N = 100\,_0^1n$ of which 40 cause fission, and 60 $_0^1n$ are absorbed without fission or escape out of the target fuel, then $40 \times \nu = 100$ $_0^1n$ are produced after the fission. In this case no $_0^1n$ are lost and the process of fissioning continues without external supply of $_0^1n$. Then the assembly is said to be

a) *Critical*, when $N = 100$.
b) *Sub-critical*, when $N < 100$.
c) *Super-critical*. For $N > 100$, and an explosion takes place.

16.5.1.2 Critical Mass

To maintain a fission chain reaction a certain minimum amount of fissile material must be present so that too many $_0^1n$ do not escape from the fuel. This amount of material is called the *critical mass*.

16.5.2 The Nuclear Reactor - Atomic Pile

A *controlled self-sustaining fission reaction* (*i.e.*, chain reaction) was first demonstrated by E. Fermi and co-workers in Dec. 2, 1942, in Chicago. A system in which the fissionable and non-fissionable materials are arranged so that the fission-chain reaction can proceed in a controlled manner is called a "nuclear reactor". Fermi et al used

$$\text{Nat U is composed of } [\,^{234}_{92}U\ (0.0057) + \,^{235}_{92}U\ (0.7204) + \,^{238}_{92}U\ (99.2739)],$$

i.e., 0.72% abundant nat U. A pile of graphite is uniformly spaced in this. This is the *uranium-graphite-pile*. Because of this type of construction the early nuclear reactors are known as "atomic piles". The cross section for thermal neutrons ($_0^1n$ of $\sqcup$ 0.025 eV) is very large ($\square$ 550 b). However, the fission neutrons have $\square$ 1 MeV energy, so have to be slowed down quickly to the thermal energies. Energy is most effectively lost by scattering from a body called moderator, which is composed of atoms with mass close to that of $_0^1n$.

A nuclear reactor is a source of the products of the fission process, viz., i) Energy, ii) Neutrons, and iii) Radio-isotopes.

16.5.3 A Natural Fission Reactor

It is quite essential, on an average, for at least one of the approximately 2.5 $_0^1n$ produced in fission event to attain a self-sustaining or chain fission reaction to cause another fission. The design of the reactor, therefore, must reduce non-fission processes that tend to absorb the fission product $_0^1n$ so as to satisfy the afore-mentioned condition. These competing processes are;

1. Non-fission capture of $_0^1n$ by the fuel (determined by the size and shape of the fuel, and the average energy of $_0^1n$ or isotopic composition, *i.e.*, enrichment, of the fuel),
2. Capture of $_0^1n$ by other materials in the reactor (the amount and type of construction materials used in the reactor) and
3. Leakage or escape of $_0^1n$ from the system (using reflector for the $_0^1n$).

Some two billion years ago, in a uranium deposit now being mined in Gabon, West Africa, a natural fission reactor went into operation and ran for perhaps several hundred thousand years before shutting itself down.

The first artificial nuclear reactor was built by Enrico Fermi *et al.* (Chicago on December 2, 1942). This reactor, which produced several kW of power, consisted of a pile of graphite blocks weighing 385 tons stacked in layers around a cubical array of 40 tons of uranium metal and uranium oxide. Spontaneous fission of $^{238}_{92}U$ or $^{235}_{92}U$ in this reactor produced a very small number of neutrons $_0^1n$. But enough uranium was present so that one of these neutrons induced the fission of a $^{235}_{92}U$ nucleus, thereby releasing an average of 2.5 $_0^1n$, which catalyzed the fission of additional $^{235}_{92}$U nuclei in a *chain reaction*, as shown in Fig 16.14. The amount of fissionable material necessary for the chain reaction to sustain itself is called the **critical mass**.

The Fermi reactor served as a prototype for larger reactors constructed in 1943 (at Oak Ridge, Tennessee, and Hanford, Washington), to produce $^{239}_{94}$Pu for one of the atomic bombs dropped on Japan at the end of World War II. As was seen, some of the neutrons released in the chain reaction are absorbed by $^{238}_{92}$U to form $^{239}_{92}$U, which undergoes decay by the successive loss of two ◆/ i>--particles to form $^{239}_{94}$Pu. $^{238}_{92}$U is an example of a *fertile* nuclide. It doesn't undergo fission with thermal $^{1}_{0}n$, but it can be converted to $^{239}_{94}$Pu, which does undergo *thermal-neutron-induced fission*.

Fission reactors can be designed to handle naturally abundant $^{235}_{92}$U, as well as fuels described as slightly enriched (2-5% $^{235}_{92}$U), highly enriched (20-30% $^{235}_{92}$U), or fully enriched (>90% $^{235}_{92}$U). Heat generated in the reactor core is transferred to a cooling agent in a closed system. The cooling agent is then passed through a series of heat exchangers in which water is heated to steam. The steam produced in these exchangers then drives a turbine that generates electrical power. There are two ways of specifying the power of such a plant: the thermal energy produced by the reactor or the electrical energy generated by the turbines. The electrical capacity of the plant is usually about one-third of the thermal power.

It takes 10^{11} fissions per second to produce $1W$ of electrical power. As a result, about 1g of fuel is consumed per day per MW of electrical energy produced. This means that 1g of waste products is produced per MW per day, which includes 0.5 g of $^{239}_{94}$Pu. These waste products must be either reprocessed to generate more fuel or stored for the tens of thousands of years it takes for the level of radiation to reach a safe limit.

Worked out Example 16.1

Determine i) the average fission cross section, ii) the average absorption cross section of $^{1}_{0}n$, and iii) the number of $^{1}_{0}n$ per fission that can be lost if a self-sustained chain reaction to occur in reactor using natU. Given, $\sigma_f(^{235}U) = 550\ b$, $\sigma_f(^{238}U) = 0\ b$, $\sigma_\gamma(^{235}U) = 101\ b$, $\sigma_\gamma(^{238}U) = 2.8\ b$.

Solution: | STEP # 1 |

Nat U is composed of [$^{234}_{92}$U (0.0057) + $^{235}_{92}$U (0.7204) + $^{238}_{92}$U (99.2739)]

| STEP # 2 | i) 3.92 b,

| STEP # 3 | ii) (0.71 + 2.8) 3.51 b,

| STEP # 4 | iii) No more than 0.5 $^{1}_{0}n$ per fission that can be lost.

16.5.4 Moderators

The fissionable nuclei in a reactor are dispersed in a matrix of a moderator. The fast $_0^1 n$ collides elastically with the nucleus of the moderator element, and the energy transfer will be largest when the two bodies of the collision are of equal mass.

<u>Case</u> 1. Water (H_2O): Hydrogen nuclei have masses almost equal with the mass of neutron. But there is discouraging factor;

$$_1^1 H + {}_0^1 n \rightarrow {}_1^2 H + \gamma \qquad (16.5.1)$$

Thus n-capture occurs!

<u>Case</u> 2. Heavy Water (D_2O): Deuterons are less likely to interact with neutrons. Further the mass deuterium atom is only two times that of neutron. Thus heavy water makes a suitable moderator material. But its availability is difficult.

<u>Case</u> 3. Graphite (C): It is composed of $_6^{12}C$ atoms, and is more readily available. $_6^{12}C$ has only very small n-capture cross section.

The Ex 16.1 reveals that the average absorption cross section of 3.51 b, is nearly equal to the average neutron fission cross section of 3.92 b, in nat U. This means that only half the slow neutrons are captured in a block of nat U inducing fissions. It is known that $\nu = 2.5$ per fission, and so no more than 0.5 neutron / fission of $_{92}^{235}U$ can be lost if a self-sustaining reaction is to occur.

16.5.5.1 The Four factor Reactor Formula

It is required to define the parameters

$$\nu, \varepsilon, p, f, \ell_f, \ell_t, \sigma_f, \sigma_{ta}, k_e, k_\infty, \eta. \qquad (16.5.2)$$

ν = # of fission $_0^1 n$ / fission,

ε = Fast $_0^1 n$ fission factor, = 1.1 for thermal reactors, and is 1 for fast neutron reactors,

p = Resonance escape probability, *i.e.*, the probability that the $_0^1 n$ slows down past the resonance region without being captured.

ℓ_f = Fast leakage probability, due to finite dimensions of the assembly, of $_0^1 n$ that will leak out during process p.

$(1 - \ell_f)$ = Chance that $_0^1 n$ will become thermalized, without leaking out.

Thus the number of $_0^1 n$ reaching the thermal region of energy

$$= \nu \varepsilon p (1 - \ell_f) \qquad (16.5.3)$$

ℓ_t = Thermal leakage probability of neutrons once they are slowed down to thermal level,

f = Thermal utilization factor, of $_0^1n$ in fuel and other materials.

The number of $_0^1n$ eventually absorbed in the fuel

$$= \nu \, \varepsilon \, (1 - \ell_f) \, p \, (1 - \ell_t) f \qquad (16.5.4)$$

Some of the $_0^1n$ will produce the reaction:

$$^{238}_{92}U + {}_0^1n \rightarrow [{}^{239}_{92}U^*] \rightarrow {}^{239}_{92}U \xrightarrow{\beta^-} {}^{239}_{93}Np \xrightarrow{\beta^-} {}^{239}_{94}Pu \qquad (16.5.5)$$

The fraction that causes fission is $\dfrac{\sigma_f(U)}{\sigma_{ta}(U)}$

σ_f = Uranium fission cross section for $_0^1n$

σ_{ta} = Total absorption cross section for $_0^1n$ in uranium.

$$\frac{\sigma_f(U\text{-}238)}{\sigma_{ta}(U\text{-}238)} = \frac{\sigma_f = 0.5}{\sigma_{ta} = 2 + 0.5}, \qquad (16.5.6)$$

Once a fission chain reaction has started, the average number of fission that results from the original fission, called the effective multiplication factor, k_e,

$$k_e = \frac{\text{Rate opf production of neutrons, P}}{\text{Rate of absorption, } A_L + \text{Rate of leakage, L}} = \frac{\#\, _0^1n \text{ in the } (n+1)\text{th generation}}{\#\, _0^1n \text{ in the nth generation}} \qquad (16.5.7)$$

$$k_e = \nu \, \varepsilon \, p f \, (1 - \ell_f)(1 - \ell_t) \, \frac{\sigma_f(U)}{\sigma_{ta}(U)} \qquad (16.5.8)$$

Defining, $\eta = f \, \dfrac{\sigma_f(U)}{\sigma_{ta}(U)}$ \hfill (16.5.9)

$$k_e = \eta \, \varepsilon \, p f \, (1 - \ell_f)(1 - \ell_t) \qquad (16.5.10)$$

When $\ell_f = 0$, $\ell_t = 0$, so that $k_e = k_\infty$,

$$k_\infty = \eta \, \varepsilon \, p f \qquad (16.5.11)$$

This is the well-known *four-factor formula in reactor physics*.

The fission reaction is indicated in Table 16.4.

Table 16.4 Fission chain reaction

1. Critical (*Steady*)	$k_e = 1$
2. Super Critical (*Divergent*)	$k_e > 1$
3. Sub Critical (*Convergent*)	$k_e < 1$

The amount of $^{235}_{92}U$ in the mass (the level of enrichment) and the shape of the mass control the criticality of the sample. You can imagine that if the shape of the mass is a very thin sheet, most of the free neutrons will fly off into space rather than hitting other $^{235}_{92}U$ atoms. A **sphere** is the optimal shape. The amount of $^{235}_{92}U$ that must be collected together in a sphere to get a critical reaction is about 0.9 *kg*. This amount is therefore referred to as the **critical mass**. For $^{239}_{94}Pu$, the critical mass is about 283 *g*.

16.5.5.2   An Alternate Reactor Formula

If F = Rate at which the fission process occurs,

$$P = \nu F \qquad (15.5.12)$$

$$k_e = \frac{\nu F}{A_L + L} = \frac{\nu F}{A_L} \cdot \frac{1}{1 + L/A_L} \qquad (16.5.13)$$

$\frac{F}{A_L}$ depends on the quantity of the fissionable and non-fissile material, and on their cross sections for fission and neutron capture.

$\frac{L}{A_L}$ depends on the ability of the reactor to contain.

Table 16.5   Size of nuclear reactor, L/A_L

	Reactor size	L/A_L	$k_e = \nu F/A_L$
1. Critical (*Steady*)	Critical size	$= 0$	$k_e = 1$
2. Super Critical (*Divergent*)		$\rightarrow$ small	$k_e > 1$
3. Sub Critical (*Convergent*)	No size	$\rightarrow$ Large	$k_e < 1$
4.	Decreases	$\rightarrow \infty$	$k_e \rightarrow 0$

16.5.5.3 Any amount of $^{238}_{92}U$ is sub-critical! Why?

$k_\infty = 0.5$ for a body of $^{238}_{92}U$. Therefore, any fission process that may be initiated will quickly die out, *i.e.*, sub-critical.

Worked out Example 16.2

Find percentage of number of fissions for $^{238}_{92}U$ to $^{235}_{92}U$ in natural Uranium as fuel.

Solution: STEP #1 $k_\infty = 2.5 \dfrac{N_{f235}\, \sigma_{f235} + N_{f238}\, \sigma_{f238}}{N_{f235}\,(\sigma_{f235} + \sigma_{f235}) + N_{f238}\,(\sigma_{f238} + \sigma_{f238})}$

$= 2.5 \dfrac{N_{f235}\,(1..3) + N_{f238}\,(1.5)}{N_{f235}\,(0 + 1.3) + N_{f238}\,(2.0 + 0.5)} \geq 40\%.$

STEP #2   Since, $k_\infty = 1$, with $N_{f238} = 0$, means $\dfrac{N_{f238}}{N_{f235}} = 1.5$).

16.5.6 Byproduct of Fission

The development of transuranic elements is the interesting byproduct of nuclear fission. In a nuclear reactor,

$$^{238}_{92}U + ^{1}_{0}n \rightarrow [^{239}_{92}U]^* \rightarrow ^{239}_{93}Np^* + \beta^- + \nu$$
$$^{239}_{93}Np^* \rightarrow ^{239}_{94}Pu\,(\text{Fissionable}) + \beta^- + \nu \qquad (16.5.14)$$

Worked out Example 16.3

How much coal is required to run a 100-W light bulb 24 hours a day for a year? Take the thermal energy content of coal is 6,150 kWh/ton.

Solution: STEP #1   0.1 kW x 8,760 hours or **876 kWh**. limited by the laws of thermodynamics, only about 40 % of the thermal energy in coal is converted to electricity.

STEP #2   So the electricity generated per ton of coal is 0.4 x 6,150 kWh or **2,460 kWh/ton**.

STEP #3   The tons of coal burned for our light bulb = 876 kWh / 2,460 kWh/ton = 0.357 tons = (0.357 tons) (2,000 pounds/ton) = **714 pounds (325 kg)** of coal.

Worked out Example 16.4

The Semi-empirical mass formula for the nuclear binding energy with coefficients of the five terms whose values are approximately (in MeV): volume, 15.5; surface, 16.8; Coulomb, 0.72; asymmetry, 23; pairing, 34. Show that the difference between the total binding energy of the isotope $^{235}_{92}U$ and the compound nucleus formed upon slow neutron absorption is 6.7 MeV while the corresponding difference for $^{238}_{92}U$ is 5.2 MeV. What relevance does this have to the fission process in natural uranium?

Solution: STEP #1

Difference	Volume	Surface	Coulomb	Assymetry	Pairing	Total
B(236) - B(235)	15.5	-1.81	1.38	-8.96	0.56	6.7 MeV
B(239) - B(238)	15.5	-1.81	1.36	-9.31	-0.56	5.2 MeV

STEP #2 Natural U = (99.3% $^{238}_{92}$U + 0.7% $^{235}_{92}$U). The latter is fissile, but the threshold for neutron induced fission of $^{238}_{92}$U = 1 *MeV* - a difference which arises, in the semi-empirical formula, due to the Pairing Term. This difference means that a neutron produced by fission in nat-U is much more likely to suffer radiative capture in $^{238}_{92}$U than it is to cause fission in $^{235}_{92}$U.

STEP #3 Consequently a chain reaction is not sustained..

16.6 NUCLEAR POWER

16.6.1 Nuclear Power Plant

Both a nuclear power plant and conventional (fossil fuel) power plant produces electricity in almost exactly the same way.

Conventionally a power plant burns fuel to create heat. The fuel used is either coal or oil. The heat is used to raise the temperature of water, thus causing it to boil. The high temperature and intense pressure results in producing steam from the boiling of the water enables turning a turbine, which then generates electricity.

In the same way, a nuclear power plant works except that the heat used to boil the water is produced by a nuclear fission reaction using $^{235}_{92}$U as fuel, not the combustion of fossil fuels. A nuclear power plant consumes much less fuel than a comparable fossil fuel plant. A rough estimate is that it takes 17,000 *kg* of coal to produce the same amount of electricity as 1 *kg* of nuclear Uranium fuel

16.6.2 Enrichment of $^{235}_{92}$U (Uranium Preparation): Why?

There is a necessity to enrich $^{235}_{92}$U for a large amount of nat U to be critical. So far discussion was mainly on the nuclear fission with $^{235}_{92}$U ($\tau_{1/2}$=703.8 *Myrs*). Actually, this will not be the only isotope of uranium present in a nuclear reactor. In naturally occurring uranium deposits, less than 1% of the uranium is $^{235}_{92}$U. The majority of the uranium is $^{238}_{92}$U ($\tau_{1/2}$= 4.468 *Byrs*). $^{238}_{92}$U is not a fissile isotope of uranium. When a free $^{1}_{0}n$ impinges on $^{238}_{92}$U, it absorbs the $^{1}_{0}n$ and does not undergo fission. Thus by absorbing free neutrons, $^{238}_{92}$U can prevent a nuclear chain reaction from occurring. This is unwanted because if a chain reaction doesn't occur, the nuclear reactions can't sustain themselves, the reactor shuts down, and no electrical power generated. In order for a chain reaction to occur, the pure uranium ore must be refined to raise the concentration of $^{235}_{92}$U. This is called *enrichment* and is primarily accomplished through a technique called *gaseous diffusion*. In this process, the uranium ore is combined with fluorine to create a chemical compound called uranium hexafluoride (UF_6). The UF_6 is heated to vaporize. The heated gas is then pushed through a series of filters. Because part of the UF_6 contains $^{238}_{92}$U and some contains $^{235}_{92}$U, there is a slight difference in the weights of the individual molecules. The molecules of UF_6 containing $^{235}_{92}$U are slightly lighter and thus pass more easily through the filters. This creates a quantity of UF_6 with a higher proportion of $^{235}_{92}U$. This is collected, the uranium is stripped from it resulting in an enriched supply of fuel. Usually, nuclear power plants use uranium fuel that is about 4% $^{235}_{92}U$.

In nuclear chemistry enrichment of fissionable materials is usually done in order to yield higher products in reactions.

Example:

Enrichment is the fraction distillation of uranium. Uranium is mixed with fluorine to form UF_6, a gas. At low pressures the compound is allowed to go through a diffusion barrier permitting the lighter $^{235}_{92}U$ to pass through faster. This is repeated hundreds of times to finally produce a 3% mixture of $^{235}_{92}U$.

16.6.3 FEATURES OF A NUCLEAR REACTOR

A large number of parameters which can vary to achieve the optimum result, are used to design any Reactor. Six of the major features describing the type of a reactor are:

1. <u>Type of fuel</u>

   a) Natural Uranium (0.73% $^{235}_{92}U$)

   b) Enriched U (greater than 0.73% $^{235}_{92}U$)

   c) Plutonium ($^{239}_{94}Pu$)

   d) Thorium ($^{233}_{90}Th$)

2. <u>Average neutron energy</u>

   a) Fast $^{1}_{0}n$

   b) Intermediate $^{1}_{0}n$ energy

   c) Thermal $^{1}_{0}n$.

3. <u>The Moderator</u>

   a) Graphite (Carbon)

   b) Heavy water (Deuterium D_2O)

   c) Water

   d) Beryllium and BeO

   e) Organic materials.

4. <u>The Fuel moderator assembly</u>

a) Heterogeneous (In moderator, fuel in lumps at regular intervals)

b) Homogeneous (Fuel and moderator as a homogeneous mixture)

5, <u>The Coolant</u>

    a) Gas

    b) Water, D_2O, organic liquid, etc.

    c) Liquid metal.

6. <u>The purpose</u>

    a) Research

    b) Power

    c) Production of fissile materials (based on the production exceeds burn up or not).

<u>Examples</u> of featured reactors:

1. Research Reactor: Enriched U, heterogeneous, thermal, heavy water moderated, and cooled. This was the research reactor at MIT, Boston (USA).

2. Nautilus submarine: Enriched-U, heterogeneous, thermal, pressurized water, power reactor.

16.6.4 Pressurized Water Reactor (PWR)

PWR is the most common of reactor types in use, both for commercial and military applications.

Parts of a Nuclear Reactor - Pressurized Water Reactor (PWR)

A typical nuclear reactor (Fig 16.13) has a few main parts.

i) The "core" where the nuclear reactions take place are the fuel rods and assemblies, the control rods, the moderator, and the coolant.

ii) Outside the core are the turbines, the heat exchanger, and part of the cooling system.

1. **Fuel**: Enriched uranium (enrichment 3.2% $^{235}_{92}U$) as uranium dioxide in zircaloy cans is used as fuel.

2. **Core layout**: The fuel assemblies are collections of fuel rods. These rods are each about 3.5 *m* long. They are each about a centimeter in diameter. These are grouped into large bundles of a couple hundred rods called fuel assemblies, which are then placed in the reactor core. Inside each fuel rod are hundreds of pellets of uranium fuel stacked end to end. Fuel pins, arranged in clusters, are placed inside a pressure vessel containing coolant & moderator

3. **Control rods**: Also in the core are control rods. These rods have pellets inside that are made of very efficient neutron capturers. An example of such a material is cadmium. These control rods are connected to machines that can raise or lower them in the core. When they are fully lowered into the core, fission cannot occur because they absorb free neutrons. However, when they are pulled out of the reactor, fission can start again anytime a stray neutron strikes a $^{235}_{92}U$ atom, thus releasing more neutrons, and starting a chain reaction.

4. **Moderator:** Ordinary water is used as coolant and moderator, because of the neutron absorption properties of H_2O (Section 16.5.4). The moderator serves to slow down the high speed neutrons "flying" all around the reactor core. If a neutron is moving too fast, and thus is at a high-energy state, it passes right through the $^{235}_{92}U$ nucleus. It must be slowed down to be captured by the nucleus and to induce fission.

5. **Coolant:** The job of the coolant is to absorb the heat from the reaction. The most common coolant used in nuclear power plants today is water. In actuality, in many reactor designs the coolant and the moderator are one and the same.

6. **Turbine:** The coolant water is heated by the nuclear reactions going on inside the core. However, this heated water does not boil because it is kept at an extremely intense pressure, thus raising its boiling point above the normal 100° C.

7. **Steam Generator:** After the hot water has passed through the turbine (steam at 317°C at 2235 psia and steam efficiency 32%), some of its energy is changed into electricity. However, the water is still very hot. It must be cooled somehow. Many nuclear power plants used steam towers to cool this water with air. These are generally the buildings that people associate with nuclear power plants. At reactors that do not have towers, the clean water is purified and dumped into the nearest body of water, and cool water is pumped in to replace it

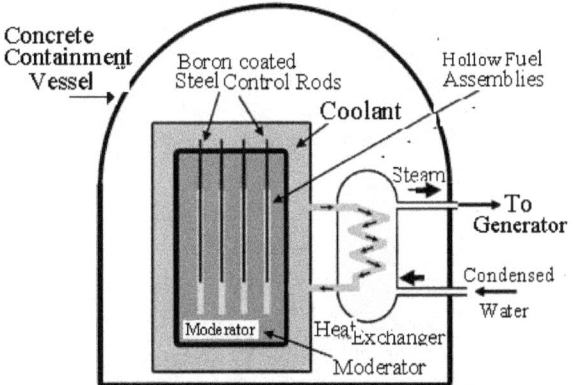

Fig 16.14   Pressurized Water Reactor (PWR)

16.7.3 Boiling Water Reactor (BWR)

Although the most common type of reactor is the Pressurized Water Reactor (PWR), many other types of reactors are also used. In the PWR, as described earlier, there are two main water cycles. One is the water inside the core that is highly radioactive. This water's heat is transferred to other, non-radioactive water inside the second loop. This water is then used to turn a turbine.

The second most popular reactor type is the Boiling Water Reactor (BRW). This type of reactor differs from the PWR in that there is only one water cycle. Radioactive water is used to turn the turbine. The major disadvantage of this is that the radioactive nuclides in the water that cause its radioactivity can be transferred to the turbine, thus causing it to become radioactive too. This produces more hazardous material that needs to be disposed of when a reactor is dismantled. However, the BWR also has a few advantages. Its core can be kept at a lower pressure, for example.

Fuel: Enriched uranium (enrichment 2.6% $^{235}_{92}U$) as uranium dioxide in zircaloy cans is used as fuel Moderator: Ordinary water is used as coolant and moderator, because of the neutron absorption properties of H_2O (Section 16.5.4). Core layout: The fuel assemblies are collections of fuel rods. These are grouped into large bundles of a couple hundred rods called fuel assemblies, which are then placed in the reactor core. Inside each fuel rod are hundreds of pellets of uranium fuel stacked end to end. Fuel pins, arranged in clusters, are placed inside a pressure vessel containing coolant & moderator.

Coolant outlet is at 286°C at 1050 psia. Schematically the BWR looks like the PWR of Fig 16.14, except that the steam generator is within the pressure vessel itself.

16.7.4 Heavy Water Reactor (HWR or PHW), like CANDU

Another type of reactor is the Heavy Water Reactor (HWR). A HWR uses heavy water as a moderator instead of normal water. Heavy water is water with deuterium, which is an isotope of hydrogen with 1 neutron. Deuterium is heavier than normal hydrogen, which has no neutrons. HWR's come in two types, pressurized and boiling, just like normal "light water" reactors. The advantage of a HWR is that un-enriched uranium fuel can be used. This is because the heavy water is a much more efficient moderator than light water. Thus, more stray neutrons can be slowed down enough to cause fission in $^{235}_{92}U$. This more efficient moderator makes up for the greater abundance of the neutron-capturing $^{238}_{92}U$. BWR is perhaps the simplest in concept of all types of reactors.

Fuel in CANDU: Natural uranium (enrichment 0.7% $^{235}_{92}U$) as uranium dioxide in zircaloy cans is used as fuel. Moderator: Heavy water D_2O is used as coolant and moderator, because of the neutron absorption properties of D_2O (Section 16.4.2.1). Core layout: The fuel assemblies are collections of fuel rods. These are grouped into large bundles of a couple hundred rods called fuel assemblies, which are then placed in the reactor core. Inside each fuel rod are hundreds of pellets of uranium fuel stacked end to end. Fuel pins, arranged in clusters, are placed inside a pressure vessel containing coolant & moderator

Coolant out let: It is at 305°C at 1285 psia. Schematically the BWR looks like the PWR of Fig 16.14, except that the steam generator is within the pressure vessel itself. The first of this type was in Ontario in 1962. (For more details, see, Hugh C. McIntyre, 'Natural Uranium HWRs', *Sci. Amer.*, Oct 1975)

In India RAPP2 is PHW type, 203 *MWe*, went critical in 1976.

SGHWR is a steam generating HWR.

Magnox gas-cooled reactors use graphite moderator.

Advance gas-cooled reactor (AGR) used graphite as moderator.

HTGR is high temperature gas-cooled reactor, used with startup Th / $^{235}_{92}$U (93% enriched); recycle Th / $^{235}_{92}$U (93% enriched / $^{233}_{92}$U (recycle)) and graphite moderator.

RBMK Reactor:

RBMK (Reactor Bolshoy Moshchnosty Kanalny) is *high-power channel reactor*, and pressurized water-cooled reactor with individual fuel channels and uses graphite as moderator. It is also known as the Light Water Graphite Reactor (LWGR). Most other power reactor designs differ from it as it is designed principally for plutonium production. It is used in Russia for both plutonium and power production. Chernobyl nuclear Power Plant used this design.

16.6.5 Breeder Reactor

Fission reactors are dependent on the supply of uranium. This is expensive and being depleted at a rate in approximately 50 years.

So it is required to design a reactor in which the ratio

$$\frac{^{239}_{94}Pu \ OR \ ^{233}_{92}U(\text{produced})}{^{235}_{92}U \ (\text{consumed})} > 1,$$

the reactor would generate more fuel than it consumed. Such reactors are known as breeder reactors.

The key to an efficient breeder reactor is a fuel that gives the largest possible number of ^1_0n released per ^1_0n absorbed. The breeder reactors being use a mixture of PuO_2 & UO_2 as the fuel and fast neutrons to activate fission. Fast neutrons carry energy of at least several keV and compared to few eV for thermal neutrons. $^{239}_{94}Pu$ in the fuel assembly absorbs one of these fast neutrons and undergoes fission with the release of 3 ^1_0n. $^{238}_{92}U$ in the fuel then captures one of these three ^1_0n to produce additional $^{239}_{94}Pu$.

This is the advantage of breeder reactors — they provide a limitless supply of fuel for nuclear reactors. There are significant disadvantages, however. Breeder reactors are more expensive to build. They are also useless without a subsidiary industry to collect the fuel, process it, and ship the $^{239}_{94}Pu$ to new reactors.

The disadvantages of $^{239}_{94}Pu$ is so dangerous as a carcinogen. The nuclear industry places a limit on exposure to Pu material that assumes workers inhale no more than 0.2 μg of plutonium over their lifetimes. Equally there is great concern that the $^{239}_{94}Pu$ produced by these reactors might be stolen and assembled to prepare bombs by terrorist organizations.

If nuclear energy is to play a dominant role in the generation of electrical energy in the 21st century, breeder reactors eventually may be essential.

The "nuclear pile" constructed by Fermi (1972) in Chicago, USA, was the first artificial nuclear reactor. In 1972, a group of French scientists discovered that uranium ore from a deposit in the Okla mine in Gabon, West Africa, contained 0.4% $^{235}_{92}U$, instead of the 0.72% abundance found in all other sources of this ore. Analysis of the trace elements in the ore suggested that the quantity of $^{235}_{92}U$ in this ore was unusually small because "natural fission reactors" operated in this deposit for a period of $6x\ 10^5$ to $8x\ 10^5$ yrs about 2 billion yrs ago.

16.6.6 Breeding Cycles

16.6.6.1 Uranium Breeding Cycle

Although the $^{238}_{92}U$ present in enriched uranium is not fissionable it can still be transmutated into fissionable $^{239}_{94}Pu$. Some, but not all $^{238}_{92}U$ reacts as follows:

$$[\text{Non-fissionable from Nat}-U] \atop 99.3\% \text{ abundant, FERTILE} \Bigg\} \; ^{238}_{92}U + ^1_0n \; \rightarrow \; [^{239}_{92}U]^* \; \xrightarrow[24\,m]{\beta^-} \; ^{239}_{93}Np^*$$

$$^{239}_{93}Np^* \; \xrightarrow[2.3\,d]{\beta^-} \; ^{239}_{94}Pu \text{ (Fissionable)}$$

(16.6.1)

$$(\text{Fissionable}) \; ^{239}_{94}Pu + ^1_0n \; \rightarrow \; ^{147}_{56}Ba + ^{90}_{38}Sr + x\,^1_0n \text{ (fast)}; \; (x > 2)$$

(16.6.2)

$$^{238}_{92}U + ^1_0n \; \rightarrow \; ^{239}_{94}Pu + 2\,^0_{-1}e$$

The Pu produced then undergoes the fission reaction

$$^{239}_{94}Pu + ^1_0n \; \rightarrow \; ^{147}_{56}Ba + ^{90}_{38}Sr + 3\,^1_0n \qquad (16.6.3)$$

The ^1_0n s produced thus are used produce more $^{239}_{94}Pu$ from $^{238}_{92}U$. Such a breeder reactor is any reactor that produces more fissionable material than it consumed at the start. This sounds great but Pu is the most toxic material known to man. It is widely held that one atom of Pu can kill one if it gets into one's lungs in the body.

16.6.6.2 Thorium Breeding Cycle

$$[\text{Non-fissionable}] \atop \text{FERTILE} \Bigg\} \; ^{232}_{90}Th + ^1_0n \; \rightarrow \; [^{233}_{90}Th]^* \; \xrightarrow[22\,m]{\beta^-} \; ^{239}_{91}Pa$$

$$^{239}_{91}Pa \; \xrightarrow[2.7\,d]{\beta^-} \; ^{233}_{92}U \text{ (Fissionable)}$$

(16.6.4)

$(^{233}_{92}U, \tau_{1/2} = 0.162\ Myrs)$.

$$(\text{Fissionable}) \; ^{233}_{92}U + ^1_0n \; \rightarrow \; \text{Pair of Fragments} + y\,^1_0n \text{ (slow)}; \; (y > 2).$$

(16.6.5)

Thus breeding is achieved by combining both the fissionable and fertile materials in the reactor core under conditions which provide not only enough ^1_0n s to propagate a chain reaction in the fissionable material but also enough to convert more fertile material into fissionable material than was originally present., *i.e.*, to produce more fissionable material than they consume. This is the figure of merit called efficiency, expressed as 'doubling time'. Design objectives are usually set to have the doubling time < 10 yrs.

16.6.7. CONCLUSIONS

After dealing with the ABCs of Fission of a nuclide, and its advantages of producing external energy, the possibility of a sustained nuclear chain reaction was described. This was followed by its potential application of harvesting nuclear energy by designing a typical nuclear reactor. Different reactor designs were included. Then the principle of breeder type of nuclear fission by which one could produce more fissile material than consumed was treated. This led to a new design type known as breeder reactor was discussed.

Even before a nuclear reactor was designed and commissioned, the first application of fission was the **atomic bomb** (**A-bomb**). It was first used in warfare at Hiroshima (August 6, 1945) (the bomb named "Little Boy") and Nagasaki (August 9, 1945) (the bomb called "Fat Man") in Japan. The two bombs played a key role in ending World War II. Created *via* the *Manhattan Project*, the A-bomb was first exploded at the top secret base of Alamogordo on July 16th, 1945. Scientists who invented it were J. Robert Oppenheimer, David Bohm, Leo Szilard, Eugene Wigner, Otto Frisch, Rudolf Peierls, Felix Bloch, Niels Bohr, Emilio Segre, James Franck, Enrico Fermi, Klaus Fuchs and Edward Teller.

From 1945 to 1947, three critical events occurred *viz*.,

1) The establishment of the **United Nations** on 26 June 1945;

2) The dramatic demonstration of the destruction of even crude nuclear weapons are capable, in August 1945; and

3) The calamities leading to partition of British India into the Republic of India and Pakistan at midnight on 14-15 August 1947

The fourth quarter of the 20th Century started deriving nuclear power, the peaceful use of nuclear fission.

REVIEW QUESTIONS

RQ.16.1 Calculate the total binding energy of $^{20}_{10}Ne$. Given, $^{20}_{10}Ne$ has a mass of 19.992439 u., $M_p = 1.007825$ u, $M_n = 1.008665$ u, .(Answer: E = 0.172461 u = 160.6 MeV. or about 8.0 MeV per nucleon).

R.Q.16.2 Calculate the total binding energy of $^{27}_{13}Al$. Given, $^{27}_{13}Al$ has a mass of 26.981541 u., $M_p = 1.007825$ u,, $M_n = 1.008665$ u,. (Answer: E = 225.0 MeV. or 8.332 MeV per nucleon).

R.Q. 16.3 Determine the excitation energy of the compound nucleus formed by n-capture by $^{235}_{92}U$ (Figure below). Given, M(236,92) = 236.0445563 u, M(235,92) = 235.043924 u, $M_p = 1.007825$ u ; $1u = 931$ MeV $/c^2$.

(Answer $E_{exc} = 6.5$ MeV)

$$^1_0n \longrightarrow \;^{235}_{92}U = \;^{236}_{92}U^*$$
fissile nucleus   Compound nucleus

R.Q. 16.4 Find the excitation energy of the compound nucleus formed by n-capture by $^{238}_{92}U$. Given, $M_n = 1.008665$ u,. (Answer $E_{exc} = 4.8$ MeV)

R.Q. 16.5 In an experiment, 1.0 g of ^{59}Co is placed in a neutron flux with an intensity f 10^{15} neutrons $s^{-1}cm^{-2}$. A handbook gives the cross section for ^{59}Co as 17 b for the reaction $^{59}Co\,(n,\gamma)\,^{60}Co$. What is the rate of producing ^{60}Co ? (Answer   Rate = $(17\, x\, 10^{15}\, cm^2) * (6.022\, x\, 10^{23}\,/ 59) * (10^{15}\, s^{-1}cm^{-2}) \approx 10^{-14}\, ^{60}Co\, s^{-1}$).

R.Q. 16.6 Find the critical energy required to produce induced fission of $^{235}_{92}U$ by n-capture. Given $E_A = 6.2$ MeV for $^{236}_{92}U$ (Answer: odd-A nucleus; $E_{Cri} = 6.2$ MeV required and thermal neutrons).

R.Q. 16.7 Find the critical energy required to produce induced fission of $^{238}_{92}U$ by n-capture. (Answer: Even-even nucleus, $E_{Cri} > 5$ MeV required and so only fast neutrons).

R.Q. 16. 8 Calculate the amount of energy available if 1 g of $^{235}_{92}U$ is completely fissioned.

(Answer $8.2\, x\, 10^{10}$ Ws ≈ 1 (MWd (Mega Watt day)), assuming energy of 200 MeV / fission.).

R.Q. 16.9 What is spontaneous fission? Why is it important?

RQ 16.10 How much energy is released in the total conversion of 10g of mass into energy? How many megatons of TNT equivalent are released? (Answer: Energy = $9\, x\, 10^{14}\, J = 22\, M\, tons$).

335

R.Q. 16.12 Consider the slowing down of fission neutrons in elastic collision with nuclei of atomic mass number **A**. Assume that these moderating nuclei are free and at rest and that the scattering is isotropic in the centre of mass (**CM**) system. If the kinetic energy in the laboratory before and after collision are T_0 and T respectively show that the average logarithmic energy loss $[\ln(T_0/T)]_{ave}$ is given by $1 + [\alpha/(1 - \alpha)]\ln(\alpha)$ where $\alpha = [(1 - A)/(1 + A)]^2$.

R.Q. 16.13 The nuclide ^{256}**Fm** decays through spontaneous fission with a half-life of 158 minutes. If the energy released is about 220 *MeV* per fission, calculate the fission power produced by 1 μg of this isotope.

R.Q. 16.14 Calculate the number of fissions required to produce 1 W of electric power from U-235..(Answer: $10^{11} s^{-1}$). How much quantity of fuel U-235 is required to produce 1*MWe* per day, and what is the amount of waste product per day produced along with this power, and of which what is the quantity of fertile produced? (Answer: 1 g;1 g; 0.5 g of Pu-239).

R.Q. 16.15 The energy liberated in the fission of a single uranium 235 atom is 3.2×10^{-11} J. Calculate the power production corresponding to the fission of one kg of uranium per day. ($N_A = 6.02 \times 10^{23}\ mol^{-1}$). (Answer 950 *MW*)

R.Q. 16.16 Explain Bohr-Wheeler theory o nuclear fission. Derive the expressions for the fissionability parameter x and the critical size.

R.Q. 16.17 What are 'prompt' and 'delayed' neutrons?

R.Q. 16.18 State the commonly used moderators in nuclear reactors. What are the important effects of a moderator in a reactor?

R.Q. 16.19 Calculate the Q-value of the fission (symmetric) reaction $^{238}_{92}U + ^1_0 n \rightarrow X + Y$, where X and Y are fragments with A values 119 and 120, respectively, and Z values 45 and 46. Given $M_n = 1.008665\ u = 939.57\ MeV$; $M_p = 1.007276\ u = 938.272\ MeV$; $(B/A)_{atA=120} = 8.5 MeV$; $(B/A)_{atA=240} = 7.6 MeV$; $M(^{238}_{92}U) = 238.650784\ u$; $1u = 931\ MeV/c^2$.

R.Q. 16.20 What is Bohr-Wheeler criterion for nuclear fission? Briefly describe the theory of fission.

R.Q.16.21 Discuss the Bohr-Wheeler theory of nuclear fission.

RQ. 16.22 Estimate the minimum rate of fuel consumption in a 500 MW nuclear reactor based on natural uranium as fuel. Given the average release of energy per fission of natural uranium is 200 MeV.

&%&%&%&%&%&%&%&

Chapter 17

NUCLEAR FUSION

(Thermo-nuclear Reaction)

Chapter 17

NUCLEAR FUSION
(Thermo-nuclear Reaction)

" I have yet to see a problem however complicated, which when you looked at it in the right way, did not become still more complicated" - Paul Anderson

"I do not know with what weapons World War 3 will be fought, but World War 4 will be fought with sticks and stones." - A. Einstein

17.1   INTRODUCTION

One of the major requirements for sustaining human progress is an adequate source of energy. The current largest sources of energy are the combustion of coal, oil and natural gas. They will last quite a while but will probably run out or become harmful in tens to hundreds of years. Solar energy will also work but is not much developed yet except for special applications because of its high cost. This high cost as a main source, *e.g.*, for central station electricity, is likely to continue, and nuclear energy is likely to remain cheaper. A major advantage of nuclear energy (and also of solar energy) is that it doesn't put carbon dioxide (CO_2) into the atmosphere. How much of an advantage depends on how bad the CO_2 problem turns out to be. Unraveling the role of THERMO-NUCLEAR REACTION (fusion) in the universe has taken almost a century since Einstein's proof of the equivalence of energy and matter in 1905. The discovery that fusion reactions are responsible for the building of the light elements in the "Big Bang" and the subsequent development of the heavier elements in the stars and in exploding supernovae is one of the field's most exciting successes. With fusions the last elements of the "Periodic table of the elements" have been created, because they are not on earth. In 1999 a few physicists thought that they have discovered the element 118 but two years later in 2001 they said that it was a mistake, so element 114 is the last know element. In stars there are also fusions

Fusion power has the following possible advantages if it can be made to work.

a) The fuel supply is potentially larger in fusion. However, for nuclear fission the uranium supply seems to be large enough.
b) Fission products are not produced, although there will be induced radioactivity in the structures of the plants.
c) No material useful for bombs is produced

17.2 WHAT IS NUCLEAR FUSION?

Nuclear energy can also be released by fusion of two light elements (elements with low atomic numbers). The power that fuels the sun and the stars is nuclear fusion. In a hydrogen bomb, two isotopes of hydrogen, deuterium and tritium are fused to form a nucleus of helium and a neutron. This fusion releases 17.6 *MeV* of energy. Unlike nuclear fission, there is no limit on the amount of the fusion that can occur.

Fusion is today possible but energy which you need for a fusion is higher than the energy you get and this is not the sense of nuclear fusions.

Because of the Coulomb repulsion between nuclei, they must have a certain KE to overcome the Coulomb potential barrier and get close enough so that nuclear forces produce the necessary consolidatingaction for fusion. This problem does not arise in fission as the neutron is electrically neutral, and so can approach a nucleus without any KE.

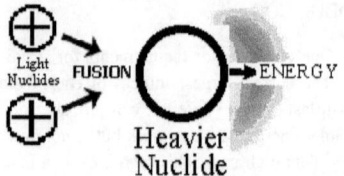

Fig 17.1 Nuclear Fusion

Each fusion reaction is characterized by a specific ignition temperature, which must be surpassed before the reaction can occur. The d-t reaction has an ignition temperature above 10^8 K.

Any substance at temperatures approaching 10^8 K will exist as a completely ionized gas, or plasma

The production of new elements is called *Nucleosynthesis*.

17.2.1 Estimation of KE required

When two nuclei of atomic numbers Z_1 & Z_2 are in contact, E_p, the mutual PE between them is

$$E_P = \frac{Z_1 Z_2 e^2}{4 \pi \varepsilon_0 r} \qquad (17.2.1)$$

where r = Sum of the radii of the two nuclei.

$r = \sim 10^{-14}$ cm

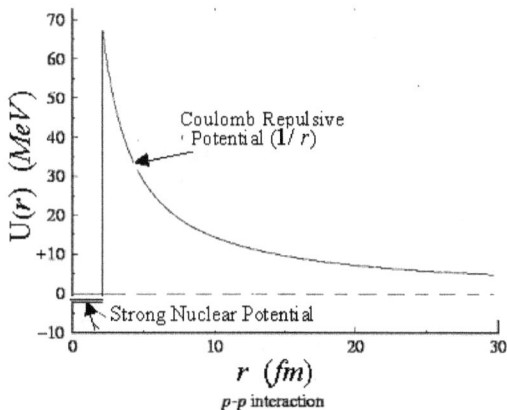

Fig 17.2 Nuclear potential and Coulomb interaction

$$E_P = \square\ 2.4 \times 10^{-14}\ Z_1\ Z_2\ J$$
$$= \square\ 1.5 \times 10^5\ Z_1\ Z_2\ eV \quad (17.2.2)$$
$$= \square\ 0.15\ Z_1\ Z_2\ MeV$$

This is the height of the potential barrier to be crossed by any particle to enter the nucleus. It is thus clear that no fusion will be possible if the E_T (KE) of the colliding nuclei

$$E_T < E_P \quad (17.2.3)$$

But barrier penetration may be possible if

$$E_T\ (= \text{slightly} < E_P) < E_P \quad (17.2.4)$$

17.2.2.1 Nuclear Fusion reactions are *Thermo-nuclear reactions* Why?

The required Temperature, T, for Fusion

The average KE of a system at temperature $T\ K$ of particles is $\square\ k_B\ T$

i.e., $1\ k_B T = 8.6 \times 10^{-5}\ T\ eV$

Thus, for $Z_1 = Z_2 = 1$, equation (17.2.2) gives

$$E_P = \square \ 0.15 \ Z_1 \ Z_2 \ MeV$$

$$0.1 \ MeV \equiv \square 1 \ k_B \ 10^9 \ eV = \square \ k_B T$$

That is, at a temperature of

$$T = \square \ 10^9 \ K \qquad (17.2.5)$$

This magnitude of temperature is unimaginable in the Earth. This means for nuclear fusion of a large number of nuclei to take place at these extreme temperatures there are major problems to face. Because of the very high temperatures involved, nuclear fusion reactions are called *thermo-nuclear reactions.*

Worked out Example 17.1

Consider the fusion reaction $d + d \rightarrow {}^3_2He + {}^1_0n + 3.2 \ MeV$ occurs with the deuterons at rest, i) What is the KE of the neutron? ii) Based on the estimate that the deuterons have to come within 100 *fm* of each other for fusion to proceed, calculate the energy that must be supplied to overcome the ES repulsion. iii) Approximately what temperature is this equivalent to?

Solution: $\boxed{STEP \ * \ 1}$ Since before the reaction there is no net momentum, the momenta of the two particles in the final state must cancel out. Let the mass of the ${}^1_0n$ as m and the mass of ${}^3_2He$ as $3m$. One gets, $3mV = m v$, i.e., $V^2 = v^2/9$,

$\boxed{STEP \ * \ 2}$ i) Now the total kinetic energy in the final state is

$$3.2 \ MeV = 3 \ m \ V^2/2 + m \ v^2/2$$

$3.2 \ MeV = (m \ v^2/2)^{4/3}$ and $(m \ v^2/2) = 2.4 \ MeV \equiv 3.84 \ x \ 10^{-13} \ J$

$\boxed{STEP \ * \ 3}$ ii) The ES energy is just $e^2/(4\pi \ \varepsilon_0 r) = 14.4 \ keV \equiv 2.3 \ x \ 10^{-15} \ J$,

$\boxed{STEP \ * \ 4}$ iii) $k_B T \equiv 2.3 \ x \ 10^{-15} \ J$ gives a temperature $T \square \ 10^8 \ K$ ..

17.2.2.2 Problem of containment of the reacting particles

No known material can contain such temperatures $T = 10^8 \ K$ as the matter will be in the plasma state (*i.e.*, collection of bare positively charged nuclei and electrons).

17.2.2.3 Magnetic fields for containment:

Containment has been attempted by means of magnetic fields. Also when the intensity of magnetic fields is rapidly increased, the plasma is adiabatically compressed and its temperature raise until fusion begins. Several ingenious devices have been built that perform these two functions of i) containment and ii) heating.

17.2.3 E_B/A for $A < 20$

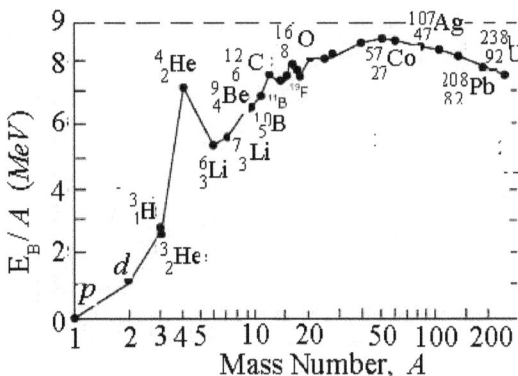

Fig 17.3 E_B/A versus A plot for lighter elements

A perusal of the E_B/A versus A plot for lighter elements $A < 20$ (Fig 17.3) shows energy can also be gained by building up an intermediate mass nuclide from lighter elements, i.e., the rest mass of an intermediate product nuclide < Sum of the masses of the two light reactant nuclei.

The abnormally high binding energy E_B for ^4_2He means the fusing of ^2_1H nuclei, to form ^4_2He, releases considerable energy. Thus in nuclear fusion, both the light nuclei and the compound nucleus (CN) lie in the left of the maximum in the

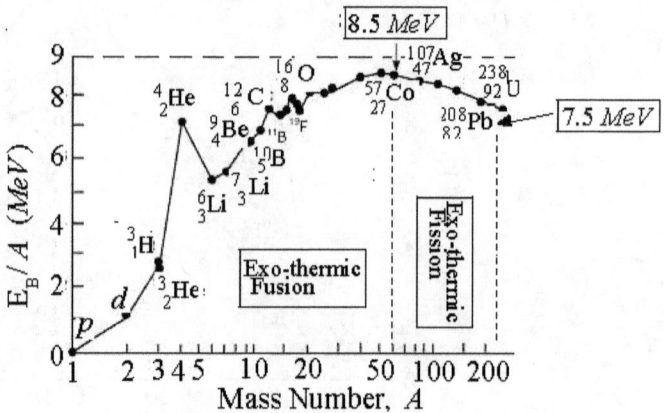

Fig 17.4   Fusion versus Fission regions

E_B / A point, viz., $^{56}_{25}$Mn of the E_B / A versus A plot (whereas in nuclear fission, normally the reactant and fragment nuclei are all lie on the right side of the $(E_B / A)_{Max}$ curve (Fig 17.4).

The various quantities have been listed in Table 17.1.

TABLE 17.1   Nuclides for Fusion and Fission - E_B / A

Nuclide	E_B / A	Type of Nuclear Reaction
1. Deuteron, $^2_1$H	4.99 MeV	$\{5.96 \frac{MeV}{nucleon}\} 4 = 24$ MeV per fusion
2. Helium, $^4_2$He	7.07 MeV	
3. Palladium, $^{Pd}_{46}$	8.4 MeV	$\{0.8 \frac{MeV}{nucleon}\} 235 = 180$ MeV per fission
4. Uranium, $^{235}_{92}$U	7.6 MeV	

17.2.4   Hydrogen Burning

Among all the nuclides in the Chart of the Nuclides, Coulomb repulsion is the least for the isotopes of hydrogen, and if these are sufficiently hot the thermal energy is sufficient to produce fusion in a proportion of the interaction. Two protons $^1_1$H fuse to form a mass 2 nucleus, and since ^{2}He is not stable, the process should be

$$^1_1\text{H} + ^1_1\text{H} \rightarrow \ ^2_1\text{H} + ^0_{+1}e + \nu_e + 0.42 \text{ MeV} \qquad (17.2.6)$$

This is generally known as *nuclear burning of hydrogen*, or as **hydrogen burning**.

$$^{0}_{+1}e + ^{0}_{-1}e \rightarrow h\nu = 1.02 \text{ MeV} \qquad (17.2.7)$$

Equation (17.2.7) is the *positron annihilation* equation.

17.3. NUCLEO-SYNTHESIS

17.3.1 The PP Cycle (Critchfield Cycle):

The burning of light hydrogen is a very *slow process*, as it goes by way of the <u>weak interactions</u>. This reaction leads to a series of processes called the <u>proton-proton cycle</u> (Critchfield cycle). The initial reaction of the PP Cycle is thus

$$\boxed{1} \quad p(p, e^+ + \nu_e) d; \; Q = +0.42 \text{ MeV} \qquad (17.3.1)$$

This reaction #1, *i.e.*,(17.3.1), **is yet to be observed in a laboratory**.

Here two protons ($p \equiv {}^{1}_{1}H$) coalesce to form a nucleus of heavy hydrogen (deuteron, $d \equiv {}^{2}_{1}H \equiv {}^{2}_{1}D$) with the emission of a positron ($e^+ \equiv {}^{0}_{+1}e$) and neutrino (ν_e). A deuteron may also be formed by means of n-capture by proton

$$^{1}_{1}H + ^{1}_{0}n \rightarrow ^{2}_{1}H + 2.226 \text{ MeV}$$

Thus *nucleo-synthesis* starts. The direct conversion of hydrogen nuclei into helium was proposed first by T.K. Fowler and Charles Christian Lauritsen through the PP cycle of hydrogen burning, of which the first reaction is given by equation (17.3.1). Once heavy hydrogen is formed, the synthesis of the heavier nuclides can go by way of the *strong interactions*.

$$\boxed{2} \quad {}^{2}H(p, \gamma) \, {}^{3}H; \; Q = +5.49 \text{ MeV} \qquad (17.3.2)$$

$$^{1}_{1}H + ^{2}_{1}H \rightarrow ^{3}_{1}H + e^+ + \nu_e + Q = 4.6 \text{ MeV} \qquad (17.3.3)$$

In contrast, the d-d reaction

$$d + d \rightarrow ^{3}_{2}He + ^{1}_{0}n + 3.2 \text{ MeV}$$

is *exo-ergic*.

$M_d = 2.014102 \, u, \, M_p = 1.007825 \, u, \, M_n = 1.008665 \, u, \, M({}^{3}_{2}He) = 3.016029 \, u$

Thus

	Mass of Reactants		Mass of Products
$M_d =$	2.014102 u		
$M_d =$	2.014102 u	$=$	4.028204 u
$M_n =$	1.008665 u		
$M(^3_2He) =$	3.016029 u	$=$	-4.024694 u
Mass difference			$+ 0.003510$ u

$$Q = (+\ 0.003510\ u)(931.5\ MeV/u) = +\ 3.270\ MeV.$$

So the reaction requires only that the acceleration of the reactants exceeds the Coulomb repulsion.

$$V = \frac{(k\,e^2)}{2\,r} = \frac{14.4\ eV\text{-nm}}{2\,(1.5\ nm)\,2^{1/3}} \approx 4.8\,MeV$$

In addition, as in Fig 17.5,

$$^1_1H + ^3_1H \rightarrow ^3_2He + ^1_0n + Q = 17.6\ MeV \qquad (17.3.4)$$

$$^2_1H + ^3_2He \rightarrow ^4_2He + ^1_1H + Q = 18.3\ MeV \qquad (17.3.5)$$

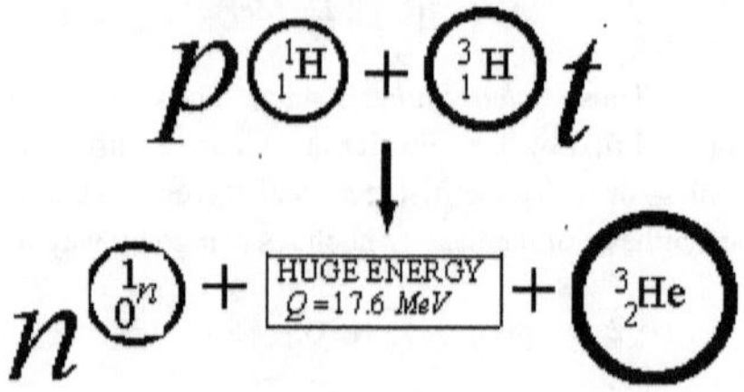

Fig 17.5 Fusion reaction of $^1_1H + ^3_1H \rightarrow ^3_2He + ^1_0n + Q = 17.6\ MeV$

Finally,

3 $\quad ^3_2He(^3He, p + p)\,^4_2He;\ Q = +\ 12.86\ MeV \qquad (17.3.6)$

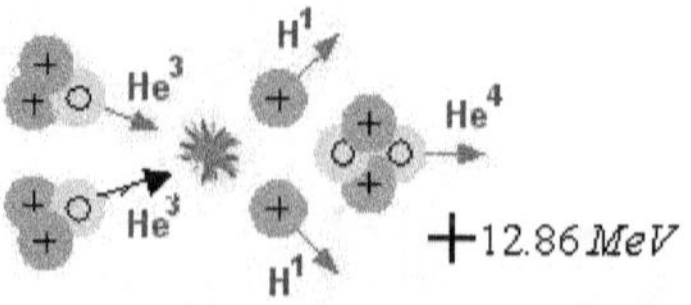

$+12.86\ MeV$

17.3.2 The PP I chain

As seen from above, the one possible PP chain of events of hydrogen burning to form ^4He is

$$\left.\begin{array}{ll} (Step\ \#1) & p(p,e^+ + v_e)d; \quad Q = +0.42\ MeV \\ (Step\ \#2) & ^2H(p,\gamma)\ ^3H; \quad Q = +5.49\ MeV \\ (Step\ \#3) & ^3_2He(^3He, p+p)\ ^4_2He; \quad Q = +12.86\ MeV \end{array}\right\} \boxed{PP\ I\ Chain}$$

(17.3.7)

$$(1)x2 + (2)x2 + (3) \Rightarrow 6\ ^1_1H \rightarrow\ ^4_2He + 2e^+ + 2v_e + 24.7\ MeV$$

(17.3.8)

Thus the total energy release resulting from the PP I chain is **24.7 MeV per helium nucleus formed**. This is equivalent to

$$6.4 \times 10^{11}\ J\ gm^{-1}\ \text{of}\ ^1_1H\ \text{consumed, to form a}\ ^4_2He\ \text{nucleus}.$$

It is seen thus that each hydrogen atom consumed in this chain leads to the emission of 6.55 MeV of EM energy from the Sun.

In the continued synthesis of the elements beyond ^4_2He, a difficulty arises in that no stable elements with mass A = 5 or 8 exist. The reaction #3 of the PP I chain leads into either PP II or PP III chain.

17.3.3 PP II Chain

PP II chain starts with the reaction #1 and #2. Then

$$\left.\begin{array}{ll} (Step\ \#3') & ^3_2He(\alpha,\gamma)\ ^7_4Be; \\ (Step\ (\#4') & ^7_4Be(e^-,v_e)\ ^7_3Li; \\ (Step\ (\#5') & ^7_3Li(p,\alpha)\ ^4_2He; \end{array}\right\} \boxed{PP\ II\ Chain} \quad (17.3.9)$$

17.3.4 PP III Chain

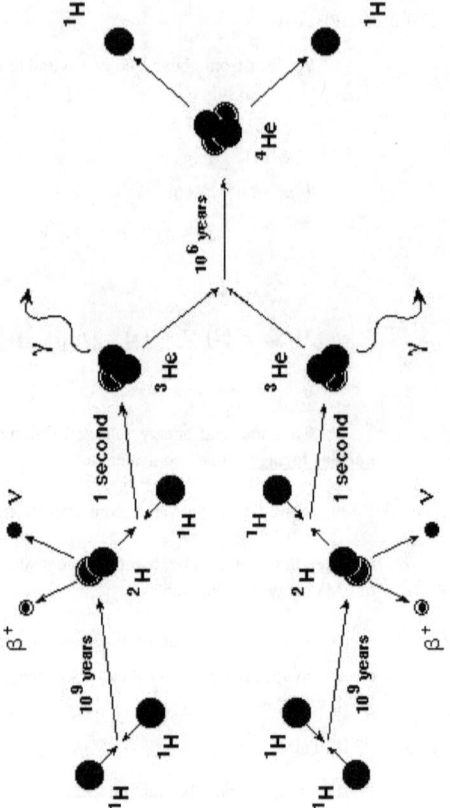

Fig 17.6 PP chain

The chain PPIII starts with the reaction #1, #2 and #3'.

$$
\left.\begin{array}{ll}
(Step\,\#4") & {}^{7}_{4}\text{Be}\,(p,\gamma)\,{}^{8}_{5}\text{B}^{*}; \\
(Step\,\#5") & {}^{8}_{5}\text{B}^{*}(\,,e^{+}\,\nu_{e}\,)\,{}^{8}_{4}\text{Be} \\
(Step\,\#6") & {}^{8}_{4}\text{Be}\,\rightarrow\,2\,\alpha;
\end{array}\right\} \boxed{\text{PP III Chain}} \qquad (17.3.10)
$$

The net result is $4p \Rightarrow 1\alpha$

17.3.5 CN Cycle (Bethe cycle or Carbon cycle, or CNO cycle)

Two α – particles can combine to form a ^{8}Be nuclide, which stays together for $\sim 10^{-15}$ s. In a sufficiently high density of helium, this results in an adequately high concentration of ^{8}Be, so that another most important fusion process takes place. This is the **Bethe or Carbon cycle**, which is equivalent to the fusion $4p \Rightarrow 1\alpha$. $\quad\quad ^8_4\text{Be} + ^4_2\text{He} \rightarrow ^{12}_6\text{C} + \gamma$.

(17.3.11)

A shown by C.W. Cook; W.A. Fowler; C.C Lauritsen and T. Lauritsen (1957), the reaction has a resonance which strongly enhances the production of ^{12}C in a helium gas of $\sim 10^8 K$. The CN cycle, in which $4p \Rightarrow 1\alpha$, involves carbon, nitrogen and oxygen, and the cycle has the following steps.

$$\left. \begin{array}{ll} (Step\ \#1) & ^{12}_6\text{C}(p,\gamma)^{13}_7\text{N}; \quad Q = +1.95\ MeV \\ & ^{13}_7\text{N}(,e^+\ v_e)^{13}_6\text{C}; \quad Q = +1.20\ MeV \\ (Step\ \#2) & ^{13}_6\text{C}(p,\gamma)^{14}_7\text{N}; \quad Q = +7.55\ MeV \\ (Step\ \#3) & ^{14}_7\text{N}(p,\gamma)^{15}_8\text{O}; \quad Q = +7.34\ MeV \\ & ^{15}_8\text{O}(,e^+\ v_e)^{15}_7\text{N}; \quad Q = +1.68\ MeV \\ (Step\ \#4) & ^{15}_7\text{N}(p,\alpha)^{12}_6\text{C}; \quad Q = +4.96\ MeV \\ \hline & \text{Total energy released} = +24.68\ MeV \end{array} \right\} \text{CN Cycle}$$

(17.3.12)

This CN cycle was suggested first and independently by Hans Bethe (1938) and by C.F. von Weizsaecker (1938). Adding all the reactions of the CN cycle,

$$4p \Rightarrow 1\alpha + 2e^+ + 2v_e + 26.7\ MeV\ (\sim +6.6 \times 10^{11}\ J-gm^{-1})$$

$$\left. \begin{array}{l} \text{The net energy liberated} \\ \text{in fusion by the CN cycle} \end{array} \right\} = \left(\begin{array}{l} +26.7\ MeV \text{ per gm of H consumed} \\ (=\sim +6.6 \times 10^{11}\ J-gm^{-1}) \end{array} \right) \quad (17.3.13)$$

In this chain carbon and nitrogen act purely as catalysts (since they are neither produced nor destroyed by the CN cycle). In the Sun for one such cycle to take place, it requires $\sim 6 \times 10^6$ yrs.

17.3.5.1 Basic Exothermic Reaction in Stars and Solar Neutrinos

The PP cycle as well as the CN cycle can take place under stellar interiors, and thus these are the basic exothermic reactions in the Sun, and hence the source of nearly all of the energy in the Universe. Self-sustaining fusion reactions can occur only under extreme temperature and pressure. The PP cycle is known to the belief of most astrophysicists the predominant process of energy generation in the '*Main Sequence Stars*', like the Sun (interior at $\Box\ 2x\ 10^6 K$) and cooler stars, whereas the CN cycle is known to the belief of most astrophysicists the predominant process of energy generation in the '*Main Sequence Stars*', like the Sun (interior at $\Box\ 2x\ 10^6 K$) and cooler stars, whereas the CN cycle is responsible for the energy output of hotter hydrogen burning stars. The **neutrinos** in PP and CN chains carry away 2 – 6 % of the energy released in the reactions.

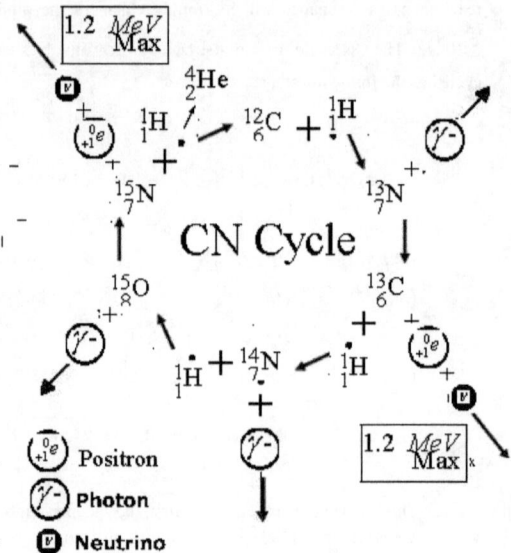

Fig 17.7 CN cycle, adapted from J.N. Bahcall, *Neutrinos from the Sun*, Scientific American, Volume 221, # 1, July 1969, pp. 28-37.

In the case of the Sun,

$$4p \Rightarrow 1\ \alpha + 2e^+ + 2\nu_e + 26.7\ MeV\ (\Box\ + 6.6 x 10^{11}\ J - gm^{-1}),$$

is at the rate of $5.64 \times 10^{11}\ kgm\text{-}s^{-1}$ of hydrogen fusing in helium, with a release of $3.7 \times 10^{25}\ W$. Of this $\Box\ 1.8 x\ 10^{14} W$ only falls on the Earth in the form of photons.

Major attributes of the Sun

Mass	$M_\odot = 1.99 \times 10^{30}$ kg
Radius	$R_\odot = 6.96 \times 10^8$ m
Luminosity	$L_\odot = 3.86 \times 10^{26}$ W
#ofHatoms	$\sim 10^{56}$

Since both PP chains and CN cycles can produce fusion energy by converting hydrogen to helium, they compete with each other. What controls which dominates in a given situation?

17.3.6 FACTORS CONTROLLING FUSION RATES

There are two points. First, there must be C, N, or O present for the CN cycle to occur, so it can only happen for stars where this is true. However, only a very small amount is required, so this condition is often fulfilled. Second, the reactions have very different temperature dependences, as illustrated in Fig 17.8.

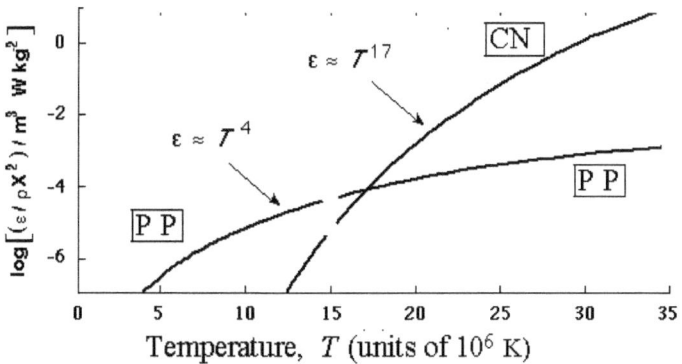

Fig 17.8   Energy production rates for the PP chain and CN cycle as a function of temperature (units of $10^6 K$).

In Fig 17.8 the horizontal axis is the temperature in units of $10^6 K$, and the vertical axis is a quantity that measures the rate of energy production plotted on a logarithmic scale.

Thus, at lower temperatures the PP chain dominates, but with rising temperatures there is a sudden transition to dominance by the CN cycle, which has an energy production rate that varies strongly with temperature. This is why the CN cycle is more important for heavier stars: their interior temperatures are higher, thus favoring the CNO cycle.

17.3.7 FUSION IN THE SUN AND STARS

17.3.7.1 The Source of Sun's Energy

In the case of the Sun, detailed considerations suggest that it is producing about 98-99% of its energy from the PP chain and only about 1% from the CN cycle. However, if the Sun were but 10-20% more massive, its energy production would be dominated by the CN cycle.

Each fusion reaction is characterized by a specific ignition temperature, which must be surpassed before the reaction can occur. The d-t reaction has an ignition temperature above 10^8 K.

Any substance at temperatures approaching 10^8 K will exist as a completely ionized gas, or plasma.

The production of new elements is called *Nucleo-synthesis*.

Table 17.2 Temperature & Fusion Reactions

	Temperature (K)	Fusion Reaction
1.	10×10^6	$^1_1H \rightarrow ^4_2He$
2.	100×10^6	$^4_2He \rightarrow ^{12}_6C$
3.	600×10^6	$^{12}_6C \rightarrow ^{16}_8O \, (^{24}_{12}Mg)$
4.	1500×10^6	$^{16}_8O \rightarrow ^{20}_{10}Ne \, (^{32}_{14}Si)$
	etc.	etc.

Probabilities of fusion reactions are quantitatively defined as the cross sections. Effective cross sections for various fusion reactions as functions of temperature are given here. For all four fusion reactions given in Table 17.2, the cross section for the reaction,

$$^2_1H + ^3_1H \rightarrow ^4_2He + ^1_0n$$

is consistently the highest at any temperature T. This reaction has been chosen in further fusion research, because of its potential for success.

Worked out Example 17.2

If free protons and neutrons could be assembled into a gram of helium nuclei, estimate the energy that would be released, (or to break up the nuclei of all the helium atoms in a gram of helium how much energy would be required?). Given: , $M_p = 1.007825 \, u$, $M_n = 1.008665 \, u$, $M_{He} = 4.003879 \, u$

Solution: $\boxed{STEP \ * \ 1}$ Total mass of the separate components of the helium nucleus

$= 2 \, (M_p = 1.007825 \, u) + 2 \, (M_n = 1.008665 \, u) = 4.032980 \, u$ But $M_{He} = 4.003879 \, u$.

STEP * 2 This, then, represents the "binding energy" of the protons and neutrons in the helium nucleus.

STEP * 3 $E = mc^2$ gives $\Delta M_{He} = 0.029101\ u \approx 4.5 \times 10^5 erg/nucleus$
$$= 2.7 \times 10^{19} erg/mole \text{ of He}$$

STEP * 4 This means that to break up the nuclei of all the helium atoms in a gram of helium would require $1.62 \times 10^{11} g-Cal \equiv 190{,}000$ kWH energy.

Worked out Example 17.3

Using the fusion reaction $^7_3Li + ^1_1H \rightarrow ^4_2He + ^4_2He$ and $M_p = 1.007825\ u$, $M_n = 1.008665\ u$, $M_{He} = 4.003879\ u$ and $M(^7_3Li) = 7.016285\ u$, show that the Einstein mass energy relation is satisfied.

Solution: **STEP * 1** On the LHS $M(^7_3Li + ^1_1H) = 8.024100\ u$,

STEP * 2 on the right $M(2 \times ^4_2He) = 8.007758\ u$

STEP * 3 so that $\Delta M = 0.0185$ units of mass had disappeared in the reaction. The experimentally determined energies of the alpha particles were approximately 8.5 *MeV* each, a figure compared to which the KE of the incident proton could be neglected. Thus $0.0185\ u$ of mass had disappeared and 17 *MeV* of KE had appeared. Now $0.0185\ u$ of mass is $3.07 \times 10^{-26} g$, 17 *MeV* is $27.2 \times 10^{-6}\ ergs$ and $c = 3 \times 10^{10}$ cm/s. If we substitute these figures into Einstein's equation, $E = mc^2$, on the left side we have $27.2 \times 10^{-6}\ ergs$ and on the right side we have $27.6 \times 10^{-6}\ ergs$, so that the equation is found to be satisfied to a good approximation. In other words, these experimental results prove that the equivalence of mass and energy was correctly stated by Einstein.

17.4 ORIGIN OF ELEMENTS

17.4.1 Carbon Burning

It is seen that, as byproduct of the PP I chain, Li and Be nuclei are produced through PP II and PP III cycles.

$$\alpha + \alpha \rightarrow ^8_4Be, \qquad (17.4.1)$$

and ^7_4Be is produced and destroyed; and the $^{11}_5B(p, \gamma)3\alpha$. Is **'helium burning'** (in 'Red Giants'). Also

$$^8_4Be(\alpha,\ )^{12}_6C. \qquad (17.4.2)$$

The byproducts of PP I chain, viz., $_1^2H$ and $_2^3He$ may be captured by the $_4^8Be$ forming $_5^{10}B$ and $_6^{11}C$ in a less probable manner at $\square\ 10^8\,K$. Once the helium is used up, temperatures increase to $10^9\,K$ and the reaction called '**Carbon burning**' starts,

$$_6^{12}C(\alpha,\gamma)\ _8^{16}O. \tag{17.4.3}$$

17.4.2 Stars and Supernovae

Nucleo-synthesis proceeds in stars as follows

Various forms of nuclear fusion take place within the stars. Stars on the *Main Sequence* radiate energy by the fusion of protons, known as *hydrogen burning*, which is followed by *helium burning*.

Later in a star's life, while it is in the *Red Giant* phase, the core temperature becomes great enough ($\square\ 10^8\,K$) to ignite further nuclear fusion reactions. First, Beryllium (Be) is created from Helium. Then Carbon (C) is created from He and Be. Finally, Oxygen (O) is created from C and He. These reactions are *exothermic* (they release energy), and thus they are spontaneous in the sense that they will continue unaided as soon as enough energy has been provided. However the activation energy in this case is too great for the reactions to occur in a Main Sequence star, so it does not occur until the Red Giant phase.

Stars which become supernovae release so much energy that they initiate *endothermic* fusion reactions: those whose products are actually less stable than the reactants. These reactions are responsible for all the elements heavier than $_{26}^{56}Fe$, which has the highest BE /A of any possible nucleus.

17.4.2 **Silicon burning** (Gravitational Confinement)

Initial gravitational collapse heats the core gas until hydrogen burning starts; and the collapse stops. When the hydrogen is exhausted, the collapse sets in again to increase the temperatures further until the 'ashes' (helium) of the previous reaction start burning. And then carbon burning occurs and so on up the Periodic Table up to iron $_{26}^{56}Fe$, which is the most stable isotope of iron. Then there is no more nuclear fuel to burn to delay further gravitational contraction, and so even higher densities and temperatures occur. Electron-capture (EC) allows forming neutron-rich nuclei. β^- - occurs which causes removal of heat by producing neutrinos which escape, as well as removing electrons.

a) Formation of Heavier Atoms

b) Further Gravitational Confinement

Then further gravitational collapse raises temperatures to several $10^9\,K$ ..

$$_{26}^{56}Fe \rightarrow 26\,\alpha + 4\ _0^1n. \tag{17.4.4}$$

The iron core suddenly becomes a helium core with a large density $\sim 10^{24}\ {}_0^1n/cc$ present. Then slow n-capture, known as "**s-process**" occurs up to the synthesis of ${}_{83}^{209}Bi$. This s-process build-up of heavy elements terminates when

$${}_{83}^{209}Bi + {}_0^1n \rightarrow \alpha + {}_{81}^{206}Tl. \qquad (17.4.5)$$

However, the trans-bismuth nuclei may be built-up through the "**r-process**", till A = 254, beyond which **nuclear fusion** occurs. The reaction proceeds in an alarmingly rapid rate so as to refer it as "**supernova explosion**". Without supernovae, there would be no Ni, Cu, Zn, Ag, Au, I, Pt, Pb, Hg, U or Pu, to name some of the most familiar elements.

17.5. MAN MADE FUSION

This aspect will be discussed in Chapter 19.

REVIEW QUESTIONS

R.Q. 17.1   It is seen thus that each hydrogen atom consumed in this chain leads to the emission of 6.55 MeV of EM energy from the Sun. Find how many hydrogen atoms are converted into helium per second. Given, the total rate of emission of photon energy (Luminosity) of the Sun $L_\odot = 3.86 \times 10^{26}\ W$ (Answer: $\frac{3.86 \times 10^{26}\ W}{6.55\ MeV} = 3.7 \times 10^{38}$).

R.Q. 17.2   As discussed in the notes the fusion reaction rate at a temperature T for two nuclei (Z_1, A_1) and (Z_2, A_2) is approximately proportional to $e^{[-1.89\,\alpha^{2/3}(k_B T)^{-1/3}]}$, where $\alpha = 31\,Z_1 Z_2 [A_1 A_2 /(A_1 + A_2)]^{1/2}\,(keV)^{1/2}$. On the basis of this expression estimate the ratio of the ${}^2H$ - ${}^3H$ to ${}^2H$ - ${}^3He$ rates at a temperature of $2 \times 10^8\ K$.

R.Q. 17.3   What is nuclear fusion? Why is energy released during this process? Why is high temperature required for fusion reactions?

R.Q. 17.4   Calculate in MeV the energy liberated when a helium nucleus is produced (i) by fusing two neutrons and two protons, and (ii) by fusing two deuterium nuclei,. Why is the quantity of energy different in the two cases ? (neutron mass = 1.00898 u: proton mass = 1.00759 u; nuclear deuterium mass = 2.01419 u; nuclear helium mass = 4.00277 u; 1 u is equivalent to 931 MeV). (Ans: 28.27477 u; 23.84291 u)

R.Q. 17.5   The Sun obtains its radiant energy from a thermonuclear fusion process. The mass of the Sun is 2×10^{30} kg and it radiates 4×10^{23} kW at a constant rate. Estimate the life of the sun in years if 0.7% of the mass of the sun is converted into radiation during the fusion process and it loses energy only by radiation. ($1 \, yr = 3 \times 10^{7} s$; $c = 3 \times 10^{10} \, ms^{-1}$) (Answer: 1.1×10^{11} yrs).

RQ. 17.6 $4 \, {}_{1}^{1}H \rightarrow {}_{2}^{4}He + 2 \, {}_{+1}^{0}e +$ Energy is the fusion reaction taking place in the Sun.. How much energy is released when 1.00 kg of hydrogen is consumed? The masses of ${}_{1}^{1}H, {}_{2}^{4}He, {}_{+1}^{0}e$ are respectively, 1.007825 u, 4.002604 u and 0.000549 u. which include the atomic electrons in both the first two values. Given, $c = 2.997925 \times 10^{8} \, ms^{-1}$ (Answer: 0.00663 kg $= 5.96 \times 10^{14} J$).

RQ. 17.7   Consider the fusion reaction $2 \, {}_{1}^{2}H \rightarrow {}_{2}^{4}He$ in reactor to produce industrial power of 150 MW, with energy efficiency 30%. How much deuterium is required each day? Tneirl atomic masses are 2.01402 u, and 4.002604 u, respectively. (Answer: 75 g/d).

RQ. 17.8   Calculate the temperature required to overcome the Coulomb barrier and the energy released during fusion if a gas consists of a) ${}_{5}^{10}B$,b) ${}_{6}^{12}C$, c) ${}_{8}^{16}O$, d) ${}_{12}^{24}Mg$.(Answers: a) r $r = 2.59$ fm, V= $1.11 \times 10^{-12} J$, $T = 5.4 \times 10^{10} K$, $Q = 31.15$ MeV / fusion,

b) $r = 2.75$ fm, V=$1.51 \times 10^{-12} J$, $T = 7.3 \times 10^{10} K$, $Q = 13.93$ MeV / fusion,

c) $r = 3.02$ fm, V $= 2.44 \times 10^{-12} J$, $T = 11.8 \times 10^{10} K$, $Q = 16.54$ MeV / fusion,

d) $r = 3.46$ fm, V=$4.79 \times 10^{-12} J$, $T = 23.1 \times 10^{10} K$, $Q = 14.95$ MeV / fusio$_n$).

RQ 17.9   Given the fusion reactions: $d + d \rightarrow {}_{2}^{3}He + n$ and $d + d \rightarrow {}_{1}^{3}H + p$, determine the Q-values. ($\Delta = 3.51 \times 10^{-3}$ u , $Q = 3.27$ MeV , $\Delta = 4.33 \times 10^{-3}$ u , $Q = 4.03$ MeV)

RQ. 17.10   Consider the thermo-nuclear reaction $d + d \rightarrow {}_{2}^{3}He + n + 3.2$ MeV .If the deuterons are at rest a) what is the kinetic energy of the neutron? b) If the deuterons are brought together within 100 fm for fusion to take place, calculate the energy of force required to apply for fusion to take place, c) What is the equivalent temperature of this energy? (Answers: a) 2.4 MeV $= 3.84 \times 10^{-13}$ J , b) 14.4 keV $= 2.3 \times 10^{-15}$ J , c) $\approx 10^{8}$ K).

&%&%&%&%&%&%&%&

Chapter 18

PARTICLES AND THEIR INTERACTIONS

Chapter 18

PARTICLES AND THEIR INTERACTIONS

	OM Namo Bhagavate Vaasudevaya	
	Ahamevaasamevaagre naanyadyat sadasatparam
Paschaadaham yadetaccha yo *f* vasishyeta so *f* smyaham	
	Riter *f* tham yatpratiyeta na pratiyeta	chaatmani
Tatvidyaatmano maayaam yatha *f f* bhaaso yatha tama:	
	Yatha mahaanti bhutaani bhuteshoocchaavacheshvanu
Pravishtaanya-pravishtaani tatha teshu na teshvaham	
	Etaavadeva jigjnaasyam Tatva jigjnaasunaa *f f* tmana:
Anvaya vyatirekaabhyaam yatsyaat sarvatra sarvadaa	
{|| Srimad *Bhaagavatam* (*Sk II*; *Chapter 9*: *Verses* 32 - 35)
Satya - Sanaa tan *a - Dharmam vijayetetaram* ||}
- Bhagavan, the Supreme GodHead, spoke to Brahma the Creator,
the Bhaagavatam in a Nutshell.

"... the nature of the perpetual things consists of small particles infinite in number... the particles are so small as to be imperceptible to us, and take all kinds of shapes and all kinds of forms and differences of size. Out of them, like out of elements (earth, air, fire, water) he now lets combine and originate the visible and perceptible bodies..." -
~ 450 B.C. Democritus

"If I could remember the names of all these particles, I'd be a botanist"

- Enrico Fermi

18.1   INTRODUCTION

In the previous Chapters, a number of elementary particles have been encountered and some features of their interactions have been described. Scientists have found that everything in the Universe is made up from a small number of basic building blocks called *elementary particles*, governed by a few *fundamental forces*. Some of these particles are stable and form the normal matter; the others live for fractions of a second and then decay to the stable ones. All of them coexisted for a few instants after the *Big Bang*.

Back in 1900, physicists believed that they had discovered more or less all the laws governing the Universe. The Laws of Classical Mechanics (due to Isaac Newton), Electromagnetism (due to Andre Marie Ampere, Karl Friedrich Gauss, Michael Faraday and James Clerks Maxwell) and Thermodynamics (due to James Prescott Joule, Lord Kelvin and Ludwig Boltzmann) seemed to explain pretty much everything.

However, at the beginning of the 20th Century two new theories emerged that revolutionized the way physicists looked at the Universe – Albert Einstein's theory of Special Relativity and Quantum Mechanics (due to such physicists as Niels Bohr, Louis deBroglie, Werner Heisenberg, Erwin Schrödinger and Paul AM Dirac). The unification of these theories led to the idea of *antimatter*. This in turn led to the mystery of why the Universe contains more matter than antimatter. It is this mystery that, at SLAC the experiment called **BaBar** experiment using the electron-positron collider, was investigated.

Until 1932, the *elementary* particles were the electron, the proton and the neutron. A perusal of the Particle Data Book will show more than 200 particles listed. They consist of the *gauge bosons* and *material particles*.

Particle physics is the study of the fundamental constituents of matter and their interactions. Modern theory, called the **Standard Model**, tries to explain all the phenomena of particle physics in terms of the properties and interactions of a small number of particles of tree distinct types, *viz.*, **leptons**, **quarks** and **gauge bosons**. The fundamental forces involved are the EM interaction, the Strong force the Weak interaction and Gravity. In this Chapter all these will be presented briefly.

18.2   SUBATOMIC PARTICLES

Until 1932, the "elementary" particles were the electron, proton, and neutron, and are listed in Table 18.1.

Table 18.1   Elementary Particles in 1913

Particle	Electron	Proton	Neutron
Symbol	$_{-1}^{0}e$	p^+ $_1^1H$	$_0^1n$
Mass (kg)	9.109×10^{-31}	1.673×10^{-27}	1.675×10^{-27}
(MeV)	0.51	938.2	939.6
Charge	$-1e$	$+1\ e$	0

One of the primary goals in modern physics is to answer the question "What is the Universe made of?". In other words, "what is matter and what holds it together?" It started from Democritus, Dalton and Rutherford. The search for the origin of matter means the understanding of elementary particles. This means their characteristics, how they interact and relate to other particles and forces of Nature.

This requires development of advanced technology, with the search for the primary constituent. Rutherford showed that the is composed of a nucleus and orbiting electrons;. Later it was understood that the nucleus is composed of neutrons and protons, which both are now known as formed from quarks.

The two most fundamental types of particle are Quarks and Leptons. Both are divided into 6 flavors corresponding to three generations of matter. Quarks and ant-quarks have electric charges in units of 1/2 and 2/3's. Leptons have charges in units of 1 or 0.

For every quark or lepton there is a corresponding anti-particle. Bosons do not have anti=particles since they are force carriers.

Quarks combine to form the basic blocks of matter, baryons and mesons. Baryons are formed from three quarks, whereas the mesons are quark pairs.

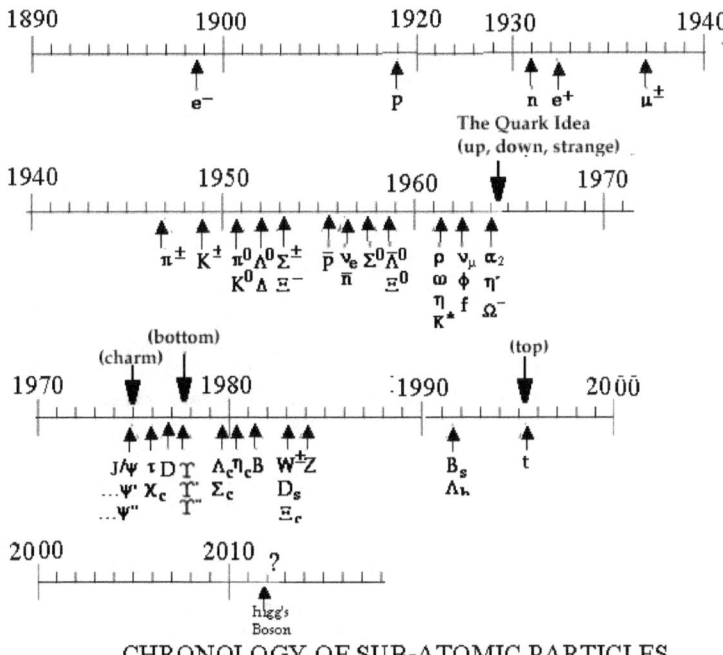

CHRONOLOGY OF SUB-ATOMIC PARTICLES

Fig 18.1 Elementary Particles 1898 – 2015

No free quarks exit. They must bind in pairs or triplets by gluons. This is called quark confinement. The exchange of gluons produces a colour force field, refers to assignment of colour charge to quarks, similar to electric charge.

Unlike in the EM force and weak forces, separation of two quarks causes the force field stronger. That energy increases until a quarks or antiquark is formed.($\Delta E = mc^2$). Quarks always travel in pairs or triplets.

18.2.1 Chronology of Elementary Particles

In 1928 Paul AM Dirac predicted that all particles should have opposites called *antiparticles*. The first of these discovered, the *positron*, was in 1932 by Carl Anderson. This was an electron with a positive electric charge (+1), the anti-electron. When an electron and positron come into contact, they mutually annihilate each other producing energy in accordance with Einstein's Equation, $\Delta E = mc^2$. Both the proton and the neutron have anti-particles. These also destroy each other if they meet with their particle. Ordinary matter is made up from particles. It appears that the Universe is made up of ordinary matter. Matter composed of anti-particles is known as *anti-matter*. Anti-matter can be created in the laboratory but does not last long as it quickly comes into contact with ordinary matter and is destroyed.

It is now known that there are many more elementary particles than the six described so far. These, with a very wide range of masses, have been created using modern high-technology equipment. These have been divided into a number of groups depending on their properties. Most of these newly discovered particles have their anti-particles. Table 18.2 lists the major discoveries related to the elementary particles

18.2.2.1 The History (1896-1930)

Henri Becquerel (1896) discovered some strange radiation coming from uranium salts. Pierre and Marie Curie isolated radium, a material much more radioactive than uranium.

Rutherford (1899) showed that two types of radiation exist, called $\alpha-$ and $\beta-$ rays. In 1902, Pierre and Marie Curie showed that $\beta-$radiation was electrons, while F. Soddy and E. Rutherford estimated that $\alpha-$, $\beta-$ and $\gamma-$ radiation are different types of radioactivity. Niels Bohr, among others, believed the sacred *energy conservation principle*, and Wolfgang Pauli (1930) verified by experiment another solution.

18.2.2.2 The Birth of Neutrino Idea(1930-1934)

Neutrinos ν appeared 15×10^9 *yrs* ago, soon after the birth of the Universe. Theoretically, they constitute a cosmic background radiation at 1.9 K ($-271.2\,^\circ C$). The other neutrinos of the Universe are produced during the life of stars and the explosion of supernovae.

J. Chadwick discovered (1932) the neutron, which did not correspond to the particle imagined by Pauli. F. Perrin showed (1933) that the neutrino mass has to be very much lower than the electron mass. The same year, Anderson discovered the positron, the first seen particle of anti-matter, verifying thus the theory of PAM Dirac and confirming the idea of neutrino hypothesized by Pauli and Fermi. Frederic Joliot Curie discovered (1933) the β^+ – radioactivity (a positron is emitted instead of an electron), while Fermi took the neutrino ν hypothesis and built his theory of $\beta-$ decay.

18.2.2.3 The Neutrino Quest

Hans Bethe and Rudolf Peierls (1934) showed that the cross section (probability of interaction) between neutrinos ν and matter should be extremely small.

18.2.3.1 First Quest of the inaccessible Star (1935-1956)

Luis Alvarez (1939) showed that tritium is radioactive. When $\beta-$ decay of tritium has been used to give the best limit on the neutrino mass. Despite of the horror os the explosion of the Atomic Bomb (1945) it inspires, Frederick Reines and Clyde Cowan performed experiment in 1956 and showed the neutrino clearly visible in the detector.

18.2.3.2 The Second Quest and Anti-Neutrino (1957-1962)

T.D. Lee and C.N. Yang (1962) verified that in

$$\mu^- \to e^- + \gamma \qquad (18.2.1)$$

the particle coming out of the interaction was identified as an e^- in some cases and in some as a μ^-.

18.2.4 Chronology of Elementary Particles discovered after 1964

Fig 18.1 gives an idea of the chronology of the particles identified from 1964.

18.2.5 Families (Collective Divisions) of Particles

Collectively those particles which experience the Strong interaction are referred to as '***hadrons***'. And this family is divided into '***baryons***' and '***mesons***'. Two other families of particles are called '***leptons***' and '***gauge bosons***'.

18.2.5.1 Leptons

The particles electron e^- (symbolically, $_{-1}^{0}e$), neutrino (ν), the muon (μ) and the tau (τ), *etc*, are referred to as *leptons*. Compared to the electron, two heavier leptons are the μ and the τ. Both are unstable and decay to simpler, more stable particles. Both have anti-particles. Muons are found in the air as cosmic rays enter the Earth's atmosphere and penetrate into atoms and molecules. There are three types of neutrino (each one associated with one of the three leptons described above), called the electron neutrino (ν_e), muon-neutrino (ν_μ), and tau neutrino (ν_t).

While neutrinos are not known to react with other types of matter they can easily pass through the Earth. They have no electric charge. Each one has its anti-particle version so there are **six types** of neutrinos. Neutrinos have a very low mass and one type can change into one of the other two types.

Nuclei of atoms do not show Leptons. They are not subject to the Strong Nuclear Force which keeps the nucleus from flying apart. They are sometimes produced in the nucleus but are quickly expelled. During β – decay a $_0^1 n$ in the nucleus breaks down to give a proton (which remains in the nucleus), an $_{-1}^{0}e$ (which emerges out and causes the radioactivity of the atom) and an ν_e (which departs at the speed of light but not usually detected).

$$_0^1 n \rightarrow p^+ + e^- + \nu_e \qquad (18.2.2)$$

The six leptons are listed in Table 18.2.

TABLE 18.2 **LEPTONS** - Participants in *Electro-weak* and *Gravitation*

Particle	Nick name	Spin ($\hbar$)	Mass (MeV/c^2)	Charge (e)	Colour Charge
1 Electron	e^-	$\frac{1}{2}$	0.511	-1	-
2 e-neutrino	ν_e	$\frac{1}{2}$	$< 10^{-6}$	0	-
3 Muon	μ	$\frac{1}{2}$	106	-1	-
4 mu-neutrino	ν_μ	$\frac{1}{2}$	< 0.17	0	-
5 Tau	τ	$\frac{1}{2}$	1.78×10^3	-1	-
6 τ-neurtrino	ν_τ	$\frac{1}{2}$	< 24	0	-

18.2.5.2 Baryons

Particles which experience the Strong interaction are collectively referred to as *hadrons*. The *fermions* with $J = half - integer$, such as the nucleons are the baryons, and the *bosons* with $J = Integer$, such as the pions form the mesons.

The proton p^+ and neutron ^1_0n (together called nucleons) are the common bosons, found in the nuclei of atoms. The Strong Nuclear Force binds them together. Table 18.3 lists them.

The proton and neutron are stable particles in the most nuclei. Outside the nucleus or in certain unstable nuclei, neutrons decay as in equation (18.2.2):

$$^1_0n \rightarrow p^+ + e^- + \nu_e \qquad (18.2.2)$$

There exist other baryons and are given in Table 18.3.

18.2.5.3 Mesons

In a nucleus, the p^+ and ^1_0n are not really separate entities, each with its own distinct identity. They change into each other by rapidly passing particles called pions (π) between themselves. Pions are the most common of the mesons' Some of the many known mesons are listed in Table 18.4.

TABLE 18.3 Nuclear Particle Zoo: (b) BARYONS $\left[j^P = \tfrac{1}{2}^+\right]$

Particle	I	S	Mass (MeV/c^2)	$\tau(s)$
p	$\tfrac{1}{2}$	0	938.28	stable
n	$\tfrac{1}{2}$	0	939.57	900
Λ	0	-1	1115.60	2.631×10^{-10}
Σ^+	1	-1	1189.37	0.800×10^{-10}
Σ^0	1	-1	1192.46	7.4×10^{-20}
Σ^-	1	-1	1197.34	1.479×10^{-10}
Ξ^0	$\tfrac{1}{2}$	-2	1314.9	2.90×10^{-10}
Ξ^-	$\tfrac{1}{2}$	-2	1321.32	1.64×10^{-10}
Charmed baryon				
Λ_C^+	0	0	2284.9	1.8×10^{-13}
Δ	$\tfrac{3}{2}$	0	1232.0	115 (*MeV*)
$\Sigma(1385)^+$	1	-1	1382.8	36 (*MeV*)
$\Sigma(1385)^0$	1	-1	1383.7	36(*MeV*)
$\Sigma(1385)^-$	1	-1	1387.2	39(*MeV*)
$\Xi(1530)^0$	$\tfrac{1}{2}$	-2	1531.8	9.1(*MeV*)
Ξ^+	$\tfrac{1}{2}$	-2	1535.0	9.9(*MeV*)
Ω^-	0	-3	1672.4	0.822×10^{-10}

Kaons (K) are short-lived mesons that decay into simpler particles. Normally, particles and anti-particles decay in a similar way. The example below shows the decay of $^1_0 n$.

$$^1_0 n \rightarrow p^+ + e^- + \nu_e \qquad (18.2.2)$$

$$^1_0 n \rightarrow p^- + e^+ + \nu_e \qquad (18.2.3)$$

The decays are mirror images of each other. Kaons are unique in that the matter and anti-matter forms occasionally decay in slightly different modes. This is referred to as a breakdown of a property called *parity* P. This breakdown of parity conservation may account for the fact that the Universe is mainly matter rather than a 50-50 mixture of matter and anti-matter. A mixed matter Universe would not last long as the matter and anti-matter would destroy each other.

18.2.5.4 Omega Particle, Ω^-

The scheme of classification into family of particles was not based on any underlying theory of fundamental structure, nor was it derived from any abstract principle. It simply provided a concise representation that exhibited symmetry and order and, additionally, predictive power. The mathematical group to fit the complex situation-*SU(3)*, the symmetric, *unitary group* of dimension 3- was proposed

TABLE 18.4 Nuclear Particle Zoo: (c) MESONS ($J^P = 0^-$)

Particle	I	S	Mass (MeV/c^2)	$\tau(s)$
π^+	1	0	139.57	2.60×10^{-8}
π^0	1	0	134.97	0.84×10^{-16}
π^-	1	0	139.57	2.60×10^{-8}
η^0	0	0	548.88	$1.08\ keV$
K^+	$\frac{1}{2}$	+1	493.65	1.237×10^{-8}
K^0	$\frac{1}{2}$	+1	497.67	$10^{-8}\ \&\ 10^{-10}$
$\eta'(958)$	0	0	957.57	$0.24(MeV)$
$\eta_c(2980)$	0	0	2979.6	$10.3(MeV)$
D^+	$\frac{1}{2}$	0	1869.3	10.7×10^{-13}
D^0	$\frac{1}{2}$	0	1864.5	4.3×10^{-13}
D_s^+	0	-1	1969.3	4.4×10^{-13}
Bottom Mesons				
B^+	$\frac{1}{2}$	0	5277.6	1.4×10^{-12}
B^0	$\frac{1}{2}$	0	5275.3	1.4×10^{-12}
$\rho(770)$	1	0	770.0	$153(MeV)$
$\omega(783)$	0	0	782.0	$8.5(MeV)$
$\phi(1020)$	0	0	1019.4	$4.41(MeV)$
$K^{*+}(892)$	$\frac{1}{2}$	+1	892.1	$51.3(MeV)$
$K^{*0}(892)$	$\frac{1}{2}$	+1	892.1	$51.3(MeV)$
$J/\psi(3097)$	0	0	3096.9	$0.068(MeV)$
$\Upsilon(9460)$	0	0	9460.3	$0.052(MeV)$
$D^{*+}(2010)$	$\frac{1}{2}$	0	2010.1	$< 2.0(MeV)$
$K^{*0}(2010)$	$\frac{1}{2}$	0	2007.1	$< 5(MeV)$

independently by Gell-Mann and Ne'eman and named by Gell-Mann the **Eightfold Way**.

The validity of *SU(3)* was demonstrated by experiment(1964). A major prediction was that a particle (named by Gell-Mann Ω^- the omega-minus), an isotopic singlet with spin = $\frac{3}{2}$, positive parity, mass of roughly 1,680 *MeV*, negative charge, baryon number B = +1, strangeness S = -3, and stable to strong decay, should exist to complete the $\frac{3}{2}^+$ baryon decuplet.

An incoming K^- meson interacts with a proton in the liquid hydrogen of the bubble chamber and produces an omega-minus Ω^-, a K^0 and a K^+ meson which all decay into other particles.

$$K^- + p \rightarrow K^0 + K^+ + \Omega^- \tag{18.2.4}$$

$$\Omega^- \rightarrow \pi^- + \Xi^0 \tag{18.2.5}$$

18.3. UNITS IN NUCLEAR PHYSICS

18.3.1. Units of Energy, Momentum and Mass

A useful unit of energy in Particle and Nuclear Physics is the eV. This is the amount of kinetic energy gained by an electron when it is accelerated through a potential difference of one Volt. Normally the energies involved in nuclear reactions are MeV and in high energy particle interactions they may be GeV (= 10^9 eV). A convenient unit for particle masses makes use of the Einstein mass-energy relationship

$$E = mc^2. \qquad (18.3.1)$$

This yields a unit mass as energy divided by the square of the velocity of light, MeV/c^2 or GeV/c^2. For example

Proton mass M_p = 938.3 MeV/c^2 $\qquad (18.3.2)$

Electron mass m_e = 0.511 MeV/c^2. $\qquad (18.3.3)$

To extend this system of units to momentum consider the relativistic relationship for the energy of a particle of rest mass m moving with momentum p,

$$E^2 = p^2c^2 + m^2c^4. \qquad (18.3.4)$$

It follows that by expressing momentum in the units of energy divided by the velocity of light (GeV/c), one gets a self-consistent system in which the velocity of light is implicitly used, but its value never explicitly appears in the calculations. Using these units,

$$E^2(GeV)^2 = p^2(GeV/c)^2 + m^2(GeV/c^2)^2 \qquad (18.3.5)$$

Kinetic energy is just expressed in GeV, so

$$E(GeV) = T(GeV) + m(GeV/c^2). \qquad (18.3.6)$$

The velocity of the particle in units of c is given by

$$\beta = \frac{p(GeV/c)}{E(GeV)} \qquad (18.3.7)$$

and the relativistic gamma factor is given by

$$\gamma = \frac{E(GeV)}{m(GeV/c^2)} \qquad (18.3.8).$$

In the non-relativistic limit $\beta \to 0$ and $p \ll m$. Thus kinetic energy

$$T = E - m = \sqrt{(p^2 + m^2)} - m$$
$$= m[1 + \sqrt{(p^2/m^2)}] - m$$

$= (p^2/2m)$ as expected.   (18.3.9)

Worked out Example 18.1

The pair annihilation can be represented by $e^- + e^+ \rightarrow 2\gamma$. What is the energy of each γ-ray and in what directions do they move? Given $m_{e^-} = m_{e^+} = 9.1 \times 10^{-31}$ kg.

Solution: Step #1 $m_{e^-} = m_{e^+} = 9.1 \times 10^{-31}$ kg

STEP * 2 Sum of rest energies of e^- and e^+ = 2 ($= 9.1 \times 10^{-31}$ kg) (3×10^8 m s^{-2}) = 1.64×10^{13} J = 1.022 MeV.

STEP * 3 Since energy is conserved, each photon has energy of 0.511 MeV.

STEP * 4 The photons must move in opposite directions to conserve momentum.

Worked out Example 18.2

Protons were accelerated in a Bevatron (proton synchrotron) to an energy of 6.4 GeV, and anti-protons were created as per the reaction: $p + p \rightarrow p + p + p + \bar{p}$. Prove that the # fermion is conserved in this reaction.

Solution: STEP * 1 The proton is a fermion and so the anti-proton. LHS has the # of protons is 2;

STEP * 2 On the RHS there are three protons and an anti-proton.

STEP * 3 However, the ant-proton is considered as a 'minus one' particle; so the total # of fermions on the RHS is 3 + (-1) = 2, which is the same for LHS.

STEP * 4 Therefore, the # of fermions is conserved in this reaction.

18.4   INTRINSIC PARTICLE PROPERTIES AND CONSERVATION LAWS

TABLE 18.5   Strength and Range of Fundamental Forces

Force	Relative Strenfth	Range	Exchanged particle
1 Strong (Nuclear)	1	$10^{-15}m$	Pions (π, K)
2 Electromgnetic	10^{-2}	Infinite	Photon, γ
3 Weak (Nuclear)	10^{-13}	$10^{-18}m$	Bosons (W$^-$, W$^+$, W^0)
4 Gravity	10^{-38}	Infinite	Graviton

Without *invariance principles*, there would be no Laws of Physics! An Invariance Principle reflects a *basic symmetry*, and is always intimately related to a Conservation Law (and to a quantity that cannot be determined absolutely). The strengths and ranges of the four fundamental interactive forces are listed in Table 18.5

18.4.1.1 Symmetry

The physical world is an undeniably complex system. Yet the Laws of Physics are remarkably simple. Gravitational force is a simple inverse square law. EM forces is described by means of concise mathematical formulae, as expressed by Maxwell's Equations. This is because the Nature displays marvelous 'symmetry' on a mathematical level.

Some mathematical operations, called *transformations*, may preserve symmetry.

In Particle Physics, non-geometric symmetries are used. Chen Ning Yang and Robert Mills (1954) made an important breakthrough, *viz.*, geometrical symmetries could be used to describe the relationship between forces and particles in terms of Quantum Field Theory (QFT). This led to a major understanding in the way nature assembles matter at the most fundamental level, including the prediction of new and exotic particles.

The standard of measurement that is applied or the gauge, does not affect the rules governing EM or gravitational fields. Particle physicists call this principle *gauge invariance*. In other words, EM, Strong, Weak and gravity are examples of gauge theories that display gauge symmetry. Many of these gauge symmetries can be manipulated by using the Group Theory of mathematics.

18.4.2 Energy, Momentum and Angular Momentum in Subatomic Physics

The mass and spin properties of particles are related to energy, momentum and angular momentum. Invariance of the Hamiltonian operator ($\hat{H}$) under displacement, translation and spatial rotation transformation characterizes the conservation laws of energy, momentum and angular momentum, respectively. (See, for example, in the book *Quantum Mechanics*, by S. Devanarayanan, 2005). These quantities are determined by applying these Conservation Laws. The laws are universal and apply to all, *i.e.*, EM, Strong and Weak interaction processes.

Some classical invariance principles are related to the nature of space-time. Invariance of the $\hat{H}$ (operator for total energy) under a translation for an isolated, multi-particle system leads directly to the conservation of the total momentum of the system. This can be demonstrated defining an operator $\hat{T}$ which produces a translation of the wave function through δx:

$$\hat{T}\ \psi(x) = \psi'(x) = \psi(x + \delta x)\ . \tag{18.4.1}$$

Then it can be shown that

$$\hat{T} \equiv e^{i\hat{p}\ \delta x/\hbar} \tag{18.4.2}$$

where $\hat{p}$ is the momentum operator.

$\hat{p}$ is said to act as a "**generator** of translations". Now since the energy of an isolated system cannot be affected by a translation of the whole system, $\hat{T}$ must commute with the Hamiltonian operator, $\hat{H}$,

i.e., $\quad [\hat{T}, \hat{H}] = 0 ;$ $\hfill(18.4.3)$

it must therefore also be true that

$$[\hat{p}, \hat{H}] = 0, \tag{18.4.4}$$

and so $\hat{p}$ has eigenvalues which are constants of the motion.

Thus there are three *equivalent* statements:

1. **Momentum is conserved in an isolated system.**
2. **The Hamiltonian is invariant under spatial translations. (Equivalently, it is impossible to determine absolute positions.).**
3. **The momentum operator commutes with the Hamiltonian ($\hat{H}$).**
   These continuous transformations cited led to additive conservation laws - **the sum of all charges or momenta is conserved.**

18.4.2.1 Electric Charge q

Another conserved quantity is electric charge (q), corresponding to an invariance of physical systems under a translation in the electrostatic potential - the physics only depends on potential differences. Quantum mechanically, we may define a charge operator $\hat{Q}$ which, when it operates on a wave function ψ_q describing a system of total charge q, returns an eigenvalue of q.

$$\hat{Q} \psi_q = q \psi_q \tag{18.4.5}$$

If q is conserved, $\hat{Q}$ and $\hat{H}$ must commute each other.

$$[\hat{Q}, \hat{H}] = 0, \tag{18.4.6}$$

this is assured by invariance under a global phase (or gauge) transformation

$$\psi'_q = e^{i\varepsilon \hat{Q}} \psi_q \, \varepsilon \tag{18.4.7}$$

where ε is an arbitrary real parameter.

Charge conservation is well known and has been verified with considerable accuracy. **Charge conservation law is universal and applies to all interactions.**

185.2.2 Noether's Theorem

Emmy Noether discovered that symmetries give rise to conservation Laws

18.5.3 Lepton number L

One can classify leptons as particles and antiparticles, associating a *lepton number L* of +1 for the particles and, -1 for the antiparticles, *i.e.* $L = +1$ for $^{0}_{-1}e$, *etc* and, while $L = -1$ for $^{0}_{+1}e$, *etc*. Table 18.5 lists the various members of the lepton family and corresponding L.

Table 18.5 Lepton number and Family of Leptons

L	e^-	ν_e	μ^-	ν_μ	τ^-	ν_τ
L_e	+1	+1	0	0	0	0
L_μ	0	0	+1	+1	0	0
L_τ	0	0	0	0	+1	+1
:	e^+	$\bar{\nu}_e$	μ^+	$\bar{\nu}_\mu$	$\bar{\tau}$	$\bar{\nu}_\tau$
L_e	-1	-1	0	0	0	0
L_μ	0	0	-1	-1	0	0
L_τ	0	0	0	0	-1	-1

It is known **that lepton number is conserved in all weak decays**, as shown in Worked out Example 18.3.

Worked out Example 18.3

Show that lepton number is conserved in the following reactions

a) $_0^1 n \rightarrow p^+ + e^- + \nu_e$, and b) $p \rightarrow _0^1 n + e^+ + \nu_e$

Solution: STEP * 1   a) $_0^1 n \rightarrow p^+ + e^- + \nu_e$

$L \qquad 0 \rightarrow 0 + 1 + (-1) = 0$

STEP * 2   b) $p \rightarrow _0^1 n + e^+ + \nu_e$
$L \qquad\quad 0 \rightarrow 0 + (-1) + 1 = 0$

Conservation of lepton number has been tested experimentally at a level of one part in 10^{12}.

Worked out Example 18.4

Given the following reactions:

i0) $\bar{\nu}_e + p \rightarrow n + e^+$
ii) $\nu_\mu + n \rightarrow \mu^- + p$
iii) $\mu^- \rightarrow e^- + \bar{\nu}_e + \nu_\mu$

Prove that lepton number for each family of lepton (*i.e.*, electron, muon, tau) must be separately conserved in reactions.

Solution STEP * 1 i) All leptons in this decay belong to the electron family.

$$\bar{\nu}_e + p \rightarrow n + e^+$$
$$L_e \quad (-1) + 0 \rightarrow 0 + (-1) = -1$$

STEP * 2 Lepton number is seen to be conserved.

STEP * 3 ii) All leptons in this decay are muon-type.

$$\nu_\mu + n \rightarrow \mu^- + p$$
$$L_\mu \quad 1 + 0 \rightarrow 1 + 0 = +1$$

STEP * 4 Lepton number is conserved.

STEP * 5 iii) This is muon decay. It contains members of electron and muon type leptons.

$$\mu^- \rightarrow e^- + \bar{\nu}_e + \nu_\mu$$
$$L_e \quad 0 \rightarrow 1 + (-1) + 0 = 0$$
$$L_\mu \quad +1 \rightarrow 0 + 0 + 1 = +1$$

STEP * 6 It can be seen that the electron lepton number and the muon lepton numbers are separately conserved.

Example 18.4 leads to the following statement:

In processes involving particle reactions, the lepton numbers for electron-type, muon-type and tau-type leptons must remain the same.

That is why neutrinos and anti-neutrinos must be observed for the conservation of lepton number in particle reactions.'

18.5.4 Baryon number, B

The antiproton was discovered by Chamberlain et al.(1955). The observation explained conservation of *baryon number*, B. Particles such as p and n are called baryons and are assigned $B = +1$, whereas the pions and kaons are mesons with $B = 0$. Antiparticles of the baryons (anti-baryons) have $B = -1$. All the particles listed in Table 18.3 are baryons.

Baryon number appears to be conserved in all reactions, thus explaining why protons, the lightest baryons, cannot decay, whereas neutrons can. Neutrons decay via

$$^1_0n \rightarrow p^+ + e^- + \nu_e \qquad (18.4.8)$$

while the only possible decays for protons would be, for example,

$$p^+ \rightarrow {}^1_0n + e^+ . \qquad (18.4.9)$$

Some Grand Unified Theories (GUTs) predict that proton decay will occur at a very low rate 10^{33} *yrs*. Experiments have failed to find it rather close to this level, which corresponds to about one decay a year in a detector of about 1,000 *tonnes*. It will become clear later that conservation of baryon number results from conservation of quarks. Non-conservation of baryon number is one of the Sakhorov criteria in a model which attempts to explain why there is no antimatter in the Universe.

Conservation of baryon number is not seen violated, except GUT suggests that the proton might decay in a manner that should violate this Law.

Worked out Example 18.5

The anti-proton was discovered by means of the following particle reaction:

$p^+ + p^+ \rightarrow p^+ + p^+ + p^+ + \bar{p}$. Check if the conserving Baryon number is obeyed..

Solution $\boxed{Step \ \#1}$ $p^+ + p^+ \rightarrow p^+ + p^+ + p^+ + \bar{p}$.

$\boxed{Step \ \#2}$ $\quad p^+ + p^+ \rightarrow p^+ + p^+ + p^+ + \bar{p}$
$\qquad\qquad B \Rightarrow \ 1 \ +1 \ = 1 \ +1 \ +1 \ +(-1) \quad \boxed{OK}$

i.e., the Fermion # is conserved.

Worked out Example 18.6

Is the following reaction possible?

$p^+ + p^+ \rightarrow p^+ + p^+ + \check{n}$.

Solution $\boxed{Step \ \#1}$ $\qquad\qquad p^+ + p^+ \rightarrow p^+ + p^+ + \check{n}$.

$\boxed{Step \ \#2}$ $\quad p^+ + p^+ \rightarrow p^+ + p^+ + \check{n}$
$\qquad\qquad B \Rightarrow \ 1 \ +1 \ = 1 \ +1 \ +(-1)$

i.e., $\boxed{LHS \neq RHS}$, the total B # is not conserved; the reaction is forbidden.

18.4.5 Isospin, I

The hadrons possess a **non-zero** quantum number called *isospin* (*isobaric* or *isotopic*), denoted by $I = \tfrac{1}{2}$, which reflects a basic symmetry found in the Strong Interaction. There are two different substates of a nucleon, viz., $I_3 = +\tfrac{1}{2}$ for the proton and $I_3 = -\tfrac{1}{2}$ for the neutron (it is conventional to use 1, 2 and 3 to indicate components in isospin space rather than $x, y,$ and z).

For each nucleon having isospin $I = \tfrac{1}{2}$ the total isotopic spin can be $I = 0$ or 1. The different 3 components of it correspond to the following configurations.

$$\begin{array}{llccc}
I_3 & = & +1 & 0 & -1 \\
I & = 1 & pp & np & nn \\
I & = 0 & & np &
\end{array} \qquad (18.5.10)$$

This is because the protons and neutron are not identical particles, *i.e*, they have different EM properties.. The triplet, $I = 1$ state is reflected, for example, in the triplet of essentially identical states in the nuclei ^{14}O, ^{14}N, ^{14}C.

In the case of pions,

$$I_3 = +1 \quad \text{signifying} \quad \pi^+$$
$$I_3 = 0 \quad , \quad \pi^0 \quad\quad\quad (18.5.11)$$
$$I_3 = -1 \quad , \quad \pi^-$$

Isospin for the hadrons and mesons are given in Table 18.3 and Table 18.4.

18.4.6 STRONG INTERACTION

18.4.6.1 Non-strange hadrons

The nucleons and pions were the only known strongly interacting particles. until the existence of **strange particles** had been established, The Feynman graph of strongly interacting particles are shown.

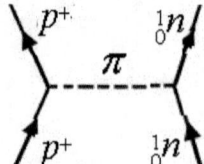

The nucleons belong to an *iso-doublet* ($I = \frac{1}{2}$, $I_3 = \pm\frac{1}{2}$), and the pions to an *iso-triplet* ($I = 1$, $I_3 = 0$, ± 1). Fermi and Yang (1949) argued that the pions can be regarded as bound states of nucleon and an anti-nucleon.

$$\pi^+ = p\bar{n}$$
$$\pi^0 = \frac{1}{\sqrt{2}} (n\bar{n} - p\bar{p}) \quad\quad\quad (18.4.12)$$
$$\pi^- = -n\bar{p}$$

This is based on the following conditions:

a) To obtain the correct spin J =0, the pion has to be in a singlet state; the lowest state of the nucleon-anti-nucleon pair will be an $S = 0$ state.
b) To have the pion in a state to have $B = 0$, the nucleon and anti-nucleon have opposite B values.
c) Since the nucleon and anti-nucleon have opposite parity, the overall parity is odd so that $J^P = 0^-$ is required.

If one applies Group Theory, the two isospin states are the simplest basic states in the Special Unitary Group in 2-D, *SU(2)* group. The corresponding *Pauli spin matrices* are the simplest representation of this group. Thus the combination of the doublet ($I = \frac{1}{2}$) isospin states for nucleon and anti-nucleon signified as 2 and $\bar{2}$ resulting in a triplet (3) $I = 1$ and the singlet (1) $I = 0$, is symbolized as

$$2 \otimes 2 = 3 \otimes 1$$

The Strong force is identical for the two, since any difference in behaviour arising only from the EM force due to the proton's charge is equivalent to the saying that I is conserved in the Strong interaction, which depends on I but is independent of I_3. There is evidence for this in nuclear energy levels for related nuclei.

Example: Consider mirror nuclei, such as ^{7}Li & ^{7}Be, ^{13}N & ^{13}C, or ^{39}Ca & ^{39}K, the energy level diagrams are identical in their energy and J^P structure, apart from small differences due to the EM interaction.

Three different nuclei can be related in a more complicated manner. For example $^8_4$Be, $^{10}_5$B, $^{12}_6$C each consists of closed p and n shells a $^8_4$Be core with nn, np and pp respectively in an outer shell.

Now two nucleons, each with $I_3 = +1$, must have a combined I_3 of +1 and hence I = 1. The different 3-components then correspond to the following configurations

$$
\begin{array}{llll}
I_z = & +1 & +1 & 0 & -1 \\
I = & 1 & pp & np & nn \\
I = & 0 & & np &
\end{array}
\qquad (18.4.13)
$$

The fact that neutron and proton are not identical particles (*e.g* they have different EM properties) and so more states are allowed them than two protons or two neutrons. Therefore two possible np states .The isospin structures of *the I* = 1, *I* = 0 neutron-proton states are

$$\psi(I=1, I_3 = 0) = \frac{1}{\sqrt{2}}(pn + np)$$
$$\psi(I=0, I_3 = 0) = \frac{1}{\sqrt{2}}(pn - np)$$
(18.4.14)

The triplet, I=1, state is reflected, for example in the tree essentially identical states in the nuclei ^{14}O, ^{14}N & ^{14}C. Now two protons, each with $I_z = +\frac{1}{2}$, must have a combined $I_3 = +1$ and hence $I = 1$; similarly two neutrons. However, an np state has $I_z = 0$ and hence can have $I = 0$ and $I = 1$. With the postulate that the strong interaction depends only on I and not on I_z, one set of energy levels with I = 1 will occur in all three nuclei, while another set with $I = 0$ will only occur in ^{10}B

The allocation of isospin can be extended to the three pions:

$I_3 = +1 \quad for \quad \pi^+$

$I_3 = 0 \quad for \quad \pi^0$ \hfill (18.4.15)

$I_3 = -1 \quad for \quad \pi^-$

Any Strong interaction H involving pions as well as nucleons must also be scalar in isospin space and the total isospin must be conserved in any process resulting from the interaction. This is seen confirmed in

$$^2H + {}^2H \rightarrow {}^4He + \pi^0$$
$$I_3 = \quad 0 + 0 \quad = \quad 0 + 0$$ \hfill (16.4.16)

But pion has $I = 1$, *i.e.*, isospin is not conserved in this process, and so not a strong interaction process. (^{2}He is a 2p bound state and ^{4}H is a 1p, 3n bound state, both having $I_3 \neq 0$, but $I = 0$).

18.4.7 EM INTERACTION

For nucleons and pions the electrical charge Q of a particle in units of e is related to the I_3 component of isospin by the following relations

Nucleons: $\quad Q = e\left(I_3 + \frac{1}{2}\right)$ \hfill (18.4.17)

Pions: $\quad Q = e(I_3)$. \hfill (18.4.18)

This can be applied for both of particles by a single relation,

For Hadrons:

$$Q = e\left(I_3 + \frac{B}{2}\right)$$ \hfill (18.4.19)

Because charge is related to I_3, EM interaction must be related to I_3. But I_3 is a vector in isospin space, and so that EM interaction cannot be invariant under spatial rotations in isospin space. So **EM interaction doesn't conserve isospin**.

To visualize how matter particles exchange messenger particles in EM interaction is, say in the case of electron and positron get annihilated creatinf a photon is obtained from the Feynman diagram shown.

18.4.8 WEAK INTERACTION

As in the case of EM interaction, it follows that Weak interaction between hadrons do not conserve isospin.

Leptons and photons are not allocated an isospin. It can therefore be stated **that isospin is only conserved in strong interaction processes.** In the Feynman graph below the neutrino – proton gives a proton.

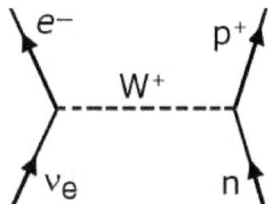

18.4.9 Strangeness, S

18.4.9.1 Strange particles

The identification of pions and muons is from the study of high energy cosmic rays. .

In 1954, Murray Gell-Mann proposed a new property of quantum particles, called strangeness, with its own quantum number S. Like electric charge, strangeness is conserved in particle decays, when the strong nuclear force is at work. Though electric charge is conserved in any interaction, strangeness is only conserved in the Strong and EM interactions. Strangeness was introduced to explain particle like K mesons and the Λ^0 (lambda) had unexpectedly long lifetimes when interacting with the strong force.

18.4.9.2 The Associated Production

The processes where strange particles are produced by ordinary particles (S = 0) is called the '**associated production**'. This concept was proposed by A. Pais (1952), and others so that particles produced by strong interactions but decay by weak interactions (*i.e.*, they are observed to live 10^5 billion times longer than they should) are called strange particles. While ready production (r-process) implies **strong interaction** with the atomic nuclei, slow decay (s-process) implies **weak interaction**, and this leads one to a dilemma

Using accelerators it was found (1953) that strange particles are 'typically formed in pairs. This mechanism, called 'associated production', is highly suggestive of an additive conserved quantity, such as charge, which was called strangeness. Further it was observed that the K meson and the Σ particle are always produced in pairs, as in

$$\pi^- + p^+ \xrightarrow{10^{-22}s} \Lambda^0 + K^0 \qquad (18.4.20)$$

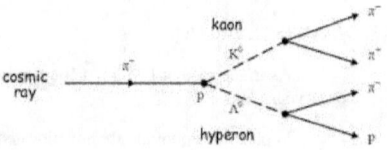

Other examples of processes of associated production are:

$$\pi^+ + p^+ \xrightarrow{10^{-22}s} \Sigma^+ + K^+ \qquad (18.4.21)$$

$$\Sigma^+ \longrightarrow n^0 + \pi^+ \qquad (18.4.22)$$

$$K^+ \longrightarrow \pi^0 + \pi^+ \qquad (18.4.23)$$

The following processes violate S-conservation, but takes place.

$$K^0 \xrightarrow{\text{slow decay}} \pi^+ + \pi^- \qquad (18.4.24)$$

$$\Xi^- \xrightarrow{\text{slow decay}} \Lambda^0 + \pi^- \qquad (18.4.25)$$

$$\Lambda^0 \xrightarrow{10^{-10}s} \pi^- + \pi^+ \qquad (18.4.26)$$

Particles are either strange or non-strange, and S is assigned to them as S = 0, S = +1, or S = -1. .

The *Law of Conservation of Strangeness* is stated as follows:

"**The total strangeness must remain constants in particle interactions governed by the strong and EM forces**".

"On the other hand, the strangeness either remains the same or $\Delta S = \pm 1$, in processes where the weak force is involved".

So Conservation of S is an experimental law.

The processes which are not allowed are:

$$n^0 + n^0 \xrightarrow{\text{NOT allowed}} \Lambda^0 + \Lambda^0$$

$$\pi^- + p^+ \xrightarrow{\text{NOT allowed}} n^0 + \Lambda^0$$

$$\pi^- + p^+ \xrightarrow{\text{NOT allowed}} \Lambda^0 + \pi^0$$

Worked out Example 18.7

Examine if the decay $\Sigma^0 \rightarrow \Lambda^0 + \gamma$ can happen?

Solution $\boxed{Step\ \#\ 1}$ $\Sigma^0 \rightarrow \Lambda^0 + \gamma$

$\boxed{Step\ \#\ 2}$
$$\Sigma^0 \rightarrow \Lambda^0 + \gamma$$
$$S = \quad -1 \quad \rightarrow \quad -1 + 0 \quad = -1 \quad \boxed{Ok}$$

$\boxed{Step\ \#\ 3}$ Yes, it happens through the strong interaction.

Worked out Example 18.8

Why is the decay $\Sigma^0 \rightarrow n + \gamma$ is not observed??

Solution $\boxed{Step\ \#\ 1}$ $\Sigma^0 \rightarrow n + \gamma$

$\boxed{Step\ \#\ 2}$
$$\Sigma^0 \rightarrow n + \gamma$$
$$S = \quad -1 \quad \rightarrow \quad 0 + 0 \quad \neq -1$$

$\boxed{Step\ \#\ 3}$ So, it violates the Law of conservation of strangeness, and the decay cannot take place when the strong interaction at play.

Worked out Example 18.9

Comment on the decay $\Xi^0 \rightarrow n + \pi^0$.

Solution $\boxed{Step\ \#\ 1}$ $\Xi^0 \rightarrow n + \pi^0$

$\boxed{Step\ \#\ 2}$
$$\Xi^0 \rightarrow n + \pi^0$$
$$S = \quad -2 \quad \rightarrow \quad 0 + 0 \quad \neq 0, \pm 1$$

$\boxed{Step\ \#\ 3}$ Since the strangeness change by more than one unit, this decay cannot happen at all.

18.4.9.3 Hypercharge, Y

For all *strange* hadrons

$$Q = e\left(I_3 + \frac{B+S}{2}\right) \quad (18.4.27)$$

This expression can be written as

$$Q = e\left(I_3 + \frac{Y}{2}\right) \quad (18.4.28)$$

where the new quantum number $Y = B + S$ is referred to as the *hypercharge*.

Strangeness (S) and *hypercharge* (Y) are the conserved in Strong interaction processes, but not always in Weak interaction decays.

For *non-strange* particle it was seen earlier that

$$Q = e\left(I_3 + \frac{B}{2}\right).$$

This can be simply modified so as to be applicable to strange hadrons by

$$Q = e\left(I_3 + \frac{B+S}{2}\right) = e\left(I_3 + \frac{Y}{2}\right) \quad (18.5.29)$$

where the new quantum number

$$Y = B + S \quad (18.5.30)$$

is called the *'hypercharge'*.

Table 18.6 – Summary

Invariance	Conserved Quantity
Gravitation, weak, electromagnetic and strong interactions are independent of:	
translation in space	linear momentum
rotations in space	angular momentum
translations in time	energy
EM gauge transformation	electric charge
CPT	(product of parities below)
Gravitation, electromagnetic and strong interactions are independent of:	
spatial inversion P	spatial parity
charge conjugation C	"charge parity"
time reversal T	"time parity"

The lifetimes are consistent with Weak interaction decays, yet that are produced in a Strong interaction process by protons or pions in the plates in the cloud chamber. This inconsistent behaviour gave rise to the name strange particles and was explained by postulating that they carry a new quantum number, called strangeness, with symbol S, which is conserved in the Strong interaction but not in the Weak interaction. These are also known as the **cascade particles**.

Table 18.6 lists summary of the conserved quantities and corresponding invariances.

18.4.10 Resonances

There are two types of particle, viz. *weakly* interacting particle (for example, μ meson) and the *strongly* interacting particle.(like K-meson). These have been produced in the decay processes taking a time range of 10^{-6} to $10^{-10}s$ in Strong interaction and $\Box\, 10^{-23}s$ in Weak interaction, respectively. However, a group of particles, christened as *resonances* or *resonant states* decay in a time $\Box\, 10^{-23}s$, (Luis W. Alvarez,1953). Resonances are specified by its symbol followed by its mass and spin within braces, example $\Lambda^*(1380, \frac{3}{2}^+)$, in the reaction:

$$K^- + p^+ \rightarrow \Lambda^0 + \pi^+$$
$$\Box$$
$$\pi^0 + \pi^- \qquad (18.4.31)$$

which is an example of baryonic resonance. If the resonant state is produced from a meson decay it is known as mesonic resonance, as shown in

$$\pi^- + p^+ \rightarrow \omega^{0+} + n$$
$$\Box$$
$$\pi^+ + \pi^0 + \pi^- \qquad (18.4.32)$$

18.4.11 Hypernuclei

A hypernucleus is formed when one of the nucleons is replaced by a hyperon a metastable nucleus. Example: Danysz and Pniewski (1957) discovered in nuclear emulsion detector exposed to cosmic radiation the hyperfragment (a nuclear fragment containing a hyperon) $^{8}_{\Lambda}Be$.

18.5 THE QUARK MODEL OF PARTICLES

18.5.1 What are Quarks?

Murray Gell-Mann and George Zweig (1964) showed that the Eightfold Way patterns could be replicated if the mesons and baryons were composed of 'furthermore' elementary particles, which Gell-Mann called **quarks**. The most striking feature of the quarks is that they have *fractional electric charges*. Table 18.7 lists the charge content of various quarks.

TABLE 18.7. QUARKS - Participants in *Electro-weak*, *Strong* and *Gravitation*

Particle	Nick name	Spin, s ($\hbar$)	Mass (MeV/c^2)	Charge (e)	Colour Charge
1 Up	u	$\frac{1}{2}$	~5	$+\frac{2}{3}$	r, g, b
2 Down	d	$\frac{1}{2}$	~10	$-\frac{1}{3}$	r, g, b
3 Strange	s	$\frac{1}{2}$	~200	$-\frac{1}{3}$	r, g, b
4 Charm	c	$\frac{1}{2}$	~1.5×10^3	$+\frac{2}{3}$	r, g, b
5 Bottom	b	$\frac{1}{2}$	~4.5×10^3	$-\frac{1}{3}$	r, g, b
6 Top	t	$\frac{1}{2}$	~180×10^3	$+\frac{2}{3}$	r, g, b

In the 1960s by the **Quark Model** which says that hadrons (Most of the more than 200 particles are mesons and baryons, or, collectively, hadrons) are made out of spin-$\frac{1}{2}$ particles called quarks.

There are 6 leptons and 6 quarks (Table 18.11). These can be subdivided further into 3 *generations*: in the quark sector, for example, the up and down quark form the *first* generation, while the charmed and strange quarks form the *second* generation. Similarly, the first lepton generation consists of the electron and its associated neutrino, the second lepton generation is comprised of the muon and its associated neutrino. Further one adds to these force carriers belonging to the 4 fundamental forces (photon, gluon, W/Z). Because quantum gravity is not yet understood, the graviton is presently left out.

The unique property of quarks is called **confinement**. This is the experimentally determined fact that except for the top quark, which decays too rapidly, individual quarks are not seen — they are always confined inside subatomic particles called **hadrons**, such as protons, neutrons, mesons. This fundamental property is expected to follow from the modern theory of strong interactions, called QCD. Although there is no mathematical derivation of confinement in QCD, it is easy to show using lattice gauge theory.

Quarks and Leptons are the two basic constituents of matter in the Standard Model of Particle Physics. Antiparticles of quarks are called antiquarks. **Quarks and antiquarks are the only fundamental particles that interact through all four of the fundamental forces.**

18.5.2 FURTHER DETAILS

18.5.2.1 What is the Eight-fold Way?

Many attempts have been made in trying to discover a system of organization that grouped elementary particles into larger groups of identification.

18.5.2.2 Spin-$\frac{1}{2}$ Baryon Multiplet

Symmetry operations on an octahedron illustrate (1964) the theory of quarks. A diagram (the Eightfold Way) is plotted that strangeness (S) is plotted horizontally and the quantum number (I_3) is plotted vertically on an x - y coordinate system, with Q (electric charge) for all the families of particles for which these are known. This a plot with a definite geometrical symmetry.

For example, the eight spin-$\frac{1}{2}$ baryons when plotted gives a geometrical symmetry shown in Fig 18.4. Look at how the particles are arranged in a hexagonal form with two baryons in the centre. Each of these baryons has a baryon number B=1, a spin of one-half and positive parity. The symmetry of charge (Q) along the diagonal of Fig 18.4 develops just by plotting strangeness (S) versus quantum number I_3. Not only does a neat geometric shape appear, but charge automatically develops into a perfect symmetry as well.

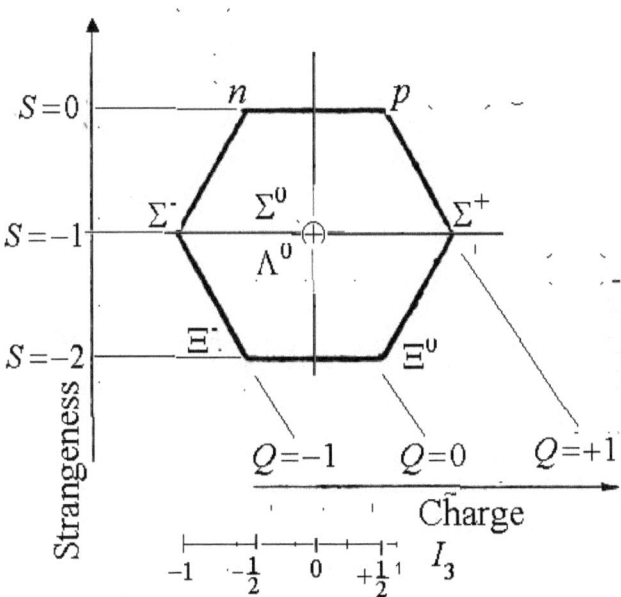

Fig 18.4. Eight-fold Way & Spin-$\frac{1}{2}$ Baryon Multiplet

These groups, from the $SU(3)$ symmetry group of the Lie groups, developed by Sophus Lie, are the result of combining smaller groups of hadrons called **multiplets**. Multiplets combine particles of like mass and like interaction capabilities that only differ in charge into a common group. Looking back at the diagram it will be seen that the $^1_0 n$ and p^+, the Σ^-, Σ^0 and Σ^+, and the Ξ^- and Ξ^0 are three different multiplets that are closely connected in the **supermultiplet** horizontally.

18.5.2.3 Spin-0 Meson Multiplet

The multiplet for spin-0 mesons is an Octet shown in Fig 18.5.

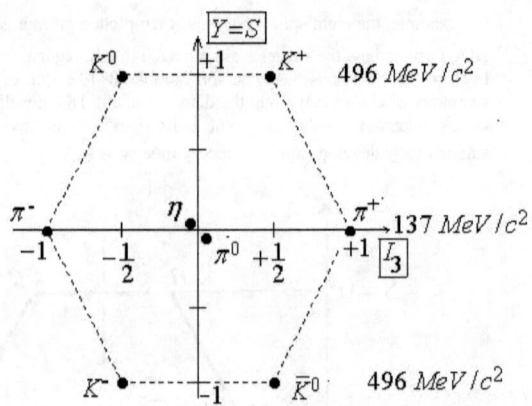

Fig 18.5 $\boxed{P=0}$ Eightfold Way & Octet spin-0 Mesons

These plots are called the **Eightfold Way patterns** by Murray Gell-Mann and Yuval Ne'eman (with the name adopted from Eight-fold path in Buddhism).

Other geometrical symmetries also emerge.

18.6.3.3 The Decuplet of Baryons.

For the ten baryons (decuplet) the Eightfold Way plot looks like in Fig 18.6.

When this plot was prepared only nine baryons were known in the family, but the predicted tenth particle Ω^-, and sooner in 1964 it was discovered.

The Eightfold Way in Particle Physics is analogous to the Periodic Table in Chemistry.

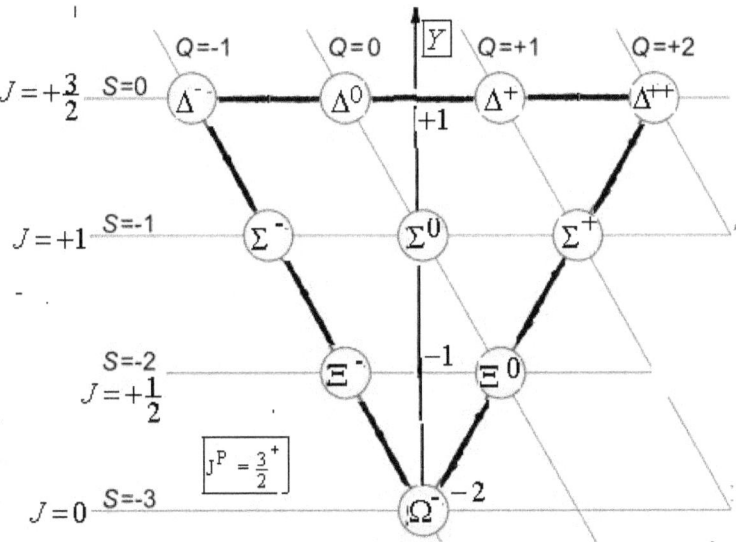

Fig 18.6   Eightfold Way & Decuplet spin-$\frac{3}{2}$ heavy Baryons

58.5.3.1 Confinement and Quark Properties

Quarks have never been found in isolation! As the force increases indefinitely as the distance between quarks increases, it is extremely difficult to extract the quark in isolation.. In a high energy collision, if a quark becomes sufficiently separated then a new quark grouping can be formed

Every subatomic particle is completely described by a small set of observables such as mass **m** and quantum numbers, such as spin **J** and parity **P**. Usually these properties are directly determined by experiments. However, confinement makes it impossible to measure these properties of quarks. Instead, they must be inferred from measurable properties of the composite particles which are made up of quarks. Such inferences are most easily made for certain additive quantum numbers called flavors.

The composite particles made of quarks and antiquarks are the hadrons. These include the mesons which get their quantum numbers from a quark and an antiquark, and the baryons, which get theirs from three quarks. The quarks (and antiquarks) which impart quantum numbers to hadrons are called **valence quarks**. Apart from these, any hadron may contain an indefinite number of virtual quarks, antiquarks and gluons which together contribute nothing to their quantum numbers. Such virtual quarks are called **sea quarks**.

18.5.3.2 What is Flavour of a Quark?

Each quark is assigned a baryon number, $B = 1/3$, and a vanishing lepton number $L = 0$.

They have fractional electric charge, Q, either $Q = +2/3$ or $Q = -1/3$. The former are called up-type (u) quarks, the latter, down-type (d) quarks.

18.5.3.3 Spin of Quark

Quantum numbers corresponding to non-Abelian symmetries like rotations require more care in extraction, since they are not additive. In the quark model one builds mesons out of a quark and an antiquark, whereas baryons are built from three quarks. Since mesons are bosons (having integer spins) and baryons are fermions (having half-integer spins), the quark model implies that quarks are fermions. Further, the fact that the lightest baryons have spin-1/2 implies that each quark can have spin $J = 1/2$. The spins of excited mesons and baryons are completely consistent with this assignment.

18.5.3.4 Isospin of Quarks

Each quark is assigned a weak isospin: $T_z = +\frac{1}{2}$ for an u-quark and $T_z = -\frac{1}{2}$ for a d-quark. Each doublet of weak isospin defines a **generation** of quarks.

There are three **generations**, and hence six flavors of quarks — the u-quark flavours are up (u), charm (c) and top (t); the d-quark flavours are down (d), strange (s), and bottom (b) (each list is in the order of increasing mass).

The number of generations of quarks and leptons are equal in the Standard Model.

The number of generations of leptons with a light neutrino is strongly constrained by experiments in CERN and by observations of the abundance of helium in the Universe. Precision measurement of the lifetime of the Z boson at LEP constrains the number of light neutrino generations to be three. Astronomical observations of helium abundance give consistent results.

Results of direct searches for a fourth generation give limits on the mass of the lightest possible 4th generation quark. The most stringent limit comes from analysis of results from the TEVATRON Collider at Fermilab, and shows that the mass of a 4th generation quark must be greater than 190 GeV. Additional limits on extra quark generations come from measurements of quark mixing performed by the experiments Belle and BaBar.

Each flavour defines a quantum number which is conserved under the strong interactions, but not the weak interactions. The magnitude of flavour changing in the weak interaction is encoded into a structure called the CKM matrix. This also encodes the CP violation allowed in the Standard Model. The flavour quantum numbers will not be described further

18.5.3.5 Colour of Quark

Since quarks are fermions, the Pauli Exclusion Principle implies that the three valence quarks must be in an anti-symmetric combination in a baryon. However, the charge $Q = 2$ baryon, Δ++ (which is one of four isospin Iz = 3/2 baryons) can only be made of three u quarks with parallel spins. Since this configuration is symmetric under interchange of the quarks, it implies that there exists another internal quantum number, which would then make the combination antisymmetric. This is given the name "colour", although it has nothing to do with the perception of the frequency (or wavelength) of light, which is the usual meaning of colour. This quantum number is the charge involved in the gauge theory called Quantum Chromo Dynamics (QCD).

The only other coloured particle is the gluon, which is the gauge boson of QCD. Like all other non-Abelian gauge theories (and unlike QED) the gauge bosons interact with one another by the same force that affects the quarks.

Colour is a gauged $SU(3)$ symmetry. Quarks are placed in the fundamental representation, 3, and hence come in three colours. Gluons are placed in the adjoint representation, 8, and hence come in eight varieties. For more on this, see the article on colour charge.

18.5.3.6 Quark Masses

Although one speaks of quark mass in the same way as the mass of any other particle, the notion of mass for quarks is complicated by the fact that quarks cannot be found free in nature. As a result, the notion of a quark mass is a theoretical construct, which makes sense only when one specifies exactly the procedure used to define it.

The approximate chiral symmetry of QCD, for example, allows one to define the ratio between various (up, down and strange) quark masses through combinations of the masses of the pseudo-scalar meson octet in the Quark Model through *chiral* Perturbation Theory, giving

$$\frac{m_u}{m_d} = 0.56, \quad \text{and} \quad \frac{m_s}{m_d} = 20.1 \tag{18.5.1}$$

The fact that $\mu \neq 0$ is important, since there would be no strong CP problem if μ were to vanish. The absolute values of the masses are currently determined from QCD Sum Rules (also called spectral function Sum Rules) and lattice QCD. Masses determined in this manner are called current quark masses. The connection between different definitions of the current quark masses needs the full machinery of renormalization for its specification.

18.5.3.7 Gell-Mann-Nishijima Mass Formula

Another method of specifying the quark masses was to use the **Gell-Mann-Nishijima Mass Formula** in the Quark Model, which connects hadron masses to quark masses. The masses so determined are called constituent quark masses, and are significantly different from the current quark masses defined above. The constituent masses do not have any further dynamical meaning. More detailed *SU(3)* theory predicts the masses of members of multiplets. For the Baryon Octet, $J^P = \frac{1}{2}^+$, it is predicted that

$$\boxed{2(m_N + m_\Xi) = 3 m_\Lambda + m_\Sigma} \tag{18.5.2}$$

or

$$\boxed{\tfrac{1}{2}(m_N + m_\Xi) - \tfrac{1}{4}[3 m_\Lambda + m_\Sigma] = 0} \tag{18.5.3}$$

This relation is call eth <u>Gell-Mann – Okubo Mss formula</u>.

where (ignoring small differences between p and n; Σ^+, Σ^0 and Σ^-; and Ξ^0, Ξ^-). This holds good for all other multiplets. This establishes quark description of hadrons.

Proceeding from general considerations and relying on his own intuition, S. Okubo (1962) obtained a formula for particle masses in iso-multiplets contained in a Unitary Multiplet

$$\boxed{m(I,Y) = m_0\{1 + aY + b\,[I(I+1) - \tfrac{1}{4}Y^2]\}} \tag{18.5.4}$$

where m_0, a and b are constants characterizing the super-multiplet as a whole.

For the Meson Multiplet

$$\boxed{m^2(I,Y) = m_0^2\{1 + b\,[I(I+1) - \tfrac{1}{4}Y^2]\}} \tag{18.5.5}$$

This for the basic Meson Octet becomes

$$\boxed{3 m_0^2 + m_1^2 = 4 m_{1/2}^2} \qquad (18.5.6)$$

Worked out Example 18.10

Arrive at the Gell-Mann-Ocubo mass formula, for the Baryon Unitary Octet, given $m(I,Y) = m_0\{1 + aY + b\,[I(I+1) - \frac{1}{4}Y^2]\}$.

Solution: $\boxed{Step\ \#\ 1}$ $m(I,Y) = m_0\{1 + aY + b\,[I(I+1) - \frac{1}{4}Y^2]\}$

$\boxed{Step\ \#\ 2}$ Substituting the values of I and Y,

I	Y	Relation
0	0	$m_\Lambda = m_0$
$\frac{1}{2}$	+1	$m_N = m_0\,(1 + a + \frac{1}{2}b)$
$\frac{1}{2}$	-1	$m_\Xi = m_0\,(1 - a + \frac{1}{2}b)$
1	0	$m_\Sigma = m_0\,(1 + 2b)$

$\boxed{Step\ \#\ 3}$ Eliminating m_0, a and b one gets

$$2\,(m_N + m_\Xi) = 3\,m_\Lambda + m_\Sigma$$

Worked out Example 18.11

Using the Gell-Mann/Okubo mass formula, together with the known masses of the nucleon N (use the average of p and n), Σ (again, use the average), and Ξ (ditto), predict the mass of the Λ. How close do you come to the observed value?

Solution: $\boxed{Step\ \#\ 1}$ The Gell-Mann-Okubo formula is $\boxed{2\,(m_N + m_\Xi) = 3\,m_\Lambda + m_\Sigma}$ (18.5.2) and $m_\Lambda = 1115.6\ MeV$.

$\boxed{Step\ \#\ 2}$ Solving Eq. (18.5.2) for m_Λ, one gets $m_\Lambda = 2\,m_\Xi - m_\Sigma/3$.

$\boxed{Step\ \#\ 3}$ Now, by averaging the masses of the p and the n (with c = 1) $m_N = 938.9\ MeV$.

$\boxed{Step\ \#\ 4}$ By similar techniques, $m_\Xi = 1318.1\ MeV$ and $m_\Sigma = 1193.1\ MeV$.

$\boxed{Step\ \#\ 5}$ Inserting these $m_\Lambda = 1107.0\ MeV$. This compares very favorably to the actual mass of the Λ, $m_\Lambda = 1115.6\ MeV$.

18.5.3.8 Heavy quark masses

From the masses of hadrons containing a single heavy quark (and one light antiquark or two light quarks) and from the analysis of quarkons, the masses of the heavy c and b quarks are obtained. Computations using the Heavy Quark Effective Theory (HQET) or Non-Relativistic Quantum Chromodynamics (NRQCD) are currently used to determine these quark masses.

Being heavy, it can be used to find the mass of the *top* quark. The best theoretical estimates of the t quark (before it was discovered in 1995) mass are obtained from global analysis of precision tests. The t quark is unique amongst quarks in that it decays before having a chance to hadronize. Thus, its mass can be directly measured from the resulting decay products. This can only be done at which is the only energetic enough to produce t quarks in abundance.

Truth and **beauty**, respectively, are the names given to the quantum numbers of the top and bottom quarks.

18.5.3.9 Antiquarks

The additive quantum numbers of antiquarks are equal in magnitude and opposite in sign to those of the quarks.

18.6   The Quark Structure of Subatomic Particles

(Hadron Multiplets)

18.6.1   Meson Multiplets

These are the 0^- octet and singlet and the 1^- octet and singlet.

So the combination of a quark and an antiquark triplet produces an octet and a singlet ($3 \otimes \bar{3} = 8 \oplus 1$) of 0^- and also of 1^- mesons.. According to this hypothesis, it should be possible to detect an octet and a singlet of 0^- and also of 1^- mesons, having similar strong interaction properties. These are shown in Fig 18.9. These are known as 'weight diagrams', which have ordinates as I_3 and abscissa as Y $(= B + S)$. Each particle is represented by a point on the diagram and bears with its symbol and quark-antiquark content. The two particles are lying at the centre of the octet. The approximate masses of the different mesons are also shown. They are the same for the different isospin multiplets, but they differ between multiplets. This is an example of symmetry breaking and that there is not complete $SU(3)$ symmetry between the three quarks.

Case a) A proton is (uud), n is (udd).

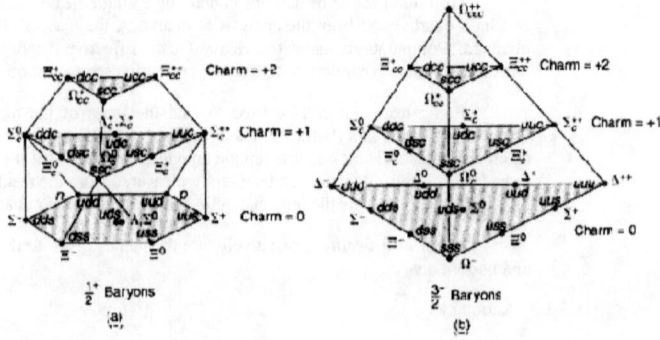

Fig 18.9 The weight diagrams for (a) lowest octet $J^P = \frac{1}{2}^+$ and (b) decuplet $J^P = \frac{3}{2}^+$.

Table 18.9 and 18.10 list the constituent quarks of the members of the Particle Zoo.

18.6.2 Baryon Multiplet:

Baryons multiplet structure, *viz.*,

. $\frac{1}{2}^-$ octet form $3 \otimes 3 \otimes 3$ and

$\frac{3}{2}^+$ decuplet $10 \otimes 8 \otimes 8 \otimes 1$.

Since there are three quarks combination, for the lowest mass states, the three quarks are expected to be in relative S-states. This gives the resultant hadron to have $J^P = \frac{1}{2}^+$ or $\frac{3}{2}^+$. The outcome of $3 \otimes 3 \otimes 3$, which is known to be $10 \otimes 8 \otimes 8 \otimes 1$ so that a decuplet and octets of baryons are expected.. They exist in nature. The weight diagrams for lowest octet $J^P = \frac{1}{2}^+$ and decuplet $J^P = \frac{3}{2}^+$ as in Fig 18.9.

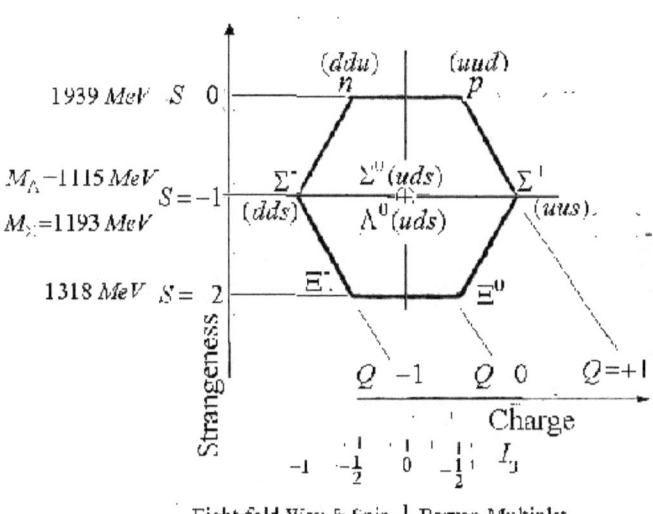

Eight-fold Way & Spin-$\frac{1}{2}$ Baryon Multiplet

Fig 18.10

Prior to 1964, the Ω^- meson was not known, only 9 members of the decuplet were known.

Worked out Example 18.12

Find the wavelength associated with the electrons with momentum 22 GeV/c used for deep inelastic scattering experiment.

Solution Step # 1 $\lambda = h/p$

$$= \frac{(6.63 \times 10^{-34} J-s)}{(22 \times 10^9 eV)(1.6 \times 10^{-19} J/eV)(3.0 \times 10^8 m-s^{-1})} = 5.7 \times 10^{-17} m.$$

Step # 2 Such energetic electrons have a wavelength in the scale $\frac{1}{100}$ th the radius of a proton.

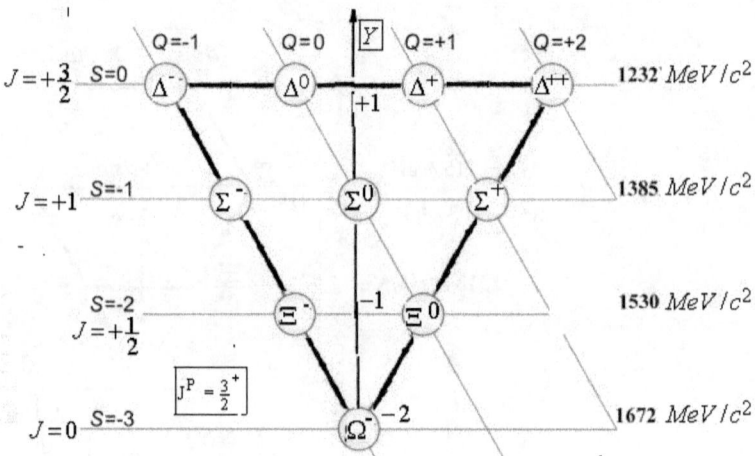

Eightfold Way & Decuplet spin-$\frac{3}{2}$ heavy Baryons

Fig 18.12

<u>Case b)</u> Experimental Evidence foe Quarks from Deep Inelastic Scattering

One can infer the internal structure of protons and neutrons by deep inelastic scattering experiments where nucleons are bombarded with high energetic (15 to 200 GeV) beams of electrons, muons, or neutrinos using particle accelerators. The first such experiment was conducted in 1970s at the SLAC using 22 GeV/c momentum electrons.

The first Eightfold Way patterns for baryons and mesons required the existence of the three quarks, u, d and s. Prediction of c, t and b quarks were made.

A meson called the J/psi (J/ψ) was discovered having a mass of 3.1 GeV/c^2 by Burton Richter and Samuel Chao Chung Ting (1974). The *b* quark was discovered by Leon Max Lederman *et al.* (1977) at the Fermilab with the discovery of the Upsilon (Y) meson, using 500 GeV protons. In 1978 the B-mesons were discovered at the DESY Electron-positron Collider and subsequently more details of *b*-quarks were found. The most massive *t* quark was found (1995) at the 1 TeV Tevatron proton synchrotron at the Fermilab using the CDF detector.

18.6.3 Quarkonium

Quark-anti-quarks ($q\bar{q}$) bound together is state of matter called *quarkonium*.

Similarly a strange and anti-strange quarks bound together ($s\bar{s}$) is called *strangeonium*.

18.6.4. Beta Decay Revisited

In terms of quarks, beta decay can be interpreted as follows:

$$\begin{pmatrix} u \\ d \\ d \end{pmatrix} \xrightarrow{\beta} \begin{pmatrix} u \\ u \\ d \end{pmatrix} + e^- + \bar{\nu}_e \qquad (18.6.4)$$

In the decay,

$$d \longrightarrow u + W^- \qquad (18.6.5)$$

followed by

$$W^- \longrightarrow e^- + \bar{\nu}_e \qquad (18.6.6)$$

Here in the weak interaction, what takes place is a quark changes its flavor! (In strong interaction quark changes to anti-quark).

Worked out Example 18.13

Confirm that Σ^- particle (spin-$\frac{1}{2}$) consists of two d and one s quarks. It is known that Σ^- has B = +1, S = -1. Also, $B = \frac{1}{3}$ for the d and s quarks, $Q = -\frac{1}{3}e$, and S = 0 for d and S = -1 for s quarks.

Solution | Step # 1 | On substituting the appropriate values of quantum numbers,

$$\Sigma^- = d + d + s$$
$$Q = -1e = (-\tfrac{1}{3}e) + (-\tfrac{1}{3}e) + (-\tfrac{1}{3}e) \quad \boxed{\text{Ok}}$$

| Step # 2 |
$$\Sigma^- = d + d + s$$
$$B = +1 = (\tfrac{1}{3}) + (\tfrac{1}{3}) + (\tfrac{1}{3}) \quad \boxed{\text{Ok}}$$

, | Step # 3 | :
$$\Sigma^- = d + d + s$$
$$S = -1 = (0) + (0) + (-1) \quad \boxed{\text{Ok}}$$

| Step # 4 |
$$\Sigma^- = d + d + s$$
$$Q = -1e = (-\tfrac{1}{3}e) + (-\tfrac{1}{3}e) + (-\tfrac{1}{3}e) \quad \boxed{\text{Ok}}$$
$$B = +1 = (\tfrac{1}{3}) + (\tfrac{1}{3}) + (\tfrac{1}{3}) \quad \boxed{\text{Ok}}$$
$$S = -1 = (0) + (0) + (-1) \quad \boxed{\text{Ok}}$$

spin-$\frac{1}{2}$ of the tree quarks can be arranged parallel or antiparallel, to yield a net spin-$\frac{1}{2}$ for the Ξ^-.

Therefore, the Σ^- particle has the quark structure (dds).

Worked out Example 18.14

Find the quark structure of the spin-$\frac{1}{2}$ particle Ξ^-; if it has S = -2, and contains three quarks involving only up or down and strange flavors. $Q = -\frac{1}{3}e$, S = 0 for the d quark and $Q = -\frac{1}{3}e$, S = -1. for the s quark; .u has S = 0, $Q = -\frac{2}{3}e$. The quarks are all spin-$\frac{1}{2}$ particle..

Solution Step # 1 Ξ^- has S = -2, means it contains two s quarks, as S = -1 for s.

Step # 2 The third quark must be either u or d.

Step # 3 The two s quarks have total $Q(s\,s) = (-\frac{1}{3}e) + (-\frac{1}{3}e) = (-\frac{2}{3}e)$ Ok.

Step # 4 The third quark must be d, since it has $Q = -\frac{1}{3}e$, and not u which has $Q = -\frac{2}{3}e$ .so that Ξ^- has $Q(d\,s\,s) = -\frac{1}{3}e) + (-\frac{1}{3}e) + (-\frac{1}{3}e) = -1e$ Ok.

Step # 3 Ξ^- has $S(d\,s\,s) = (0) + (-1) + (-1) = -2$ Ok.

Step # 5 spin-$\frac{1}{2}$ of the tree quarks can be arranged parallel or antiparallel, to yield a net spin-$\frac{1}{2}$ for the Ξ^-.

New theories called Superstrings, Twisters and M- Theory are attempting to link relativity (especially gravity) and predict the properties of all the sub-atomic particles and the forces of nature.

TABLE 18.8 Conservation Rules in Particle Physics

Interaction	Conserved?	P,E, L.q.	Lepton L_e, L_μ, L_τ	Baryon B	Strangeness S	Isospin I	Parity P	Conjugation C	Time reversal T
1. Strong		Yes	Yes	Yes	Yes	Yes	Yes	Yes	Yes
2. EM		Yes	Yes	Yes	Yes	No	Yes	Yes	Yes
3. Weak		Yes	Yes	Yes	No	No	No	No	No

Table 18-5 lists strength and range of the four fundamental forces between two protons. Note that the strong force acts between quarks by an exchange of gluons. The residual strong force between two protons can be described by the exchange of a neutral pion. Note, the $W^\pm$ is not included as an exchange particle for the weak interaction because it is not exchanged in the simplest proton-proton interaction.

The potential energy associated with each force acting between two protons is characterized by both the strength of the interaction and the range over which the interaction takes place. In each case the strength is determined by a coupling constant, and the range is characterized by the mass of the exchanged particle. The potential energy, U, between two protons a distance r apart is written as

$$U = \frac{C^2}{r} e^{-r/R},\qquad(18.6.7)$$

where $R = \frac{h}{mc}$ is the range of the interaction, and C^2, is the strength of the interaction. In each case the interaction is due to the exchange of some particle whose mass determined the range of the interaction, R. The exchanged particle is said to mediate the interaction.

The range of the gravitational and EM forces is infinite, while that of the strong and weak forces is very short. Also, the strength of the interaction depends on the separation between the two protons. Both gravitation and electromagnetism are of infinite range and their strengths decrease as the separation, r, increases – falling off as $1/r^2$. Table 18.8 gives an useful listing of the conserved quantities in an interaction process.

18.6.5.1 Carriers of the Basic Forces

Each force is carried by an elementary particle. The electromagnetic force, for instance, is mediated by the *photon*, the basic quantum of EM radiation. The strong force is mediated by the *gluon*, the weak force by the *W* and *Z* particles, and gravity is thought to be mediated by the *graviton*. QFT applied to the understanding of the EM force is called QED, and applied to the understanding of strong interactions is called QCD. In 1979 Sheldon Glashow, Steven Weinberg, and Abdus Salam were awarded the Nobel Prize in Physics for their work in demonstrating that the EM and weak forces are really manifestations of a single *electroweak* force. A *Unified Theory* that would explain all the *four* forces as manifestations of a single force is being sought.

18.6.5.2 The Quark Structure of Hadrons

In order to include the strange particles in the ideas of the previous section, one has to reject the concept of the nucleons as the building blocks. M. Gell-Mann & Ne'eman (1961) suggested the $SU(3)$ symmetry for the hadrons.

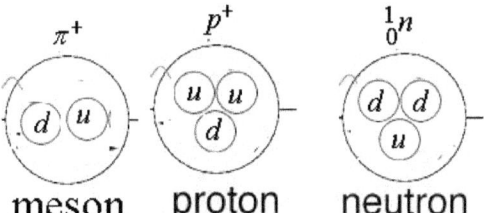

Fig 18.10 $SU(3)$ scheme of the mesons and the nucleons

Gell-Mann & Zweig (1964) suggested that the hadrons are states of three fundamental particle called 'quarks' which have spin $\frac{1}{2}$. Accordingly the $SU(3)$ scheme the mesons are built up from quark and anti-quark; the baryons from three quarks (Fig 18.10).as shown in Table 18.9 and Table 18.10.

18.6.5.3 Antisymmetry and Colour

TABLE 18.9   MESONS :- Quark Structure

Particle		Intrinsic Spin $J^P(\hbar)$	Mass (MeV/c)	Strangeness S	Isospin I, I_3	Life time $\hbar/\Gamma$
1	π^+ ($u\bar{d}$)	0^-	140	0	1, +1	$2.6 \times 10^{-8}s$
2	π^0 ($u\bar{u}+d\bar{d}$)	0^-	135	0	1, 0	$8.4 \times 10^{-17}s$
3	π^- ($d\bar{u}$)	0^-	140	0	1, -1	$2.6 \times 10^{-8}s$
4	K^+ ($u\bar{s}$)	0^-	494	+1	$\frac{1}{2}, +\frac{1}{2}$	$1.2 \times 10^{-8}s$
5	K^0_S ($d\bar{s}+s\bar{d}$)	0^-	498	mix	$\frac{1}{2}$, mix	$8.9 \times 10^{-11}s$
6	K^0_L ($d\bar{s}+s\bar{d}$)	0^-	498	mix	$\frac{1}{2}$, mix	$5.2 \times 10^{-8}s$
7	K^{--} ($s\bar{u}$)	0^-	494	+1	$\frac{1}{2}, -\frac{1}{2}$	$1.2 \times 10^{-8}s$
						Line width
8	ρ^+ ($u\bar{u}$)	1^-	769	0	1, +1	151 MeV
9	ρ^0 ($u\bar{u}+d\bar{d}$)	1^-	769	0	1, 0	151 MeV
10	ρ^- ($d\bar{u}$)	1^-	769	0	1, -1	151 MeV
11	K^{*+} ($u\bar{s}$)	1^-	892	+1	$\frac{1}{2}, +\frac{1}{2}$	50 MeV
12	K^{*0} ($d\bar{s}$)	1^-	896	+1	$\frac{1}{2}, -\frac{1}{2}$	51 MeV
13	$\bar{K}^{*0}$ ($s\bar{d}$)	1^-	896	-1	$\frac{1}{2}, +\frac{1}{2}$	51 MeV
14	$\bar{K}^{*-}$ ($s\bar{u}$)	1^-	892	-1	$\frac{1}{2}, -\frac{1}{2}$	50 MeV

The Pauli Exclusion Principle insists that the antisymmetry of the intrinsic particle wavefunction under the exchange of identical quarks manifests most strikingly for Δ^-(ddd), Δ^{++}(uuu) and Ω^-(sss). Since the overall wavefunction for these particles is symmetrical (say, for $J_z = \frac{3}{2}$ state, each quark has $s = \frac{1}{2}$, and the overall wavefunction for these particles is symmetrical under the exchange of quarks). This is not allowed, and so a further attribute called '**colour**' was proposed by Greenberg (1964), as there should be three varieties- red, yellow and blue. The colour couples the carrier of the inter-quark interaction, called the 'gluon' to a quark.

The three colour states of a quark are regarded as a 3-D representation of an $SU(3)$ 'colour group'. This leads to a singlet state

$$3 \otimes 3 \otimes 3 = 1 \oplus 8 \oplus 8 \oplus 10$$

as does the combination of quark and antiquark

$$3 \otimes \bar{3} = 1 \oplus 8.$$

18.6.5.4 Charm Quarks

The first definite evidence (1974) of the existence of hadrons whose properties could not be accounted for in terms of the three basic quarks, viz., u, d, and s. The bombardment of Be by 30 GeV protons led to the production of the $e^- - e^+$ pair with specific energy (3097 MeV) in the CM system suggested that this was the result of the decay a particle or resonance, called the J/ψ

meson. The SLAC experiment with colliding beams of 2 *GeV* electrons and positrons also revealed the resonance J/ψ.

TABLE 18.10   HADRONS - Quark Structure

BARYONS		Intrinsic Spin $J^P(\hbar)$	Mass (MeV/c)	Strangeness S	Isospin I, I_3		Life time
1 p	(uud)	$\frac{1}{2}^+$	938	0	$\frac{1}{2}$, $+\frac{1}{2}$		$> 10^{32}$ yrs
2 n	(udd)	$\frac{1}{2}^+$	940	0	$\frac{1}{2}$, $-\frac{1}{2}$		889 s
3 Σ^+	(uus)	$\frac{1}{2}^+$	1189	-1	1, +1		8.0×10^{-11} s
4 Σ^0	(uds)	$\frac{1}{2}^+$	1193	-1	1, 0		7.4×10^{-20} s
5 Σ^-	(dds)	$\frac{1}{2}^+$	1197	-1	1, -1		1.5×10^{-10} s
6 Λ^0	(uds)	$\frac{1}{2}^+$	1116	-1	0, 0		2.6×10^{-10} s
7 Ξ^0	(uss)	$\frac{1}{2}^+$	1315	-2	$\frac{1}{2}$, $+\frac{1}{2}$		2.9×10^{-10} s
8 Ξ^-	(dss)	$\frac{1}{2}^+$	1321	-2	$\frac{1}{2}$, $-\frac{1}{2}$		1.6×10^{-10} s
							Line width
9 Δ^{++}	(uuu)	$\frac{3}{2}^+$	1232	0	$\frac{3}{2}$, $+\frac{3}{2}$		120 MeV
10 Δ^+	(uud)	$\frac{3}{2}^+$	1232	0	$\frac{3}{2}$, $+\frac{1}{2}$		120 MeV
11 Δ^0	(udd)	$\frac{3}{2}^+$	1232	0	$\frac{3}{2}$, $-\frac{1}{2}$		120 MeV
12 Δ^-	(ddd)	$\frac{3}{2}^+$	1232	0	$\frac{3}{2}$, $-\frac{3}{2}$		120 MeV
13 Σ^{*+}	(uus)	$\frac{3}{2}^+$	1383	-1	1, +1		~40 MeV
14 Σ^{*0}	(uds)	$\frac{3}{2}^+$	1384	-1	1, 0		~40 MeV
15 Σ^{*-}	(dds)	$\frac{3}{2}^+$	1387	-1	1, -1		~40 MeV
16 Ξ^{*0}	(uss)	$\frac{3}{2}^+$	1532	-2	$\frac{1}{2}$, $+\frac{1}{2}$		~10 MeV
16 Ξ^{*-}	(dss)	$\frac{3}{2}^+$	1535	-2	$\frac{1}{2}$, $-\frac{1}{2}$		~10 MeV
17 Ω^-	(sss)	$\frac{3}{2}^+$	1672	-3	0, 0		8.2×10^{-11} s

This led to the knowledge that J/ψ has the structure $(c,\bar{c})$. This is due to the involvement of a new quark *flavour* called *charm*.(denoted by *c*). The $(c,\bar{c})$ is referred to as the '*charmonium*' just like the $e^- - e^+$ is called *positronium*. In a similar way, *bottom*, denoted by *b* quark was identified in 1977.

The charges of all quarks, in particular, and all hadrons, in general, is led to the relationship

$$Q = e\{I_3 + \frac{1}{2}(B + S + \Box + \wp + \Im)\} \qquad (18.6.8)$$

18.6.5.5 Interaction between a Pair of Quarks

The exchange particle between say a proton and an electron in the EM potential is the photon, as calculated by QED. This has three main components:

$$V_{EM} = V_C + V_{LS} + V_{SS} \qquad (18.6.9)$$

The last two terms are responsible for fine structure and hyperfine structure effects in the H-atom.

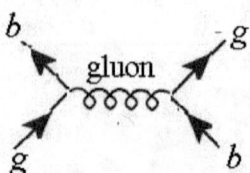

In the Strong interaction between b and g quarks, the exchange of massless particles called **gluons** (which couple to the colour charge of a quark just as the massless photon couples to electric charge) is attributed (See the Feynman diagram). The theory, called a **Gauge Theory,** invariant under certain gauge transformations, is the QCD.

In the case of Gravitation interaction the Feynman graph is drawn below.

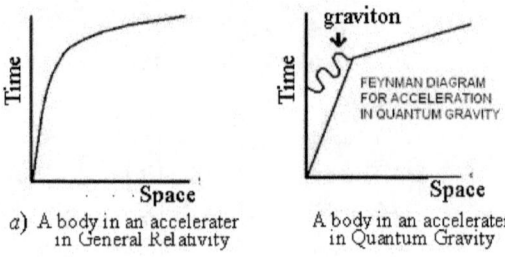

a) A body in an accelerater in General Relativity

A body in an accelerater in Quantum Gravity

18.6.6 THE WEAK INTERACTION AND UNIFICATION

18.6.6.1 Introduction

There are three types of decays:

a) Decays characterized by a width Γ, rather than a lifetime t, measured in *MeV* corresponding to $\tau \Box 10^{-21} s$ to $10^{-23} s$, and such decays are Strong interaction processes in which all Conservation Laws are satisfied.
b) Decays with somewhat longer lifetimes characterize EM interactions, in which the isospin conservation is violated.
c) Many decays which have lifetimes greater than $\tau \approx 10^{-13} s$ are attributed to Weak interaction, where parity is violated. In this section this weak decays will be dealt with.

18.6.6.2 The Electro-weak Theory

After Maxwell's unification of electric and magnetic forces was the first made by Clerk Maxwell. Broken symmetry in the weak interaction led (1967) Steven Weinberg, Sheldon Lee Glashow and Abdus Salam to the *Electroweak Theory*. This involves combining together the group representing the EM interaction, *viz.*, the U(1), and the group representing the weak, called the *SU(2)*, to form a composite group $SU(2) \times U(1)$, which predicted the existence of the three bosons that carry the weak force.. Richard Feynman diagrams of the electro-weak force is shown in Fig 18.11 Thus the EM and the weak forces are different manifestations of a single electroweak force.

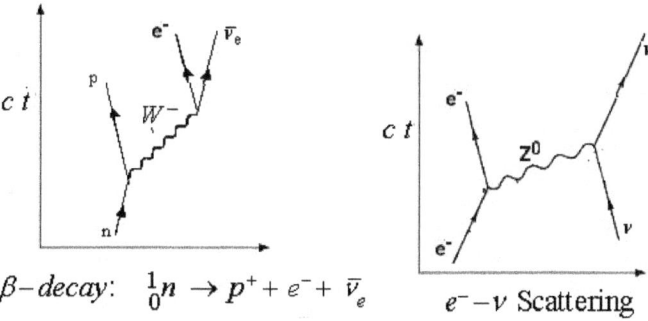

Fig 18.11   Feynman diagrams for β^- –decay and e^- –ν scattering

18.6.6.3 Leptonic Decays

The decays of the μ^- & μ^+ belong to leptonic decays, in which the conservation of lepton numbers, L_- and L_+ are only involved.

$$\mu^- \rightarrow \nu_\mu + e^- + \bar{\nu}_e$$
$$\mu^+ \rightarrow \bar{\nu}_\mu + e^+ + \nu_e \qquad (18.6.10)$$

18.6.6.4 Non-Leptonic Decays

Both hadrons and leptons are involved in these decays.

For $\Delta S = 0$,

$$^1_0 n \rightarrow p^+ + e^- + \nu_e \qquad (18.6.11)$$

$$\pi^- \to \mu^- + \bar{\nu}_\mu$$
$$\to e^- + \bar{\nu}_e \qquad (18.6.12)$$

$$\mu^- + {}^A_Z X \to {}^A_{Z-1} Y + \nu_\mu \qquad (18.6.13)$$

For $\Delta S = \pm 1$,

$$K^+ \to \pi^0 + e^+ + \nu_e \qquad (18.6.14)$$

$$\Lambda^0 \to p + e^- + \bar{\nu}_e \qquad (18.6.15)$$

18.6.6.5 An Exotic Five-quark particle

All subatomic particles were fit into two categories for about 4 decades: quark-triplet baryons, like protons and neutrons; or mesons, made up of a quark pair. Then the first five-quark particle was detected in 2003 (Fig 18.12).

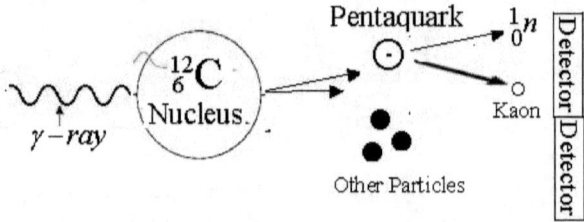

Fig 18.12   Penta-quark Experiment scjhamatics

The new particle is a sort of baryon-meson hybrid with five quarks--or, more precisely, four quarks and one anti-quark. The **pentaquark** is a member of the baryon family, but it is remarked to be **exotic** because the anti-quark has a different "flavor" to the other quarks.

One could think of many possible combinations of the six quarks and six antiquarks. But not all of them can exist, according to the rules of Quantum Chromodynamics (QCD), the theory that describes the Strong interactions between quarks. For instance, QCD forbids four-quark configurations, while the pentaquark is a permitted state. From conservation laws the only possible configuration for this new particle, dubbed Theta-plus, is two up quarks, two down quarks, and an anti-strange quark

Pentaquark = $(uudd\bar{s})$.

However, Dmitri Diakonov et al. (1997) predicted by theoretical means the existence of a pentaquark with a mass about 50 % heavier than that of a hydrogen atom. A pentaquark contains an exotic antiquark (the so-called strange antiquark). The composite particle also contains two up quarks and two down quarks—the same ones found in ordinary matter.

18.7 The STANDARD MODEL

Particle physicists mean the Standard Model as the combination of QED, the Electro-weak unification, the Quark model and QCD together.

$$\boxed{\text{STANDARD MODEL}} = \boxed{\text{Q E D}} + \boxed{\text{Electroweak Theory}} + \boxed{\text{Quark Model}} + \boxed{\text{Q C D}} \;.$$
(18.7.1)

Both the leptons and the quarks are structureless (point-like) particles which are the most and ultimate fundamental particle. It can be stated that

"All matter can be thought of as composed of combinations of six types of quark and their antiquarks and six types of lepton and their antiparticles.".

The fundamental building blocks of matter in the Universe are quarks and leptons. The Standard Model is a Gauge theory that explains all the known particles and their interactions as far as the energies of the current particle accelerators will allow, but it is incomplete as 't' only partially unify together the forces in nature. On the other hand it leaves many unanswered questions about the fundamental nature of the universe. The goal of modern theoretical physics has been to find a "unified" description of the universe. This has historically been a very fruitful approach.

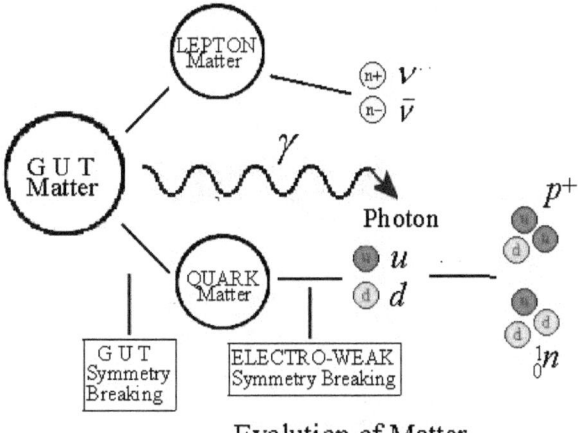

Fig 18.13   Evolution of matter

For example Einstein-Maxwell theory unifies the electric and magnetic forces into the EM force. In QED involves only the particle photon. In QCD the quark forces are mediated by eight types of gluon. The work of Weinberg, Glashow and Salam successfully showed that the EM and Weak forces can be unified into a single Electro-weak force, with the discovery of the W^+, W^- and Z^0 particles. There is actually some strong evidence that the forces of the Standard Model should all unify as well. Attempts are being made to unify the Electro-weak theory and

QCD into a Grand Unified Theory (GUT), which embraces electricity, magnetism, the weak and the strong force. See Fig 18.13. As the Universe expanded and cooled, matter began to condense starting with massive GUT matter. Each *symmetry breaking* produces a phase change and different

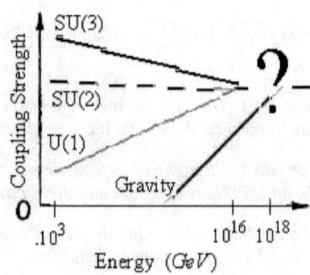

Fig 18.14 Coupling strength *versus* Energy of the Unitary Symmetries

forms of matter appear. Here the quarks and leptons are treated as some equivalent particles, called massive 12 types of X boson. On examination on how the relative strengths of the Strong force and Electro-weak force behave at higher energies, they become the same at energy of about 10^{16} GeV.

Further the Gravitational force should become equally important at an energy of about 10^{19} GeV (Fig 18.14). The Standard Model cannot explain the fruit falling from a tree. It can describe the existence of the fruit and of the tree, but the net force between them, including only the forces in the Standard Model, is zero. All EM, Weak and Strong forces cancel out. The force responsible for the fruit falling on the ground is the fourth force known to everyone, the Gravitational force. That force is simply not part of the Standard Model.

18.7.1 Proton Decay

In a proton (*uud*) an *u* quark emits a virtual X boson and then gets transformed into an anti-quark. A *d* quark absorbs the X boson to become a $^0_{+1}e$, and the remaining quark-anti-quark pair convert to become a π– meson. Such an event takes place once in 10^{30} yrs, which is 10^{20} times the age of the Universe!!!

18.7.2 Parity ($\hat{P}$)

Parity inversion changes a right hand into a left hand, while keeping same the direction of an angular momentum. Parity operator $\hat{P}$ is defined by

$$\hat{P}(x,y,z) = (-x,-y,-z) \qquad (1`8.7.8)$$

Since two successive parity transformations leave the system invariant, it implies that

$$\hat{P} = +1, -1 \tag{18.7.9}$$

For the possible values of $\hat{P}$. In both Strong and EM interactions

$$[\hat{P}, \hat{H}] = 0, \tag{18.7.10}$$

means that **parity is conserved in the decays involving Strong and EM forces**. But it is found that **parity is violated in Weak interaction decays**.

18.7.3 Charge Conjugation ($\hat{C}$)

Charge conjugation $\hat{C}$ replaces all particles with their anti-particles. Momentum and spin are opposite for a neutrino.

For the EM and Strong interactions only $\hat{C}$ is conserved, *i.e.*

$$[\hat{C}, \hat{H}] = 0 \tag{18.7.11}$$

Charge conjugation is violated in Weak interaction processes.

18.7.4 $\hat{C}\hat{P}$-invariance

$\hat{C}\hat{P}$ violation has been found in the decay

$$K_L^0 \rightarrow \pi^+ + \pi^- \tag{18.7.12}$$

and $\quad K_L^0 \rightarrow \pi^0 + \pi^0 \tag{18.7.13}$

18.7.5 Time Reversal ($\hat{T}$)

Time-reversal invariance or $\hat{T}$-invariance, is defined as invariance under the transformation

$$t \rightarrow t' = -t \tag{18.7.14}$$

leaving all position vectors unchanged. It is a symmetry of the Strong and EM interactions, but is violated in Weak interactions. Although $\hat{T}$-invariance does not give rise to a conservation law, *i.e.*, a quantum number, it does lead to equation (18.7.14).

18.7.6 $\hat{C}\hat{P}\hat{T}$ Theorem

The combined operation $\hat{C}\hat{P}\hat{T}$ converts a particle at rest to a anti=particle at rest, and invariance under $\hat{C}\hat{P}\hat{T}$ operation requires particles and anti-particles to have the same masses and lifetimes. This is in accord with experiment.

18.7.7 The Higgs boson (God particle)

The Higgs boson, nicknamed the **God particle**, is a hypothetical passive elementary particle that is predicted to exist by the Standard Model (SM) of Particle physics. The Higgs boson is an integral part of the theoretical Higgs mechanism. The salient characteristic, instruments. Starting with the atom smasher, in 1990s the FERMILAB (Fermi National Accelerator Lab) in Batavia, Illinois, USA, established the TEVATRON, 4 miles around, smashes protons and anti-protons together with unprecedented energies. This monopoly in energy frontier was broken in 2000. The SSC (Superconducting Super Collider) is 54 miles around at Texas, USA. Not satisfied with the energies, at CERN, near Geneva, Switzerland, the LHC (Large Hadron Collider) became operational in Sep 2008. This is a 17 *miles* (27 *km*) long was built deep underground, straddling the French and Swiss borders. Its primary aim has been to provide the conditions that existed immediately after the Big Bang If shown to exist, it would help explain why other particles can have mass It is the only predicted elementary particle yet to be confirmed experimentally. Theories that do not need the Higgs boson also exist and would be considered if the existence of the Higgs boson were ruled out. They are described as Higgsless Models.

If shown to exist, the Higgs mechanism would also explain why the W and Z bosons, which mediate Weak Interactions, are massive whereas the related photon, which mediates Electro-magnetism, is massless. The Higgs boson is expected to be in a class of particles known as <u>scalar bosons</u>. (Bosons are particles with integer spin, and scalar bosons have spin 0*).

Experiments attempting to find the particle are currently being performed using the Large Hadron Collider (LHC) at CERN (near Geneva), and were performed at Fermilab's Tevatron until its closure in late 2011. Some theories suggest that any mechanism capable of generating the masses of elementary particles must be visible at energies above 1.4 *TeV*, therefore, the LHC is expected to be able to provide experimental evidence of the existence or non-existence of the Higgs boson.

A Higgs boson of mass ~ 125 *GeV* has been tentatively confirmed by CERN on 14 March 2013, although unclear as yet which model the particle best supports or whether multiple Higgs bosons exist.

18.8   STRING THEORY AND UNIFICATION OF FORCES

18.8.1 Introduction

The goal of *String Theory* is to explain the "?" in the diagram of Fig 18.10. The characteristic energy scale for quantum gravity is called the Planck Mass, and is given in terms of Planck constant, the speed of light, and Newton's constant (G_N),

$$M_{Pl} = \sqrt{\tfrac{\hbar c}{G_N}} = 1.22 \times 10^{19} \ GeV/c^2 \qquad (18.8.1)$$

Physics describes the Universe at this high energy scale as it existed during the first moments of the Big Bang. These high energy scales are completely beyond the range which can be created in the current particle accelerators. Most of the physical theories that are in use to understand the <u>U</u>niverse also break down at the Planck scale. However, String Theory shows unique promise in being able to describe the physics of the Planck scale and the Big Bang. In its final form String Theory should be able to provide answers to explain

i) Where do the four forces come from?

ii) Why do the various types of particles that are observed?

iii) Why do particles have the masses and charges that are known?

iv) Why do we live in 4 space-time dimensions?

v) What is the nature of space-time and gravity?

18.8.2 General Theory of Relativity

The General theory describes the attraction of massive objects. Indeed, it describes the falling of the massive fruit from a tree towards the very massive Earth. Very massive objects are difficult to move. It will turn out that the description involves new concepts, most importantly that the space-time

18.8.3 Mass and curvature

The way in which the mathematics describes the attractive force between two masses, is by describing *space-time to be curved* around each massive object separately (whether a second body is present or not). The curved space-time caused by one massive body, causes a second massive body in its vicinity not to travel in a straight line. The trajectory of the second massive body is bend towards the first massive body. Thus the net effect is that the second massive body is attracted towards the first and that is exactly what is observed.

Most importantly, the precise mathematical description of the attraction of massive bodies in General Relativity is more accurate than the Newtonian one, which did not take into account the dynamics of space-time itself. The light rays of stars gets bend when pass right next to the Sun, and confirmed the theory of General Relativity. For the precise angle of deflection to be explained the bending of space-time is taken into account

A more concrete question is that the forces of the Standard Model are transmitted by photons, vector bosons, or gluons, for the electromagnetic, the weak and the strong force respectively. Now that we have a fourth force, we have another messenger particle, the graviton, it has spin 2. There could be a quantum field theory that would describe the interactions of all elementary matter particles and messenger particles with the graviton That is the most obvious unification

18.8.4 What is String Theory?
String theory is the most promising theory of the 22nd Century for a unified description of the fundamental particles and forces in nature including gravity. String Theory is beyond experiment

What are then the new principles that string theory is based on? Firstly, importantly, it replaces the basic principle *of point like* particles that underlie our intuition for QFT and General Relativity with the idea that the elementary excitations of our Universe are not particles but *strings*.

It turns out that a QM theory of **open string** can be formulated, and that it automatically incorporates a lot of excitations that look like particles. One of those excitations is a vector particle that is exactly massless. Open strings can merge into closed loops of energy Not only it is needed to accept open strings as basic ingredients, but also closed loops of energy, closed strings. Now, a quantum mechanical theory of closed strings also incorporates a lot of massless particle like *excitations*. Further, it is *a massless spin 2 particle*.

As a theory of quantum gravity String Theory is at present the best hope to give concretely computable answers to fundamental questions such as the underlying symmetries of nature, the quantum behaviour of black holes, the existence and breaking of supersymmetry, and the quantum treatment of *singularities*. It might also shed light upon larger issues such as the nature of QM and space and time. In String theory all the forces and particles emerge in an elegant geometrical way, realizing Einstein's dream of building everything from the geometry of space-time.

String theory is based on the simple premise that at Planckian scales, where the quantum effects of gravity are strong, Particles are actually 1-D extended objects. Just as a particle that moves through space-time sweeps out a curve (the worldline) (Fig 18.15),

Fig 18.15

string will sweep out a surface (the world-sheet)

In contrast with particle theories, String theory is highly constrained in the choice of interactions, super-symmetries and gauge groups. Actually, all the usual particles emerge as excitations of the string and the interactions are simply given by the geometric splitting and joining of these strings..

By merely studying the theory of open and closed strings, or string theory, in its quantum mechanical formulation, one got everything one needed to formulate a Unified Theory that incorporates the Standard Model and General Relativity.

In this way the usual Feynman diagrams of QFT (quantum field theory) are generalized by arbitrary Riemann surfaces.

Much recent interest has been focused on D-branes. A D-brane is a sub-manifold of space-time with the property that strings can end or begin on it.

18.8.5 Extra Dimensions

1) Space is 3-D

2) **Time is** the fourth dimension.

3) **More than 3+1**

4) For years it was thought that String theory needs precisely six extra spatial dimensions.

9+1 = 10

Thus according to String theory humans live in 10-Dimensions. The six extra dimensions are curled up, and in such a way that they are extremely tiny. Since nobody has seen them yet, in particle accelerators, it means that they are smaller than $10^{-15}\,m$, a tiny 6-D ball!

18.8.5.1 Branes and Dualities

The idea of *duality* refers to the possibility that there might be two descriptions for the same thing, as the wave-particle duality of matter.

T-duality could allow one to examine properties of a theory that would otherwise be very difficult to access. **S-duality** is another duality that allows for such a drastic simplification.

It is known that the open strings are now restricted to live on objects called **branes**. The open strings cannot freely move everywhere in space anymore, but their endpoints are confined to move on these branes. Since a brane is an object on which open strings live, it is possible to find a Gauge field living on a brane. Further, if only closed string modes live in the whole of the space, gravitons can be expected in the whole of the space, on the brane and in the rest of the space. This leads to a new possibility for the answer to the question: where are the six extra dimensions of String theory ?

A brane that is 3-D, just like our space, may be assumed. Then one would have gauge fields living on the brane, in three *spatia*l dimensions. That brane could look exactly like our world if the gauge fields and particles would be like the ones in the Standard Model. In other words, the world just might be such a 3-D brane. If the photon, say, only lives on the brane, all the other 6-D would be literally dark. Indeed, no photons, or light, would come from these other dimensions, and no light would leak into these dimensions, since the photons (gauge fields) can only live on the brane. Actually, these extra dimensions can be as large as a tenth of a *mm* and invisible at this moment. Experiments are now being done that would allow one to view extra dimensions having size a tenth of a *mm*, but since they involve the very weak gravitational force, the experiments are technically very involved -- everybody would be thrilled if one detects deviations in the gravitational force due to the existence of extra dimensions.

18.8.5.2 Black Holes

Black holes are called *black*, as General Relativity does not allow anything to escape from a black hole, not even light. Therefore a black hole will look like a black object -- there's no light coming from a black hole. It is to be remembered that that a body closer than a certain distance will never be able to escape anymore. The critical distance to the black hole is marked by the horizon. In fact anything behind the horizon will never be seen again by an observer far from the black hole. Of course, he might decide to travel past the horizon himself and might meet all kinds of aliens that were hidden from his view before.

A black hole radiates nothing in General Relativity, but when looked at quantum-mechanically, it does..

A natural question is what is the statistical mechanical description of black holes? String theory is a quantum mechanical description of gravity, so it should provide us with a statistical mechanical explanation of all thermo-dynamical properties of black holes.

18.8.5 M-Theory (Maps between String theories)

There are in fact five seemingly different String theories, known as Type I, Type IIA and Type IIB and Heterotic E8 and Heterotic SO32. All these types of string theories are linked by chains of dualities. They seem to be part of a bigger whole, called M-theory, which requires a better description. M-theory remains a challenge for physicists. Unified picture of all string theories is shown the diagram below.

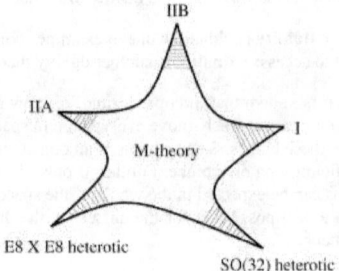

Fig 18.16 Unified picture of all string theories

18.8.6 Unified Theories.

A Unified Theory would be a mathematical frame work in which all the different kinds of forces and particles occur naturally. There is no need to fix the masses and charges of particles from experiment; rather the theory should fix the right values automatically. Why does the electron weigh as much as it does? Why do particles interact with a given strength and not any other? In the Standard Model the measured values are assumed in experiments, but in a Unified Theory these values should be predicted. Clearly this is an ambitious goal.

This suggests that the theory should possess a great degree of mathematical elegance and consistency. To discover the Unified Theory, only those physical models which broadly resemble nature may be considered and in addition satisfy the above criteria. Then check whether it describes the living world.

18.8.7 Quantum gravity: Super-string Theory

The ultimate "*Theory of Everything*" must include a quantum theory of gravity. A promising candidate for such a theory (so far the only one) is "Superstring Theory" which predicts that the quarks and leptons are not really point-like but consist of incredibly tiny strings of size $10^{-35} m$, much smaller than the size of a proton ($10^{-15} m$). Although this theory seemingly much appealing because it incorporates gravity into the Standard Model, it must be pointed out so far there is no evidence so far any of its model predictions.

REVIEW QUESTIONS

R.Q. 18.1   Explain what are meant by an elementary particle. What are the differences among leptons, baryons and mesons?

R.Q. 18.2   By emitting a an electron $_{10}^{23}Ne$ undergoes decay giving $_{11}^{23}Na$ and an anti-neutrino.. Write down the nuclear equation for this decay. Estimate the maximum KE with which the electron is emitted.   Given: Mass of $_{10}^{23}Ne = 22.994\ 465\ u$ mass of $_{11}^{23}Na = 22.989\ 768\ u$. [Answer: $_{10}^{23}Ne \longrightarrow\ _{11}^{23}Na +\ _{-1}^{0}\beta + \bar{v}$;. $Q = +0.004\ 697\ u$; The maximum KE of the emitted electrons = 4.37 MeV].

R.Q. 18.3   What is spin? Explain how spin can be used to classify subatomic particles.

R.Q. 18.4   Find out the angular momentum of the Earth ($R_E = 6400\ km$, $M_E = 6 \times 10^{24}\ kg$); and a proton ?

[Answer: $L = I\ \omega = \{\frac{2}{5}(6 \times 10^{24}\ kg)(6.4 \times 10^6\ m)^2 = 9.8 \times 10^{37}\ kg\ m^2\}\ [2\pi / 86400\ rad\ s^{-1}] = 7 \times 10^{33}\ N\ ms$;
$L = 0.5\ (1.05 \times 10^{-34}\ Js) = 5.25 \times 10^{-35}\ N\ ms$].

R.Q. 18.5   How do the angular momenta at the atomic and sub-nuclear level differ from angular momentum in everyday large objects? [The L at both atomic and sub-nuclear levels is quantized].

R.Q. 18.6   Write an equation for the decay of a neutron. Describe briefly without experimental details the principle behind the detection of neutrinos.

[Answer: $_{-1}^{0}n \longrightarrow\ _{+1}^{1}p +\ _{-1}^{0}\beta + \bar{v}$].

R.Q. 18.7   What is the nature of a β^+ particle? Name another particle emitted during β^+ decay. [Answer: A e^+ having same mass of an e^- but of opposite charge; a v neutrino].

R.Q. 18.8   Distinguish between leptons, baryons, hadrons and mesons. Name the six leptons associated with ordinary matter. Which of these four categories of particles are fermions and which are bosons?

R.Q. 18.9   What is strangeness? In which interactions is strangeness always conserved? [Answer: Any hadron having a strange quark is a particle having property strangeness. Strangeness is always conserved in strong and EM interactions].

R.Q. 18.10   By examining strangeness (S), determine which of the following decays or reactions proceed via the strong interaction   i) $K^o \longrightarrow \pi^+ + \pi^-$

ii) $\Lambda^o + p \longrightarrow \Sigma^+ + n$ iii) $\Lambda^o \longrightarrow p + \pi^-$

iv) $K^- + p \longrightarrow \Lambda^o + \pi^o$. Given: S = 0, for p, n, π^o, π^+, π^-; S = +1, for K^o; S = -1, K^-, Λ^o, Σ^+
.[Answers: i) S is not conserved, it cannot proceed by strong interaction. ii) S is conserved, so it proceeds via strong interaction; iii) S is not conserved; it cannot proceed by strong interaction. iv) S is conserved, so it proceeds via strong interaction].

R.Q. 18.11 Indicate, with an explanation, whether the following interactions proceed through the strong, EM or weak interactions, or whether they do not occur: (1) $K^+ \to \mu^+ + \nu_\mu$, (2) $\tau^+ \to \mu^+ + \bar{\nu}_\tau$, (3) $\Omega^0 \to \Lambda^0 + \pi^-$, (4) $\pi^0 \to \gamma + \gamma$, (5) $\Lambda^0 \to p + \pi^-$, (6) $\pi^- \to n + K^0 + \Sigma^-$, (7) $\pi^- + n \to K^- + \Sigma^0$, (8) $\bar{p} + p \to n + \pi^+ + \pi^-$

R.Q. 18.12 Draw the Feynman diagrams for the decays $\pi^- \to \pi^0 + e^- + \bar{\nu}_e$ and

$$K^- \to \pi^0 + e^- + \bar{\nu}_e.$$

R.Q. 18.13 List the fundamental fermions, showing how they may be divided into families or generations. Discuss the different properties of the classes of particles you have named.

R.Q. 18.14 Explain what is meant by a virtual particle. How are space-like and time-like virtual photons distinguished?

R.Q. 18.15 Use the position-momentum Uncertainty Relation, estimate the minimum momentum of an electron contained to a nucleus From the relativistic energy-momentum relation, determine the corresponding energy, and compare it with that of an electron emitted in, say, the beta decay of tritium (Answers: $\Delta x = 10^{-15}$ m implies $\Delta p = 197.3 MeV/c$;, which, since the mass of the electron is so much less than the energy = kE.. Thus, the electrons confined to the nucleus have much higher kinetic energy than those observed coming from a nucleus in beta decay).

R.Q. 18.16 What is the Gell-Mann-Nishijima formula? Can it be generalized?

R.Q. 18.17 Using four quarks (u, d, s, and c), construct a table of all the possible baryon species. How many combinations carry a charm of +1? How many carry charm +2, and +3?.

R.Q. 18.18 Is there sufficient experimental and observational data to unify the forces?

R.Q. 18.19 Discuss sub-structure of the pions'

R.Q. 18.20 If the neutron β^- – decay involves only a proton, an electron, the momentum and energy conservations dictate that the electrons must emerge with a fixed momentum / energy. But experimentally they do not, one observes a range of electron kinetic energies. Discuss. (Answer: Neutrno discovery).

R.Q. 18.21 The anti-proton was discovered in 1955 using the Bevatron that could accelerate protons up to energy of 6.4 GeV. Anti-protons $\bar{p}$ were produced by the reaction: $p + p \to p + p + p + \bar{p}$. Show that the number of fermions in the reaction is conserved.

R.Q. 18.22 Which of these, leptons, baryons and mesons are elementary particles? Why? (Answers: Leptons only, elementary particle are the ones which apparently do not have internal structure).

R.Q. 18.23 By examining strangeness S, in which of the following decays or reactions proceed via the strong interactions. i) $K^0 \to \pi^+ + \pi^-$, ii) $\Lambda^0 + p \to \Sigma^+ + n$, iii) $\Lambda^0 \to p + \pi^-$, iv) $K^- + p \to \Lambda^0 + \pi^0 \pi^0$ [Given: S = 0 for p, n, π^0, π^+, π^-; S = +1 for K^0; S = -1 for K^-, Λ^0, Σ^+]. (Answers: i) and iii) cannot proceed by strong interactions; ii) and iv) proceed via strong interaction.).

R.Q. 18.24   The reaction $\mu^- \to e^+ + \nu_e + \bar{\nu}_\mu$ takes place by conserving momentum and energy. Show that lepton number for electron type and muon type leptons must each remain constant. (Answer: L_e is conserved for electron family; L_μ conserved for muon family).

R.Q. 18.25   Using four quarks (u, d, s, and c), construct a table of all the possible baryon species. How many combinations carry a charm of +1? How many carry charm +2, and +3? (Answers: three combinations with *ccu*, *ccd*, *ccs with C=+2;* six combinations with $C = +1$:;*cuu, cud, cus, cdd, cds, css;* and ten combinations with $C = 0$: *uuu, uud, udd, ddd, uus, uss, sss, dds, dss, uds*)

R.Q. 18.26   The anti-proton was discovered by means of the following particle reaction:

$p + p \to p + p + \bar{p}$, by conserving Baryon number. Is the following reaction possible? $p + p \to p + p + \bar{n}$.(Answer: B # is not conserved; the reaction is forbidden).

R.Q. 18.27   Confirm that the η (eta) particle, which has a life-time of 8.0×10^{-19} s, can decay via the strong interaction through the process, $\eta \to \pi^+ + \pi^- + \pi^0$.(Answer: The life-time of 8.0×10^{-19} s, of η makes this a strong interaction process. The η has odd parity (-1); The pi-mesons also have odd parity, so η (parity) = (-1). This decay is allowed by conservation of parity.)

R.Q. 18.28   By checking to see whether Q, B and S are each conserved, decide which one of these strong interaction can occur a), b) $K^- + p \to \bar{K}^0 + n$, c) $\pi^- + p \to p + n$ (Answers: a) satisfied Q conservation; B not conserved, S Ok; b) satisfied Q conservation, B conserved, S okayed This strong interaction process is allowed; c) Can not occur as Q is not conserved).

R.Q. 18.29   Which of the conservation law {s} are violated in each case? i) $\nu_e + p \to n + e^+$; ii) $p + p \to p + n + K^+$; iii) $p + p \to p + n + \Lambda^0 + K^0$; iv) $K^- + p \to n + \Lambda^0$ W (Answers: (i) Lepton # violated (ii) S is violated (iii) B is violated (iv) B is violated).

R.Q. 18.30   The strong interaction $K^- + p \to K^0 + K^+ + X$ was first observed in 1964; where X is a hadron, when until then had not been observed. Deduce (i) the charge of X in Coulombs; (ii) whether X is a meson or baryon, (iii) the strangeness of X. Use conservation of Q, B, S. (Answers: Q(X) to have charge -1e, B must have +1, S conservation requires, S = -3)

R.Q. 18.31   The Feynman diagram shows the interaction mechanism of an electron and positron. What type of interaction is represented by it?

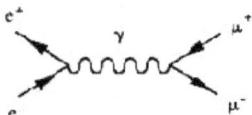

Draw the Feynman diagram that represents the exchange of a □⁺ between a proton and a neutron. (Answers: a) EM interaction, b) e⁺ and e⁻ pair has annihilated to produce a μ⁺ and μ⁻ pair)

R.Q. 18.32   The Σ^- particle is a baryon with Q, B and S ; –1e, +1, and -1, respectively. Confirm that it has the quark structure dds. (Answer: B = 1, Q (dds) = - 1 e , S(dds) = -1 .So Σ^- particle is a baryon with structure (dds).

R.Q. 18.33 The Ξ^- particle has spin ½ and charge – e and S = -2. It is thought to be made up of three quarks involving only up, down and strange flavors. What must the combination be? (Answer: Since Ξ has S = -2, it contains two s quarks, each of S = -1. The third quark mu8st be d., with S = - 0; satisfies Q (dss) = -1 e. S (dss) = -2.).

R.Q. 18.34. Which of the following reactions are allowed? If forbidden, state the reason.

a) $\pi^- + p^+ \to K^- + Z^+$, b) $d + d \to {}^4He + \pi^0$,

c) $K^- + p^+ \to \Xi^- + K^-$.(Answer: a) $\Delta I_3 = \frac{1}{2}-(-\frac{1}{2})=1 \neq 0$, the reaction is forbidden.

b) $\Delta I_3 = 1-0 = 1$, the reaction is forbidden., c) $\Delta I_3 = 1-0 = 1$, the reaction is forbidden).

R.Q. 18.35 Verify if the reaction shown in the diagram can proceed.

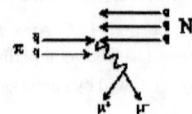

R.Q. 18.36 Complete the following decay process, $K^- + p \to \bullet + \bullet + \Omega^-$.

(Answers K^0, K^+). $\Sigma^0 \to \Lambda^0 + \bullet$).

R.Q. 18.37 State the conservation laws for nuclear reactions.

R.Q. 18.38 Outline the concept of 'strangeness'.

R.Q. 18.39 Explain with reasons which of the following reactions are allowed:

a) $\pi^+ + n \to \Lambda^0 + K^+$, b) $\pi^- + p \to \Lambda^0 + K^0$.(b) allowed, as S is balanced).

R.Q. 18.40 Give an account of the symmetry schemes of elementary particles with special reference to SU(3) theory.

R.Q. 18.41 Give a detailed account of the classification and properties of elementary particles with special reference to strong, weak, and Electromagnetic interactions.

R.Q. 18.42 For the Baryon Unitary Decuplet, obtain the Gell-Mann-Ocubo mass formula, given $m(I,Y) = m_0 \{1 + aY + b\ [I(I+1) - \frac{1}{4}Y^2]\}$.

(Answer: $(m_\Delta - m_\Sigma) = m_\Sigma - m_\Xi = m_\Xi - m_\Omega$ }

R.Q. 18.43 Describe SU(3) classification of elementary particles.

RQ. 18.44 i) Give an account of the various conservation laws which determine the properties and classification of elementary particles. Ii) What do you understand by 'associated production' ? Is the process: $n^0 + n^0 \to \lambda^0 + \lambda^-$ allowed? Iii) Why is a proton stable?

RQ. 18.45   i) Discuss the basis of the SU(2) symmetry classification of strongly interacting particles. ii) Why do you consider SU(2) is broken in EM interactions? iii) What are quarks? Describe the unusual properties of various quarks. Explain how mesons and baryons are composed of quarks.

RQ. 18.46   State the Gell-Mann-Nishijima relation. What is SU(3) symmetry?

RQ. 18.47   Find the charge and strangeness of all mesons that can be built from quark-antiquark pairs with flavors u, d and s.

(Answers: $\pi^+(u\bar{d})$, $\pi^-(d\bar{u})$, $K^+(u\bar{s})$, $K^-(s\bar{u})$, $K^0(s\bar{d})$, and $K^0(d\bar{s})$).

&&*&*&*&*&*&*

Chapter 19

NUCLEAR POWER

Chapter 19

NUCLEAR POWER

"For the present I believe that the war will be over long before the first atom bomb is built" — W. Heisenberg, (1939)

"Atomic power can cure as well as kill. It can fertilize and enrich a region as well as devastate it. It can widen man's horizons as well as force him back into the cave."

- Alvin M. Weinberg, 1944

"Peace, progress, human rights – these three goals are insolubly linked to one another; it is impossible to achieve one of these goals if the other two are ignored"

A. Sakharov, Nobel lecture

19.1 INTRODUCTION

It is known that generation of electricity on an industrial scale dates back to 1881. Popular electric power generation methods include: hydro-electric, coal fired, geothermal, natural gas, oil, biogas, biomass (which includes firewood), nuclear, wind, wave, solar thermal, solar photovoltaic and bio-voltaic.

One of the most interesting phenomena in Nuclear Physics with far reaching consequences has been the fission of atomic nuclei Chapter 16). Discovered in 1934, it paved the way for the modern nuclear reactor, a bountiful source of energy. From the fundamental point of view it is a beautiful illustration of Einstein's $E = mc^2$ formula.

Enrico Fermi *et al.* (1934) tried to produce elements of Z > 92 by bombarding with neutrons the Uranium nuclei. According to previous beliefs when Uranium is bombarded with $_0^1 n$ *transuranic* elements must be produced.

$$_{92}U + _0^1 n \rightarrow \text{Transuranic elements} \qquad (19.1.1)$$

But Paul Harteck Otto Hahn and Fritz Strassmann (1938) discovered that 'alkaline earth' metals ($_{56}^{139}Ba$ and $_{57}^{140}La$) are produced when uranium is irradiated with slow $_0^1 n$, in radio-chemical experiments..

$$_{92}U + _0^1 n \rightarrow \text{Alkaline rare earth metals}. \qquad (19.1.2)$$

An important and special type of nuclear reaction phenomenon is *nuclear fission*.

Lise Meitner and Otto R. Frisch (1939) suggested that by absorption of a neutron (*a*), the uranium nucleus, X becomes sufficiently excited to split nearly in half into a pair of fragments, *b* & Y of approximately equal mass (Fig 16.1). Thus the process

$$X\left(a, b\right) Y \qquad (19.1.3)$$

is called *fission*, if b and Y are comparable in masses.

To provide the power for a dynamo-electric machine, or electric generator, nuclear power plants rely on the process of nuclear fission. In this process, the nucleus of a heavy element, such as uranium, splits when bombarded by a free neutron in a nuclear reactor. The fission process for uranium atoms yields two smaller atoms, one to three free neutrons, plus an amount of energy, as stated by eq (19.1.3). Because more free neutrons are released from a uranium fission event than are required to initiate the event, the reaction can become self-sustaining--a chain reaction--under controlled conditions, thus producing a tremendous amount of energy

Naturally occurring Uranium ($nat\ U$) consists of approximately 99.28 % $^{238}_{92}U$ and 0.71% $^{235}_{92}U$. $^{235}_{92}U$ is said to be a *fissile* isotope. produced from $^{232}_{90}Th$ The nucleus of a $^{238}_{92}U$ atom, on the other hand, instead of undergoing fission when struck by a free $^{1}_{0}n$, will nearly always absorb the neutron and yield an atom of the isotope $^{239}_{92}U$. This isotope then undergoes natural radioactive decay to yield $^{239}_{94}Pu$, which, like $^{235}_{92}U$, is a fissile isotope. $^{238}_{92}U$ is said to be *fertile*, because, through neutron irradiation in the core, some eventually yield atoms of fissile $^{239}_{94}Pu$.

It takes 10^{11} **fissions per second to produce one watt of electrical power**. As a result, about **1 g of fuel is consumed per day per *MWe* energy produced.** This means that 1 g of waste products is produced per *MW* per day, which includes $0.5g\ ^{239}_{94}Pu$. These waste products must be either reprocessed to generate more fuel or stored for the tens of thousands of years it takes for the level of radiation to reach a safe limit.

19.1.1. Turning Points in Nuclear India

The prime objective of India's nuclear energy programme is the development and use of nuclear energy for **peaceful purposes** such as power generation, applications in agriculture, medicine, industry, research and other areas. India is today recognized as one of the countries most advanced in nuclear technology including production of source materials. India is self-reliant. It has mastered the expertise covering the complete nuclear cycle - from exploration and mining to power generation and waste management.

Accelerators and research & power reactors are now designed and built indigenously. The sophisticated Variable Energy Cyclotron (VEC) at Kolkatta and a medium energy heavy ion accelerator 'Pelletron' set up at Mumbai are national research facilities in the frontier areas of science.

It was the farsightedness of Dr. **Homi Jahangir Bhabha** to start nuclear research in India at a time following the discovery of nuclear fission phenomena by Otto Hahn and Fritz Strassman and soon after Enrico Fermi *et al* from Chicago reporting the feasibility of sustained nuclear chain reactions. At that time very little information was available to the outside world about nuclear fission and sustained chain reactions and nobody was willing to subscribe to the concept of power generation based on nuclear energy.

India's indigenous efforts in nuclear science and technology were established remarkably early. In Mar 1944, the first step was taken up by the visionary Dr. Homi J. Bhabha, who submitted a proposal to the Sir Dorab Tata Trust, to establish a nuclear research institute This led to the creation of the **Tata Institute of Fundamental Research (TIFR)** on 19 Dec 1945 with Dr.

Bhabha as its first Director. The new Government of India passed the Atomic Energy Act, on 15 April 1948, leading to the establishment of the **Indian Atomic Energy Commission (IAEC)**.

The IAEC decided to set up a new facility - the **Atomic Energy Establishment**, Trombay (AEET), in 3 Jan 1954. Followed on 3 Aug 1954 the **Department of Atomic Energy (DAE)** was created with Dr. Bhabha as the Secretary. On 12 Jan 1967 the AEET got the present name – **Bhabha Atomic Research Centre (BARC)** when Indira Gandhi (then the Prime Minister) renamed it in memory of Dr. Bhabha (who died in an airplane crash on Mt. Alps on 24 Jan 1966).

In May 2000, the Atomic Energy Regulatory Board (**AERB**).was set up.

19.2 NUCLEAR BATTERY

Devices for converting natural radioactive decay directly into electricity are nothing new. Nuclear battery technology began in 1913, when Henry Moseley first demonstrated the Beta Cell. Nuclear batteries were essential to the success of the long distance space probes. Solar cells would not have provided enough power to operate the required equipment because the sun was too far away. Chemical batteries can provide enough power to run the instruments for a short period of time, but their total energy storage capacity is many times less than that of a nuclear battery. Imagine how many rechargeable NiCad batteries would have to be replaced during 10 years of continuous use The terms **atomic battery**, **nuclear battery** and **radioisotope battery** are used to describe a device which uses the charged particle emissions from a radioactive isotope to directly generate electricity

Scientists in the USA have developed a new fabrication technique that will lead to nuclear batteries that could last for decades. The researchers, from the University of Rochester, claim that the technique is already ten times more efficient than current nuclear batteries, and has the potential to outstrip them nearly 200 times.

A battery that converts the energy of particles emitted from atomic nuclei into electric energy. Two basic types have been developed: (1) a high-voltage type, in which a beta-emitting isotope is separated from a collecting electrode by a vacuum or a solid dielectric, provides thousands of volts but the current is measured in pA; (2) a low-voltage type gives about 1 V with current in microamperes

The Voyager space probes carried devices called Radioisotope Thermal Generators. (RTGs) These devices are simply amazing. They work on a simple principle.

19.2.1 Optoelectric nuclear battery

The battery would consist of an excimer of argon, xenon, or krypton (or a mixture of two or three of them) in a pressure vessel with an internal mirrored surface, finely-ground radioisotope, and an intermittent ultrasonic stirrer, illuminating a photocell with a band gap tuned for the excimer. When the beta active nuclides (*e.g.*, $^{85}_{36}Kr$ or $^{39}_{18}Ar$) are excited, their own electrons in the narrow exciter band that this radiation is converted in a high band gap photovoltaic layer (*e.g.* in p-n diamond) very efficiently into electricity. The electric power per weight compared with existing radionuclide batteries can then be increased by a factor 10 to 50 and more betavoltaics.

Shielding requirements in a RTG are the second lowest of all possible isotopes; only $^{238}_{94}Pu$ requires less.

Most RTGs use $^{238}_{94}Pu$ which decays with a $\tau_{1/2} = 87.7\ yrs$. RTGs using this material will therefore lose $1 - 0.5$ or 0.787% of their capacity per year. 23 years after production, such an RTG would produce 83.4% of its starting capacity. Thus, with a starting capacity of 470 W, after 23 years it would have a capacity of $0.834\ x\ 470\ W = 392\ W$.

Dynamic generators, unlike thermo-electrics, use moving parts to mechanically convert heat into electricity. Unfortunately, those moving parts can wear out and need maintenance, which may not be possible for certain applications like space probes. Dynamic power sources also cause vibration and RF noise. Even so, development by NASA on a next generation RTG called a Stirling Radioisotope Generator (SRG) that uses Free-Piston Stirling engines to produce power. SRG prototypes demonstrated an average efficiency of 23%, and higher efficiency can be achieved with the use of greater temperature differentials between the hot and cold ends of the generator

19.3 BASIC PRINCIPLE OF A REACTOR.

A detailed account on nuclear fission process has been provided in Chapter 16.

19.3.1 Fission Chain Reaction

In an induced fission, for each neutron ^1_0n absorbed by the fissile nucleus, $v > 1$, so more than one ^1_0n are emitted (Table 16.3). This suggests a '**self-sustaining**' sequence of fissions; called *chain fission reaction* is in principle possible (See Chapter 16). Whether the chain reaction (Fig 19.1) remains steady, builds up, or dies down, depends on the competing processes, *viz.*,

1. Non-fission capture of the ^1_0n by the fissionable material (fuel),
2. ^1_0n - capture by other materials in the system; (n, γ) reactions.
3. Leakage of neutrons through the surface of the system.

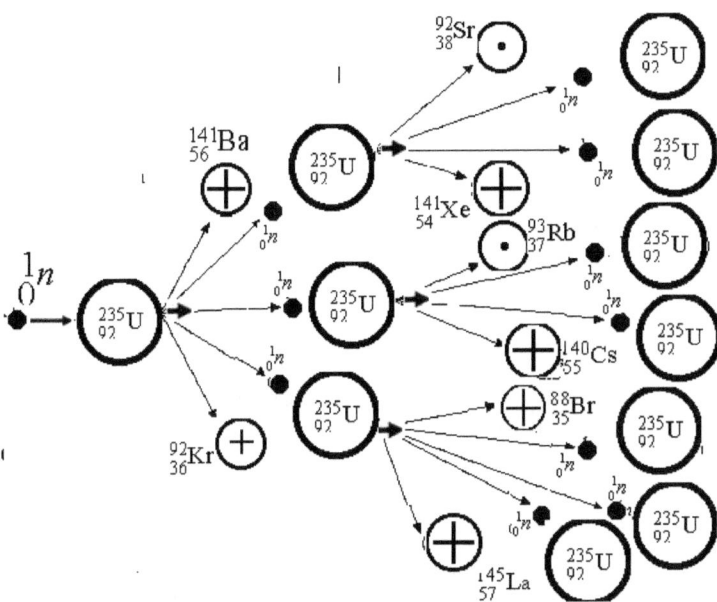

Fig 19.1. Neutron induced fission chain reaction

Suppose a small amount of the fuel is bombarded with $_0^1n$ numbering 100 of which 40 cause fission, and 60 are absorbed without fission or escape out of the target fuel, then ($40 \ x \ \nu = 100$) $_0^1n$ are produced after the fission. In this case none of the $_0^1n$ is lost, and the fission process continues without external supply of $_0^1n$. Then the assembly is said to be

a) *Critical*, when N = 100.
b) *Sub-critical.*, when N < 100.
c) *Super- critical.* For N > 100, and an explosion takes place.

19.3.2 The Nuclear Fuel Cycle

For typical light-water reactor the nuclear fuel cycle consists of

a) "*front end*" steps that lead to the preparation of Uranium for use as fuel for reactor operation and

b) "*back end*" steps that are necessary to safely manage, prepare, and dispose of the highly radioactive spent nuclear fuel.

Chemical processing of the spent fuel material to recover the remaining fractions of fissionable products, $^{235}_{92}U$ and $^{239}_{94}Pu$, for use in fresh fuel assemblies is technically feasible.

The *front end* of the nuclear fuel cycle commonly is separated into the following steps

| Step # 1 | **Exploration.** By geophysical techniques, say a deposit of Uranium, is discovered, and it is first evaluated, and then sampled to determine the amounts of uranium ore materials that are extractable from the deposit at specified costs.

| Step # 2 | **Mining.** The methods used are similar to the ones used for mining metals. Uranium ore can be extracted through conventional mining in open pit and underground methods applied for other metals..

| Step # 3 | **Milling.** Mined uranium ores normally are processed by grinding the ore materials to a uniform particle size and then treating the ore to extract the uranium by chemical leaching. The milling process commonly yields dry powder-form material consisting of natural uranium, "yellowcake", is uranium oxide which is sold on the uranium market as U_3O_8.

| Step # 4 | **Uranium conversion.** Milled U_3O_8, must be converted to uranium hexafluoride, UF_6, which is the form required by most commercial uranium enrichment facilities currently in use. A solid at room temperature, UF_6 can be changed to a gaseous form at moderately higher temperatures. The UF_6 conversion product contains only natural uranium, not enriched.

| Step # 5 | **Enrichment.** The concentration of the fissionable isotope, $^{235}_{92}U$ (0.71 % in natural uranium) is less than that required to sustain a nuclear chain reaction in light water reactor cores. Natural UF_6 thus must be "enriched" in the fissionable isotope for it to be used as nuclear fuel. The different levels of enrichment required for a particular nuclear fuel application are specified by the customer: light-water reactor fuel normally is enriched up to about 4 % $^{235}_{92}U$, but uranium enriched to lower concentrations also is required. Gaseous diffusion and gas centrifuge are the commonly used uranium enrichment technologies.

| Step # 6 | **Fabrication.** For use as nuclear fuel, enriched UF6 is converted into uranium dioxide (UO_2) powder which is then processed into pellet form. The pellets are then fired in a high temperature sintering furnace to create hard, ceramic pellets of enriched uranium. The cylindrical pellets then undergo a grinding process to achieve a uniform pellet size. The pellets are stacked, according to each nuclear core's design specifications, into tubes of corrosion-resistant metal alloy. The tubes are sealed to contain the fuel pellets: these tubes are called fuel rods. The finished fuel rods are grouped in special fuel assemblies that are then used to build up the nuclear fuel core of a power reactor.

The *back end* of the cycle is divided into the following steps:

a) **Interim Storage.** After its operating cycle, the reactor is shut down for refueling. The fuel discharged at that time (spent fuel) is stored either at the reactor site or, potentially, in a common facility away from reactor sites. The spent fuel rods are usually stored in water, which provides both cooling (the spent fuel continues to generate heat as a result of residual radioactive decay) and shielding (to protect the environment from residual ionizing radiation).

b) **Reprocessing.** Light-water reactors discharge spent fuel which contains appreciable quantities of fissile ($^{235}_{92}U$, $^{239}_{94}Pu$), fertile ($^{238}_{92}U$), and other radioactive materials. These fissile and fertile materials can be chemically separated and recovered from the spent fuel.

c) **Waste Disposal.** A current concern in the nuclear power field is the safe disposal and isolation of either spent fuel from reactors or, if the reprocessing option is used, wastes from reprocessing plants. These materials must be isolated from the biosphere until the radioactivity contained in them has diminished to a safe level.

19.3.3 Enrichment of $^{235}_{92}U$ (Uranium Preparation):

Earlier discussion was mainly on the nuclear fission with $^{235}_{92}U$ ($\tau_{1/2}$ =703.8 *Myrs*). In reality, this will not be the only isotope of Uranium present in a nuclear reactor. In naturally occurring Uranium deposits, less than 1% is $^{235}_{92}U$. The major part of the Uranium is $^{238}_{92}U$ ($\tau_{1/2}$ = 4.468 *Byrs*). $^{238}_{92}U$ is not a fissile isotope of uranium. When a loose $^{1}_{0}n$ strikes $^{238}_{92}U$, it absorbs the $^{1}_{0}n$ and does not undergo fission. Thus, by absorbing neutrons, $^{238}_{92}U$ can prevent a nuclear chain reaction to take place. This is unwanted because if a chain reaction doesn't occur, the nuclear reactions cannot sustain themselves, the reactor shuts down. In order for a chain reaction to occur, the pure Uranium ore must be refined to raise the concentration of $^{235}_{92}U$. This is called *enrichment* and is primarily accomplished through a technique called *gaseous diffusion*. In this process, the uranium ore is combined with fluorine to create a chemical compound called UF_6 (uranium hexafluoride) which is heated and vaporizes. The heated gas is then pushed through a series of filters. Because some of the UF_6 contains $^{238}_{92}U$ and some contains $^{235}_{92}U$, there is a slight difference in the weights of the individual molecules. The molecules of UF_6 containing $^{235}_{92}U$ are slightly lighter and thus pass more easily through the filters. This creates a quantity of UF_6 with a higher proportion of $^{235}_{92}U$. This is collected, the uranium is stripped from it, and the result is an enriched supply of fuel. Usually, nuclear power plants use uranium fuel that is about 4% $^{235}_{92}U$.

Enrichment is the process of purifying a specific isotope. In nuclear chemistry enrichment of fissionable materials is usually done in order to yield higher products in reactions. One example of enrichment is the fraction distillation of uranium. Uranium is mixed with fluorine to form UF_6, a gas. At low pressures the compound is allowed to go through a diffusion barrier permitting the lighter $^{235}_{92}U$ to pass through faster. This is done hundreds of times to finally produce a 3% mixture of $^{235}_{92}U$.

19.3.4 Moderators

In a matrix of a moderator, the fissionable nuclei in a reactor are dispersed. The fast $^{1}_{0}n$ collides elastically with the nucleus of the moderator element, and the energy transfer will be largest when the two bodies of the collision are of equal mass.

Case 1. Water (H_2O) as Moderator:

Hydrogen nuclei have masses almost equal with the mass of neutron. But there is discouraging factor;

$$^{1}_{1}H + ^{1}_{0}n \rightarrow ^{2}_{1}H + \gamma \tag{19.3.1}$$

Thus n-capture occurs!

Case 2. Heavy Water (D_2O) as Moderator

Deuterons are less likely to interact with neutrons. Further the mass deuterium atom is only two times that of neutron. Thus heavy water makes a suitable moderator material. But its availability is difficult.

Case 3. Graphite (C) Moderator:

It is composed of $^{12}_{6}C$ atoms, and is more readily available. $^{12}_{6}C$ has only very small n-capture cross section.

It obvious from the *Worked out Example* 16.1 that the average absorption cross section of 3.51 b, is nearly equal to the average neutron fission cross section of 3.92 b, in *nat* U. This means that only half the slow neutrons are captured in a block of *nat* U inducing fissions. It is known that $\nu = 2.5$ per fission, and so no more than 0.5 neutron / fission of $^{235}_{92}U$ can be lost if a self-sustaining reaction is to occur.

19.3.5 The Four factor Reactor Formula

Define the parameters ν, ε, p, f, ℓ_f, ℓ_t, σ_f, σ_{ta}, k_e, k_∞, η.

$\nu = $ # of fission $^{1}_{0}n$ per fission,

$\varepsilon = $ Fast $^{1}_{0}n$ fission factor, $= 1.1$ for thermal reactors, and is 1 for fast neutron reactors,

$p = $ Resonance escape probability, *i.e.*, the probability that the $^{1}_{0}n$ slows down past the resonance region without being captured.

$\ell_f = $ Fast leakage probability, due to finite dimensions of the assembly, of $^{1}_{0}n$ that will leak out during process *p*.

$(1 - \ell_f) = $ Chance that $^{1}_{0}n$ will become *thermalized*, without leaking out.

Thus the number of $^{1}_{0}n$ reaching the thermal region of energy

$$= \nu \, \varepsilon \, p \, (1 - \ell_f) \qquad (19.3.2)$$

$\ell_t = $ Thermal leakage probability of neutrons once they are slowed down to thermal level,

$f = $ Thermal utilization factor, of $^{1}_{0}n$ in fuel and other materials.

The number of $^{1}_{0}n$ eventually absorbed in the fuel

$$= \nu \, \varepsilon \, p \, (1 - \ell_f) \, (1 - \ell_t) f \qquad (19.3.3)$$

Some of the $^{1}_{0}n$ will produce the reaction:

$$^{238}_{92}U + ^{1}_{0}n \to [^{239}_{92}U^*] \to ^{239}_{92}U \xrightarrow{\beta^-} ^{239}_{93}Np \xrightarrow{\beta^-} ^{239}_{94}Pu \quad (19.3.4)$$

The fraction that causes fission is $\dfrac{\sigma_f(U)}{\sigma_{ta}(U)}$

σ_f = Uranium fission cross section for $^1_0 n$

σ_{ta} = Total absorption cross section for $^1_0 n$ in uranium.

$$\frac{\sigma_f(^{238}_{92}U)}{\sigma_{ta}(^{238}_{92}U)} = \frac{\sigma_f(=0.5)}{\sigma_{ta}=(2+0.5)}, \quad (19.3.5)$$

Once a fission chain reaction has started, the average number of fission that results from the original fission, called the effective multiplication factor, k_e,

$$k_e = \frac{\text{Rate opf production of neutrons, P}}{\text{Rate of absorption, } A_L + \text{Rate of leakage, L}} = \frac{\#\ ^1_0 n \text{ in the } (n+1)\text{th generation}}{\#\ ^1_0 n \text{ in the nth generation}} \quad (19.3.6)$$

$$k_e = \nu\,\varepsilon\,pf\,(1-\ell_f)(1-\ell_t)\frac{\sigma_f(U)}{\sigma_{ta}(U)} \quad (19.3.7)$$

Defining, $\quad \eta = f\,\dfrac{\sigma_f(U)}{\sigma_{ta}(U)} \quad (19.3.8)$

$$k_e = \eta\,\varepsilon\,pf\,(1-\ell_f)(1-\ell_t) \quad (19.3.9)$$

When $\ell_f = 0$, $\ell_t = 0$, so that $k_e = k_\infty$,

$$k_\infty = \eta\,\varepsilon\,pf \quad (19.3.10)$$

This is the well known *four-factor formula in reactor physics*.

The fission reaction is listed in Table 19.1.

Table 19.1 Fission chain reaction

1. Critical (*Steady*)	$k_e = 1$
2. Super Critical (*Divergent*)	$k_e > 1$
3. Sub Critical (*Convergent*)	$k_e < 1$

Both the amount of $^{235}_{92}U$ in the mass (the level of enrichment) and the shape of the mass control the criticality of the sample. One can imagine that if the shape of the mass is a very thin sheet, most of the free neutrons will fly off into space rather than hitting other $^{235}_{92}U$ atoms. A **sphere** is the optimal shape. The quantity of $^{235}_{92}U$ that one must collect together in a sphere to get a critical reaction is about 2 *pounds* (0.9 kg). This quantity is referred to as the **critical mass**. For $^{239}_{94}Pu$, the critical mass is about 10 *ounces* (283g).

19.3.6 An Alternate Reactor Formula

If F = Rate at which the fission process occurs,

$$P = \nu F \qquad (19.3.11)$$

$$k_e = \frac{\nu F}{A_L + L} = \frac{\nu F}{A_L} \cdot \frac{1}{1 + L/A_L} \qquad (19.3.12)$$

F / A_L depends on the quantity of the fissionable and non-fissile material, and on their cross sections for fission and neutron capture.

L / A_L depends on the ability of the reactor to contain (Table 19.2).

Table 19.2 Size of nuclear reactor, L / A_L

Reactor size	L / A_L		$k_e = \nu F / A_L$
1. Critical (*Steady*)	Critical size	$= 0$	$k_e = 1$
2. Super Critical (*Divergent*)		$\to$ small	$k_e > 1$
3. Sub Critical (*Convergent*)	No size	$\to$ Large	$k_e < 1$
4.		Decreases $\to \infty$	$k_e \to 0$

19.3.6.1 Any amount of $^{238}_{92}U$ is sub-critical! Why?

$k_\infty = 0.5$ for a body of $^{238}_{92}U$. Therefore, any fission process that may be initiated will quickly die out, *i.e.*, sub-critical.

19.3.6.2 Enrichment of $^{235}_{92}U$, Why?

Enrichment of $^{235}_{92}U$ is required for a large quantity of *nat* U to be critical.

Worked out Example 19.1

Find percentage of number of fissions for $^{238}_{92}U$ to $^{235}_{92}U$ in natural Uranium as fuel.

Solution: $\boxed{STEP\ \#1}$ $k_\infty = 2.5 \dfrac{N_{f235}\,\sigma_{f235} + N_{f238}\,\sigma_{f238}}{N_{f235}\,(\sigma_{a235} + \sigma_{f235}) + N_{f238}\,(\sigma_{a238} + \sigma_{f238})}$

$\boxed{STEP\ \#2}$ $k_\infty = 2.5 \dfrac{N_{f235}\,(1.3) + N_{f238}\,(1.5)}{N_{f235}\,(0 + 1.3) + N_{f238}\,(2.0 + 0.5)} \geq 40\%$.

$\boxed{STEP\ \#3}$ Since, $k_\infty = 1$, *with* $N_{f238} = 0$, means $\dfrac{N_{f238}}{N_{f235}} = 1.5$,

19.3.7 Byproduct of Fission

The development of transuranic elements is the interesting byproduct of nuclear fission. In a nuclear reactor,

$$^{238}_{92}U + ^{1}_{0}n \rightarrow [^{239}_{92}U^*] \rightarrow ^{239}_{93}Np^* + \beta^- + \nu$$
$$\searrow ^{239}_{94}Pu \text{ (Fissionable)} + \beta^- + \nu \qquad (19.3.13)$$

Worked out Example 19.2

How much coal is required to run a 100 W light bulb 24 hours a day for a year? Given: The thermal energy content of coal is 6,150 kWh/ton.

Solution: | STEP #1 | 0.1 kW x 8,760 hours or **876 kWh**.

| STEP #2 | Only about 40% of the thermal energy in coal is converted to electricity. So the electricity generated per ton of coal is 0.4 x 6,150 kWh or **2,460 kWh/ton**.

| STEP #3 | Amount of coal burned for the light bulb = (876 kWh)/ (2,460 kWh/ton) = (0.357 tons) (2,000 *pounds/ton*) = **714 *pounds* (325 *kg*)** of coal.

19.4.1 A Natural Fission Reactor

Francis Perrin discovered in 1972 a uranium deposit (known for to be two billions of *yrs* old) now being mined in Gabon, West Africa. This natural fission reactor has been operational for at least several hundred thousand years before shutting itself down.

19.4.2 The First Artificial Nuclear Reactor (Atomic Pile)

A *controlled self-sustaining fission reaction* (*i.e.*, chain reaction) was first demonstrated by E. Fermi *et al.* in 2 Dec.,1942, in Chicago, USA. A nuclear system in which the fissionable and non-fissionable materials are prepared so that the fission-chain reaction can proceed in a controlled manner is called a "nuclear reactor". Fermi *et al* used.

$$\text{Nat U composed of } [^{234}_{92}U \ (0.0057) + ^{235}_{92}U \ (0.7204) + ^{238}_{92}U \ (99.2739)] \quad (19.4.1)$$

i.e., 0.72% abundant *nat* U. A pile of graphite is uniformly spaced in this. This is the uranium-graphite-pile. Due to this type of construction the early nuclear reactors are known as "atomic piles". The cross section for thermal neutrons ($^{1}_{0}n$ of $\sim 0.025 \ eV$) is very large ($\sim 550 \ b$). On the other hand, the fission neutrons have $\sim 1 \ MeV$ energy, which requires bringing down quickly to the thermal energies.

It is well known that energy is most effectively lost to the incident particle by scattering from a body called moderator, which contains atoms having mass close to that of $^{1}_{0}n$.

A nuclear reactor is a source of the products of the fission process, *viz.*,

i)   *a*) Energy, *b*) Neutrons, and *c*) Radio-isotopes,

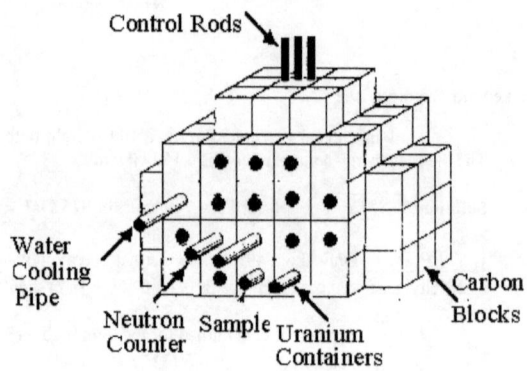

Fig 19.2   The Atomic Pile

ii)   The moderator is graphite block that slows down $_0^1n$,
iii) The fuel is fissionable $_{92}^{235}U$,
iv)   Control rods of Cadmium regulates the rate of reaction by absorbing $_0^1n$,
v)   Coolant is water, that removes heat of reaction, to make steam for turbine generator,
vi)   The shielding, concrete, keeps the radiation keeps the radiation within the reactor..

This Fermi reactor at Chicago served as a prototype for larger reactors constructed (1943) at 3 locations in USA, *viz.*, Oak Ridge, Tennessee, and Hanford, Washington, to produce $_{94}^{239}Pu$ for one of the atomic bombs dropped on Japan at the end of World War II. As we have seen, some of the neutrons released in the chain reaction are absorbed by $_{92}^{238}U$ to form ^{239}Pu, which undergoes decay by the successive loss of two $_0^1n$ *particles to form* $_{94}^{239}Pu$. $_{92}^{238}U$ *is an example of a fertile nuclide. It doesn't undergo fission with thermal neutrons, but it can be converted to* $_{94}^{239}Pu$, *which does undergo thermal-neutron-induced fission.*

19.5   TYPES OF REACTORS

In Chapter 16 different types of design of nuclear reactor have been dealt with.

19.5.1 THE FIRST REACTOR OF INDIA

AEET (Atomic Energy Establishment Trombay) employed over one thousand scientists and engineers by 1959 . Construction began in 1955 on India's first reactor, *viz,*. 1 *MW* **APSARA** research reactor, with British assistance. After more than a year of negotiation, in Sep 1955, Canada agreed to supply India with a powerful research reactor - the 40 *MW* Canada-India Reactor (**CIRUS**). Under the Eisenhower Administration's "Atoms for Peace" programme the USA, in Feb 1956, supplied 21 *tons* of heavy water for this reactor.

APSARA, fueled by enriched Uranium from the UK, went *critical* on 4 August 1957, becoming the first operating reactor in Asia outside of the then Soviet Union. CIRUS achieved criticality at BARC on 10 July 1960. This reactor produced the Plutonium used in the first nuclear test by India in 1974; provided the design prototype for India's more powerful **DHRUVA** Plutonium production "research" reactor; and is directly responsible for producing nearly half of the weapons grade Plutonium.

19.5.2 NUCLEAR POWER PLANT

It is known that conventionally, a Power Plant (fossil fuel) burns fuel to generate heat. The fuel is generally coal, but oil is also sometimes used. The heat is used to causing water to boil, resulting intense pressure of the steam generated at high temperature. This is used to turn a turbine, which then generates electricity.

A nuclear power plant produces electricity in almost exactly the same way that a conventional power plant does, except that the heat used to boil the water is produced by a nuclear fission reaction using $^{235}_{92}U$ as fuel. A nuclear power plant uses much less fuel than a comparable fossil fuel plant. For example, a rough estimate is that Conventional power plant takes 17,000 *kg* of coal to produce the same amount of electricity as 1 *kg* of nuclear uranium fuel

19.5.3 **Pressurized Water Reactor** (PWR)

Commercial as well as military applications Pressurized Water Reactor (PWR) is the popular type of reactor in use.

The essential parts of a typical nuclear reactor are shown Fig 19.3.

i) The nuclear reactions take place in the "core". The core comprises the fuel rods and assemblies, the control rods, the moderator, and the coolant.

ii) Outside the core are the turbines, the heat exchanger, and part of the cooling system.

1) Fuel:

Enriched uranium (enrichment 3.2% $^{235}_{92}U$) as UO_2 in zircaloy cans is used as fuel.

2. Core layout:

Each fuel assembly is a collection of fuel rods. Each fuel rod is of about 1 *cm* diameter and about 3.5 *m* long. A couple of hundred such rods called are grouped and bundled as fuel assemblies, which are then placed in the reactor core. Inside each fuel rod are hundreds of pellets of uranium fuel stacked end to end. Fuel pins, arranged in clusters, are placed inside a pressure vessel containing coolant & moderator

3. Control rods:

Control rods are also arranged in the core. These rods enclose pellets formed of very efficient neutron capturing materials such as cadmium. These control rods are connected to machines that can raise or lower them in the core. When they are fully lowered into the core, fission cannot occur because they absorb free neutrons. On the other hand, when they are fully lifted out of the reactor, fission can start again as and when a stray neutron strikes a $^{235}_{92}U$ atom, thus releasing more neutrons, and starting a chain reaction.

4. Moderator:

Ordinary water is used as coolant and moderator. The reason is the neutron absorption properties of H_2O (Section 19.3.4). The moderator serves to slow down (thermalize) the fast neutrons $^{1}_{0}n$ moving all around the reactor core. If a $^{1}_{0}n$ is moving too fast, and thus is at a high-energy state, it passes right through the $^{235}_{92}U$ nucleus. It must be slowed down to be captured by the nucleus and to induce fission.

5. Coolant:

The coolant absorbs the heat released from the nuclear reaction. The most common coolant used in nuclear power plants today is water. Actually, many reactors are designed such that the coolant and the moderator are one and the same.

6. Turbine:

The coolant water absorbs and gets heated due to the nuclear reactions taking place inside the core. Being kept at an extremely intense pressure, the boiling point of the water coolant is raised above the normal value of 100° C. Thus the water coolant only gets heated and does not boil. It passes through the turbine.

7. Steam Generator:

After the hot water has passed through the turbine (steam at 317°C at 2235 *psia* and steam efficiency 32%), a part of its energy is changed into electricity. However, the water is still very hot. It must be cooled somehow. Many nuclear power plants used steam towers to cool this water with air. These are generally the buildings that workers associate with nuclear power plants. At reactors that do not have towers, the clean water is purified and dumped into the nearest body of water, and cool water is pumped in to replace it.

Example:

In India RAPP2 is PHW type, 203 *MWe*, went critical in 1976.

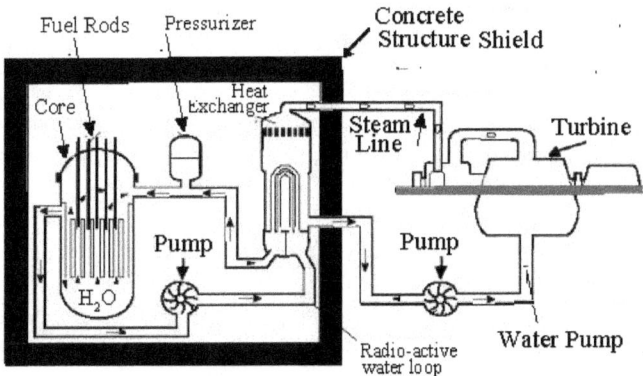

Fig 19.3 Pressurized Water Reactor (PWR) [RAPP2 is PHW type, 203 MWe, critical in 1976]

19.5.3 Boiling Water Reactor (BWR)

PWR is the most common type of reactor, yet few other types of reactors are also used. The second most popular type of reactor is the Boiling Water Reactor (BRW). This differs from the PWR in that there is only one water cycle. Radioactive water is used to turn the turbine, which is the major disadvantage of this. Hence radioactive nuclides in the water cause radioactivity that can be transferred to the turbine, thus spreading radioactivity. This produces more hazardous material that needs to be disposed of when a reactor is dismantled. However, the BWR has a few advantages. Its core can be kept at a lower pressure.

a) Fuel:

Enriched uranium (enrichment 2.6% $^{235}_{92}U$) as UO_2 in zircaloy cans is used as fuel

b) Moderator:

Ordinary water is used as coolant and moderator, because of the neutron absorption properties of H_2O (Section 19.3.4).

c) Core layout:

Each fuel assembly is a collection of fuel rods. Each fuel rod is of about 1 *cm* diameter and about 3.5 *m* long. A couple of hundred such rods called are grouped and bundled as fuel assemblies, which are then placed in the reactor core. Inside each fuel rod are hundreds of pellets of uranium fuel stacked end to end. Fuel pins, arranged in clusters, are placed inside a pressure vessel containing coolant & moderator.

Coolant outlet is at 286°C at 1050 *psia*. Schematically the BWR looks like the PWR of Fig 19.3, except that the steam generator is within the pressure vessel itself.

19.5.4 Heavy Water Reactor (HWR or PHW), like CANDU

A HWR uses heavy water (D_2O) as a moderator instead of normal water. Deuterium is heavier than normal hydrogen. HWR's come in two types, pressurized and boiling, just like normal *light water* reactors. The advantage of a HWR is that un-enriched Uranium fuel can be used. This is because the heavy water is a much more efficient moderator than light water. Thus, more stray neutrons can be slowed down enough to cause fission in $^{235}_{92}U$. More efficient moderator makes up for the greater abundance of the neutron-capturing $^{238}_{92}U$. BWR is perhaps the simplest in concept of all types of reactors.

a) Fuel in CANDU:

Natural uranium (enrichment 0.7% $^{235}_{92}U$) as UO_2 in zircaloy cans is used as fuel.

b) Moderator:

D_2O is used as coolant and moderator, because of the neutron absorption properties of D_2O (Section 19.3.4).

c) Core layout:

Each fuel assembly is a collection of fuel rods. Each fuel rod is of about 1 *cm* diameter and about 3.5 *m* long. A couple of hundred such rods called are grouped and bundled as fuel assemblies, which are then placed in the reactor core. Inside each fuel rod are hundreds of pellets of uranium fuel stacked end to end. Fuel pins, arranged in clusters, are placed inside a pressure vessel containing coolant & moderator.

d) Coolant out let:

It is at 305°C at 1285 *psia*. Schematically the BWR looks like the PWR of Fig 19.3, except that the steam generator is within the pressure vessel itself. The first of this type was designed and established in Ontario (Canada) in 1962.

Table 19.3   Research Reactors of INDIA

Name of Reactor	Type	Power	Modertor	Fuel	Location	Supplier	Date of Criticality
1 APSARA	PWR	1 *MWt*	Water	enriched U	BARC	UK	Aug 1957
2 CIRUS	PHWR(Candu)	40*MWt*	D_2O	Nat UO_2	BARC	Canada	Jul 1960
3 ZERLINA		0.1 *kW*					1961
4 DHRUVA	PHWR	100 *MW*	D_2O	Nat UO_2	BARC		Aug 1985
5 PURNIMA			Na cooled	$^{239}_{94}Pu$	DAE		May 1972
6 FBTR		40 *MWt*	Na cooled	U-Pu carbide	DAE		Oct 1985
7 KAMINI		30 *kW*	Na cooled	$^{233}_{92}U$	Kalpakkom	DAE	Oct 1996
8 Prtotype FBR			Na Cooled	$^{239}_{94}Pu$	Kalpakkam	DAE	(2009)

19.5.4.1 SGHWR is a steam generating HWR.

19.5.5 Magnox Gas-cooled Reactors (MGR) use graphite moderator.

19.5.6 Advance Gas-cooled Reactor (AGR) used graphite as moderator.

19.5.7 HTGR is High Temperature Gas-cooled Reactor, used with startup Th/$^{235}_{92}U$ (93% enriched); recycle Th/$^{235}_{92}U$ (93% enriched/ $^{233}_{92}U$ (recycle)) and graphite moderator.

TABLE 19.4 OPERATING POWER REACTORS OF INDIA

	Reactor	Type	MWe net	Year
1	Tarapur 1 (TAPS) (MH)	BWR	160	Oct 1969.
2	Tarapur 2 (TAPS) (MH)	BWR	160	Oct 1969
3	Tarapur 3 TAPS) (MH)	PHWR	540	Sep 2005
4	Tarapur 4 (TAPS) (MH)	PHWR	540	Aug 2006
5	Kaiga 1 (KA)	PHWR	220	Mar 2000
6	Kaiga 2 (KA)	PHWR	220	Nov 2000
7	Kaiga 3 (KA)	PHWR	220	May 2007
8	Kaiga 4 (KA)	PHWR	220	(Dec 2010)
9	Kakrapar 1 (GUJARAT)	PHWR	220	May 1993
10	Kakrapar 2 (GUJARAT)	PHWR	220	Sep 1995
11	Kakrapar 3 (GUJARAT)	PHWR	700	(Jun 2015)
12	Kakrapar 4 (GUJARAT)	PHWR	700	(Dec 2015)
13	Kalpakkam 1 (MAPS)	PHWR	170	Jan 1984
14	Kalpakkam 2 (MAPS)	PHWR	220	Mar 1986
15	Narora 1 (UP)	PHWR	220	Jan 1991
16	Narora 2 (UP)	PHWR	220	Jul 1992
17	Rawatbhata 1 (Rajasthan)	PHWR	90	Dec 1973
18	Rawatbhata 2 (RAUJASTHAN)	PHWR	187	Apr 1981
19	Rawatbhata 3 (RAJASTHAN)	PHWR	202	Jun 2000
20	Rawatbhata 4 (Rajasthan)	PHWR	202	Dec 2000
21	Rawatbhata 5 (RAJASTAN)	PHWR	202	Dec 2009
22	Rawatbhata 6 (RAJASTHAN)	PHWR	202	Mar 2010
23	Rawatbhata 7 (Rajasthan)	PHWR	700	(Jun 2016)
24	Rawatbhata 8 (Rajasthan)	PHWR	700	(Dec 2016)
25	Kudankulam 1 (TN)	PHWR	1000	2013
26	Kudankulam 2 (TN)	PHWR	1000	2013
27	Kalpakkam (MAPS)	PFBR	470	2013
28	Kudankulam 3 (TN)	PHWR	1000	(TBD)

19.5.8 Research Reactor of India

In India various research reactor have been designed and commissioned since 1957. They are listed in Table 19.3.

19.5.9 Power Plants of India

In February 2001 India had 14 small nuclear power reactors in commercial operation. Further two larger ones are under construction and ten more planned

Table 19.4 lists the power reactors operating by 2010 in India.

19.5.10 BREEDER REACTORS

Breeder reactor is efficient because the key factor in its fuel that gives the largest possible number of $_0^1n$ released per neutron absorbed. Such a reactor is being built use a mixture of $PuO_2 + UO_2$ as the fuel and fast $_0^1n$ (~1 MeV) activate fission. Fast neutrons carry energy of at least several keV and therefore travel more than 10^4 times faster than thermal (~1 keV) neutrons. $_{94}^{239}Pu$ in the fuel assembly on absorbing one such fast neutrons undergoes fission with the release of 3 neutrons. Through $_0^1n$ capture process $_{92}^{238}U$ in the fuel then produce additional $_{94}^{239}Pu$.

Obviously breeder reactors have the advantage that they provide limitless supply of fuel for nuclear reactors. This is known as the breeding cycle.

Disadvantages:

i) The primary disadvantage of a breeder reactor is that it is more expensive to build than the other types of reactors.
ii) They are also useless without a subsidiary industry to collect the fuel, process it, and transport the $_{94}^{239}Pu$ to new reactors.
iii) It is the reprocessing of $_{94}^{239}Pu$ that concerns most of the scientists of breeder reactors.
iv) $_{94}^{239}Pu$ is so dangerous because carcinogen that the nuclear industry places a limit on exposure to this material that assumes those workers inhale no more than 0.2 μg of Pu over their lifetimes.
v) The $_{94}^{239}Pu$ produced by these reactors might be stolen and assembled into bombs by terrorist organizations.

But fission reactors depend on the supply of uranium, which is quite expensive and gets depleted at a rate which will deplete the supply in approximately 50 years.

19.5.9.1. Breeding Cycles

There are two breeding cycles.

i) Uranium Breeding Cycle

The $_{92}^{238}U$ present in enriched uranium is not fissionable. But it can be made to undergo transmutation into $_{94}^{239}Pu$, a fissionable material. Some $_{92}^{238}U$ (Non-fissionable from *nat* U; 99.3% abundant, but FERTILE) give through the following reaction, the fissionable $_{94}^{239}Pu$:

$$\text{(Fertile)} \, _{92}^{238}U + {_0^1}n \rightarrow [{_{92}^{239}U}^*] \xrightarrow[24\,m]{\beta^-} {_{93}^{239}Np}^* \xrightarrow[23\,d]{\beta^-} {_{94}^{239}Pu} \, \text{(Fissionable)}$$

$$(19.5.1)$$

The $_{94}^{239}Pu$ produced then undergoes the fission reaction

(Fissionable) $^{239}_{94}Pu + ^{1}_{0}n \rightarrow ^{147}_{56}Ba + ^{90}_{38}Sr + x\ ^{1}_{0}n$ (fast); $(x > 2)$ (19.5.2)

The neutrons produced thus are then used to make more $^{239}_{94}Pu$ from $^{238}_{92}U$. Reactors that are specially designed to harvest the Pu are breeder reactors. By definition a breeder reactor is any reactor that creates more fissionable matter than it started out with. This is palatable but Pu is the most toxic material known to humanity. It is widely known that **one atom of Pu can kill a man if it gets into his lungs**!

ii) Thorium Breeding Cycle

(Fertile) $^{232}_{90}Th + ^{1}_{0}n \rightarrow [^{233}_{90}Th^{*}] \xrightarrow[22\ m]{\beta^{-}} ^{233}_{91}Pa^{*} \xrightarrow[27\ d]{\beta^{-}} ^{233}_{92}U$ (Fissionable)

(19.5.3)

($^{233}_{92}U$; $\tau_{1/2} = 0.162\ MYrs$).

(Fissionable) $^{233}_{92}U + ^{1}_{0}n \rightarrow$ Pair of Fragments $+ y\ ^{1}_{0}n$ (slow); $(y > 2)$

(19.5.4)

Thus breeding is achieved by combining both the fissionable and fertile materials in the reactor core under conditions which

a) Provide enough $^{1}_{0}n$ s to sustain a chain reaction in the fissionable material
b) Enough to convert more fertile material into fissionable material than was originally present., *i.e.*, to produce more fissionable material than they consume.

This is the figure of merit called efficiency, expressed as 'doubling time'. Design objectives are usually set to have the doubling time < 10 *yrs*.

19.5.9.2. Liquid Metal Fast Breeding Reactors (LMFBR)

It is based on

i) Fuel: A mixture of ($UO_2 + ^{239}Pu\ O_2$) in stainless steel container.
ii) Moderator: None
iii) Core layout: Assemblies of fuel elements are immersed in a tank containing the liquid sodium coolant. The core is surrounded by a "blanket" of uranium carbide in stainless steel container.
iv) Heat extraction: The sodium is heated by the core and pumped through an intermediate heat exchanger, where it heats sodium in a separate secondary circuit. The sodium (M.P. 208°F ; B.P. 1616°F) in the secondary circuit transfers its heat to ordinary water in a steam generator. The steam drives a turbine coupled to an electric generator.
v) Fuel enrichment: 20% Pu.
Coolant outlet temperature: 620°C, at coolant pressure of 5 *psia*.

19.5.9.3 Light Water Breeder Reactor (LWBR)

LWBR uses $^{232}_{90}\text{Th} \rightarrow\ ^{233}_{92}\text{U}$ fuel cycle and employs PWR. In PWR, using $^{235}_{92}\text{U}$ fuel, has a conversion ratio of ▢ 0.6, *i.e.*, during operation▢ 0.6 atom of fissionable $^{239}_{94}\text{Pu}$ are formed to every atom of $^{235}_{92}\text{U}$ consumed. But in LWBR, this ratio is made 1.0, at which point the reactor makes as much new fuel as it consumes.

19.5.9.4 Molten Salt Breeder Reactor (MSBR) (Thermal Breeder Reactor)

It uses $^{232}_{90}\text{Th} \rightarrow\ ^{233}_{92}\text{U}$ cycle.

Fuel salt: $^7\text{Li} - \text{BeF}_2 - \text{ThF}_4 - \text{UF}_4$

Moderator: Unclad, sealed graphite

19.5.9.5 Integral Fast Reactor (IFR) –

Argonne National Laboratory near Chicago (USA) has a new fast reactor designed to have a highly compact core with enriched fuel. The enriched fuel allows for a longer "burn" period between shutdowns.

19.5.9.6. World's Safest Nuclear Reactor **ATBR** (2005)

India unveiled before the international community on August 25, 2005 its revolutionary design of 'A Thorium Breeder Reactor' (**ATBR**) that can produce 600 *MW* of electricity for two years 'with no refueling and practically no control maneuvers. Designed by scientists of the Mumbai-based BARC, the ATBR is claimed to be far more economical and safer than any power reactor in the world. Most significantly for India, ATBR does not require natural or enriched Uranium which the country is finding difficult to import. It uses Thorium -which India has in plenty in Kerala - and only requires Plutonium as 'seed' to ignite the reactor core initially. Eventually, the ATBR can run entirely with Thorium and fissile ^{233}U bred inside the reactor (or obtained externally by converting fertile Thorium into fissile U-233 by neutron bombardment). The uniqueness of the ATBR design is that there is almost a perfect 'balance' between fissile depletion and production that allows in-bred ^{233}U to take part in energy generation thereby extending the core life to two years. The ATBR annually requires 2.2 *tons* of Plutonium as 'seed'. Although India has facilities to recover plutonium by reprocessing spent fuel, it requires plutonium for its Fast Breeder Reactor programme as well.

19.6 NUCLEAR WEAPONS.

Two types of atomic explosions that can be achieved by

a) $^{235}_{92}\text{U}$ fission and

b) Fusion.

Fission is a nuclear reaction in which a heavy atomic nucleus splits into usually two fragments of comparable mass, and releasing 100 *MeV* to several hundred *MeV* of energy. This energy is expelled explosively and violently in the Atomic Bomb. A fusion reaction usually requires a fission reaction to start. But unlike the fission (atomic) bomb, the fusion (hydrogen) bomb derives its power as a result of the fusion of nuclei of various hydrogen isotopes into helium nuclei.

19.6.1 ATOMIC BOMB (A-Bomb)

19.6.1.1 A Working Thermonuclear Device.

To build an atomic bomb, the requirements are:

(i) A source of fissionable fuel,

(ii) A triggering device, a way to allow the majority of fuel to fission or fuse before the explosion occurs (otherwise the bomb will fizzle out),

There are three things about this *induced* fission process, the probability of a $^{235}_{92}U$ atom capturing a $^{1}_{0}n$ as it passes by is fairly high. In a bomb to work properly, more than one $^{1}_{0}n$ released from each fission causes a second fission to occur. This condition is known as *super-criticality, t*he process of capturing the $^{1}_{0}n$ and splitting happens very quickly, on the order of picoseconds ($1 \times 10^{-12} s$) and an incredible amount of energy is released, in the form of heat and γ-radiation, when an atom splits. The energy released by a single fission is due to the fact that the fission products and the neutrons, together, weigh less than the original $^{235}_{92}U$ atom. The difference in mass is converted to energy at a rate governed Einstein formula $\Delta E = m\ c^2$.

19.6.1.2 Fabrication

i) First, say about 50 *Lbs* (110 *kgs*) of weapons grade Plutonium, is taken. One has to pay much attention since Pu, especially pure & refined Pu, is somewhat dangerous, ii) Then wash the hands with soap and warm water after handling the material in a lead box. iii) Next, Bind together a metal enclosure to house the device, by sheet metal, other than tinfoil. iv) Then arrange the Pu into two hemispherical shapes, separated by about 4 *cm*. v) Use rubber cement to hold the Pu dust together. vi) Now get about 100 *Lbs* (220 *kgs*) of trinitrotoluene (TNT) (or Gelignite is much better. vii) Pack the TNT around the hemispheric arrangement constructed in step 4 packed in with any modeling clay, and viii) enclose the structure from step 6 into the enclosure made in step 3. ix) Use a strong glue such as "Araldite" to bind the hemispheric arrangement against the enclosure to prevent accidental detonation, which might result from vibration or mishandling. x) To detonate the device, obtain a radio controlled (RC) servo-mechanism, as found in toys. A remote plunger can be made that will strike a detonator cap to effect a small explosion.

Nuclear devices have been known to spontaneously detonate in the unstable conditions, like temperature and high humidity (Hall closet or under the kitchen sink suits well. .

It can be used for **National Defense purposes**.

The minimum amount to start a chain reaction as described above is known as super critical mass.

$^{235}_{92}U$, $^{238}_{92}U$ and $^{239}_{94}Pu$

19.6.2 The Hydrogen Bomb (Thermonuclear Bomb)

When a fission bomb, called the primary, explodes a flood of radiation including a large number of neutrons are released. This radiation impinges on the thermonuclear portion of the H-bomb, known as the secondary. The secondary consists largely of lithium deuteride. The neutrons react with the lithium in this chemical compound, producing tritium and helium

$$^{6}_{3}Li + ^{1}_{0}n \rightarrow ^{4}_{2}He + ^{3}_{1}H \qquad (19.6.1)$$

This reaction produces the tritium ($^{3}_{1}H$) on the spot, so there is no need to include tritium in the bomb itself. In the extreme heat which exists in the bomb, the $^{3}_{1}H$ fuses with the deuterium in the lithium deuteride.

Design: The shock waves produced by the primary (A-bomb) would propagate too slowly to permit assembly of the thermonuclear stage before the bomb get itself split apart. Edward Teller and Stanislaw Ulam found a solution to this problem. They introduced a high energy γ-ray absorbing material (Styrofoam) to capture the energy of the radiation. As high energy γ-radiation from the primary (fission bomb) is absorbed, radial compression forces are exerted along the entire cylinder at almost the same instant. This produces the compression of the lithium deuteride core. Additional neutrons are also produced by various components and reflected towards the core of lithium deuteride. When the compressed lithium deuteride core is bombarded with neutrons, tritium $^{3}_{1}H$ is formed and the fusion process begins.

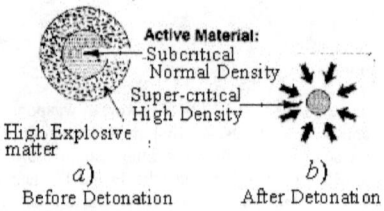

a) Before Detonation   b) After Detonation

Fig 19.6   Implosion Assembly Principle

The yield of a H-bomb is controlled by the amounts of lithium deuteride and of additional fissionable materials. $^{238}_{92}U$ is usually the material used in various parts of the bomb's design to supply additional neutrons for the fusion process. This additional fissionable material also produces a very high level of radioactive fallout.

19.6.3 The Neutron Bomb (ERW)

The neutron bomb is a small H-bomb. It differs from standard nuclear weapons insofar as its primary lethal effects come from the radiation damage caused by the neutrons it emits. It is also known as an E*nhanced Radiation Weapon* (ERW).

The augmented radiation effects mean that blast and heat effects are reduced so that physical structures including houses and industrial installations are less affected. Because neutron radiation effects drop off very rapidly with distance, there is a sharper distinction between areas of high lethality and areas with minimal radiation doses.

19.6.4 Chronology of Developments and Fabricated Devices

1) On June 27, 1954, the World's first nuclear power plant generated electricity, at Obnisk near Moscow (Russia). The capacity of the generator was only 5 *MWe*.

2) In 1954: the World's first nuclear powered submarine, the USS *Nautilus*, was launched. It was the first vessel to complete a submerged transit to the North Pole on 3 August 1958.

3) In 1955, Arco (Idaho, USA) was the first town to be lit entirely by nuclear power. The BORAX II reactor(BWR) prototype, was used.

4) Launched on December 5, 1957, the First Nuclear-Powered Surface Ship, *Lenin* by the USSR during the Cold War. The ice breaker never fired a shot as it had no guns, depth charges or weapons of any kind. I was built in the Admiralty Shipyards in the then Leningrad, .

5) 18 Dec 1957, '*Shippingpor*t', the first U.S. Nuclear Power Plant, at Beaver, near Pittsburgh, PA (USA). Started production of electricity. It was a light water moderator thermal Breeder Reactor..

6) By late 1950s, AECL, Canada had developed the first nuclear reactor (CANDU, *i.e.*, CANADA Deuterium Uranium) environmentally-sensitive reactors consistently lead the world in productivity, safety, and ease of use. In Ontario, it was a pressurized heavy water (D_2O) (Deuterium oxide) coolant and moderator with fuel, UO_2 + nat U (0.7% $^{235}_{92}U$).

7) A-Bomb: On July 16, 1945, the first $^{235}_{92}U$ -fueled *atomic bomb* was detonated at Alamogordo, New Mexico, and on August 6, 1945, the USA dropped an A- bomb on Hiroshima (Japan), killing more than 100,000 people, and the $^{239}_{94}Pu$ -fueled bomb dropped on Nagasaki (Japan)

8) I Nov 1952: The first artificial fusion reaction occurred when the hydrogen bomb was tested, at Eniwetok, Marshall Islands (USA). It is the Teller-Ulam design for the H-Bomb.

9) 21 Jul 1959, the First U.S. Nuclear-Powered Cargo passenger (commercial) vessel, *viz.*, the Nuclear Ship 'Savannah', was substantial

10) India joined the nuclear club. On May 18, 1974, Indian Army detonates a 12-*kiloton* nuclear explosive (a peaceful Bomb) in the Pokhran, Rajasthan desert. It was built using Plutonium from a research reactor

11) India's first 40 *MWt* Fast Breeder Test Reactor (FBTR) at IGCAR, Kalpakkam, attained criticality on 18th October 1985. It uses the Thorium Reactor design lasting for 100 years

(compared to the 40 yers life of all the other reactor designs)). The fuel used is Pu-U Mono-Carbide. India has abundant supply of Thorium. Thorium provides 60% of the reactor power. It uses $^{233}_{92}U$ only for neutron radiography. India becomes the sixth nation having the BOT (build and operate technology) a FBTR besides USA, UK, France, Japan and the then USSR.

Worked out Example 19.4

The dominant energy-releasing reaction in the Sun is $4\,(^1_1H) \rightarrow {}^4_2He + 2\,{}^0_{-1}e + 2\,\nu + \gamma$. Given that the Sun is fully composed of hydrogen atoms. If the total power output of the Sun is assumed to remain constant at $3.9 \times 10^{26}\,W$, estimate the time period that will take for burning up all of the hydrogen. The mass of the Sun is $M_{\Box} = 1.99 \times 10^{30}\,kg$.

Solution: $\boxed{STEP\ \#1}$ Total number of hydrogen atoms in the sun by calculating = $(M_{\Box} = 1.99 \times 10^{30}\,kg)/(1.67 \times 10^{-27}\,kg/atom) = 1.192 \times 10^{57}$ atoms.

$\boxed{STEP\ \#2}$ Now, the reaction quoted has a mass difference of $4 \times 1.007825 - (4.002603 + 2 \times 0.000549) = 0.027599\,u$,

$\boxed{STEP\ \#3}$ which corresponds to an energy

$E = (0.027599\,u)\,(931.5\,MeV/u\,c^2) = 25.71\,MeV$

$\boxed{STEP\ \#4}$ Each individual reaction consumes 4 H-atoms, the total energy available in the Sun $= (1.192 \times 10^{57}\,atoms)\,(25.71\,MeV/4\,atoms)\,(10^6\,eV/MeV)\,(1.6 \times 10^{-19}\,J/eV) = = 1.225 \times 10^{45}\,J$

$\boxed{STEP\ \#5}$ The lifetime of the Sun =

$[(1.225 \times 10^{45}\,J)/(3.9 \times 10^{26}\,Js^{-1})]\,(1\,hr/3600\,s)\,(1\,d/24\,h)\,(1\,yr/365\,d) = 9.96 \times 10^{10}\,yrs$

$\boxed{STEP\ \#6}$ Estimated lifetime = $\Box$ 99.6 Byrs.

19.7   POWER FROM CONTROLLED FUSION (APPLICATION OF PLASMA)

19.7.1.1 Types of Fusion

It is known from Chapter 17 that the fusion of two deuterium nuclei produces helium or tritium s from the reaction (19.7.1).

$${}^2_1H + {}^2_1H \rightarrow {}^3_2He + {}^1_0n + 3.27\,MeV \qquad (19.7.1)$$

$${}^2_1H + {}^2_1H \rightarrow {}^3_1H + {}^1_1H + 4.03\,MeV \qquad (19.7.2)$$

The tritium, and helium produced this way react with deuterons to produce finally $_{2}^{4}He$ (α-) particle as follows:

$$_{1}^{2}H + _{1}^{3}H \rightarrow _{2}^{4}He + _{0}^{1}n + 17.58 \text{ MeV} \qquad (19.7.3)$$

$$_{1}^{2}H + _{2}^{3}He \rightarrow _{2}^{4}He + _{1}^{1}H + 18.34 \text{ MeV} \qquad (19.7.4)$$

All the above reactions are *exo-thermic*.

The number of these reactions inside the plasma depends on

i) the confinement time τ of the plasma, and
ii) the number density n of nuclei taking part in the fusion. It can be seen that for high probability of fusion, one must have

$$n \, \alpha \, \bar{v} \, \tau \geq 1 \qquad (19.7.5)$$

where $\bar{v} = \sqrt{\frac{3 k_B T}{m}}$. Assuming T = 300 keV, then $\bar{v}$ = 6.5 x $10^6 \, m \, s^{-1}$, for the D-D reaction. The fusion cross section, $\alpha \simeq 10^{-29} \, m^{-2}$ will be obtained. The condition for the high possibility of fusion o occur

$$n \, \tau \geq \frac{1}{\bar{v} \, \alpha} \qquad (19.7.6)$$

If n is in m^{-3} and τ in seconds, J.D. Lawson gave

$$n \, \tau \geq 1.5 \times 10^{22} \, m^{-3} \, s \qquad (19.7.7)$$

This is known as **the Lawson Criterion**, for the D-D reaction to take place. For the D-T reaction, it can be verified that the criterion becomes $\tau \geq 10^{20} \, m^{-3} \, s$.

Further the temperature of the plasma should be very high.

Among the methods adopted to heat the plasma to high temperatures, one among the following methods may be adopted, *viz.*,

a) Ohmic heating,

b) Cyclotron heating,

c) Shock heating,

d) Magnetic heating, and

e) Heating by neutral particles.

19.7.1.2 Confinement Methods

Several magnetic confinement systems have been developed. They are:

a) Magnetic confinement: Tokomak - Stellarator – Reverse field pinch –Field Reversed Configuration – b) b) Levitated Dipole

c) Using pinch effects - Z-pinch and Θ-pinch,.

d) Stellarator

e) Tokomak.

It is known that the fusion reactions

To enable one deuterium ion to fuse with the other deuterium ion it must physically touch the other deuterium ion. This causes ES repulsion. It is the effort of overcoming this repulsion between the nuclei that is a cost of doing business. There were two kinds of *magnetic bottles*;

i) the doughnut-shaped arrangement which in time evolved to the Tokomak, front-runner in fusion, and

ii) the *mirror machine,* with the configuration or "magnetic field-line pattern" resembled a long cylinder squeezed in near its ends. These two very different shapes represented different responses to a fundamental problem of magnetic containment:

Persa Raymond (1957) became instrumental in the design and operation of the DCX-2 mirror machine experiment.

Consider the *containment* of water in a hose. A hose is effective at *containing* the water within itself, but is ineffective in keeping it from flowing out the ends. A cylindrical bundle of magnetic field lines resembles a water hose.

19.7.2 The Mirror Machines (DCX Machine)

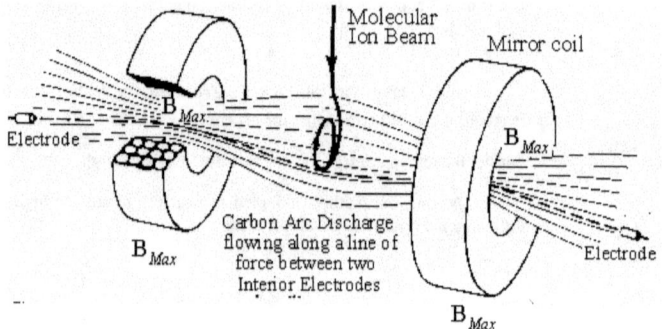

Fig 19.8   Principles of the DCX Mirror Machine

The first attempt at fusion containment was the DCX machine, built in 1950s. Basically it was a large water cooled discharge tube. Its principle is shown in Fig 19.8. In discharge tubes that are at a non-conducting state, the resistance for flow exists. The working environment of the tube is vacuum and so the resistance. But as soon as the tube starts to create ions the resistance vanishes.

The magnetic field is much weaker in the middle of the machine than at the ends (B_{Max}). If the magnetic field is strong it will curl the paths of the charged ions and electrons of a hot plasma into corkscrew-shaped orbits, thereby <u>containing</u> the ions and electrons (Fig 19.9). But such a field will do nothing to keep the plasma from escaping out the ends. *Mirror machines*, solve the problem of the ends by taking advantage of the "magnetic-mirror" effect. That is, a region of increased magnetic field (the "mirror") will tend to reflect and turn back tight-spiraling ions or electrons that approach that region. This is because a charged particle that has a momentum vector making an angle ϑ with field line will spiral along the line, as it enters the highest $\vec{B}$ field at the end, creating a field component of $\vec{F} = q\,(v \wedge \vec{B})$ in a direction parallel to the axis of the machine, but opposite the velocity direction, as shown in Fig 19.9. Thus confinement begins.

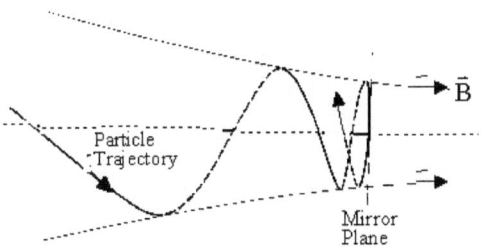

Fig 19.9 The spiral path of a charged particle as it enters a region of converging

magnetic induction.

if the angle ϑ is above a critical value and less than 90^0. The condition of this confinement is determined by the magnetic field gradient. Thus a **magnetic bottle** can be formed by adding coils that increase the field strength to B_{Max} at the ends of a long cylindrical bundle of magnetic field lines. This was the principle of early mirror machines. The concept devised to control mirror end leakage, called the "tandem mirror", was put forward, in 1976, by Dimov (in Russia), and Fowler and Logan at Livermore (in U.A).

The huge advantage that mirror machines (such as the tandem mirror) have is the possibility of containing the plasma in a non-turbulent state, with major gains, both in simplification and in economics. These gains can only be realized if the tandem mirror is reconfigured to the original simple cylindrical form. This implies that new means must be found to prevent the sideways drift that plagued mirror machines before the invention of the magnetic well.

19.7.3 Tokomak

Tokomak is the Russian name for 'toroidal chamber with axial magnetic field'. Experimental arrangement for controlled nuclear fusion in a *Tokomak* machine, is shown in Fig 19.10. Two superimposed magnetic fields enclose the plasma: this is the toroidal field generated by (a) external coils and (b) the field of a flow in the plasma. In the combined field, the field lines run helicoidally around the doughnut shaped vacuum container centre. (Fig 19.10). This first (toroidal) field is produced by a set of coils equally spaced around the doughnut chamber and keeps the plasma away from the walls. However, this field is not enough to confine the plasma by itself. The necessary twisting of the field lines and the structure of the magnetic areas are achieved in this manner. Apart from the toroidal field generated, a third (poloidal) field is added to counteract the natural pressure inside the plasma which tries to make it expand. The third vertical field (poloidal field), fixes the position of the flow in the plasma container. In a Tokomak, the poloidal field is generated by the large plasma current which is also used to heat the plasma.

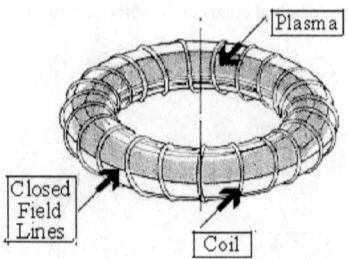

Fig 19.10 Schematic showing working principle of a Tokamak.

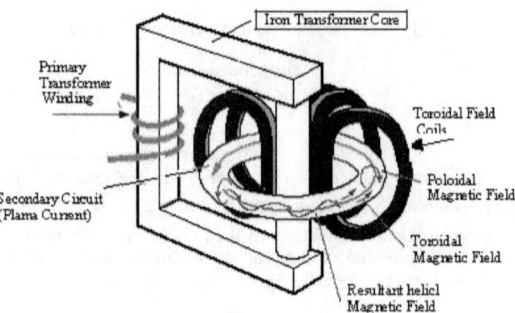

Fig 19.11 Schematics of the Ohmic heating principle to increase the energy of plasma

Though Tokomaks work in a pulsed mode they are capable of reaching steady state conditions by using additional current drive systems. Once the fusion burn is initiated external heating can be reduced as the desired operating value for power amplification Q (= fusion power out / input power to the plasma) is reached. The heating systems can then be used to 'push' the plasma electrons in the toroidal direction, creating a current independent of the transformer.

When fusion conditions are reached, α – particles resulting from the fusion reaction will also heat the plasma. In a commercial power reactor, nearly all of the heating needed to compensate plasma energy losses will come from the α – particles, with just a low level of external heating to fine-tune the plasma performance.

The Tokomak design, developed by Andrei D Sakharov and Igor EY Tamm in Moscow in the 1960s, has been most successfully demonstrated in JET (Joint European Torus experiment).

JET has a plasma volume of approx. 85 m^3, the central part of which reaches fusion temperatures of $1-2 \times 10^8$ $°C$. In 1997, over 16 MW of fusion power was generated by JET. . In a large device like JET, this toroidal current is about 3 to 6 MA and is induced by transformer action – the plasma acts as a secondary winding. The resulting total magnetic field is 'helical' around the toroidal direction. Other magnetic field components are generated by further coils to shape and position the plasma in the reactor.

The world's largest Tokomak is ITER (International Nuclear Fusion Project), built in France in July 2010, in which D-T mixture is used, heating is at temperature greater than 160×10^6 $°C$. Superconducting magnet is used. ITER is expected to show that fusion process is the energy source for the future.

High-Beta Tokomak of the Columbia University is being under construction (2012).

EFDA (European Fusion Development Association) has been working for the energy supply for the future by fusion research by their Tokomak.

19.7.4 The Stellarator Machine

A *Stellarator* is a device used to confine hot plasma with magnetic fields in order to sustain a controlled nuclear fusion reaction. It was invented by in 1951 by Lyman Spitzer of Princeton University. The name was given to this early fusion concept because of the possibility of harnessing the power source of the Sun.

The Stellarator solves issues faced by Tokomak fusion reactors where the windings of an electromagnet's wiring around a torus are less dense on the outside of the loop than on the inside, which makes it difficult for magnetic torus to contain plasma. The Stellarator addresses this issue by using a toroid bent into a figure-eight shape.

In standard torus plasma particles (ions) on the inner portion of the tube are subjected to a greater magnetic force than those at the outside. Only particles near the middle receive the optimum amount. Since magnetic forces are generally at right angles to motion (Fig 19.12), non-centered plasma moving around the toroid would be forced up or down until it hit the edges of the tube (Fig 19.13).

In a Stellerator, when a particle orbits the tube, it spends half the time on the inside of the tube and half on the outside. This helps equalizes the forces and the particle is subject to a much smaller drifting force.

The earliest Stellarators were figure-eights, consisting of two sides of a torus connected together with crossed straight tubes. To allow the tubes to cross without hitting, the torus sections on either end were rotated slightly. This arrangement was less than perfect, however, as a particle on the inner portion at one end would not end up at the outer portion at the other, but at some other point rotated from the perfect location due to the tilt of the two ends.

Different geometries were tried to address these problems, starting with simple changes to allow the ends to lie flat at different levels and placing symmetrical bends in the arms instead. A later version solved the problem more convincingly by introducing a peanut-shaped tube instead of a figure-eight, the in-bent sides offsetting the out-bend toroidal sections on either end.

The solution was magnetic rather than mechanical. By rotating the magnetic windings themselves as they were wrapped around the chamber, the plasma would be rotated around a simple torus, slowly moving from inside to outside.

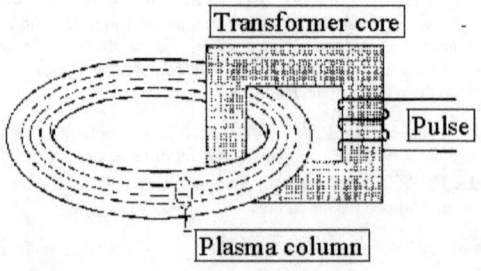

Fig 19.12 The Stellarator

For a fusion power plant, the Stellarators could provide a technically simpler solution than Tokomak. Some important Stellarator experiments are Wendelstein (in Germany), and the Large Helical Device (in Japan). A new Stellarator, NSCX, is currently being built at the Princeton Plasma Physics Lab.

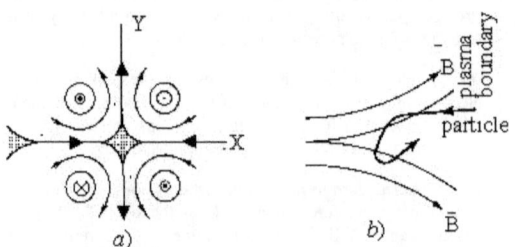

Fig 19.13 The sweep field created by parallel conductors of opposite current., and the motion of charged particle in a sweep field.

19.7.5 Laser Fusion

A second technique for confining the plasma is called **inertial confinement**. It involves compressing a fuel pellet by "zapping" it from all sides by laser beams (or particle beams), thus compressing it and increasing its temperature and particle density so that thermonuclear fusion can occur. By comparison with devices such as the Tokomak, inertial confinement involves working with much higher particle densities for much shorter times.

Laser fusion is being investigated in many laboratories in the United States and elsewhere. At the Lawrence Livermore Laboratory, the laser pulses are designed to deliver, in total, some 200 kJ of energy to each fuel pellet in less than a nanosecond. This is a delivered power of about 2×10^{14} W during the pulse, which is roughly 100 times the total sustained electric power generating capacity of the world! The feasibility of laser fusion as the basis of a thermonuclear power

19.8 COLD FUSION

So far using high temperatures to increase this energy is called *thermonuclear fusion*, or **hot fusion**. This type of fusion is used in the hydrogen bomb, and in the Sun and other stars.

If the fusion reaction takes place at normal room temperatures, using an electric current or something besides high speeds or temperatures to cause the nuclei to fuse, it is called *cold fusion*.

An amazing discovery was announced by B. Stanley Pons and Martin Fleischmann's (1989) at the University of Utah: they could make nuclear fusion work at room temperature instead of millions of K. The reactions involved deuterium (^2_1H). It was known that deuterium could undergo the fusion reactions

$$^2_1H + ^2_1H \rightarrow ^3_2He + ^1_0n + 3\, MeV \qquad (19.7.1)$$

$$^2_1H + ^2_1H \rightarrow ^3_1H + p^+ + 4\, MeV \qquad (19.7.2)$$

When two deuterons collide, the reaction (19.7.1) occurs about half the time, the (19.7.2) about half the time. The **chemists** held the view that the deuterium in palladium rods, making use of the fact that palladium absorbs hydrogen.

They reported that more energy was produced in their apparatus than was used to force the deuterium into the rod. They attributed the excess energy to nuclear fusion. If this were true it would imply a very fundamental discovery in physics had been made:

*the distances between deuterium nuclei in the palladium is approximately $10^{-10}\, m$.

*the distance between the deuterium nuclei in order for fusion to happen needs to be $10^{-15}\, m$.

*the deuterium nuclei repel each other.

*this discovery would have enormous economic consequences.

Further experiments by many groups were not able to confirm the reported result.

But scientists have said: "Where there is nuclear fire there have to be nuclear ashes," and everyone knows that the *ashes* of a nuclear reaction is radiation. Cold fusion produces intense alpha radiation, which is totally different: it takes only a few atoms to shield you from it, whereas it takes three feet of lead to shield you from the nuclear radiation produced by an atom bomb explosion. In sum, because cold fusion does not produce energetic, penetrating nuclear radiation, scientists have disclaimed it - as not being nuclear. In fact, several research institutions now believe that the Spherical Tokamak is the solution.

REVIEW QUESTIONS

R.Q. 19.1 Calculate the number of fissions required to produce 1 W of electric power from U-235. (Answer: $10^{11} s^{-1}$). How much quantity of fuel U-235 is required to produce 1MWe per day, and what is the amount of waste product per day produced along with this power, and of which what is the quantity of fertile produced? (Answer: , 1 g;1 g; 0.5 g of Pu-239).

R.Q. 19.2 Calculate the amount of energy available if 1 g of $^{235}_{92}U$ is completely fissioned. (Answer 8.2 x 10^{10} W s ≈ 1 MWD (MegaWattDay), assuming energy of 200 MeV/fission).

R.Q 19.3 Find i) the average fission cross section, ii) the average absorption cross section of $^1_0 n$, and iii) the number of $^1_0 n$ per fission that can be lost if a self-sustained chain reaction to occur in reactor using nat U. Given, $\sigma_f(^{235}U) = 550\ b$, $\sigma_f(^{238}U) = 0\ b$ $\sigma_\gamma(^{235}U) = 101\ b$ $\sigma_\gamma(^{238}U) = 2.8\ b$. (Answer: i) 3.92 b, ii) (0.71 + 2.8) 3.51 b, iii) No more than 0.5 $^1_0 n$ per fission that can be lost.).

R.Q 19.4 In a mixture of hydrogen and helium at a temperature of 10^7 K a particular pair of one each of these nuclei has both moving at their most probable speed $(2kT/m)^{1/2}$. Calculate the maximum and minimum kinetic energies the pair may have due to their **relative** motion. (Answer: 1554 eV, 170 eV).

R.Q 19.5 Given the data listed in the table below calculate the Q-values for the following reactions $^{12}C + ^4He \rightarrow ^{16}O + \gamma$; $^{16}O + ^4He \rightarrow ^{20}Ne + \gamma$; $^{20}Ne + ^4He \rightarrow ^{24}Mg + \gamma$. Given: The masses of nuclides are $M(^4He) = 4.002603\ u$, $M(^{12}C) = 12.000000\ u$, $M(^{16}O) = 15.994915\ u$, $M(^{20}Ne) = 19.992436\ u$, and $M(^{24}Mg) = 23.985042\ u$. (Answers: 7.16 MeV, 4.73 MeV, 9.31 MeV).

R.Q 19.6 The combination of four protons and two electrons to form an alpha particle

(^4He), two neutrinos, and six gamma rays. Thus, the overall equation is $4\ ^1_1H + 2\ ^0_{-1}e \rightarrow\ ^4_2He + 2\nu + 6\gamma$. where $1.007825u$ is the mass of a hydrogen atom and $4.002603u$ is the mass of a helium atom; neutrinos and gamma-ray photons have no mass and thus do not enter into the calculation of disintegration energy. Show that the energy release in this reaction is $\Delta E = 26.7 MeV$.

R.Q. 19.7 (a) Explain the term nuclear binding energy. Sketch a graph showing the variation of binding energy per nucleon with nucleon number (mass number) and show how both nuclear fission and nuclear fusion can be explained from the shape of this curve. Calculate in MeV the energy liberated when a helium nucleus is produced (i) by fusing two neutrons and two protons, and (ii) by fusing two deuterium nuclei. Why is the quantity of energy different in the two cases? (neutron mass = 1.00898 u; proton mass = 1.00759 u; nuclear deuterium mass = 2.01419 u; nuclear helium mass = 4.00277 u; 1 u is equivalent to 931 MeV).

(b) Give the sketch showing the essential features of a thermal fission reactor. What is its principle of functioning? What is the function of (i) the moderator, (ii) the control rods and (iii) the coolant?

R.Q. 19.8 (a) What is the function of a moderator in a nuclear reactor? Name any one substance which is commonly used as moderator.

RQ 19.9 A reactor using U-235 has an output of 700 MW and is 20% efficient a) How many uranium atoms are burnt every day? B) What is the mass of Uranium consumed each day? (Given, Energy liberated per fission of Uranium is 200 MeV, $N_A = 6.0225 \times 10^{23} \, mol^{-1}$).

(Answer: a) $9.5 \times 10^{24} \, d^{-1}$, b) $3.7 \, kg$)

%%*%*%*%*%*

Chapter 20
NUCLEAR INDIA

Chapter 20

NUCLEAR INDIA

"The high destiny of the individual is to serve rather than to rule"

-Albert Einstein

"When Nuclear Energy has been successfully applied for power production in, say a couple of decades from now, India will not have to look abroad for its experts but will find them ready at hand"-   in 1944, H. J. Bhabha

"We must have the capability. We should first prove ourselves and then talk of Gandhi, non-violence and a world without nuclear weapons."- ,   H. J. Bhabha

20.1   INTRODUCTION

The **atomic bomb** (**A-bomb**) was first used in warfare at Hiroshima (August 6, 1945) and Nagasaki (August 9, 1945) in Japan. The two bombs played a key role in ending World War II. Created via the *Manhattan Project*, the A-bomb was first exploded at the top secret base of Alamogordo on July 16th, 1945. Scientists who invented it were J. Robert Oppenheimer, David Bohm, Leo Szilard, Eugene Wigner, Otto Frisch, Rudolf Peierls, Felix Bloch, Niels Bohr, Emilio Segre, James Franck, Enrico Fermi, Klaus Fuchs and Edward Teller.

The bomb that dropped on Hiroshima was code-named **'Little Boy'**, on explosion, generated an amount of energy equivalent to a 15 kiloton TNT explosion. The second one called **'Fat Man'** (dropped on Nagasaki) differed from 'Little Boy' which was a 'gun-type' weapon shooting a piece of sub-critical ^{235}U into another -cup-shaped piece to create a super-critical mass - and a nuclear explosion. 'Little Boy', weighed about 9,000 *lbs*, and diameter 28 *in* and length 10 *ft* . The reported estimated temperature was $7,000\,^\circ F$ directly beneath the centre of the explosion. 'Fat Man' used an implosion method, with a ring of 64 detonators shooting segments of Plutonium together to obtain the super-critical mass that would create a nuclear explosion. 66,000 people were killed and 69,000 injured by a 10-kiloton atomic explosion by 'Little Boy'. On the other hand, in a split second, Nagasaki's population dropped from 422,000 to 383,000 and over 25,000 people injured when 'Big Brother' exploded. Studies on these two atomic explosions gave an estimate that the bombs utilized only $\frac{1}{10}$ th of 1 % of their respective explosive capabilities.

Upon witnessing the explosion, Isidor Rabi felt that the equilibrium in nature had been upset -- as if humankind had become a threat to the world it inhabited. Oppenheimer, though ecstatic about the success of the project, quoted a remembered fragment from the Bhagavad Gita. "I am become Death," he said, "the destroyer of worlds". Ken Bainbridge, the test director, told Oppenheimer, "Now we are all sons of bitches".

The end of World War II marked a revolution in world affairs - the recasting of nations and relations between nations, and the emergence of a new technology which fundamentally changed the role of warfare. Within the span of two years and two months, from 1945 to 1947, three critical events occurred *viz*.,

1) The establishment of the **United Nations** on 26 June 1945;

2) The dramatic demonstration of the destruction of even crude nuclear weapons are capable, in August 1945; and

3) The calamitous partition of British India into the Republic of India and Pakistan at midnight on 14-15 August 1947.

Further the legacy of partition is a key driving force behind the nuclear standoff that now exists between India and Pakistan. India's acquisition of nuclear weapons is because of the potential threat and regional challenge presented by the nuclear-armed state of China, which faces India along much of its northern border. Disputes covering about 80,000 km^2 of this border region exist:

The prime objective of India's nuclear energy programme is the development and use of nuclear energy for **peaceful purposes** such as power generation, applications in agriculture, medicine, industry, research and other areas. India is today recognized as one of the countries most advanced in nuclear technology including production of source materials. India is self-reliant. It has mastered the expertise covering the complete nuclear cycle - from exploration and mining to power generation and waste management.

Accelerators and research & power reactors are now designed and built indigenously. The sophisticated Variable Energy Cyclotron (VEC) at Kolkatta and a medium energy heavy ion accelerator 'Pelletron' set up at Mumbai are national research facilities in the frontier areas of science.

20.2. ESTABLISHMENT OF THE **DEPARTMENT OF ATOMIC ENERGY (DAE)**

It was the farsightedness of Dr. **Homi Jahangir Bhabha** to start nuclear research in India at a time following the discovery of nuclear fission phenomena by Otto Hahn and Fritz Strassman and soon after Enrico Fermi *et al.* from Chicago reporting the feasibility of sustained nuclear chain reactions. At that time very little information was available to the outside world about nuclear fission and sustained chain reactions and nobody was willing to subscribe to the concept of power generation based on nuclear energy.

India's indigenous efforts in nuclear science and technology were established remarkably early. In March 1944, the first step was taken up by the visionary Dr. Homi J. Bhabha, who submitted a proposal to the Sir Dorab Tata Trust, to establish a nuclear research institute This led to the creation of the **Tata Institute of Fundamental Research (TIFR)** on 19 December 1945 with Dr. Bhabha as its first Director. The new Government of India passed the Atomic Energy Act, on 15 April 1948, leading to the establishment of the **Indian Atomic Energy Commission (IAEC)**.

On 3 January 1954 the IAEC decided to set up a new facility - the **Atomic Energy Establishment**, Trombay (AEET). On 3 August 1954 the **Department of Atomic Energy (DAE)** was created with Dr. Bhabha as the Secretary. On 12 January 1967 the AEET got the present name – **Bhabha Atomic Research Centre (BARC)** when Indira Gandhi (then the Prime Minister) renamed it in memory of Dr. Bhabha (who died in an airplane crash on Mt. Alps on 24 January 1966).

The DAE, while performing a key role in the scientific and technological scenario of the country, has also been vital to the overall nation-building exercise. This strategy has accorded India the status of a 'Developed Nation' amongst the *'Developing Nations'*, a fact made clear by

India getting elected as the Chairman of the Board of Governors of the International Atomic Energy Agency (IAEA), in September 1994, where it occupies a position as Permanent Member since its inception.

20.3. THE FIRST REACTOR OF INDIA

AEET employed over one thousand scientists and engineers by 1959 . In 1955 construction began on India's first reactor, *viz.* 1 *MW* **APSARA** research reactor, with British assistance. In September 1955, after more than a year of negotiation, Canada agreed to supply India with a powerful research reactor - the 40 *MW* Canada-India Reactor (**CIRUS**). Under the Eisenhower Administration's "Atoms for Peace" program the US supplied 21 *ton*s of heavy water for this reactor in February 1956.

APSARA, fueled by enriched Uranium from the UK, went *critical* on 4 August 1957, becoming the first operating reactor in Asia outside of the Soviet Union. CIRUS achieved criticality at BARC on 10 July 1960. This reactor produced the Plutonium used in India's first nuclear test in 1974; provided the design prototype for India's more powerful **DHRUVA** Plutonium production "research" reactor; and is directly responsible for producing nearly half of the weapons grade Plutonium currently believed to be in India's stockpile.

20.4. NUCLEAR POWER

a) India's civilian nuclear program was also established during this period. Negotiations with American firms to construct India's first nuclear power plant at Tarapur were held in 1960-61. In April 1965 L.B. Shastri (PM) gave Bhabha formal approval to move ahead with nuclear explosive development. In April 1965 Bhabha initiated the effort by setting up the nuclear explosive design group.

b) Study of Nuclear Explosions for Peaceful Purposes (SNEPP). Bhabha selected Raja Ramanna - Director of Physics Group at AEET - to lead the effort. In 1966 **Vikram A. Sarabhai** who was chosen by Indira Gandhi to succeed Bhabha as the Chairman of the IAEC, and Secretary of the DAE.

In 1966 India's diplomatic policy towards nuclear weapons made a fateful shift. International interest in non-proliferation focused on restricting the spread of nuclear weapons to *any* additional states; India would not eschew nuclear arms unless the existing nuclear states did also. This fundamental logic led India to refusing to sign the **Nuclear Non-Proliferation Treaty** and voting against it in June 1968, and has informed Indian nuclear diplomacy ever since.

Dr. P'K. Iyengar set about developing and got the approval for building up the reactor, called **PURNIMA** (Plutonium Reactor for Neutron Investigation in Multiplying Assemblies),in March 1969. It was designed to use a hexagonal core of 177 stainless steel pencil shaped rods containing 18 *kg* of Plutonium as 21.6 *kg* of plutonium oxide pellets with a nominal average power of 1 *W*. M. Srinivasan developed a sophisticated physics model for criticality calculations in 1970. He became the reactor's Chief physicist. PURNIMA went critical on 18 May 1972. It was a test bed and the Indian physicists were able to refine their understanding of the **physics of fast fission and fast neutrons.**

DHRUVA went critical on 8 August 1985, and achieved full power on 17 January 1988, becoming the workhorse of the Indian nuclear weapons production program. The first test flight of the 150 *km* Prithvi tactical missile with a 1000 *kg* payload had occurred on 18 February 1988. The Agni MRBM had its first test flight on 22 May 1989 a successful launch in which it carried a 1000 *kg* payload 800 *km*. India described the AGNI at this stage as a "technology demonstrator.

20.5 INDIA'S FIRST BOMB: POKHRAN I (1974)

India's nuclear weapons program moved in to full swing and in1997 led to the "Peaceful Nuclear Explosive" (PNE) program. This detonation test was on 18 May 1974, at 8:05 (IST), and was located at Pokhran (27.1 N, 71.8 E), Thar Desert, Rajasthan, India, with test height underground, 107 *m*, of yield 8 *kTonne*. It has been known since its public announcement as "**Smiling Buddha**". The crater produced by this detonation of a Plutonium implosion device has been reported to have a radius of 47 *m* with a crater depth of 10 *m*. The device was manufactured (in 1972 with the consent from the Prime Minister Indira Gandhi) from Plutonium produced at the CIRUS reactor at BARC, but the explosive lenses were made by the Defence Research and Development Organization (DRDO). The neutron initiator was a ^{210}Po/Beryllium type bearing code-name.

20.5.1 Missile Programme

Both the new missile effort (based at DRDL and led by **APJ Abdul Kalam**), in 1983 and the Pokhran I had put together an ambitious development Programme for five missiles based on a family of related technologies. In August 1983 the Integrated Guided Missile Development Programme (IGMDP) was born with a budget of Rs 38.8 billion ($370 million). Anticipating the tightening of international restrictions on the export of missile technology the Missile Technology Control Regime was signed in 1987. Both the **PRITHVI** and the **AGNI** were intended to carry nuclear warheads. The Prithvi was a 150 *km* liquid fueled missile derived from "Project Devil", and the Agni was a long range (1500 *km*) two stage solid fuel (first stage) / liquid fuel (upper stage) design derived both from the SLV-3 and "Project Devil."

What is this ICBM programme that India is working on?

It is a three-stage ballistic missile that DRDO, along with a number of defence agencies in the country, is working hard on.The missile will have solid fuel rockets in the first and second stages, and a liquid propellant rocket in the third stage. The launch weight of the missile may reach 270 to 275 tonnes. The missile could have a 5,480 pound to 7,680 pound releasable front section with two to three warheads of 15 kilo tonne to 20 kilo tonnes each. The ICBM is being developed by combining the technology of the Agni II with that of the Polar Satellite Launch Vehicle. It is expected to have a range of more than 8,000 km.".

THE INDIAN MISSILES

Name	Type/propulsion	Warhead	Payload	Range	Status
Prithvi-1/ SS-150	Ballistic/ Single-stage/ Liquid-engine	Conventional/ nuclear	1,000 kg	150 km	Operational
Prithvi-2/ SS-250	Ballistic/ Single-stage/ Liquid-engine	Conventional/ nuclear	500 kg	250 km	Undergoing user trials
Dhanush/ Prithvi-3/ SS-350	Ballistic/ Single-stage/ Liquid-engine	Conventional/ nuclear	Undisclosed	350 km	Undergoing flight tests
Agni Technology Demonstrator	Ballistic/ Two-stage hybrid/ solid-motor/ Liquid-engine	Nuclear	1,000 kg	1,200 km to 1,500 km	Small number available to army
Agni-I	Ballistic/ Single-stage/ Solid-motor	Nuclear	1,000 kg	700 km to 800 km	Undergoing flight tests
Agni-II	Ballistic/ Two-stage/ Solid-motor	Nuclear	1,000 kg	2,000 km to 2,500 km	Completed flight tests
Agni-III	Ballistic	Nuclear	Undisclosed	3,000 km to 4,000 km	Flight tests expected this year
BrahMos/ PJ-10	Cruise/ Two-stage/ Solid-booster/ Liquid-sustainer engine	Conventional	200 kg to 300 kg	280 km to 300 km	Serial production to begin this year
Sagarika	Class contested	Conventional/nuclear	Undisclosed	Undisclosed	Expected to be operational by 2010

India's nuclear capable missiles

Name	Class	Range	Payload	Status
Agni-I	SRBM	700 km	1,000 kg	Operational
Agni-II	MRBM	2,000 km - 3,000 km	500 kg - 1,000 kg	Operational
Agni-II Prime	MRBM	2,750 km - 3,000 km	500 kg - 1,500 kg	Under development
Agni-III	IRBM	3,500 km	2,490 kg	Inducted
Agni-V	ICBM	5,000 km - 6,000 km	3,000 kg+	Under development
Surya-I	ICBM	5,200 km - 11,600 km	700 kg - 1,400 kg	Under development
Dhanush	SRBM	350 km	500 kg	Operational
Nirbhay	Subsonic Cruise Missile	1,000 km	?	Under development
Brahmos	Supersonic Cruise Missile	290 km	300 kg	Operational
P-70 Ametist	Anti-shipping Missile	65 km	530 kg	Operational
P-270 Moskit	Supersonic Cruise Missile	120 km	320 kg	Operational
Popeye	ASM	78 km	340 kg	Operational
Prithvi-I	SRBM	150 km	1000 kg	Operational
Prithvi-II	SRBM	250 km	500 kg	Operational
Prithvi-III	SRBM	350 km	500 kg	Operational
Sagarika (missile)	SLBM	700 km - 2,200 km	150 kg - 1000 kg	Under development
Shaurya	TBM	700 km - 2,200 km	150 kg - 1,000 kg	Under development

20.5.2 Nuclear Weapon State

India became a **nuclear weapon state** in reality in May 1994, the nuclear powers pushed two treaties that provided for restrictions on specific proliferation activities - the Comprehensive Test Ban Treaty (**CTBT**) to prohibit all nuclear tests, and a treaty to cut-off the production of fissile materials for weapons. Then on 15 December 1995 the *New York Times* disclosed that US satellites had detected test preparations underway at Pokhran, during the time of PM Narasimha Rao. "India is now a nuclear weapons state."

India has the capacity for a big bomb now. Ours will never be weapons of aggression. "according to PM AB Vajpayee 14 May 1998). With the multiple tests India marched towards an openly declared nuclear capability.

20.6 The POKHRAN II (SHAKTI Test Devices) and AERB

POKHRA II consisted of three test devices and emplaced in the test shafts: i) a lightweight pure fission bomb (12 *kiloton*); ii) a boosted fission bomb; and iii) a sub-*kiloton* experimental test device. On Monday, 11 May 1998, as measured by international seismic monitors, India declared itself a **full-fledged nuclear armed state**. The devices tested were a thermonuclear device, a fission device, and a low-yield device. Just two days later on 13 May 12.21 PM. local Indian time India detonated two more sub-*kiloton* nuclear devices underground before declaring that the test series was completed. The principal purposes of this technique are to minimize the ability of other nations to collect intelligence data about the tests.

In May 2000, the Atomic Energy Regulatory Board (**AERB**).was set up.

20.7 NUCLEAR POWER PLANTS

The three stage Programme, charted by Bhabha, aimed at establishing nuclear power with resources comprises the following guidelines:

a) **First stage** - use of natural Uranium in pressurized heavy water reactors (PHWR) and production of power and Plutonium;

b) **Second stage** - use of Plutonium produced in Fast Breeder Reactors (FBR) and production of additional Plutonium /^{233}U and power; and

c) **Third stage** - use of Thorium /^{233}U in an advanced fuel cycle and reactor system (under development).

The other important multi-disciplinary R&D Centre, Indira Gandhi Centre for Atomic Research (**IGCAR**), at Kalpakkom is dedicated to fast reactor technology and associated fuel cycle, material sciences, fuel reprocessing and sodium technology. The Centre is also engaged in basic research relating to material science, radiochemistry, and applied research in the sphere of non-destructive technology, advanced instrumentation and materials.

India's first 40 *MWt* **Fast Breeder Test Reactor (FBTR)** attained criticality on 18th October 1985. India becomes the sixth nation having the technology to built and operate a FBTR besides USA, UK, France, Japan and the then USSR.

20.7.1 Heavy Water

High purity heavy water is used in PHWRs for serving as the moderator and the primary coolant.

The first heavy water plant was set up in India at Nangal in 1962. Other Heavy water plants are at Baroda, Tuticorin, Kota, Thal, Hazira and Manuguru. The hydrogen sulphide - water process used at Kota and Manuguru plants is based on the expertise developed through indigenous R&D. The technology for upgrading heavy water was developed at BARC. The current research is directed towards the development of alternative, more cost-effective, technologies for heavy water production, such as hydrogen-water exchange process.

20.7.2. Nuclear Fuels Complex (NFC)

The Nuclear Fuel Complex (NFC) was established in Hyderabad in the early 70s. It is for making nuclear fuel assemblies and core structural components for the entire nuclear power programme of the country. The activities at NFC includes Processing of Uranium ore concentrate and zircon sand brought from Bihar and Kerala, through a series of indigenously developed chemical and metallurgical operations, making seamless tubes of stainless steel, carbon steel, titanium and other special alloys of Nickel, Magnesium, *etc*.by hot extrusion and cold pilfering process

20.7.3. **Kamini**, a 30 *kWt* reactor at the IGCAR,

This achieved criticality in October 1996 for providing neutron radiography facilities and is a small but significant step towards utilization of our vast Thorium reserves. It is the only operating reactor in the world using ^{233}U fuel. Some of the larger facilities built by DAE are now available to university researchers through Inter-university Consortium for DAE facilities.

20.7.4. World's Safest Nuclear Reactor **ATBR** (2005)

India unveiled before the international community on August 25, 2005 its revolutionary design of 'A Thorium Breeder Reactor' (**ATBR**) that can produce 600 *MW* of electricity for two years 'with no refueling and practically no control maneuvers'. Designed by scientists of the Mumbai-based BARC, the ATBR is claimed to be far more economical and safer than any power reactor in the world. Most significantly for India, ATBR does not require natural or enriched Uranium which the country is finding difficult to import. It uses Thorium -which India has in plenty in Kerala - and only requires Plutonium as 'seed' to ignite the reactor core initially. Eventually, the ATBR can run entirely with Thorium and fissile ^{233}U bred inside the reactor (or obtained externally by converting fertile Thorium into fissile U-233 by neutron bombardment). The uniqueness of the ATBR design is that there is almost a perfect 'balance' between fissile depletion and production that allows in-bred ^{233}U to take part in energy generation thereby extending the core life to two years. The ATBR annually requires 2.2 *tons* of Plutonium as 'seed'. Althouth India has facilities to recover plutonium by reprocessing spent fuel, it requires plutonium for its Fast Breeder Reactor programme as well.

20.7.5 Indian Rare Earths Limited (IREL)

IREL is a public sector undertaking unit of DAE; it process minerals to products which are not only of value to the Department, but also useful to other users in the country and outside. Nuclear Fuel Complex (NFC), Hyderabad fabricates fuel and structural components for all operating power reactors, thorium blankets and structural components for FBTR.

20.7.6. Nuclear Agriculture

In the sphere of agriculture, the Department of Atomic Energy caters to crop improvement programmes, fertilizer and pesticide related studies, food preservation by irradiation and water management, thus contributing immensely to the country's rural development programme.

20.7.7. Water Management

The DAE has been involved in the process of water management all over the country on different aspects like artificial recharge of ground water system, the study of silt movement in ports, flow measurements in rivers and seepage identification in dams

20.7.8. Health Care

The Board of Radiation and Isotope Technology formed to commercialize the use of isotopes supplies radiopharmaceuticals, labeled compounds and radioimmunoassay kits to over 400 medical institutions all over the country, to enable early diagnosis and treatment of diseases like cancer. The Radiation Medicine Centre of BARC and the Tata Memorial Centre have been performing a vital role in the research and technology involved in fighting cancer in the country through nuclear energy. The gamma sterilization plants set up in various parts of the nation have provided an added impetus to the process of growth of better health care facilities.

TABLE 20.1   India's Operationg Power Reactor

	Reactor	Type	MWe net	Year
1	Tarapur 1 (TAPS) (MH)	BWR	160	Oct 1969.
2	Tarapur 2 (TAPS) (MH)	BWR	160	Oct 1969
3	Tarapur 3 TAPS) (MH)	PHWR	540	Sep 2005
4	Tarapur 4 (TAPS) (MH)	PHWR	540	Aug 2006
5	Kaiga 1 (KA)	PHWR	220	Mar 2000
6	Kaiga 2 (KA)	PHWR	220	Nov 2000
7	Kaiga 3 (KA)	PHWR	220	May 2007
8	Kaiga 4 (KA)	PHWR	220	(Dec 2010)
9	Kakrapar 1 (GUJARAT)	PHWR	220	May 1993
10	Kakrapar 2 (GUJARAT)	PHWR	220	Sep 1995
11	Kakrapar 3 (GUJARAT)	PHWR	700	(Jun 2015)
12	Kakrapar 4 (GUJARAT)	PHWR	700	(Dec 2015)
13	Kalpakkam 1 (MAPS)	PHWR	170	Jan 1984
14	Kalpakkam 2 (MAPS)	PHWR	220	Mar 1986
15	Narora 1 (UP)	PHWR	220	Jan 1991
16	Narora 2 (UP)	PHWR	220	Jul 1992
17	Rawatbhata 1 (Rajasthan)	PHWR	90	Dec 1973
18	Rawatbhata 2 (RAUJASTHAN)	PHWR	187	Apr 1981
19	Rawatbhata 3 (RAJASTHAN)	PHWR	202	Jun 2000
20	Rawatbhata 4 (Rajasthan)	PHWR	202	Dec 2000
21	Rawatbhata 5 (RAJASTAN)	PHWR	202	Dec 2009
22	Rawatbhata 6 (RAJASTHAN)	PHWR	202	Mar 2010
23	Rawatbhata 7 (Rajasthan)	PHWR	700	(Jun 2016)
24	Rawatbhata 8 (Rajasthan)	PHWR	700	(Dec 2016)
25	Kudankulam 1 (TN)	PHWR	1000	Mar 2011
26	Kudankulam 2 (TN)	PHWR	1000	Dec 2011
27	Kalpakkam (MAPS)	PFBR	470	2010
28	Kudankulam 3 (TN)	PHWR	1000	(TBD)

20.8.1 THE INDO -US NUKE DEAL

Indicating a watershed development on the trail of bilateral cooperation between India and the US, the US Congress on December 8, 2006 passed the legislation seeking approval to the landmark Indo-US civilian nuclear deal, shrouded in uncertainties regarding its fate ever since it was clinched n 2005. The US House of Representatives cleared the legislation with a thumbing majority by a margin of 350 – 59 votes after long debate, though the Chairman of the House of International Relations (Mr. Henry Hyde), Mr. Tom Lantos supported it while MA Governor Mr. Edward Markey was against it. Both the US Senate and House of Representatives approved this. It is christened as **the Henry J. Hyde US-India Peaceful Atomic Energy Cooperation Bill 2006**. Accordingly the US Administration will provide nuclear technology and material to India's civilian programme. This giant leap forward with Congressional Approval is to bring an end to the boycott required to the *Non-Proliferation Treaty* (NPT) regime. Legally, it exempts India from fulfilling *Section 123 (a) 2of the US Atomic Energy Act*.

The agreement for the Indo-US civilian nuke deal was reached between Prime Minister Dr. Man Mohan Singh and the US President Mr. George W. Bush on July 18, at US and later in March 2, 2006 in India.2005. The 123 agreement is a civil nuclear deal, therefore, it will have no bearing on India's strategic and military programme and India can make a bomb. It is completely out of the ambit of the deal.

India and the United States on Oct 11, 2008 seem to have signed the 123 agreement on the peaceful use of the nuclear energy, vowing to use their nascent strategic partnership in shaping a new world order. The agreement bringing into reality the accord envisioned by Indian Prime Minister Manmohan Singh and US President George W. Bush over three years ago, was signed by the two ministers at a ceremony in the Benjamin Franklin Room of the State Department. With NSG waiver for India from the Nuclear Suppliers' Group for trade in the atomic energy, the US today said its next aim is to make New Delhi a "full partner" in the nuclear cartel."President Bush, Secretary of State and the entire administration had worked tirelessly to ensure that India reached the stage where it has today in the Nuclear Suppliers Group (NSG Assistant Secretary (Market Access and Compliance) in the US Department of Commerce David Bohigian said here.

The next step for the US (administration) will be working through the Congress and the Hyde Act and make sure that business opportunities will enable the US firms to stay in what is estimated to be 100 billion dollar market," the official said at a CII seminar.

Of India's 22 nuclear plants, 14 classified for civilian use would be subject to new and permanent international inspections under the deal. The country's eight other reactors, as well as future ones designated for military use, would be off- For India, which faces dwindling supplies of indigenous uranium, the deal would allow it to import uranium to fuel its civilian program and free up its local supplies to fuel the weapons program. India, Washington says, has demonstrated that it is a responsible nuclear power

20.8.2 INDO-JAP NUKE DEAL

After six years of wrangling and intense negotiations, India and Japan on Friday, 11 November 2016 signed a landmark Civil Nuclear Agreement in Tokyo, where the Prime Minister Narendra Modi and his Japanese counterpart Shinzo Abe held talks. "The Agreement for Cooperation in Peaceful Uses of Nuclear Energy marks a historic step in India's engagement to build a clean energy partnership", said PM Modi. The agreement will allow Japan to supply nuclear reactors, fuel and technology to India, which will be the first country that has not signed the NPT to have such a deal with Tokyo. Hope this deal will make it easier for US based Companies to setup atomic plants in India.

20.9 CONCUSION

In pursuit of the peaceful uses of Atomic Energy, power generation based on nuclear energy assumes first and foremost place and India has achieved many milestones in this area. A well planned programme for the progressive expansion for the tapping of atomic energy for electricity keeping in view of the country's future requirements for increased power generation capacity and available resources has been under implementation. A strong R&D base has been established and functions as a back bone for the smooth transition of the research and development activities to the deployment phase and thereby realizing the Department of Atomic Energy's mandate. Many technologies of strategic importance have been mastered to meet developmental needs. Indigenous technology development in the areas of fuel reprocessing, enrichment, production of special materials, computers, lasers, accelerators represents a whole spectrum of activities necessary for realizing full potential of our energy resources to meet future energy needs. Radiation Technology and Isotope Applications represents another prominent area of the peaceful uses of Atomic Energy in health care, agriculture, industries, hydrology and food preservation where self- reliance has been accomplished.

In 1950 India was producing a meagre 1,800 MW power but in 1998-99 we generated about 90,000 MW. Almost all of this was thermal and hydropower and the share of nuclear power was an insignificant 1,840 MW -- a ridiculously low 2 % of the total energy production. As of today, the Indian nuke Dom claims, their energy output has increased to 2,770 MW. It is hardly 3 % even if we keep the total energy output at the stagnant level of 90,000 MW.

India is not a signatory to NPT and has opposed the treaty as discriminatory to non-weapons states. India has previously taken the position that a world-wide ban on nuclear testing, and the production of fissionable material for weapons is called for. Except for China, which continues testing, there is now a de facto halt to testing worldwide, as well as the production of weapons grade plutonium and uranium by the US and Russia. India has shown no interest so far in restricting its own activities despite these changes in the world situation. India has also rejected offers at bilateral negotiation with Pakistan, but in December 1988 the two nations signed an agreement prohibiting attacks on each other's nuclear installations and informing each other of their locations (though not their purposes).

20.9.1 Major Nuclear Power Plant Accidents

A calendar of nuclear accidents shows the threat that humanity faces from the atom bomb and the nuclear fuel cycle. Almost 18 major nuclear power plant accidents have occurred the world over during 1952 – 2011.It demonstrates how technological failures coupled with human error risk public health and the environment on an almost daily basis.

Chalk River, Ottawa (Dec 12, 1952), Sellafield, Britain), Oct 1957, Kyshtym, Urals (1957-58); Idaho Falls USA (Jan 1961); nuclear powered submarine of USSR (July 1961); Detroit, USA (Oct 1968); Lucens Vad, Switzerland, Jan 1969) ; Lubmin nuclear power complex , Baltic coast, East Germany (Dec 1975); Harrisburg, PA, USA, (Mar 1979); Sequoya Plant 1, TN, USA, (Feb 1981); Tsuruga, Japan,(Apr 1981); **Chernobyl Plant**, USSR (Apr 26, 1986); Sosnovy Bar, St Petersburg, Russia (Mar 1992); Forbach, France (Nov 1992); Monju FBR, Japan (Nov 1995); Tokamura, Japan (Mar 1997); Tokamura, Japan (Sep 1999); **Fukushima Daiichi**, Japan (11 Mar 2011).

Of these the Chernobyl and the Fukushima Daiichi are the worst disasters in nuclear crisis.

%%*%*%*%*%*

Appendix A
BIBLIOGRAPHY

(For further reading)

Other Texts:

1. Burcham, W.E. & M. Jobes (1995) *"Nuclear and Particle Physics"* (Longman, N.Y).

2. Enge, H.A. (1975) *"Introduction to Nuclear Physics"* (Addison Wesley).

4. Fermi, E. (1974) *"Nuclear Physics"* (Reprint) (Univ. Chicago Press).

3. Krane, Kenneth S. (1988) *"Introductory Nuclear Physics"* (John Wiley, N.Y.).

References

1. Alfassi, Z.B. (1998) '*Chemical Analysis by Nuclear Methods*'. (John Wiley: New York, NY).

2. Alonso, M & EJ Finn (1968) *"Fundamental University Physics"* Vol III (Addison-Wesely, Reading)

3. Bais, Sandor (2005) *"The Equations: Icons of Knowledge"* (Harvard Univ Press, Cambridge, MA).

4. Barron, John D. (2007) *"New Theories of Everything"* (OUP, NY).

5. Bethe, Hans A. & Morrison, Philip (19) *"Elementary Nuclear Theory"* (John Wiley,)

6. Bishop, Christopher (2002) "Particle Physics" (John Murray, London).

7. Blatt, John M. & Victor F. Weisskopf, (1979) *"Theoretical Nuclear Physics"* (Dover, N.Y.).

8. Burcham, W.E. (1988) *"Elements of Nuclear Physics"* (ELBS / Longman, N.Y.).

9. Blin-Stoyle, R.I. (1991) *"Nuclear and Particle Physics"* (Chapman & Hall, London).

10. Chown, Marcus (2006) *"Quantum Zoo – A Tourists Guide to the Neverending Universe"* (Joseph Henry Press, Wash DC).

11. Close, Frank (2004), *"Particle Physics"* (OUP,)

12. Cohen, B.L. (1977), *Sci. Amer.* Vol. 236, p 21 ; Rev. Mod. Phys., Vol 49, p 1.

13. Cohen, B.L. (1971) *"Concepts of Nuclear Physics"* (TMH, Mumbai)

14. Cottingham, W.N. & D.A. Greenwood (2001) *"An Introduction to Nuclear Physics"* (CUP, UK)

15. Cowan, George A.,(1976) "A Natural Fission Reactor", Scientific American 235, July 1976, p 36.

16. Das, A. & T. Ferbel (1994) *"Introduction to Nuclear and Particle Physics"* (Wiley, N.Y.).

17. Das, T.P. & E.L. Hahn (1958) *"NQR"* (AP, N.Y.).

18. DeGroot, Gerard J, (2004) *"The Bomb- A Life"* (Harvard Univ Press, Harvard, Cambridge, MA).

19. Devanarayanan, S. (1974) *"Mossbauer Spectroscopy"*, J. Sci. Ind. Res., (CSIR, N.D.)

20. Devanarayanan, S. (2005) '*Quantum Mechanics*' (SciTech Publ, Chennai).

21. Devanarayanan, S. (2008) '*Quantum Chemistry'* (SciTech Publ, Chennai).

22. Davies, Paul(1993) *"About Time"* (Simon & Schuster, NY).

23. Davies, Paul (2007) *"Cosmic Jackpot"* (Houghton Mifflin Co, Boston).

24. Davies, Paul (2007) *"The Mind of God"* (Houghton Mifflin Co, Boston).

25. Dyson, N.A. (2005) '*X-RAYS in Atomic and Nuclear Physics'* (CUP, Cambridge). TR49381; 539.7222DYS

26. Enge, H.A. (1975) *"Introduction to Nuclear Physics"* (Addison Wesley,)

27. Evans, R.D. (1969) *"The Atomic Nucleus"* (MGH, N.Y.).

28. Frauenfelder, H. & E.M. Hensley (1974) *"Sub-atomic Physics"* (Prentice Hall, N.J.).

29. Garry McCracken and Peter Stott (2005) *"Fusion - The Energy of the Universe"* (Academic Press)

30. Gasiorowicz, S. (1974) *"Quantum Physics"* (John Wiley).

31. Gibson, W.M & B.R. Pollard (1976)*"Symmetry Principles in Elementary Particle Physics"* (CUP, Cambridge).

32. Glasstone, (1979) *"Source Book on Atomic Energy"* (D Van Nostrand, Princeton).

33. Glascock, M.D. (1996), *Nuclear and Radiochemistry* (John Wiley & Sons, New York)

34. Golding, R.M. (19) *"Applied Wave Mechanics"* (D Van Nostrand, Princeton).

35. Goswami,SN, (1995) *"Elements of Plasma Physics"* (New Central Book Agency, Kolkatta).

36. Graham Woan (2000) *"The Cambridge Handbook of Physics Formulas"* (Cambridge Univ. Press, Cambridge).

37. Green, Alex E.S., (1955) *"Nuclear Physics"* (Kogakusha-MGH).

38. Hales, PB (1997) *"Atomic Spaces: Living on the Manhattan Project"*.

39. Hawking, Stephen (2005) *"The Theory of Everything – The Origin of the Universe"* (Phoenix Books, CA)

40. Hawking, Stephen (1988) ,*"A Brief History of Time – From the Big Bang to Black Holes"* (Bantam , London))

41. Heyde, K. (1994) *"Basic Ideas and Concepts in Nuclear Physics"* (Institute of Physics, London).

42. Ingram, J.E. (1976) *"Radio and Microwave Spectroscopy"* (Butterworths, London).

43. Jelley, N.A. (1990) *"Fundamentals of Nuclear Physics"* (CUP, Cambridge).

44. Kaplan, I. (1975) *"Nuclear Physics"*, 7th Print (Addison Wesely,)

45. Keepin, GR(1965) *"Physics of Nuclear Reactors"* (Addison-Wesley Publishing Co., Massachusetts).

46. Kenyon, IR. (1987), *"Elementary Particle Physics"* (Routlege).

47. Lederman, Leon (2006) *"The God Particle"* (Houghton Mifflin Co, Boston)

48. Lederer, C.M. and Shirley, V. (Edtd) (1978), *"Table of Isotopes"* (Wiley, NY).

49. Leo, William R., "Techniques for Nuclear and Particle Physics Experiments A How-to Approach"(Springer-Verlag, NY)

49. Levi, B G (1986) *"Soviets assess cause of Chernobyl accident"*, Physics Today, Dec 86, vol.39.

50. Littlefield, T.A. & N. Thorley (1979) *"Atomic and Nuclear Physics: an Introduction"* (Van Nostrand Rinehart, London; also ELBS)

51. Martin, B.R. & G. Shaw (1992) *"Particle Physics"* (John Wiley, Chichester).

52. Marmier, P. & E. Sheldon (1969) *"Physics of Nuclei and Particles"*, Vol. 1 (AP, N.Y.).

53. Marmier, P. & E. Sheldon (1970) *"Physics of Nuclei and Particles"*, Vol. 2 (AP, N.Y.).

54. Marmier, P. & E. Sheldon (1970) *"Physics of Nuclei and Particles"*, Vol. 3 (AP, N.Y.).

55. Mason, Thomas (2006) "Pulsed Neutron Scattering for the 21st Century ", **Physics Today**, p 44 – 49, May 2006 (AIP, Washington).

56. Meyerhof, Walter E. (1967) *"Elements of Nuclear Physics"* (MGH, N.Y.).)

57. McIntyre, Hugh C(1975) 'Natural Uranium HWRs', *Sci Amer.*, Oct 1975)

58. '*Nuclear Structure- Selected Reprints*' (1965) AIP Resource Letters (for AAPT by AIP, NY)

59. Roy, RR & BP Nigam (1967) *"Nuclear Physics"* (Wiley).

60. Perkins, Donald H. (2000) *"Introduction to High Energy Physics"* (CUP, UK).

61. Preston, M.A. & R.K. Braduri (1975) *"Structure of the Nucleus"* (Addison-Wesley).

62. Roy, RR. & BP. Nigam (1979) "Nuclear Physics – Theory & Experiment" (Wiley Eastern, N.D.).

63. Segre, E. (1982) *"Nuclei and Particles: An Introduction to Nuclear and Subnuclear Physics, 2nd ed.,"* (W. A. Benjamin)

64. Seiden, Abraham (2005) *"Particle Physics: A Comprehensive Introduction"* (Addison Wesley, San Francisco).

65. Seife, Charles (2003) *"Alpha and Omega : The Search for the Beginning and End of the Universe"* (Penguin, NY)

66. Siegbahn, Kai (Ed.) (1966) *" α-, β- and γ- Ray Spectroscopy"* (North-Holland, Amsterdam).

67. *Physics Today*, (1997) Special Issue on Radioactive Waste; 5 Topics, p 22 – 62 (APS, N.Y.).

68. Puri, RK & VK Babbar (1996) "Introductory Nuclear Physics" (Narosa, New Delhi).

69	Singru, RM (1972) "*Introductory Experimental Nuclear Physics*" (Wiley Easetern).
70	Srinivasan. MR (), "*Fission to Fusion: The Story of India's Atomic Energy Programme*" (Viking, New Delhi) ISBN 0-67-004924-7.
71	Sundaram, CV; L.V. Krishnan & TS Iyengar, (1998) "*Atomic energy in India fifty years*" (Department of Atomic Energy, Mumbai).
72	Semat, H. (1959), "*Introduction to Atomic & Nuclear Physics*" (Rinehart & Co, N.Y.)
73	Stacey, WM (2001): "*Nuclear Reactor Physics*" (John Wiley &Sons Inc., 2001)
74.	Seshagiri, N. (1975) "*The Bomb*" (Vikas, Delhi)
75	Spent Fuel Stores, Reprocessing, and Waste Disposal in *SCOPE 50 - Radioecology after Chernobyl.*
76	White, H.E. (1968), "*Introduction to Atomic and Nuclear Physics*" (Affiliated East-West Press, N.D.).
77	Wong, S.S.M. (1996) "*Introduction to Nuclear Physics*" (PHI,).
78.	Weinberg, Steven (2003) "*The Discovery of Subatomic Particles*", Revised Edition (CUP, Cambridge).
79	Wolf, Fred Alan (2001) " *Mind into Matter*" (Moment Point Press, Needham, MA).
80	Booth, C. N. and Combley, F. H., *Nuclear Physics* 303.
81	*CODATA*, (1998).
82	*J. Phys. Chem. Ref. Data*,(1999) Vol. 28 (6).
83	*Rev. Mod Phys*. 72,(2) 2000.
84	Gell-Mann, M & Rossenbaum (1975) '*Elementry Particles*'(*Scientific American* ,197, 72-88).

%^*%*%*%*%*%

Appendix B

Values of Physical Constants

[Ref: *CODATA*, 1998; *J. Phys. Chem. Ref. Data*, Vol. 28 (6) 1999; *Rev. Mod Phys.* 72,(2) 2000]

Planck constant	$h = 6.6256 \times 10^{-34} J-s$;
	$\hbar = (h/2\pi) = 1.054 \times 10^{-34} J-s$
Permittivity of free space	$\varepsilon_0 = 8.8542 \times 10^{-12} F\ m^{-1}$;
	$\varepsilon_0 = 8.8542 \times 10^{-12} C^2 N^{-1}\ m^{-2}$
The Coulomb constant	$k = (1/4\pi\ \varepsilon_0) = 8.9875 \times 10^9\ Nm^2 C^{-2}$
	$k = (1/4\pi\ \varepsilon_0) = 8.9875 \times 10^9\ F^{-1} m$
Permeability of free space	$\mu_0 = 4\pi \times 10^{-7} T\ m\ A^{-1}$,
	$\mu_0 = 4\pi \times 10^{-7} N\ A^{-2}$.
Gravitational constant	$G = 6.67 \times 10^{-11} Nm^2 kg^{-2}$
Speed of light in Free space	$c = 2.997925 \times 10^8\ ms^{-1}$
Boltzmann constant	$k_B = 1.3805 \times 10^{-23} JK^{-1}$
Avogadro's constant	$N_A = 6.0225 \times 10^{23}\ mol^{-1}$
Stefan's constant	$\sigma = 5.67 \times 10^{-8} Wm^{-2} K^{-4}$,
Universal Gas constant	$R = 8.314\ J\ K^{-1} mol^{-1}$
Electronic Charge	$e = 1.6021 \times 10^{-19} C$
Electron Rest mass	$m_e = 9.1094 \times 10^{-31}\ kg$
	$m_e = 5.4858 \times 10^{-4} u = 0.5110\ MeV$
Electron Volt	$1\ eV = 1.6021 \times 10^{-19} J$
Proton Rest mass	$M_p = 1.6726 \times 10^{-27} kg$;
	$M_p = 1.007276\ u\ = 938.272\ MeV$

Mass of Neutron	$M_n = 1.67493 \times 10^{-27}$ kg
	$M_n = 1.008665\ u = 939.57$ MeV
Mass of Deuteron	$M_d = 2.013553\ u$
Mass of Alpha particle	$M_\alpha = 4.002602\ u$
Mass of H-atom	$M_H = 1.67493 \times 10^{-27}$ kg ;
	$M(^1_1H) = 1.007825\ u$
Mass of Tritium	$M(^3_1H) = 3.016049\ u$
Mass of He	$M(^4_2He) = 4.00387\ u$
Mass of Nitrogen	$M(^{14}_7N) = 14.00752\ u$
Mass of Oxygen	$M(^{17}_8O) = 17.00453\ u$
Mass of U-238	$M(^{238}_{92}U) = 238.650784\ u$
Mass of Th 234	$M(^{234}_{90}Th) = 234.043593\ u$
Mass of Th-232	$M(^{232}_{90}Th) = 232.038124\ u$
Mass of Ra-228	$M(^{228}_{88}Ra) = 228.031139\ u$
Unified atomic mass unit	$1u = \frac{1}{12}$ (mass of $^{12}_6C$) $= 931$ MeV$/c^2$
$\hbar c$	$\hbar c = hc/2\pi = 1.973 \times 10^{-11}$ MeV $-$ cm
Molar volume at 1 atm, 0°C	$= 2.2414 \times 10^{-2}\ m^3 mol^{-1}$
Specific electronic charge,	$e/m_e = 1.7588 \times 10^{11}\ Ckg^{-1}$
Rydberg constant,	$R_H = 1.09737 \times 10^7\ m^{-1}$
Bohr first Radius,	$a_0 = \hbar^2/m_e e^2 = 0.529167 \times 10^{-10}$ m
	$a_0 = \hbar^2/m_e e^2 = 0.5292$ nm
Radius of Proton	$r_0 = 1.2 \times 10^{-15}$ m
Fine Structure constant	$\alpha = e^2/hc = 7.2973 \times 10^{-3} = 1/137.036$
Bohr magneton,	$\mu_B = 9.2740 \times 10^{-24}\ A\ m^2\ (= JT^{-1})$
	$\mu_B = 5.788383 \times 10^{-11}\ MeVT^{-1}$
Nuclear magneton,	$1\ \mu_N = 1\ nm = 5.9505 \times 10^{-27}\ A\ m^2$
	$\mu_N = 1\ nm = 3.152452 \times 10^{-14}\ MeV\ T^{-1}$
Magnetic dipole moment, electron	$\mu_e = 1.0011597\ \mu_B$
Proton	$\mu_p = 2.792847\ \mu_N$
Neutron	$\mu_n = -1.910426\ \mu_N$
Barn	$1\ b = 1 \times 10^{-24}\ cm^{-2} = 100\ fm^2$
Hydrogen Ground state	$R_y = 13.6057 eV$.
Calorie	$1\ Cal = 4.1840\ J$
Tesla (the unit of magnetic field)	$1\ T = 1.0\ Wb\ m^{-2} = 10^4\ G$
Atomic weight	$= N M_n + Z (M_p + m_e) - $ (binding energy)
Year in seconds	$1\ yr = 3.16 \times 10^7 s$

Atm. Pressure	$1\ atm = 1.01 \times 10^5\ N\ m^{-2}$
Acceleration due to gravity	$1\ g = 9.81\ ms^{-2}$
Unit for energy absorption	, rad $1\ rad = 0.1\ J/kg$
Curie *Ci)	$1\ Ci = 3.7 \times 10^{10}\ s^{-1}$

%%*%*%*%*%*%*%*

APPENDIX C

List of Elements

LIST OF ELEMENTS

ATOMIC NUMBER	SYMBOL	NAME	ATOMIC WEIGHT	ATOMIC NUMBER	SYMBOL	NAME	ATOMIC WEIGHT
0	n	neutron		52	Te	tellurium	127.60
1	H	hydrogen	1.0079	53	I	iodine	126.9045
2	He	helium	4.00260	54	Xe	xenon	131.30
3	Li	lithium	6.941	55	Cs	cesium	132.9054
4	Be	beryllium	9.01218	56	Ba	barium	137.34
5	B	boron	10.81	57	La	lanthanum	138.9055
6	C	carbon	12.011	58	Ce	cerium	140.12
7	N	nitrogen	14.0067	59	Pr	praseodymium	140.9077
8	O	oxygen	15.9994	60	Nd	neodymium	144.24
9	F	fluorine	18.99840	61	Pm	promethium	
10	Ne	neon	20.179	62	Sm	samarium	150.4
11	Na	sodium	22.98977	63	Eu	europium	151.96
12	Mg	magnesium	24.305	64	Gd	gadolinium	157.25
13	Al	aluminum	26.98154	65	Tb	terbium	158.9254
14	Si	silicon	28.086	66	Dy	dysprosium	162.50
15	P	phosphorus	30.97376	67	Ho	holmium	164.9304
16	S	sulfur	32.06	68	Er	erbium	167.26
17	Cl	chlorine	35.453	69	Tm	thulium	168.9342
18	Ar	argon	39.948	70	Yb	ytterbium	173.04
19	K	potassium	39.098	71	Lu	lutetium	174.97
20	Ca	calcium	40.08	72	Hf	hafnium	178.49
21	Sc	scandium	44.9559	73	Ta	tantalum	180.9479
22	Ti	titanium	47.90	74	W	tungsten	183.85
23	V	vanadium	50.9414	75	Re	rhenium	186.2
24	Cr	chromium	51.996	76	Os	osmium	190.2
25	Mn	manganese	54.9380	77	Ir	iridium	192.22
26	Fe	iron	55.847	78	Pt	platinum	195.09
27	Co	cobalt	59.9332	79	Au	gold	196.9665
28	Ni	nickel	58.71	80	Hg	mercury	200.59
29	Cu	copper	63.546	81	Tl	thallium	204.37
30	Zn	zinc	65.38	82	Pb	lead	207.2
31	Ga	gallium	69.72	83	Bi	bismuth	208.9804
32	Ge	germanium	72.59	84	Po	polonium	
33	As	arsenic	74.9216	85	At	astatine	
34	Se	selenium	78.96	86	Rn	radon	
35	Br	bromine	79.904	87	Fr	francium	
36	Kr	krypton	83.80	88	Ra	radium	
37	Rb	rubidium	85.4678	89	Ac	actinium	
38	Sr	strontium	87.62	90	Th	thorium	232.0381
39	Y	yttrium	88.9059	91	Pa	protactinium	
40	Zr	zirconium	91.22	92	U	uranium	238.029
41	Nb	niobium	92.9064	93	Np	neptunium	
42	Mo	molybdenum	95.94	94	Pu	plutonium	
43	Tc	technetium		95	Am	americium	
44	Ru	ruthenium	101.07	96	Cm	curium	
45	Rh	rhodium	102.9055	97	Bk	berkelium	
46	Pd	palladium	106.4	98	Cf	californium	
47	Ag	silver	107.868	99	Es	einsteinium	
48	Cd	cadmium	112.40	100	Fm	fermium	
49	In	indium	114.82	101	Md	mendelevium	
50	Sn	tin	118.69	102	No	nobelium	
51	Sb	antimony	121.75	103	Lr	lawrencium	

About the Author

Prof. (Dr.) S. DEVANARAYANAN, Ph.D. (IISc); D.Sc. (USA), Dip (Uppsala)

Dr. S. Devanarayanan was educated at the University College, Thiruvananthapuram (1961 – 63), Indian Institute of Science, Bangalore (1963 – 70), and Institute of Physics, Uppsala, Sweden (1970 – 71). He had a brilliant academic career throughout. He was in the Faculty of the University of Kerala (1971 – 2000) and was the Professor & HOD of the Department of Physics, University of Kerala (1993 – 2000); and has 37 years of teaching / research experience in Physics and Materials Science.. He was Professor in the University of Puerto Rico (1989 - 91), Some 21 students have completed Ph.D. and M.Phil. under his supervision. He has to his credit over 80 published research papers in standard scientific periodicals. A monograph entitled THERMAL EXPANSION OF CRYSTALS (Pergamon, Oxford, 1979, also Kindle edition) and books "QUANTUM MECHANICS: Principles and Applications" (SciTech, 2005, 2010), QUANTUM CHEMISTRY (SciTech, 2013), PHYSICS IN A NUTSHELL (Amazon, 2016), A TEXT BOOK ON NUCLEAR PHYSICS (Amazon, 2016) and FERROELCTRICITY IN CRYSTALS (Amazon, 2016) were authored by him. He has served as a Professor in Physics at the University of Puerto Rico, USA, (1989 –91). He was awarded the SIDA Fellowship and worked at The Institute of Physics, Uppsala, Sweden (1970 – 71).

A Life Member of various academic bodies like the IAPA Indian Physics Association, American Physical Society, American Chemical Society, and Founder Fellow of the Indian Cryogenic Council, his biography has found place a number of times in the publications of Marquis' Who's Who (USA) (2013)(2014), IBC (UK)(2011), ABC(USA) (2005), etc. The Govt. of Kerala appointed him as a member of the Commission of Enquiry on the working of the University of Kerala, in Oct 2000 – Mar 2001.

He was in several Committees, Boards of Studies, Examinations and Selection Boards in different Universities, Faculty of Science, Academic Council of University of Kerala, Expert in State Level and National, UPSC competitive Tests.

Prof. Devanarayanan has found honour in finding a name in the Star Chart, - Cat #TYC-7882-99-1- Scorpio Constellation- NASA- Feb 2013 –png.

He had the special honour of being invited by the Royal Swedish Academy to submit proposals for the award of the Nobel Prize in Physics for 1995. Devanarayanan believes in Sir C.V. Raman's advice that one can become a good scientist only when one takes up research along with teaching at a University. He has made academic visits in Sweden, Finland, Leningrad (USSR), The Netherlands, Germany, France, Australia, Czech Republic, Hungary, Austria, England, and USA.

-%-%-%-%-

www.ingramcontent.com/pod-product-compliance
Lightning Source LLC
Chambersburg PA
CBHW071408180526
45170CB00001B/18